Queridos Felipa y Gerardo,
Con el cariño de siempre,
Besitos y bendiciones,
Yvette
abril del 2020

Caring for Latinxs with Dementia in a Globalized World

Behavioral and Psychosocial Treatments

Hector Y. Adames • Yvette N. Tazeau
Editors

Caring for Latinxs with Dementia in a Globalized World

Behavioral and Psychosocial Treatments

Editors
Hector Y. Adames
The Chicago School of Professional Psychology
Chicago, IL, USA

Yvette N. Tazeau
Independent Practice
San Jose, CA, USA

ISBN 978-1-0716-0130-3 ISBN 978-1-0716-0132-7 (eBook)
https://doi.org/10.1007/978-1-0716-0132-7

This Springer imprint is published by the registered company Springer Science+Business Media, LLC part of Springer Nature.
The registered company address is: 233 Spring Street, New York, NY 10013, U.S.A.

To the Latinx immigrant community that continues to survive and thrive—with the indomitable spirit of our ancestors, full of joy, hope, and wisdom, amidst the many struggles and often dehumanizing experiences in this land. And to mis padres, Rosa y Angelmanuel, y a mi abuelita Mamá Bache por darle a nuestra familia lo que ellos nunca tuvieron. Ustedes son mi todo.

Hector Y. Adames

To Latinxs with dementia, their families, and their caregivers, and to the practitioners and researchers who endeavor to help them. And to my maternal families, Posada and Melhado, sobretodo a mi madre, Noemy, quien inspira el gran aprecio y amor que siento hacia mi origen Hispano-Latino.

Yvette N. Tazeau

Preface

Many are the motivations that brought this book into being, not the least of which is the ever present clarion call to recognize the growing elderly population around the world and its health needs regarding dementia. Readers of this book will find that the chapter authors indeed have much to say about this important demographic trend, particularly regarding Hispanics/Latinos/Latinas/Latinxs in the United States (U.S.) and internationally. Those involved in the creative and scientific journey of contributing to this book will, without a doubt, look back with gratification on its trajectory that yielded a magnificent representation of 58 writers across eight countries or territories. The authors include academicians, clinicians and other practitioners/interventionists, and researchers from multidisciplinary fields inclusive of demography, geriatrics, gerontology, medicine, mental health, neurology, neuropsychology, nursing, occupational therapy, pharmacology, psychiatry, psychology, rehabilitation, social work, sociology, and statistics, which collectively bear on the problem and the solutions for better care for Latinxs affected by dementia. Momentarily we summarize the contents of this volume which describe the importance of behavioral and psychosocial treatments of dementia, while also making the case that these services be at the forefront of the minds of all health providers.

For our international readers, we digress in order to address the descriptive use of the terminology "Latinx" in the book's title. Many readers will be familiar with the terms "Hispanic" and "Latino/a" that have been used in scientific literature, as well as the popular lexicon. These terms also speak to the United States' historically rooted explorations of self and sociopolitical identification. Fast forward, "Latinx" now comes to represent, for many, a reparative justice and liberation psychology perspective (Santos, 2017) which seeks to encompass Spanish language, culture, and customs, that is all-inclusive of identity (e.g., gender, disability status, Indigenous and African roots). This volume's first co-editor is among the recognized experts in this area and readers are encouraged to explore his other contributions to the literature on the topic (see Adames & Chavez-Dueñas, 2017; Adames, Chavez-Dueñas, & Organista, 2016; Adames, Chavez-Dueñas, Sharma, & La Roche, 2018).

The volume *Caring for Latinxs with Dementia in a Globalized World: Behavioral and Psychosocial Treatments* is an addition to the few texts available that focus on Latinxs with Alzheimer's Disease and related dementias (ADRD), in both the United States and Spanish-speaking countries around the world including Central America, the Caribbean, South America, Mexico, and Spain. In the United States, there continues to be significant demographic shifts. The population is growing older while also becoming more racially, ethnically, and linguistically diverse than ever before. Together, these changes have important implications for the delivery of health services for Latinx and Asian populations, both of which are growing at a faster rate than other U.S. racial and ethnic groups, including non-Latinx Whites, Blacks, and American Indian/Alaskan Natives (National Center for Education Statistics, 2017).

The chapters are written by a group of international multidisciplinary experts, scholars, and practitioners who address the unique cultural ways in which patients, families, and caregivers of Latinx patients with dementia experience, describe, understand, and manage ADRD. Unlike other texts that provide a broad coverage of Latinx populations, *Caring for Latinxs with Dementia in a Globalized World: Behavioral and Psychosocial Treatments* notably provides a broad, biopsychosocial presentation of Latinxs while underscoring and contextualizing the heterogeneity that exists within this population, including nationality, immigration, race, sexual orientation, gender, and the like. No other book covering topics related to Latinxs living with ADRD, and their caregivers, singularly addresses the complexities, within-group differences, and racial heterogeneity characteristic of Latinxs across different geographical locations around the globe.

The volume is comprised of 22 chapters organized into five sections. Part-I, *Characterizing the Context: Latinxs and Dementia*, is designed to orient readers to demographics and epidemiological risk factors for dementia in Latinx populations (Chap. 1). The section also provides elements that outline Latinx culture and details the contextual, systemic challenges, and sociopolitical factors (e.g., immigration status, racial biases, xenophobia, institutional barriers) that impact the care of Latinx patients with dementia and their families (Chap. 2). The section concludes with a meta-analysis of dementia rates in Central and South America, as well as the Spanish-speaking countries of the Caribbean (Chap. 3).

Part-II, *Evaluations, Assessment, and Diagnosis*, provides practical approaches to evaluate medical comorbidities associated with dementia among Latinxs (Chap. 4) and to assess functioning with a focus on culture (Chap. 5). The section also provides a review of the challenges related to assessing dementia and depression in Latinxs (Chap. 6). The last chapter in Part-II reviews common neuropsychiatric symptoms, their expression among Latinx individuals with dementia, and their effect on caregiver burden (Chap. 7).

Part-III of the book introduces *Supportive Interventions and Services* to consider when caring for Latinxs with dementia and their families. This section walks readers through existing models that address the various behavioral issues and placement problems Latinx families face (Chap. 8) and presents specific examples of educational, diagnostic, and supportive psychosocial interventions for Latinxs with ADRD (Chap. 9). Contributors in this section also discuss topics related to cognitive rehabilitation for maintenance of function in Latinx patients (Chap. 10), pharmacological, nonpharmacological, and ethnocultural healthcare interventions (Chap. 11), and practical and effective healthcare solutions to address the unique challenges and multiple forms of oppression that Latinxs with dementia often encounter because of sexual and gender identities (Chap. 12). The section concludes with the presentation of conceptual models for understanding the role of stress and coping in dementia among Latinxs (Chap. 13), and the ways in which grief, loss, and depression are manifested in Latinx caregivers and families affected by ADRD (Chap. 14).

As a unique feature of the book, Part-IV centers on *International Perspectives from the Americas and Beyond* regarding the care of Latinxs with dementia. Specifically, this section is comprised of several chapters that contextualize the current state of dementia in different countries, including Mexico (Chap. 15), Guatemala, Central America (Chap. 16), and the Spanish-speaking countries of the Caribbean, inclusive of Cuba, the Dominican Republic, and Puerto Rico (Chap. 17). Additionally, one of the chapters addresses dementia treatments and health disparities among Puerto Ricans, with attention paid to the unique cultural and political history of Puerto Rico vis-a-vis the United States (Chap. 18). The last two chapters focus on dementia care in Colombia, South America (Chap. 19), and Spain (Chap. 20).

The book concludes with Part-V titled *Social/Public Policy and Community Perspectives*. Chapter 21 describes the pivotal role that civil society organizations, e.g., community partnerships, nongovernmental groups, and faith-based organizations, can play in the care of Latinxs with dementia. Lastly, Chap. 22 generates several policy options to reduce risk factors for ADRD among Latinxs in an effort to meet the unique needs and systemic challenges of this community.

No book of this magnitude and scope comes to life without the wonderfully synchronous gathering of many key individuals. We are indebted to Springer Publications' longtime Senior Editor of Health and Behavior, Janice Stern, and her assistant Christina Tuballes, for their steadfast encouragement, guidance, and support. We hope Ms. Stern will find time to peruse our book in her retirement! We are also thankful to current Springer Senior Editor of Health and Behavior, Katherine Chabalko, and her team Sara Yanny-Tillar and Lilith Dorko, for taking us to the "finish line." We also express our appreciation to Dale Glaser, PhD, for his excellent review and suggestions, and to graduate students Karen Bugarin, Carla da Cunha, Denisha Maddie, Claire Manley, Radia Mchabcheb, and Jessica G. Perez-Chavez, all from the *Immigration, Critical Race, And Cultural Equity (IC-RACE) Lab* in Chicago, who helped with reviews and suggestions.

Yvette N. Tazeau, PhD, is very grateful to her colleague and friend Paula Hartman-Stein, PhD, clinical geropsychologist, who introduced her to Janice Stern. It was Ms. Stern who invited Dr. Tazeau to create this book. Yet, this book would not have happened were it not for Dr. Hector Y. Adames. Utmost gratitude is expressed to Dr. Adames for being instrumental to this book's creation, and for the high level of collaboration that resulted in its completion. The trajectory of the work together was one of an extraordinary partnership and of true joy. Dr. Adames is a distinctly talented psychology scholar and standout of his generation, and a genuine humanist—a believer in human freedom and progress. Dr. Tazeau also wants to express her appreciation to the other inspiring individuals in her life whose mentoring and friendships across time helped realize the dream of developing a book such as this, including Herr Gerald E. Logan (QEPD), Juanita R. Mendoza, LMFT (QEPD), Deborah A. DiGilio, MPH, and Allen Calvin, PhD. Last, but never least, the greatest of appreciations is to God, Creator of All, and to the Tazeau family for steadfast support throughout this project, and always.

Hector Y. Adames, PsyD, expresses thanks to Helen A. Neville, PhD, Joseph L. White, PhD (QEPD), and his Academic *Madrina* (Godmother) Lillian Comas-Diaz, PhD, for their mentoring, unwavering support, and for inspiring Dr. Adames to write about his Latinx roots with love and pride. Dr. Adames also thanks his soul sisters Janette Rodriguez, PsyD, Nayeli Y. Chavez-Dueñas, PhD, Shweta Sharma, PsyD, Arianne E. Miller, PhD, and Wendy Y. Blanco, BA. Life for Dr. Adames would not be the same without these fierce, brilliant, and powerful Women of Color. He also thanks his family: Rosa A. Adames, Angelmanuel H. Adames, Angel Adames, Jayden R. Adames, Mia Angelina Adames, Dulce Matos, Luis Ney Matos, Luis Matos, Rosanna Matos, Deiby Marte, Alexandra De La Cruz, Jeancarlos Paulino, Carmen De La Cruz, Nancy Adames, and Zunilda Adames for their support throughout the decades. Ultimately, Dr. Adames thanks Yvette N. Tazeau, PhD for, together, undertaking this formidable and exciting book enterprise. Dr. Adames will forever cherish their collegial and interpersonal connection that has grown from the work. The gratifying experience of this project underscores Dr. Adames' esteem for Dr. Tazeau as a first-rate professional; she is a gem in all ways!

Researching and treating the seemingly intractable issues associated with dementia conditions are truly incremental efforts, yet as Mother Theresa said, "There are no great deeds; there are just small deeds done with great love." The time and effort dedicated to this book has been all about love. We are grateful for the camaraderie of all our multinational colleagues who contributed to this book and, in closing, our expectation and hope is that this volume assists those interested in the study and care of Latinxs with dementia to envision and advance a more nuanced, critical, and contextual understanding of this population, along with their families and caregivers affected by Alzheimer's disease and related dementias. Our desire is that this volume motivate the embodiment of a practice that delivers optimal healthcare, comfort, dignity, and the best quality of life possible for Latinxs with ADRD, for their families, and their caregivers. We assert that access to quality healthcare is, fundamentally, a human right to which all are entitled.

Hector Y. Adames
Chicago, IL, USA

Yvette N. Tazeau
San Jose, CA, USA

References

Adames, H. Y., & Chavez-Dueñas, N. Y. (2017). *Cultural foundations and interventions in Latino/a mental health: History, theory, and within group differences*. New York, NY: Routledge.

Adames, H. Y., Chavez-Dueñas, N. Y., & Organista, K. C. (2016). Skin color matters in Latino/a communities: Identifying, understanding, and addressing Mestizaje racial ideologies in clinical practice. *Professional Psychology: Research and Practice, 47*(1), 46–55. https://doi.org/10.1037/pro0000062

Adames, H. Y., Chavez-Dueñas, N. Y., Sharma, S., & La Roche, M. J. (2018). Intersectionality in psychotherapy: The experiences of an AfroLatinx queer immigrant. *Psychotherapy, 55*(1), 1–7. https://doi.org/10.1037/pst0000152

National Center for Education Statistics. (2017). Status and trends in the education of racial and ethnic groups. Retrieved from https://nces.ed.gov/programs/raceindicators/indicator_raa.asp

Santos, C. E. (2017). The history, struggles, and potential of the term Latinx. *Latino/a Psychology Today, 4*(2), 7–14.

Contents

Part III Supportive Interventions and Services

Part IV International Perspectives: The Americas and Beyond

Contributors

Daisy Acosta, MD Universidad Nacional Pedro Henriquez Ureña, Santo Domingo, Dominican Republic

Hector Y. Adames, PsyD The Chicago School of Professional Psychology, Chicago, IL, USA

Paola Alejandra Andrade Calderón, PhD Universidad del Valle de Guatemala, Guatemala City, Guatemala

Erin E. Andrews, PsyD, ABPP Central Texas Veterans Health Care System, University of Texas at Austin Dell Medical School, Texas A&M Health Science Center College of Medicine, Austin, TX, USA

Jacqueline L. Angel, PhD The University of Texas at Austin, Austin, TX, USA

Ronald J. Angel, PhD The University of Texas at Austin, Austin, TX, USA

Lee Ashendorf, PhD, ABPP-CN University of Massachusetts Medical School; VA Central Western Massachusetts, Worcester, MA, USA

Laura Barba-Ramírez, MSc National Council to Prevent Discrimination, Mexico; El Colegio de la Frontera Norte, Tijuana, Mexico

Gregory Benson-Flórez, PhD The Chicago School of Professional Psychology, Chicago, IL, USA

Cristalís Capielo Rosario, PhD Arizona State University, Tempe, AZ, USA

Nayeli Y. Chavez-Dueñas, PhD The Chicago School of Professional Psychology, Chicago, IL, USA

Jhokania De Los Santos, MA University of Georgia, Athens, GA, USA

Leticia E. Fernandez, PhD U.S. Census Bureau, Washington, DC, USA

Milton A. Fuentes, PsyD Montclair State University, Montclair, NJ, USA

Joshua T. Fuller, MA Massachusetts General Hospital, Harvard Medical School, Boston, MA, USA; Boston University, Boston, MA, USA

Samuel C. Gable, PhD University of Massachusetts at Boston, Boston, MA, USA

Dolores Gallagher-Thompson, PhD Stanford University, School of Medicine, Stanford, CA, USA

Margaret Gatz, PhD University of Southern California, Los Angeles, CA, USA

Mackenzie T. Goertz, MA Marquette University, Milwaukee, WI, USA

Julie E. Horwitz, PhD, ABPP-CN Memorial Hospital, University of Colorado, Colorado Springs, CO, USA

Ivonne Z. Jiménez-Velázquez, MD University of Puerto Rico, School of Medicine, San Juan, Puerto Rico

Norman J. Johnson, PhD U.S. Census Bureau, Washington, DC, USA

Tedd Judd, PhD, ABPP-CN Universidad del Valle de Guatemala, Guatemala City, Guatemala

Juan Llibre-Rodríguez, MD, PhD Universidad de Ciencias Medicas, La Habana, Cuba

Francisco Lopera, MD Grupo de Neurociencias de Antioquia, Medellín, Colombia

Mariana López-Ortega, MPP, PhD National Institutes of Health, Mexico, National Institute of Geriatrics, Cuernavaca, Mexico

Claire R. Manley, MA The Chicago School of Professional Psychology, Chicago, IL, USA

Silvia Mejía-Arango, PhD El Colegio de la Frontera Norte, Department of Population Studies, Tijuana, Mexico

Julian Montoro-Rodriguez, PhD University of North Carolina at Charlotte, Charlotte, NC, USA

Alexander Navarro, RN Grupo de Neurociencias de Antioquia, Medellín, Colombia

Maureen K. O'Connor, PsyD, ABPP-CN Bedford Veterans Hospital, Center for Translational Cognitive Neuroscience, Boston University School of Medicine, Boston, MA, USA

Paula Ospina-Lopera, BA Grupo de Neurociencias de Antioquia, Medellín, Colombia

Viviana Padilla-Martinez, PhD Bay Pines VA Healthcare System, Lee County VA Healthcare Center, Cape Coral, FL, USA

Shawneen R. Pazienza, PhD Central Texas Veterans Health Care System, Austin, TX, USA

Jessica G. Perez-Chavez, BA University of Wisconsin-Madison, Madison, WI, USA

Francisco J. Piedrahita, BA Grupo de Neurociencias de Antioquia, Medellín, Colombia

Catherine V. Piersol, PhD, OTR/L Thomas Jefferson University, Philadelphia, PA, USA

Yakeel T. Quiroz, PhD Massachusetts General Hospital, Harvard Medical School, Boston, MA, USA

Carlos A. Rodriguez, PhD Spectrum Health, Grand Rapids, MI, USA

Janette Rodriguez, PsyD Miami VA Healthcare System, Miami, FL, USA

Dinelia Rosa, PhD Teachers College, Columbia University, New York City, NY, USA

Caroline Rosenthal Gelman, LCSW, PhD Silberman School of Social Work, Hunter College, City University of New York, New York, NY, USA

Silvia P. Salas, MA University of Wisconsin-Milwaukee, Milwaukee, WI, USA

Amanda Saldarriaga, BA Grupo de Neurociencias de Antioquia, Medellín, Colombia

Azara L. Santiago-Rivera, PhD Merrimack College, North Andover, MA, USA

Philip Sayegh, PhD, MPH University of California, Los Angeles, Los Angeles, CA, USA

Amber Schaefer, MEd Arizona State University, Tempe, AZ, USA

Christian E. Schenk-Aldahondo, MD University of Puerto Rico School of Medicine, San Juan, Puerto Rico

Juan P. Serrano Selva, PhD Universidad de Castilla-La Mancha, Albacete, Spain

Shanna N. Smith, MA The Chicago School of Professional Psychology, Chicago, IL, USA

Karla Steinberg, LMSW The Icahn School of Medicine at Mount Sinai, New York City, NY, USA

Yvette N. Tazeau, PhD Independent Practice, San Jose, CA, USA

Valeria L. Torres, MA Massachusetts General Hospital, Harvard Medical School, Boston, MA, USA; Florida Atlantic University, Boca Raton, FL, USA

Mari Umpierre, LCSW, PhD The Icahn School of Medicine at Mount Sinai, New York City, NY, USA

Valentine M. Villa, PhD California State University, Los Angeles, CA, USA; UCLA Fielding School of Public Health, Los Angeles, CA, USA

Steven P. Wallace, PhD UCLA Fielding School of Public Health, Los Angeles, CA, USA

Kathleen D. Warman, EdM, MA The Chicago School of Professional Psychology, Chicago, IL, USA

Diana Lynn Woods, PhD, GNP-BC Azusa Pacific University, Azusa, CA, USA; University of California Los Angeles, Los Angeles, CA, USA

About the Editors

Hector Y. Adames, PsyD received his doctorate in Clinical Psychology from the American Psychological Association's (APA) accredited program at Wright State University in Ohio. He completed his APA predoctoral internship at the Boston University School of Medicine's Center for Multicultural Training in Psychology (CMTP), and APA-accredited postdoctoral fellowship in Clinical Neuropsychology at Boston University School of Medicine, Bedford Veterans Hospital. Currently, he is an Associate Professor at The Chicago School of Professional Psychology (TCSPP), Chicago campus. He co-founded and co-directs the *Immigration, Critical Race, And Cultural Equity Lab* (IC-RACE Lab), and has a small private practice in neuropsychology. Dr. Adames is the co-author of a textbook on Latinx Psychology titled *Cultural Foundations and Interventions in Latino/a Mental Health: History, Theory and within Group Differences* published by Routledge Press (2017). His scholarship focuses on topics related to racism and colorism, Latinx Psychology, intersectionality, and cognitive health among People of Color. His academic work is published in many leading journals including the *American Psychologist*, *Journal of Palliative & Supportive Care*, *Psychotherapy*, *Hispanic Journal of Behavioral Sciences*, *Professional Psychology: Research and Practice*, *Journal of Cross-Cultural Psychology*, and *The Counseling Psychologist.* He is the editor of *Latinx Psychology Today* (*LPT*), serves on the editorial board of *The Counseling Psychologist*, and *Professional Psychology: Research and Practice*, and is ad hoc reviewer for several journals, including the *Archives of Clinical Neuropsychology*, among others. He has earned numerous awards and recognitions, including the 2018 Distinguished Emerging Professional Research Award from The Society for the Psychological Study of Culture, Ethnicity, and Race, a Division of the American Psychological Association (APA).

Yvette N. Tazeau, PhD is a Licensed Psychologist in independent practice in Silicon Valley. A third-generation Californian of Western-European and Salvadoran heritage, she is a psychology graduate of Palo Alto University (PGSP) and University of California at Davis, and has been serving the San Francisco (SF) Bay Area in private practice since 1998. As a native speaker of the Spanish language, she provides bilingual (Spanish/English) services. She works across the developmental life-span, with specialties including Geropsychology, Clinical Neuropsychology, Child Psychology, and Industrial/Organizational Psychology. Her clinical psychology psychotherapy services focus on Cognitive-Behavioral Therapy (CBT) and Interpersonal Therapy (IPT). Her clinical neuropsychology practice includes Cognitive Rehabilitation in addition to testing. She uses digital technologies for clinical treatment and incorporates mobile applications and virtual reality tools in her work. She has also created bilingual CBT mobile apps. Her interests and scholarship regarding Gerodiversity topics include clinical and biopsychosocial/behavioral interventions with Latino/a populations, as well as organizational development issues of geriatric mental health workforce planning. She has worked at SF Bay Area hospitals and clinics/agencies, for public school districts and private schools, has taught psychology graduate courses at local universities, and has published studies in scientific journals and chapters in scientific books. Her clinical training and work experiences include Stanford University School of Medicine's Department of Psychiatry and Behavioral Sciences, Veterans Affairs Palo Alto Health Care System, The Children's Health Council (Palo Alto), Gardner Family Health Network/Centro de Bienestar (San Jose), Alexian Hospital/Regional Medical Center, Alexian Brothers Senior Health Center, Santa Clara University, and Notre Dame de Namur University.

List of Figures

List of Tables

Part I

Characterizing the Context: Latinxs and Dementia

1 Demographics and the Epidemiological Risk Factors for Dementia in Hispanic/Latino Populations

Leticia E. Fernández and Norman J. Johnson

Abstract

The aging of the Hispanic/Latino population, their projected gains in life expectancy, and their high prevalence of chronic health conditions raise concerns about the well-being of their elder in the coming decades. This chapter examines demographic and epidemiologic risk factors for dementia in Hispanics/Latinos as well as provides estimates of the prevalence of dementia and associated mortality risks in a sample of elderly Hispanics/Latinos. Aging trends come from the 2014 National Population Projections by the Census Bureau. Their prevalence and mortality rates are estimated using data from the National Longitudinal Mortality Study (NLMS). The NLMS sample in this study consists of Hispanics/Latinos of any race and non-Hispanic White individuals in selected Current Population Study (CPS) surveys who were enrolled in Medicare and who reached age 65 between 1991 and 2011. The findings confirm that male and female Hispanics/Latinos have a higher prevalence of dementia than non-Hispanic White individuals in the same age groups for ages 60–64 and 65–69 and for females ages 70–74. At older ages, Hispanics/Latinos have similar or lower rates of dementia than non-Hispanic/Latino White individuals. We find differences in the prevalence of dementia by Hispanic country of origin and by educational attainment, but not by nativity (U.S.-born vs. foreign-born). In terms of mortality, Hispanics/Latinos afflicted with dementia are less likely to die at each age group compared to non-Hispanic White individuals with dementia. These patterns could be idiosyncratic to our Medicare-CPS linked sample, warranting further research.

Introduction

Previous studies report that the prevalence of dementia among elder Hispanics is about one and a half times higher than for non-Hispanic Whites and that these differences seem to remain throughout the oldest (85+) age groups (Demirovic et al., 2003; Gurland et al., 1999; Manly & Mayeux, 2004; Semper-Ternent et al., 2012). Tabulations using 2014 Medicare data estimate that 6.9% of non-Hispanic Whites ages 65 and older have a dementia diagnosis compared

This chapter is released to inform interested parties of research and to encourage discussion. Any views expressed on statistical, methodological, technical, or operational issues are those of the authors and not necessarily those of the U.S. Census Bureau. All results have been reviewed by the Census Bureau's Disclosure Review Board to ensure that no confidential information is released.

L. E. Fernández (✉) · N. J. Johnson
U.S. Census Bureau, Washington, DC, USA
e-mail: leticia.esther.fernandez@census.gov; norman.j.johnson@census.gov

H. Y. Adames, Y. N. Tazeau (eds.), *Caring for Latinxs with Dementia in a Globalized World*,
https://doi.org/10.1007/978-1-0716-0132-7_1

to 11.5% of Hispanics in the same age group (Alzheimer's Association, 2016a). Researchers suggest that about one-third of Hispanics that survive dementia-free to age 65 could develop dementia in their later years (Haan et al., 2003; Mayeda, Glymour, Quesenberry, & Whitmer, 2016). The risk of dementia among Hispanics increases severalfold if comorbidities such as Type 2 diabetes mellitus and other cardiovascular diseases are present (Haan et al., 2003). Differences in the risk of dementia have been observed among Hispanic subgroups. Caribbean Hispanics tend to have higher risk than Mexican Americans (Haan et al., 2003; Tang et al., 2001). However, there are additional factors such as lifestyle, prevalence of cardiovascular disease, genetics, and socioeconomic status that are associated with differences in risk of developing dementia (Lines & Wiener, 2014).

This chapter provides an overview of the arguments supporting increased concern for risk of dementia in the Hispanic population, including shifts toward an older age distribution in the coming decades. Using data from the National Longitudinal Mortality Study (NLMS), we link data from the Current Population Survey (CPS) to Medicare claims and to mortality outcomes from the National Death Index to estimate the prevalence of dementia among Hispanics ages 65 and older who are enrolled in Medicare, and we provide estimates of their mortality risks.

Literature Review

Dementia describes a group of neurodegenerative diseases that cause progressive cognitive decline, destroying a person's memory, ability to reason, and independence to perform daily activities (Alzheimer's Association, 2016a). This cognitive decline is the result of damaged or destroyed neurons (brain nerve cells) in the areas involved in cognitive and other functions (Alzheimer's Association, 2016a, 2016b). In most cases, symptoms of dementia become apparent as individuals reach their mid-60s or later, and its progression can be devastating for individuals, caregivers, and the healthcare system (Alzheimer's Association, 2016b; Chatterjee et al., 2016; Ninomiya, 2014; Plassman et al., 2007; van den Berg & Splaine, 2012). In the advanced stage, dementia causes severe functional limitations and may become an underlying or a contributing factor in a person's death (Alzheimer's Association, 2016a). As an underlying cause, dementia may lead to difficulty eating or swallowing, thus giving rise to complications that may result in death (Wilkins, Parsons, Gentleman, & Forbes, 1999). Alternatively, dementia may indirectly increase risks of mortality by reducing a person's physical resistance to illnesses (Aguero-Torres, Fratiglioni, Guo, Viitanen, & Winblad, 1999; Newcomer, Covinsky, Clay, & Yaffe, 2003; Wilkins et al., 1999).

Several subtypes of dementia exist. The most prevalent type of dementia is Alzheimer's disease accounting for 60 to 80% of cases (this percentage increases with age), followed by vascular dementia at about 17% to 20% (Alzheimer's Association, 2016b; Barnes & Yaffe, 2011; Plassman et al., 2007). Other forms of dementia account for the remainder and include mixed dementia (dementia of the Alzheimer's type and vascular dementia), Parkinson's disease, dementia with Lewy bodies, physical brain injury, Huntington's disease, Creutzfeldt–Jakob disease, frontal temporal dementia/Pick's disease, and normal pressure hydrocephalus (Alzheimer's Association, 2016b; Chatterjee et al., 2016; Chen, Lin, & Chen, 2009; Plassman et al., 2007). In this chapter, we refer to dementia in general, including all subtypes, except when specifically noted otherwise.

In the United States (U.S.), the prevalence of dementia among individuals over 70 years old is approximately 14.0%. A breakdown of this estimate shows that 5.0% of individuals aged 71 to 79 suffer dementia, that its prevalence increases to 24.2% among individuals aged 80 to 89, and to 37.4% among individuals aged 90 and older (Plassman et al., 2007). The prevalence of Alzheimer's disease, the most common form of dementia, is about 3% among individuals 65 to 74, but it increases to 18% for those 75 to 84, and to 33% among individuals 85 or older (Hebert, Weuve, Scherr, & Evans, 2013). In 2016, it was estimated that 5.2 million individuals ages 65 and older had Alzheimer's disease in the United States, about 11% of the elder population. The number is expected to reach 7.1 million by 2025 and 13.8 million by 2050 (Alzheimer's Association, 2016a; Hebert et al., 2013).

Growth and Aging of the Hispanic/Latino Population

In 2015, there were 56.6 million individuals of Hispanic origin, representing 17.6% of the U.S. population. Hispanics are a young, fast-growing, and diverse population. They are projected to reach 105 million in 2050 and 119 million in 2060, accounting for 26.5% and 28.6% of the nation's population, respectively (Colby & Ortman, 2015; U.S. Census Bureau, 2014a). Past and continuous immigration and gains in life expectancy suggest that in the coming decades Hispanics will be increasingly represented in older age groups in the U.S. population (Alzheimer's Association, 2013; Colby & Ortman, 2015; Haan et al., 2003). Table 1.1 shows the U.S. Census Bureau's 2014 National Population Projections of total and Hispanic populations for the next few decades. Although Hispanics have a younger age distribution than the country's population as a whole, the percentage for people age 65 and older is projected to more than double between 2015 and 2050, reaching 16% (U.S. Census Bureau, 2014a). This projection reflects expected gains in life expectancy for all groups in the coming decades (Ortman, Velkoff, & Hogan, 2014).

As shown in Table 1.1, the life expectancy at birth for Hispanic males is projected to increase

Table 1.1 Estimated and projected age distribution of total and Hispanic U.S. population 2015, 2030, and 2050

Projected population by age group and Hispanic origin	2015[a]	2020	2030	2040	2050
Total U.S. population (in 1000s)	321,419	334,503	359,402	380,219	398,328
Percent 0 to 17 years	22.9	22.2	21.2	20.6	20.1
Percent 18 to 64 years	62.2	61.0	58.2	57.8	57.9
Percent 65 years and older	14.9	16.9	20.6	21.7	22.1
Total	100.0	100.0	100.0	100.0	100.0
U.S. Hispanic origin population[b] (in 1000s)	56,593	63,551	77,463	91,626	105,550
Percent 0 to 17 years	32.1	30.0	26.8	25.4	24.2
Percent 18 to 64 years	61.3	62.2	62.3	61.0	59.9
Percent 65 years and older	6.7	7.9	10.9	13.6	15.9
Total	100.0	100.0	100.0	100.0	100.0
Life expectancy at birth					
Hispanic female	84.2	84.3	84.5	85.3	86.4
Hispanic male	79.6	80.0	80.6	81.7	83.2
Life expectancy at age 65					
Hispanic female	22.4	22.4	22.5	22.9	23.7
Hispanic male	20.0	20.0	20.0	20.0	21.2
Population age 65 and older (in 1000s)	47,761	56,441	74,107	82,344	87,996
Percent non-Hispanic	92.2	91.1	88.6	84.9	80.9
Percent native Hispanic	3.50	3.8	4.6	5.7	7.6
Percent foreign-born Hispanic	4.30	5.0	6.8	9.4	11.4
Total	100.0	100.0	100.0	100.0	100.0
Population age 85 and older (in 1000s)	6287	6727	9132	14,634	18,972
Percent non-Hispanic	93.5	92.1	90.9	90.0	86.9
Percent native Hispanic	3.1	3.7	3.8	4.2	5.2
Percent foreign-born Hispanic	3.4	4.2	5.2	5.8	7.9
Total	100.0	100.0	100.0	100.0	100.0

Source: U.S. Census Bureau, 2014a (National Projections)

[a]The population age distribution for 2015 comes from the Annual Estimates of the Resident Population by Sex, Age, Race, and Hispanic Origin for the United States and States: April 1, 2010 to July 1, 2015 (U.S. Census Bureau, 2016). All other figures are from the 2014 National Projections developed by the U.S. Census Bureau Population Division (U.S. Census Bureau, 2014a, 2014b)

[b]Native and foreign-born Hispanics combined

from 79.6 years in 2015 to 83.2 years by 2050 and for females, from 84.2 to 86.4 for the same period. For Hispanics who attain the age of 65, gains in life expectancy are more modest but still positive. Hispanic male life expectancy at 65 is projected to increase from 20.0 years in 2015 to 21.2 years by 2050 and for females, from 22.4 to 23.7 (U.S. Census Bureau, 2014b). Table 1.1 also shows that the nation's population 65 and older is projected to increase by nearly 18 million people between 2020 and 2030, as the last generation of the baby boomer cohort (born between 1946 and 1964) reaches age 65 (U.S. Census Bureau, 2014a). In the following decades, Hispanics will become an increasing share of the fast-growing 65 and older population, reaching 20% in 2050; that is, one out of five elderly will be of Hispanic/Latino descent (U.S. Census Bureau, 2014a).

Aging is the main risk factor for dementia. With an ever-increasing elderly population having a longer life expectancy, there will be more individuals likely to live well into their 80s and 90s when the risk of Alzheimer's disease and other dementias are highest (Hebert et al., 2013). In addition, the increasing representation of Hispanics in the elderly population may bring further increases in the number of people with dementia if elderly Hispanics are more likely than non-Hispanic to develop dementia (Alzheimer's Association, 2016a). Furthermore, because of past migration trends, among Hispanics 65 and older, the foreign-born are projected to outpace the share of the native-born. In 2050, six out of every ten elderly Hispanics will be foreign-born, with implications for language preference and need for culturally appropriate healthcare services.

Epidemiology of Dementia in the Hispanic/Latino Population

Recent studies report that Hispanics tend to experience higher rates of cognitive decline and dementia than non-Hispanic Whites, while also having more comorbidities (Alvarez, Rengifo, Emrani, & Gallagher-Thompson, 2014; Clark et al., 2005; Haan et al., 2003; Manly & Mayeux, 2004; Novak & Riggs, 2004; Tang et al., 2001). In addition to age, the risks of dementia are higher for individuals with low levels of education, cardiovascular diseases, and genetic predisposition (Alzheimer's Association, 2016a, 2016b). Cultural practices and lifestyle may also influence risks, as well as the progression of the disease (Chen et al., 2009). We discuss each of these as they pertain to the Hispanic population.

Age

After age 65, the risk of dementia doubles for each 5 years of age (Novak & Riggs, 2004). As summarized in the previous section, in the coming decades, Hispanics will increasingly be overrepresented in the age groups most at risk for dementia. In particular, in 2015, 6.5% of the individuals 85 and older were Hispanic, but this share is expected to double to 13.1% by 2050.

Low Education/Socioeconomic Status

Having more years of formal education and regularly engaging in mentally stimulating leisurely activities are both associated with lower risk of dementia (Barnes & Yaffe, 2011; Chen et al., 2009). In part, the association between education and dementia may be mediated by socioeconomic status. Low socioeconomic status has been associated with a broad range of adverse health outcomes (Galea, Tracy, Hoggatt, DiMaggio, & Karpati, 2011; Haan et al., 2003; Marmot, 2005).

Hispanics tend to have lower formal education levels than any other group in the United States. In 2015, about one-third of Hispanics ages 25 and older had less than a high school education, and only 16% had a college degree or higher education (Ryan & Bauman, 2016). As a group, foreign-born Hispanics/Latinos have much lower levels of education than their native counterparts; the educational attainment of native Hispanics/Latinos has improved in recent decades. In 2015, about 20% of native Hispanics had a college degree or higher, twice the percentage from 1994 (Administration for Community Living, 2016; Ryan & Bauman, 2016). For the older generations of Hispanics, low levels of education may translate into higher risk of developing dementia. Some studies do confirm that elder Hispanics with low socioeconomic status are more likely to have symptoms of cognitive impairment than Hispanics with higher socioeconomic status (Alzheimer's Association, 2016a; Schneiderman et al., 2014).

Cardiovascular Risk Factors

Cardiovascular diseases, such as Type 2 diabetes mellitus, hypertension in midlife, high cholesterol, obesity in midlife, and a history of stroke have been associated with increased risk of dementia in the older ages, and these risks are heightened in the presence of comorbidities (Barnes & Yaffe, 2011; Chatterjee et al., 2016; Ninomiya, 2014). Hispanics present a high prevalence of cardiovascular disease. For instance, a study documenting self-reported health among Hispanics ages 18–74 found that half of the men and nearly 40% of the women reported high cholesterol, and obesity was reported by over one-third of the men and over 40% of the women (Daviglus et al., 2012). In addition, for both men and women, one in four reported hypertension and 17% reported diabetes (Daviglus et al., 2012; Schnaider Beeri et al., 2004). Among Hispanics ages 45–64, one-quarter had three or more of these risk factors. This percentage increased to 40% for ages 65–74. Hispanics with low education and low income, native Hispanics, and immigrant Hispanics who have lived in the United States for over 10 years were the most likely to report multiple risk factors (Daviglus et al., 2012; Schneiderman et al., 2014). It should be noted that some of the differences in the prevalence of morbidities might be due to undiagnosed conditions in recent immigrants (Barcellos, Goldman, & Smith, 2012).

In addition, as mentioned earlier, there seem to be differences in the prevalence of risk factors by country of origin or ancestry. For example, Mexican, Puerto Rican, or Dominican Hispanics tend to have a higher prevalence of Type 2 diabetes mellitus than Cuban and South Americans. In contrast, the prevalence of hypertension is reported to be higher for Dominicans, Cubans, and Puerto Ricans and lower for South Americans and Mexicans (Daviglus et al., 2012; Schnaider Beeri et al., 2004). The prevalence of Type 2 diabetes mellitus in Hispanics/Latinos, estimated at 13 to 17%, raises major concerns about future increases in the prevalence of cognitive impairment and dementia (Centers for Disease Control and Prevention, 2014; Daviglus et al., 2012; Haan et al., 2003; Noble, Manly, Schupf, Tang, & Luchsinger, 2012).

Genetic Factors

Having one or more parents or siblings with Alzheimer's disease increases an individual's risk of developing the disease. This may be due to a combination of shared heredity, lifestyle, or environmental factors (Green et al., 2002; Lautenschlager et al., 1996). In addition, there is a well-documented association between higher risk of dementia, specifically Alzheimer's disease, and the presence of an identified gene on chromosome 19, the apolipoprotein E-ε4 (APOE-ε4) allele. There are three forms of the APOE gene (ε2, ε3, and ε4) from each parent. Having one copy of the APOE-ε4 gene triples the risk of developing Alzheimer's disease, and having two copies multiplies the risk by 8–12 times compared to those without the ε4 form (Alzheimer's Association, 2016a; Holtzman, Herz, & Bu, 2012; Loy, Schofield, Turner, & Kwok, 2014). However, this does not mean that every carrier of this gene will develop dementia. Estimates vary, but studies report that about half the individuals with Alzheimer's do not have the APOE-ε4 allele (National Institute on Aging, 2015; Pastor & Goate, 2004). Hispanic genetic ancestry is a mixture of Amerindian (American Indian), African, and European genomes, and the percentage contribution of each tends to vary by country of origin. Generally, for example, Mexicans and Central Americans have a higher Amerindian genetic ancestry, while Caribbean groups tend to have a higher African ancestry (Campos, Edland, & Peavy, 2013). For Hispanics of Amerindian ancestry, the frequency of APOE-ε4 allele is relatively low, and the association between Alzheimer's and the APOE-ε4 allele seems to be weaker or attenuated (Campos et al., 2013; Farrer et al., 1997; Gamboa et al., 2000). Moreover, some researchers suggest that regardless of APOE genotype, Hispanics tend to have higher risks of Alzheimer's than non-Hispanic White individuals of the same age. This may be due to their higher prevalence of Type 2 diabetes mellitus, particularly in Mexican and Puerto Rican groups (Farrer et al., 1997; Haan et al., 2003; Plassman et al., 2007; Tang et al., 1998).

Health-Impacting Factors Associated with Lifestyle and Culture

Close to 40% of adult Hispanics are obese (Body Mass Index of 30 or higher), a risk factor for diabetes, hypertension, and other conditions associated with dementia (Daviglus et al., 2012; Flegal, Carroll, Ogden, & Curtin, 2010). Smoking has also been associated with increased risks of cognitive decline and dementia. Puerto Ricans and Cubans show higher rates of smoking than other Hispanic groups (Baumgart et al., 2015; Daviglus et al., 2012).

Some studies suggest that prescription drugs to manage cardiovascular conditions such as diabetes, high cholesterol, or hypertension may be protective of cognitive functions through their anti-inflammatory mechanisms, so that early detection and treatment of these chronic conditions may have positive impact on preventing dementia in later years (Baumgart et al., 2015; Chen et al., 2009; Ninomiya, 2014). However, Hispanics are less likely than non-Hispanics to seek regular checkups. This can result in Hispanics going undiagnosed, experiencing late diagnoses or mismanagement of chronic conditions and cognitive decline symptoms (Alzheimer's Association, 2007; Center on Aging Society, 2003; Cooper, Tandy, Balamurali, & Livingston, 2010; Espino et al., 2001; Novak & Riggs, 2004). Early diagnosis of dementia could help individuals obtain the most benefits from available treatments to improve cognitive function and delay institutionalization, including opportunities to participate in promising clinical trials (Prince, Bryce, & Ferri, 2011). Some of the barriers to healthcare utilization among elderly Hispanics include lack of health insurance, poverty, low levels of formal education and poor health literacy, limited English proficiency, and underreporting of cognitive impairment due to the perception that memory loss is a normal part of aging (Lines & Wiener, 2014; Novak & Riggs, 2004).

Prevalence of Dementia and Mortality Risks in the Hispanic/Latino Medicare Population

Using data from the NLMS, we estimate the prevalence of dementia and risks of mortality for each 5-year age group starting with age 60. The sample consists of N = 19,832 Hispanics and N = 304,173 non-Hispanic White individuals. These individuals were enrolled in Medicare, reached age 65 between 1991 and 2011, and were matched to their individual-level responses in the CPS. We linked the sample to their Medicare claims to identify those afflicted with dementia and determined mortality outcomes by using the National Death Index (NDI).

We focus on differences in dementia prevalence and mortality among Hispanics by sex, education, and place of birth (foreign-born vs. U.S.-born) as reported in the CPS. Dementia diagnoses, which included Alzheimer's disease and related disorders or senile dementia, were identified from the Medicare inpatient and outpatient claims records based on the International Classification of Diseases, Ninth Revision (ICD-9; Centers for Medicare & Medicaid Services, 2014a) and Tenth Revision (ICD-10; Centers for Medicare & Medicaid Services, 2014b) codes, as recommended by the Centers for Medicare and Medicaid Services (CMS) Chronic Conditions Data Warehouse (CCW) algorithms. The three-digit ICD-9 codes included mild cognitive impairment, Alzheimer's disease, dementia with Lewy bodies, and other cerebral degeneration (331), presenile and senile dementia, including vascular dementia (290), senility without mention of psychosis (797), and other and non-specified brain disorders (294). The corresponding three-digit ICD-10 codes included Alzheimer's disease, dementia with Lewy bodies, other dementias (F02, G30), vascular dementia (F01), senile dementia (G31.1, R54), frontotemporal dementia (G31.01, G31.09), and other non-specified brain disorders (F03, F04, F05, F06.1, F06.8, G13.2 G13.8, G31.2, G94, R41.81).

A limitation for this type of study is that the diagnosis of dementia is influenced by differential rates of underdiagnosis and underreporting across Hispanic groups of varying socioeconomic characteristics (Kotagal et al., 2015; Lines & Wiener, 2014). In addition, the sample combines individuals who were diagnosed in a 20-year span, and changes over the years in the disease itself, its coding in the ICD system, and its criteria for a diagnosis might influence our findings. Table 1.2 compares the prevalence of dementia by 5-year age group among non-Hispanic White male and female Medicare beneficiaries with those of their similarly aged Hispanic counterparts and to Hispanics from specific countries of origin or ancestry. Our estimates show that Hispanics as a group have a higher prevalence of dementia in the younger-old groups, ages 60 to 64 and 65 to 69 (and for women, also ages 70 to 74), confirming other findings of early-onset dementia in Hispanics (Clark et al., 2005; Gurland et al., 1999; Manly & Mayeux, 2004). At older ages, our findings for this Medicare sample suggest that Hispanics have similar or lower rates of dementia than non-Hispanic White individuals.

Looking at particular countries of origin or ancestry, Puerto Rican male Medicare beneficiaries ages 60 to 64 have rates that are twice as high as those of other Hispanics/Latinos in the same age group (11.0%), which could be specific to the sample and require further research. We observe that Puerto Rican males have higher prevalence of dementia than other groups for ages 80 to 84 and Puerto Rican females have higher prevalence of dementia than other groups for ages 70–74, 75–79, and 80–84, although the differences are not always statistically significant (the confidence intervals overlap). These patterns could be idiosyncratic to the Medicare sample, warranting further research. Although the differences are not always statistically significant, Cuban and Mexican men, as well as Cuban, Mexican, and Central and South American women have a prevalence of dementia that is higher than for non-Hispanic White individuals prior to age 70, but are similar or lower for older ages. The lowest prevalence of dementia is found in the other Hispanics for both males and females.

Table 1.3 shows differentials in the prevalence of dementia among Hispanics/Latinos by educational attainment. As reported in the literature, males that are more educated have lower prevalence of dementia, at least before age 70, but the pattern is not as clear for females. Similarly, Table 1.4 shows that the prevalence of dementia among Hispanics does not follow a clear trend for place of birth. Finally, Table 1.5 shows the percentage of Hispanics/Latinos who died within each 5-year interval by sex and dementia diagnosis. There are three findings shown in this table that should be emphasized. First, on average, although not for every age group, individuals with dementia are at higher risk of death than those with no dementia. Second, as a group, Hispanics afflicted with dementia are less likely to die at each age group compared to non-Hispanic White individuals with dementia. Third, in general, educated Hispanics with dementia are less likely to die in each age interval than their less-educated counterparts. Education seems to be associated not only with lower prevalence of dementia among Hispanics but also with lower mortality among those who develop dementia.

Discussion

Our research shows that Hispanics/Latinos ages 60–64 tend to have higher prevalence of dementia than similarly aged non-Hispanic White individuals, which confirms patterns reported elsewhere of early-onset dementia. At the same time, mortality rates of Hispanics with dementia

Table 1.2 Prevalence of dementia among White non-Hispanics and Hispanics and by detailed Hispanic origin groups, Medicare beneficiaries aged 65 and older[a] (confidence interval in parenthesis)

Males, % with dementia diagnosis in each age interval							
Age	Non-hispanic white ($N = 127{,}644$)	All Hispanic ($N = 8274$)	Mexican ($N = 4330$)	Puerto Rican ($N = 940$)	Cuban ($N = 1042$)	Central/ South American ($N = 766$)	Other hispanic ($N = 1196$)
60–64	4.0 (3.8, 4.2)	5.6 (4.6, 6.6)	5.2 (3.9, 6.6)	11.0 (7.1,14.9)	5.3 (2.2, 8.3)	4.4 (1.6, 7.2)	3.1 (1.1, 5.2)
65–69	13.0 (12.6, 13.4)	14.6 (13.0, 16.1)	15.0 (12.9,17.1)	15.8 (11.2, 20.4)	15.7 (10.8, 20.5)	12.4 (7.9, 17.0)	12.4 (8.6, 16.3)
70–74	24.1 (23.5, 24.6)	24.8 (22.6, 26.9)	25.0 (22.1, 27.9)	25.8 (19.6, 32.0)	25.9 (19.8, 32.0)	20.8 (14.2, 27.5)	24.5 (18.9, 30.2)
75–79	40.9 (40.2, 41.6)	39.3 (36.6, 42.0)	38.9 (35.1, 42.7)	39.7 (31.0, 48.4)	41.0 (34.3, 47.6)	37.6 (28.5, 46.7)	39.5 (32.7, 46.3)
80–84	57.3 (56.5, 58.1)	49.8 (46.4, 53.3)	52.9 (48.2, 57.7)	59.0 (48.5, 69.6)	42.9 (33.4, 52.3)	45.5 (34.3, 56.6)	40.9 (31.9, 49.9)
85+	73.9 (73.1, 74.7)	63.0 (58.9, 67.0)	62.9 (57.1, 68.7)	61.0 (48.6, 73.5)	65.4 (56.2, 74.5)	54.8 (37.3, 72.4)	64.5 (54.8, 74.2)
Total	28.3 (28.0, 28.5)	24.8 (23.8, 25.7)	24.6 (23.4, 25.9	26.3 (23.5, 29.1)	28.3 (25.6, 31.0)	20.5 (17.6, 23.4)	23.7 (21.3, 26.2)
Females, % with dementia diagnosis in each age interval							
Age	Non-Hispanic white ($N = 176{,}529$)	All Hispanic ($N = 11{,}558$)	Mexican ($N = 5592$)	Puerto Rican ($N = 1379$)	Cuban ($N = 1544$)	Central/south American ($N = 1378$)	Other hispanic ($N = 1665$)
60–64	3.9 (3.7, 4.1)	4.8 (4.0, 5.6)	4.3 (3.2, 5.4)	5.8 (3.4, 8.3)	5.5 (2.8, 8.1)	5.7 (3.2, 8.3)	4.4 (2.3, 6.5)
65–69	13.2 (12.8, 13.5)	15.4 (14.0, 16.8)	15.6 (13.6, 17.5)	19.9 (15.5, 24.3)	18.1 (13.5, 22.8)	13.5 (10.0, 16.9)	10.5 (7.2, 13.8)
70–74	23.3 (22.8, 23.8)	27.2 (25.3, 29.1)	27.1 (24.4, 29.8)	40.2 (34.1, 46.4)	25.7 (20.6, 30.8)	25.8 (20.4, 31.1)	20.1 (15.8, 24.5)
75–79	40.9 (40.3, 41.5)	41.2 (38.9, 43.4)	42.3 (39.0, 45.6)	50.2 (43.3, 57.2)	38.1 (32.6, 43.7)	39.7 (32.5, 46.8)	34.6 (28.8, 40.4)
80–84	58.9 (58.3, 59.5)	56.9 (54.3, 59.5)	55.7 (51.9, 59.5)	62.9 (54.6, 71.1)	60.4 (54.2, 66.7)	54.8 (46.4, 63.2)	54.1 47.1, 61.1)
85+	77.1 (76.6, 77.6)	71.7 (69.1, 74.4)	71.6 (67.6, 75.7)	71.8 (63.9, 79.7)	69.6 (63.3, 75.9)	74.8 (66.8, 82.7)	72.4 (66.1, 78.7)
Total	33.9 (33.6, 34.1)	29.4 (28.6, 30.3)	28.6 (27.5, 29.8)	33.1 (30.6, 35.6)	34.3 (31.9, 36.6)	26.6 (24.2, 28.9)	27.0 (24.9, 29.2)

Source: Authors' computations, NLMS

[a]The NLMS sample includes only Hispanics/Latinos of any race and non-Hispanic White individuals enrolled in Medicare who reached age 65 between 1991 and 2011 and who matched to their individual-level responses in the Current Population Survey (NLMS website: https://www.census.gov/nlms).

within each 5-year interval seem to be lower than for non-Hispanic White individuals. With the aging of the Hispanic population and the expected gains in life expectancy, the number of elderly Hispanics/Latinos with dementia may be higher than expected, and they may spend a greater number of years afflicted by dementia. There are no treatments to cure dementia, slow down its progression, or stop the damage to brain cells. However, drug treatments can lessen at least for a limited time the cognitive and behavioral symptoms of the disease (memory loss, confusion, problems with thinking and reasoning, etc.) (Alzheimer's Association, 2016b; O'Brien et al., 2017). Responses to treatment as well as side effects vary by individual, and in general, the use of anti-dementia drugs must be balanced with the risk and severity of side effects (O'Brien et al., 2017). The hope is that in the future therapies will be developed that will stop the disease or at

Table 1.3 Prevalence of dementia among Hispanics by educational attainment, Medicare beneficiaries aged 65 and older[a] (confidence interval in parenthesis)

Hispanic males, % with dementia diagnosis in each age interval			
Age	Less than high school ($N = 5077$)	High school diploma/equivalent ($N = 1628$)	More than high school ($N = 1566$)
60–64	6.3 (4.9, 7.7)	5.4 (3.4, 7.5)	3.3 (1.5, 5.0)
65–69	15.6 (13.5, 17.6)	15.0 (11.6, 18.4)	11.2 (8.2, 14.3)
70–74	24.8 (22.1, 27.6)	27.7 (22.7, 32.7)	21.6 (17.0, 26.2)
75–79	38.4 (35.0, 41.7)	40.4 (34.0, 46.9)	41.6 (35.1, 48.1)
80–84	50.1 (45.9, 54.3)	49.2 (40.2, 58.1)	49.3 (40.8, 57.7)
85+	65.7 (61.0, 70.3)	52.1 (40.5, 63.7)	59.0 (48.5, 69.6)
Total	26.6 (25.4, 27.8)	22.2 (20.2, 24.3)	21.3 (19.3, 23.4)
Hispanic females, % with dementia diagnosis in each age interval			
Age	Less than high school ($N = 7562$)	HS diploma/equivalent ($N = 2427$)	More than high school ($N = 1568$)
60–64	4.3 (3.3, 5.3)	5.8 (3.9, 7.7)	5.3 (3.2, 7.5)
65–69	16.5 (14.7, 18.3)	15.7 (12.7, 18.6)	10.8 (7.8, 13.8)
70–74	26.4 (24.1, 28.7)	30.3 (26.2, 34.4)	25.9 (20.7, 31.0)
75–79	41.4 (38.7, 44.1)	41.9 (36.6, 47.2)	38.6 (32.0, 45.2)
80–84	56.8 (53.7, 60.0)	56.4 (50.3, 62.5)	58.3 (50.3, 66.4)
85+	70.7 (67.6, 73.8)	76.8 (70.6, 83.1)	70.8 (61.7, 79.9)
Total	31.0 (29.9, 32.0)	28.4 (26.6, 30.2)	23.7 (21.6, 25.8)

Source: Authors' computations, NLMS

[a]The NLMS sample includes only Hispanics/Latinos of any race enrolled in Medicare who reached age 65 between 1991 and 2011 and who matched to their individual-level responses in the Current Population Survey (NLMS website: https://www.census.gov/nlms).

least improve the quality of life for people with dementia (Alzheimer's Association, 2016b).

Therefore, strategies to improve the quality of life of elder Hispanics/Latinos in the coming years should emphasize multimodal lifestyle interventions to slow down cognitive decline and the onset of dementia (Barnes & Yaffe, 2011; Manuel et al., 2016). Seven modifiable risk factors have been identified that could reduce the prevalence of dementia in the United States and worldwide. According to Barnes and Yaffe (2011), if rates of physical inactivity were reduced, and the prevalence of midlife obesity, diabetes, midlife hypertension, depression, smoking, and low education were lowered by 10%, there would be 184,000 fewer Alzheimer's cases in the United States (a 3.5% reduction). If the negative risk factors were lowered by 25%, 492,000 cases of Alzheimer's could be averted (9.3% reduction) in the coming decades.

These suggested changes are likely to benefit from multilevel public health interventions that tailor messages to specific at-risk subpopulations in promoting weight loss and healthier diets,

Table 1.4 Prevalence of dementia among Hispanics by nativity, Medicare beneficiaries aged 65 and older[a] (confidence interval in parenthesis)

Hispanic males, % with dementia diagnosis in each age interval			
Age	Born in the United States (*N* = 3898)	Born in outlying U.S. territories (*N* = 624)	Foreign-born (*N* = 3359)
60–64	4.9 (3.5, 6.3)	8.3 (4.2, 12.5)	5.1 (3.6, 6.6)
65–69	14.0 (11.8, 16.2)	14.1 (8.6, 19.6)	15.1 (12.7, 17.5)
70–74	24.4 (21.3, 27.5)	23.2 (15.8, 30.6)	25.8 (22.4, 29.1)
75–79	38.1 (34.3, 41.9)	36.6 (26.2, 47.0)	40.1 (35.8, 44.3)
80–84	49.9 (45.0, 54.7)	50.9 (37.7, 64.1)	47.7 (42.1, 53.3)
85+	64.9 (59.1, 70.7)	44.7 (28.9, 60.5)	62.2 (55.9, 68.6)
Total	24.9 (23.5, 26.2)	22.4 (19.2, 25.7)	24.6 (23.1, 26.0)
Hispanic females, % with dementia diagnosis in each age interval			
Age	Born in the US (*N* = 5237)	Born in outlying U.S. territories (*N* = 963)	Foreign-born (*N* = 4815)
60–64	5.0 (3.7, 6.3)	4.5 (2.0, 7.0)	4.2 (3.0, 5.4)
65–69	14.0 (12.0, 16.0)	18.3 (13.4, 23.2)	15.9 (13.8, 18.1)
70–74	25.5 (22.8, 28.2)	38.6 (31.4, 45.8)	25.6 (22.7, 28.5)
75–79	39.2 (35.9, 42.4)	42.4 (34.0, 50.9)	42.1 (38.5, 45.7)
80–84	53.5 (49.6, 57.3)	58.5 (47.9, 69.2)	60.3 (56.2, 64.4)
85+	75.1 (71.4,78.9)	65.8 (54.9, 76.6)	68.9 (64.7, 73.1)
Total	29.2 (27.9, 30.4)	28.6 (25.7, 31.4)	29.2 (28.0, 30.5)

Source: Authors' computations, NLMS

[a]The NLMS sample includes only Hispanics/Latinos of any race enrolled in Medicare who reached age 65 between 1991 and 2011 and who matched to their individual-level responses in the Current Population Survey (NLMS website: https://www.census.gov/nlms).

smoking cessation, engaging in frequent physical activity, and keeping regular healthcare visits to identify and manage cardiovascular diseases (Golden et al., 2012; Lopez & Golden, 2014; Middleton & Yaffe, 2010; van den Berg & Splaine, 2012). In addition, given the evidence of early-onset dementia among Hispanics, healthcare providers could suggest more frequent evaluations for the onset of cognitive disorders, as well as adopt culturally relevant instruments (Haan et al., 2003; Lee, 2010; Swedish Council on Technology Assessment in Health Care, 2008). In the past, Hispanics have been less likely than non-Hispanic Whites to be screened for cognitive decline (Alzheimer's Association, 2013; Kotagal et al., 2015; Lines & Wiener, 2014). Furthermore, tests to assess cognitive decline might be confusing if they are culturally, educationally, and linguistically biased given that at least half of the elderly Hispanics are foreign-born with diverse levels of English language proficiency (Lines & Wiener, 2014).

Conclusion

The shift toward an older age distribution in the Hispanic population, together with projected longer lifespans in the coming decades, and relatively high prevalence of risk factors for dementia suggest that the coming years will bring increases in the number of elderly Hispanics suffering from dementia. Dementia is a devastating progressive disease that destroys a person's cognitive abilities and eventually causes severe functional limitations that may contribute or lead to death. Aging is the main risk factor for dementia, but risks are higher in the presence of cardiovascular disease, such as diabetes and hypertension. Strategies are needed to both address and minimize the impact of this potential public health crisis.

Elderly Hispanics and their caregivers likely will need culturally and linguistically appropriate emotional, informational, and tangible support, as well as greater access to healthcare. Younger Hispanics may benefit from prevention, early detection, and treatment of a host of chronic conditions associated with dementia in later life. Behavioral modification is complex and challenging, but researchers have identified risk factors that could be targeted with culturally appropriate interventions. The challenge in the

Table 1.5 Percent of individuals who died within each 5-year interval by sex, age group, and whether they had a dementia diagnosis[a] (confidence interval in parenthesis)

Age	Non-hispanic white (N = 127,644)		Hispanic (N = 8274)		Hispanic, less than high school (N = 5077)		Hispanic, high school Diploma/equivalent (N = 1628)		Hispanic, more than high school (N = 1566)	
Males, % who died with and without dementia diagnosis in each age interval										
	With dementia	No dementia	With dementia	No dementia	With dementia	No dementia	With dementia	No dementia	With dementia	No dementia
60–64	57.0	52.2	42.9	33.3	45.8	37.4	42.3	31.9	30.8	23.8
65–69	46.3	56.2	38.7	37.0	44.6	43.5	31.3	31.7	25.5	24.5
70–74	56.1	62.6	44.8	41.0	47.3	48.8	41.2	35.6	40.3	22.2
75–79	63.0	64.0	49.0	47.1	50.8	54.3	45.1	37.3	46.7	28.7
80–84	68.7	67.2	57.9	55.9	59.8	60.7	61.0	52.5	47.0	39.7
85+	77.5	70.2	67.1	60.8	69.2	63.2	67.6	50.0	55.1	61.8
Total	64.9	58.5	51.2	39.9	54.4	46.3	46.4	34.5	43.1	26.1
Females, % who died with and without dementia diagnosis in each age interval										
Age	Non-hispanic white (N = 176,529)		Hispanic (N = 11,558)		Hispanic, less than HS (N = 7562)		Hispanic, HS diploma/ equivalent (N = 2427)		Hispanic, more than high school (N = 1568)	
	With dementia	No dementia	With dementia	No dementia	With dementia	No dementia	With dementia	No dementia	With dementia	No dementia
60–64	50.4	44.2	37.9	25.3	43.9	31.3	34.3	18.3	26.1	13.2
65–69	41.5	50.4	27.3	28.0	27.3	33.3	33.0	21.8	15.9	17.0
70–74	48.6	58.6	34.2	32.9	35.6	37.7	32.7	27.5	30.6	18.0
75–79	56.5	61.9	44.4	43.7	46.8	51.1	38.1	28.5	39.5	24.8
80–84	63.3	64.4	50.7	41.7	52.0	43.9	46.1	39.4	50.0	30.0
85+	73.7	68.3	63.4	52.6	65.8	56.8	61.0	43.9	48.5	28.6
Total	62.0	54.2	46.3	32.1	48.7	38.0	42.2	24.2	38.2	17.6

Source: Authors' computations, NLMS

[a]The NLMS sample includes only Hispanics/Latinos of any race and non-Hispanic White individuals enrolled in Medicare who reached age 65 between 1991 and 2011 and who matched to their individual-level responses in the Current Population Survey (NLMS website: https://www.census.gov/nlms).

coming decades will be to turn research findings into successful programs and practices that improve the life of Hispanic elderly and their families, both those already suffering from cognitive decline or dementia and those at risk of developing dementia.

References

Administration for Community Living. (2016). *A statistical profile of Hispanic older Americans aged 65+*. Washington, DC: US Department of Health and Human Services.

Aguero-Torres, H., Fratiglioni, L., Guo, Z., Viitanen, M., & Winblad, B. (1999). Mortality from dementia in advanced age: A 5-year follow-up study in incident dementia cases. *Journal of Clinical Epidemiology, 52*(8), 737–743.

Alvarez, P., Rengifo, J., Emrani, T., & Gallagher-Thompson, D. (2014). Latino older adults and mental health: A review and commentary. *Clinical Gerontologist, 37*(1), 33–48.

Alzheimer's Association. (2007). *Alzheimer's disease and type 2 diabetes: A growing connection*. Chicago, IL: Alzheimer's Association. Retrieved from https://www.alz.org/national/documents/latino_brochure_diabetes.pdf

Alzheimer's Association. (2013). *2013 Alzheimer's disease facts and figures*. Chicago, IL: Alzheimer's Association. Retrieved from http://www.sciencedirect.com/science/article/pii/S1552526013000769

Alzheimer's Association. (2016a). *2016 Alzheimer's disease facts and figures*. Chicago, IL: Alzheimer's Association.

Alzheimer's Association. (2016b). *What is dementia?* Chicago, IL: Alzheimer's Association. Retrieved from http://www.alz.org/what-is-dementia.asp

Barcellos, S. H., Goldman, D. P., & Smith, J. P. (2012). Undiagnosed disease, especially diabetes, casts doubt on some of reported health 'advantage' of recent Mexican immigrants. *Health Aff (Millwood), 31*, 2727–2737.

Barnes, D. E., & Yaffe, K. (2011). The projected effect of risk factor reduction on Alzheimer's disease prevalence. *The Lancet Neurology, 10*(9), 819–828.

Baumgart, M., Snyder, H. M., Carrillo, M. C., Fazio, S., Kim, H., & Johns, H. (2015). Summary of evidence on modifiable risk factors for cognitive decline and dementia: A population-based perspective. *Alzheimer's & Dementia, 11*(6), 718–726.

Campos, M., Edland, S. D., & Peavy, G. M. (2013). An exploratory study of APOE-E4 genotype and risk of Alzheimer's disease in Mexican Hispanics. *Journal of the American Geriatrics Society, 61*(6), 1038–1040.

Center on Aging Society. (2003). *Older Hispanic Americans: Less care for chronic conditions*. Data profile, 9. Washington, DC: Georgetown University Health Policy Institute. Retrieved from https://hpi.georgetown.edu/agingsociety/pubhtml/hispanics/hispanics.html

Centers for Disease Control and Prevention (CDC). (2014). *National diabetes statistics report: Estimates of diabetes and its burden in the United States* (p. 2014). Atlanta, GA: U.S. Department of Health and Human Services.

Centers for Medicare & Medicaid Services (CMS). (2014a). ICD-9-CM Diagnosis and Procedure Codes: Abbreviated and Full Code Titles, Version 31, Effective October 1, 2013. Retrieved from https://www.cms.gov/Medicare/Coding/ICD9ProviderDiagnosticCodes/codes.html

Centers for Medicare & Medicaid Services (CMS). (2014b). 2015 ICD-10-CM and GEMs. Retrieved from https://www.cms.gov/Medicare/Coding/ICD10/2015-ICD-10-CM-and-GEMs.html

Chatterjee, S., Peters, S. A., Woodward, M., Mejia Arango, S., Batty, G. D., Beckett, N., … Huxley, R. R. (2016). Type 2 diabetes as a risk factor for dementia in women compared with men: A pooled analysis of 2.3 million people comprising more than 100,000 cases of dementia. *Diabetes Care, 39*(2), 300–307.

Chen, J.-H., Lin, K.-P., & Chen, Y.-C. (2009). Risk factors for dementia. *Journal of the Formosan Medical Association, 108*(10), 754–764.

Clark, C. M., DeCarli, C., Mungas, D., Chui, H. I., Higdon, R., Núñez, J., … van Belle, G. (2005). Earlier onset of Alzheimer disease symptoms in Latino individuals compared with Anglo individuals. *Archives of Neurology, 62*(5), 774–778.

Colby, S. L., & Ortman, J. M. (2015). Projections of the size and composition of the U.S. population: 2014 to 2060. U.S. Census Bureau current population reports, P25–1143. Retrieved from https://www.census.gov/content/dam/Census/library/publications/2015/demo/p25-1143.pdf

Cooper, C., Tandy, A. R., Balamurali, T., & Livingston, G. (2010). A systematic review and meta-analysis of ethnic differences in use of dementia treatment, care, and research. *The American Journal of Geriatric Psychiatry, 18*(3), 193–203.

Daviglus, M. L., Talavera, G. A., Avilés-Santa, L., Allison, M., Cai, J., Criqui, M. H., … Stamler, J. (2012). Prevalence of major cardiovascular risk factors and cardiovascular diseases among Hispanic/Latino individuals of diverse backgrounds in the United States. *JAMA, 308*(17), 1775–1784.

Demirovic, J., Prineas, R., Loewenstein, D., Bean, J., Duara, R., Sevush, S., & Szapocznik, J. (2003). Prevalence of dementia in three ethnic groups: The South Florida program on aging and health. *Annals of Epidemiology, 13*(6), 472–478.

Espino, D. V., Moulton, C. P., del Aguila, D., Parker, R. W., Lewis, R. M., & Miles, T. P. (2001). Mexican American elders with dementia in long term care. *Clinical Gerontologist, 23*(3/4), 83–96.

Farrer, L. A., Cupples, L. A., Haines, J. L., Hyman, B., Kukull, W. A., Mayeux, R., ... van Duijn, C. M. (1997). Effects of age, sex, and ethnicity on the association between apolipoprotein E genotype and Alzheimer disease. A meta-analysis. APOE and Alzheimer disease Meta Analysis Consortium. *JAMA, 278*(16), 1349–1356.

Flegal, K. M., Carroll, M. D., Ogden, C. L., & Curtin, L. R. (2010). Prevalence and trends in obesity among US adults, 1999–2008. *JAMA, 303*(3), 235–241.

Galea, S., Tracy, M., Hoggatt, K. J., DiMaggio, C., & Karpati, A. (2011). Estimated deaths attributable to social factors in the United States. *American Journal of Public Health, 101*(8), 1456–1465.

Gamboa, R., Hernandez-Pacheco, G., Hesiquio, R., Zuñiga, J., Massó, F., Montaño, L. F., ... Vargas-Alarcón, G. (2000). Apolipoprotein E polymorphism in the Indian and Mestizo populations of Mexico. *Human Biology, 72*(6), 975–981.

Golden, S. H., Brown, A., Cauley, J. A., Chin, M. H., Gary-Webb, T. L., Kim, C., ... Anton, B. (2012). Health disparities in endocrine disorders: Biological, clinical, and nonclinical factors – An Endocrine Society scientific statement. *The Journal of Clinical Endocrinology and Metabolism, 97*(9), E1579–E1639.

Green, R. C., Cupples, L. A., Go, R., Benke, K. S., Edeki, T., Griffith, P. A., ... MIRAGE Study Group. (2002). Risk of dementia among white and African American relatives of patients with Alzheimer disease. *JAMA, 287*(3), 329–336.

Gurland, B. J., Wilder, D. E., Lantigua, R., Stern, Y., Chen, J., Killeffer, E. H., & Mayeux, R. (1999). Rates of dementia in three ethnoracial groups. *International Journal of Geriatric Psychiatry, 14*(6), 481–493.

Haan, M. N., Mungas, D. M., Gonzalez, H. M., Ortiz, T. A., Acharya, A., & Jagust, W. J. (2003). Prevalence of dementia in older Latinos: The influence of type 2 diabetes mellitus, stroke and genetic factors. *Journal of the American Geriatrics Society, 51*, 169–177.

Hebert, L.E., Weuve, J., Scherr, P.A., & Evans, D.A. 2013. Alzheimer disease in the United States (2010–2050) estimated using the 2010 census. *Neurology, 80*(19): 1778–1783.

Holtzman, D. M., Herz, J., & Bu, G. (2012). Apolipoprotein E and apolipoprotein E receptors: Normal biology and roles in Alzheimer disease. *Cold Spring Harbor Perspectives in Medicine, 2*(3), a006312.

Kotagal, V., Langa, K. M., Plassman, B. L., Fisher, G. G., Giordani, B. J., Wallace, R. B., ... Foster, N. L. (2015). Factors associated with cognitive evaluations in the United States. *Neurology, 84*(1), 64–71.

Lautenschlager, N. T., Cupples, L. A., Rao, V. S., Auerbach, S. A., Becker, R., Burke, J., ... Farrer, L. A. (1996). Risk of dementia among relatives of Alzheimer's disease patients in the MIRAGE study: What is in store for the oldest old? *Neurology, 46*(3), 641–650.

Lee, Y. (2010). Systematic review of health behavioral risks and cognitive health in older adults. *International Psychogeriatrics, 22*(2), 174.

Lines, L. M., & Wiener, J. M. (2014). *Racial and ethnic disparities in Alzheimer's disease: A literature review*. Washington, DC: U.S. Department of Health and Human Services Assistant Secretary for Planning and Evaluation office of Disability, Aging and Long-Term Care Policy.

Lopez, L., & Golden, S. H. (2014). A new era in understanding diabetes disparities among U.S. Latinos—All are not equal. *Diabetes Care, 37*, 2081–2083.

Loy, C. T., Schofield, P. R., Turner, A. M., & Kwok, J. B. (2014). Genetics of dementia. *Lancet, 383*, 828–840.

Manly, J. J., & Mayeux, R. (2004). Ethnic differences in dementia and Alzheimer's disease. In N. B. Anderson, R. A. Bulatao, & B. Cohen (Eds.), *Critical perspectives on racial and ethnic differences in health in late life* (pp. 95–141). Washington, DC: The National Academies Press National Research Council Panel on Race, Ethnicity, and Health in Later Life.

Manuel, D. G., Garner, R., Finès, P., Bancej, C., Flanagan, W., Tu, K., ... Bernier, J. (2016). Alzheimer's and other dementias in Canada, 2011 to 2031: A microsimulation population health modeling (POHEM) study of projected prevalence, health burden, health services, and caregiving use. *Population Health Metrics, 14*, 37.

Marmot, M. (2005). Social determinants of health inequalities. *The Lancet, 365*(9464), 1099–1104.

Mayeda, E. R., Glymour, M. M., Quesenberry, C. P., & Whitmer, R. A. (2016). Inequalities in dementia incidence between six racial and ethnic groups over 14 years. *Alzheimer's & Dementia, 12*(3), 216–224.

Middleton, L. E., & Yaffe, K. (2010). Targets for the prevention of dementia. *Journal of Alzheimer's Disease, 20*(3), 915–924.

National Institute on Aging. (2015). *Alzheimer's disease genetics, fact sheet*. NIH Publication #15-6424. Bethesda, MD: U.S. National Institutes of Health. Retrieved from https://www.nia.nih.gov/alzheimers/publication/alzheimers-disease-genetics-fact-sheet

Newcomer, R., Covinsky, K. E., Clay, T., & Yaffe, K. (2003). Predicting 12-month mortality for persons with dementia. *Journal of Gerontology, 58B*(3), S187–S198.

Ninomiya, T. (2014). Diabetes mellitus and dementia. *Current Diabetes Reports, 14*, 487.

Noble, J. M., Manly, J. J., Schupf, N., Tang, M.-X., & Luchsinger, J. A. (2012). Type 2 diabetes and ethnic disparities in cognitive impairment. *Ethnicity & Disease, 22*(1), 38–44.

Novak, K., & Riggs, J. (2004). *Hispanics/Latinos and Alzheimer's disease*. Chicago, IL: Alzheimer's Association. Retrieved from https://www.alz.org/national/documents/reporthispanic.pdf

O'Brien, J. T., Holmes, C., Jones, M., Jones, R., Livingston, G., McKeith, I., ... Burns, A. (2017).

Clinical practice with anti-dementia drugs: A revised (third) consensus statement from the British Association of Psychopharmacology. *Journal of Psychopharmacology, 31*(2), 147–168.

Ortman, J. M., Velkoff, V. A., & Hogan, H. (2014). A*n aging nation: The older population in The United States.* Current population reports, P25-1140. Washington, DC: U. S. Census Bureau.

Pastor, P., & Goate, A. M. (2004). Molecular genetics of Alzheimer's disease. *Current Psychiatry Reports, 6*, 125–133.

Plassman, B. L., Langa, K. M., Fisher, G. G., Heeringa, S. G., Weir, D. R., Ofstedal, M. B., ... Wallace, R. B. (2007). Prevalence of dementia in the United States: The aging, demographics, and memory study. *Neuroepidemiology, 29*, 125–132.

Prince, M., Bryce, R., & Ferri, C. (2011). *World Alzheimer report 2011: The benefits of early diagnosis and intervention.* London: Alzheimer's Disease International. Retrieved from https://www.alz.co.uk/research/world-report-2011

Ryan, C. L., & Bauman, K. (2016). Educational attainment in the United States: 2015 *Population characteristics.* Current population reports, P20-578. Washington, DC: U.S. Census Bureau. Retrieved from https://www.census.gov/content/dam/Census/library/publications/2016/demo/p20-578.pdf

Schnaider Beeri, M., Goldbourt, U., Silverman, J. M., Noy, S., Schmeidler, J., Ravona-Springer, R., ... Davidson, M. (2004). Diabetes mellitus in midlife and the risk of dementia three decades later. *Neurology, 63*, 1902–1907.

Schneiderman, N., Llabre, M., Cowie, C. C., Barnhart, J., Carnethon, M., Gallo, L. C., ... Avilés-Santa, M. L. (2014). Prevalence of diabetes among Hispanics/Latinos from diverse backgrounds: The Hispanic community health study/study of Latinos (HCHS/SOL). *Diabetes Care, 37*(8), 2233–2239.

Semper-Ternent, R., Kuo, Y. F., Ray, L. A., Ottenbacher, K. J., Markides, K. S., & Snih, S. A. (2012). Prevalence of health conditions and predictors of mortality in oldest old Mexican Americans and non-Hispanic whites. *Journal of the American Medical Directors Association, 13*(3), 254–259.

Swedish Council on Technology Assessment in Health Care. (2008). *Dementia etiology and epidemiology – A systematic review.* Volume 1 June 2008 SBU, Stockholm.

Tang, M.-X., Cross, P., Andrews, H., Jacobs, D. M., Small, S., Bell, K., ... Mayeux, R. (2001). Incidence of AD in African-Americans, Caribbean Hispanics, and Caucasians in northern Manhattan. *Neurology, 56*(1), 49–56.

Tang, M.-X., Stern, Y., Marder, K., Bell, K., Gurland, B., Lantigua, R., ... Mayeux, R. (1998). The APOE-epsilon4 allele and the risk of Alzheimer disease among African Americas, whites and Hispanics. *JAMA, 279*(10), 751–755.

U.S. Census Bureau. (2014a). 2014 National Projections. Washington, DC: U.S. Census Bureau Population Division. Retrieved from https://www.census.gov/data/datasets/2014/demo/popproj/2014-popproj.html

U.S. Census Bureau. (2014b). *Methodology, assumptions, and inputs for the 2014 national projections.* Washington, DC: U.S. Census Bureau Population Division. Retrieved from https://www2.census.gov/programs-surveys/popproj/technical-documentation/methodology/methodstatement14.pdf

U.S. Census Bureau. (2016). Annual estimates of the resident population by sex, age, race, and Hispanic origin for the United States and states: April 1, 2010 to July 1, 2015. Washington, DC: U.S. Census Bureau Population Division. Retrieved from https://factfinder.census.gov/faces/tableservices/jsf/pages/productview.xhtml?src=bkmk

van den Berg, S., & Splaine, M. (2012). *Policy brief, risk factors for dementia.* UK: Alzheimer's Disease International.

Wilkins, K., Parsons, G. F., Gentleman, J. F., & Forbes, W. F. (1999). Deaths due to dementia: An analysis of multiple-cause-of-death data. *Chronic Diseases in Canada, 20*(1), 26–35.

2 Contextual, Cultural, and Sociopolitical Issues in Caring for Latinxs with Dementia: When the Mind Forgets and the Heart Remembers

Nayeli Y. Chavez-Dueñas, Hector Y. Adames, Jessica G. Perez-Chavez, and Shanna N. Smith

Abstract

Latinxs in the USA continue to be disproportionately impacted by dementia. Despite its prevalence within the Latinx community, there is a dearth of literature that explicitly addresses the unique cultural ways in which patients and families of Latinx descent experience, describe, understand, and manage brain-related diseases such as dementia. This chapter provides a nuanced and complex understanding of the contextual and sociopolitical factors (e.g., racial biases, xenophobia, institutional barriers) that impact the care of Latinx patients and their families. The central elements that characterize the Latinx culture are described, followed by a section about within-group Latinx differences, which underscores the heterogeneity (e.g., nationality, immigration, race) that exists within this population. Systemic challenges and their impact on diagnosis and symptom management of dementia are provided. The chapter concludes with a brief discussion on how Latinx culture and context impacts the caregiving of Latinxs with dementia.

Contextual, Cultural, and Sociopolitical Issues in Caring for Latinxs with Dementia: When the Mind Forgets What the Heart Remembers

Latinxs[1] are currently the largest ethnic minority group, comprising 18% of the U.S. population (U.S. Census Bureau, 2016). The Bureau of the Census reports that, as of 2016, there were approximately 57 million Latinxs living in the country. The growth of the Latinx population is expected to continue in the upcoming decades with projections calculating that by 2030, 20% of the total population will be of Latinx descent (U.S. Census Bureau, 2004). Latinxs also represent the fastest-growing aging population in the U.S., with 3.6 million individuals over the age of

[1] To include and center the broad range of gender identities present among individuals of Latin American descent, the term Latinx is used throughout the chapter.

N. Y. Chavez-Dueñas (✉) · H. Y. Adames
S. N. Smith
The Chicago School of Professional Psychology, Chicago, IL, USA
e-mail: NChavez@thechicagoschool.edu; HAdames@thechicagoschool.edu; shannanicoles@gmail.com

J. G. Perez-Chavez
University of Wisconsin, Madison, WI, USA
e-mail: jperezchavez@wisc.edu

H. Y. Adames, Y. N. Tazeau (eds.), *Caring for Latinxs with Dementia in a Globalized World*,
https://doi.org/10.1007/978-1-0716-0132-7_2

65 (Administration on Aging, 2015). The Administration on Aging reports that between 2008 and 2030, the Latinx aging population will increase by 224% compared to a 65% increase for non-Latinx Whites. Thus, it estimated that by 2060 there will be 21.5 million Latinx elders, totaling 22% of the older U.S. population. In addition to being the fastest-growing aging population in the U.S., Latinxs also have a longer life expectancy than their African American and non-Latinx White American counterparts (Centers for Disease Control and Prevention, 2015). In fact, Latinx life expectancy is projected to increase from 80 to 87 by 2050 (Centers for Disease Control and Prevention, 2015).

With the long life expectancy projected for Latinxs, they will undoubtedly face changes related to age-associated conditions including the loss of cognitive functioning, for example, memory, language, comprehension, and thinking. One of the most pernicious cognitive diseases related to old age is dementia: an umbrella term used to define a group of incurable brain diseases that cause loss of memory and cognitive functioning which impacts daily life and is oftentimes progressive. While there are many different types of dementia (e.g., vascular, Lewy bodies, frontotemporal), dementia of the Alzheimer's type (AD) is the most common, accounting for 60–80% of all dementia cases (Alzheimer's Association, 2016a). The statistics of people with AD are overwhelming. Currently, more than five million individuals are living with the disease, one in three dying with AD or another dementia, and every 66 seconds someone develops AD (Alzheimer's Association, 2016a). These alarming figures make AD the sixth leading cause of death in the U.S., with more deaths than prostate and breast cancer combined (Alzheimer's Association, 2016a).

Over time, dementia results in marked behavioral changes that limit the patient's ability to problem-solve and perform simple, everyday tasks and activities (Wu, Vega, Resendez, & Jin, 2016). With regard to Latinxs, research indicates that this population is disproportionately impacted by dementia. In fact, the incidence of Alzheimer's disease and related dementias has been documented as being much higher among African Americans and Latinxs compared to non-Latinx White Americans (Manly & Mayeux, 2004). For instance, according to figures reported by the Alzheimer's Association (2016b), Latinxs are 1.5 times more likely to develop dementia of the Alzheimer's type (7.5% between ages 65 and 74; 27.9% between ages 75 and 84; 62.9% over age 85) than non-Latinx Whites (2.9% between 65 and 74; 10.9% between ages 75 and 84; 30.2% over age 85).

Despite the prevalence of dementia within the Latinx community, there is a dearth of literature that explicitly addresses the unique ways in which patients and families of Latinx descent experience, describe, understand, and manage brain-related diseases such as dementia. Hence, there is an urgent and sustained need for healthcare providers to integrate culture into healthcare service delivery for Latinxs. This chapter seeks to provide a more nuanced and complex understanding of the contextual and sociopolitical factors that impact the care of Latinx patients and their families. The central elements that characterize the Latinx culture are described, followed by a section about within-group Latinx differences, which underscores the heterogeneity that exists within this population. Systemic challenges and their impact on diagnosis and symptom management of dementia are also provided. The chapter concludes with a brief discussion on how Latinx culture and context impact the caregiving of Latinxs with dementia.

Understanding Latinx Culture in Dementia Care

Latinxs are a heterogeneous group, rich in diversity manifested through differences in generational status, race, socioeconomic status, ethnicity, degree of bilingualism, nationality, level of acculturation, and stage of ethnic identity. Despite these differences, Latinxs are united by a common culture, a history of colonization (Chavez-Dueñas, Adames, & Organista, 2014), and traditional cultural values focusing on interdependence, collectivism, and a group orientation to family unity (Adames, Chavez-Dueñas, Fuentes, Salas, & Perez-Chavez, 2014; Romero,

Cuéllar, & Roberts, 2000; Saetermoe, Beneli, & Busch, 1999; Santiago-Rivera, Arredondo, & Gallardo-Cooper, 2002). This section briefly describes the roots of the common traditional cultural values practiced by Latinxs in contemporary society and how such values may impact dementia care.

The Latinx culture is a complex constellation of ancestral Indigenous and African traditions, mores, beliefs, and cultural values combined with European practices and belief systems imposed through the process of conquest and colonization (Adames & Chavez-Dueñas, 2017). Latinx culture serves as the general foundation upon which traditional cultural values are formed. Defined as beliefs and practices, traditional cultural values are considered important among members of an ethnic group (Kluckhohn & Strodtbeck, 1961). In general, cultural values are based on collective experiences believed to influence thoughts, goals, and behaviors, impacting preferred styles of communication (Adames et al., 2014; Añez, Silva, Paris, & Bedregal, 2008; Chandler, 1979; Inclan, 1985; Szapocznik, Scopetta, & King, 1978). Among healthcare providers, it is imperative to understand cultural values since these drive patients' preferences for practices that are considered culturally congruent (Adames & Chavez-Dueñas, 2017). They also play a key role in shaping a person's internal frame, informing attitudes, and promoting behaviors (Miller & Rollnick, 2013). Cultural values have been found to shield against the adverse effects of immigration, acculturation, and oppression, which are common experiences among many Latinxs in the U.S. (Miller & Rollnick, 2013; Villarruel et al., 2009). Research has also revealed that cultural values related to interpersonal relationships (e.g., *simpatía, respeto*) and family (e.g., *familismo*) play an important role in the provision of health services of U.S. Latinxs. Specifically, interventions that take into account a client's cultural values result in better treatment outcomes among individuals from this population (Bernal, Jiménez-Chafey, & Domenech Rodríguez, 2009; Carter, 1991). The integration of traditional Latinx cultural values can also increase treatment retention (Kalibatseva & Leong, 2014). Furthermore, among Latinx families and patients affected by dementia, traditional cultural values may impact how they make sense of the condition; understand, describe, and interpret symptoms; communicate with healthcare providers; and respond to treatment options. Below is a brief review of the literature regarding the most common Latinx cultural values, followed by a discussion on how such values interact with traditional gender role ideologies to produce dementia care that is highly gendered.

Latinx Cultural Values

Familismo/familism continues to be the traditional Latinx cultural value in the dementia literature that is often underscored (Talamantes, Trejo, Jiminez, & Gallagher-Thompson, 2006). The concept of *familismo* involves broad networks of support that extend beyond the nuclear family to include aunts, uncles, grandparents, godparents, and other close family members (Adames et al., 2014). Family structures, processes, and interactions are informed by the collectivistic norms of the Latinx culture (Falicov, 1989), emphasizing obligation, affiliation, and cooperation (Adames & Chavez-Dueñas, 2017). Within Latinx families, individual identity is commonly secondary to family identity, requiring individuals to prioritize family needs over individual ones. Such family-centered socialization creates a strong sense of connectedness and interdependence within its members. *Familismo* also promotes and maintains loyalty, attachment, and a sense of responsibility toward the members of one's family (Falicov, 1989).

Given the strong sense of obligation, loyalty, and responsibility that familism engenders, it is often believed to serve as a protective factor for patients with dementia. Overall, Latinxs are more likely to be cared for by family members when faced with chronic and terminal conditions (Adames et al., 2014). However, research on familism and caregiving has produced mixed results. For instance, some studies demonstrate that familism conveys a perceived availability of support and caregiving satisfaction while assisting patients with dementia (John, Resendiz, & de Vargas, 1997; Scharlach et al., 2006; Wallace & Facio, 1987). Thus, higher scores on familism are associated with lower scores on caregiving burden (Losada et al., 2006). However, other studies have not found a relationship between familism and caregiving burden (Crist et al., 2009). Additional studies have concluded that familism is also associated with worse psychological outcomes (Shurgot & Knight, 2005), high levels of stress (Youn, Knight, Jeong, & Benton, 1999), lower levels of perceived support (Schurgot & Knight, 2005), and depression (Rozario & DeRienzis, 2008) among Latinx caregivers. Overall, the role of familism in the care of Latinxs with dementia appears to function as a double-edged sword. On one side, familism serves as a determinant factor in the decision to become the caregiver of a family member with dementia (John et al., 1997; Scharlach et al., 2006). On the other side, familism decreases the perceived burden associated with such care. Such strong sense of loyalty to one's family may also decrease the probability that supportive resources will be utilized, thus increasing the objective burden on the caregiver.

Dementia Care: A Gendered Practice Caregiving is a highly gendered practice wherein the expectation is on women to become the caretaker of patients with dementia (Mendez-Luck, Applewhite, Lara, & Toyokawa, 2016). These findings can also be extended to Latinxs where traditional gender roles including *machismo* and *marianismo* interact and influence caregiving expectations and practices. *Machismo*, defined as a "socially constructed, learned, and reinforced set of behaviors comprising [the] male gender in Latino/a society" (De La Cancela, 1986, p. 291), is comprised of two aspects, including *traditional machismo* and *caballerismo*. The former is often associated with negative behaviors (e.g., firm, domineering, aggressive, stoicism), while *caballerismo* includes more positive aspects of Latinx male behavior socialization (e.g., honor, respect, reliability, courage). The socially constructed, learned, and reinforced behaviors comprising the content of the female gender role in traditional Latinx culture is known as *marianismo*. Latinas who conform to this ideal of womanhood are expected to be long suffering, nurturing, humble, and spiritually stronger than men (Gil & Vasquez, 1996; Santiago-Rivera et al., 2002). When considering how familism and gender ideologies interact with dementia, it becomes clear that while the value of familism holds that "caregiving is a family obligation ... it is predominantly women who enact the care" (Shurgot & Knight, 2005, p. 1057). Moreover, women are expected to accept and fulfill the caregiving role without complaints (Magaña, Schwartz, Rubert, & Szapocznik, 2006). For example, in a qualitative study of caregivers of elderly family members, Mendez-Luck et al. (2016) reported that when it came to caregiving, women of Mexican descent sacrifice themselves to follow the idealized principles of familism (e.g., obligation, loyalty, responsibility). The devotion and sense of responsibility that Latina women experience may propel them to leave their jobs, postpone their personal goals, and sacrifice their needs in order to care for a family member with dementia. However, less research exists examining the relationship between *machismo/caballerismo*, familism, and caregiving for patients with dementia. Nonetheless, given the available literature on the male gender roles in Latinx culture, we can speculate that when Latinx men do find themselves as the only viable caregiver, they may face particular challenges that need to be considered by health professionals. In particular, Latinx men may be less likely to express feeling overwhelmed, stressed, or ask for help when unable to provide the level of care required by their ailing family member.

As discussed throughout this section, Latinx cultural values and traditional gender role socialization influence an individual's decision to become the caregiver of the patient with dementia, as well as how they understand such responsibility. Thus, it is essential that healthcare providers become aware of how Latinxs give cultural meaning to caregiving experiences and how such experiences are informed by traditional gender role expectations. However, two additional factors need to be considered when working with Latinxs. These factors include ethnic identity development (Atkinson, Morten, & Sue, 1989) and acculturation (Kohatsu, Concepcion, & Perez, 2010), both of which serve as variables that moderate the extent to which individuals adhere to traditional cultural values and Latinx gender role expectations. Acculturation is defined as the process through which individuals adjust to a new culture, which involves the incorporation of the new culture's beliefs, values, norms, language, and behaviors. Ethnic identity is an aspect of the self that includes a sense of acceptance and congruence regarding one's membership in a socially constructed ethnic group. Furthermore, it involves an individual's perceptions and feelings about members of his/her/their own ethnic group as well as members of the dominant group. Both ethnic identity and acculturation are influenced by the current sociohistorical context and within-group Latinx differences such as nationality, immigration, legal status, and race. While an in-depth review of the literature on acculturation and ethnic identity is beyond the scope of this chapter, readers are encouraged to review several foundational and contemporary publications on acculturation (see Berry, 1990; Capielo Rosario, Adames, Chavez-Dueñas, & Renteria, 2019; Cuéllar, Arnold, & Maldonado, 1995) and ethnic identity theory (see Adames & Chavez-Dueñas, 2017; Atkinson et al., 1989; Helms, 1990; Phinney, 1989). Being well versed in this body of literature will set the groundwork to help providers develop culturally congruent dementia care for Latinx patients.

Considering Nationality, Immigration, and Race: Diversity Within Latinxs

While Latinxs share elements of a common culture including traditional values and beliefs, considerable diversity exists within this heterogeneous population. To illustrate, Latinxs can differ in their nation of origin, history of immigration, legal status, skin color, and phenotype (Adames & Chavez-Dueñas, 2017; Chavez-Dueñas et al., 2014). These differences may impact their sense of connection to the traditional Latinx culture, adherence to cultural values, connection to the dominant culture, as well as their perceptions of themselves and others. Additionally, these areas of diversity are likely to affect access to life-enhancing resources such as formal education and health insurance, factors that are key for early detection and treatment management of dementia (Alzheimer's Association, 2016b). Hence, effective healthcare delivery for Latinx patients with dementia requires providers to develop a nuanced understanding on how Latinxs' intersecting identities impact dementia care.

Nationality

Statistics regarding the incidence of dementia in Latin America are varied. Though it appears that Latin American countries may have approximately half the number of dementia diagnoses than the U.S. (1.8 million compared to 3.4 million, respectively), it is projected that these numbers may become even by the year 2040 (Ferri et al., 2005; Sousa et al., 2010). Overall, urbanized areas in Latin America appear to have a higher prevalence of dementia diagnoses than rural regions (Rodríguez, Ferri et al., 2008; Sousa et al., 2010). Moreover in Cuba, the prevalence of Alzheimer's disease and related dementias is reported to be as high as in developed countries (Rodríguez, Valhuerdi et al., 2008).

In the U.S., Latinxs can trace their origins to 19 countries with their own unique cultures and traditions, with the three biggest groups being Mexicans (64%), Puerto Ricans (9.5%), and Cubans (3.7%; Lopez & Patten, 2015). Each Latinx country has its own history of contact, settlement, and immigration to the U.S. To illustrate, individuals from some Latinx groups have deep-seated roots in the U.S. that date back several generations. This is the case for Latinxs of Mexican descent whose ancestors lived in the southwest part of today's U.S. prior to the Mexican-American War (1846–1848). After the war, Mexico lost over half of its territory, and Mexicans who were living in the new U.S.

territory became foreigners in their own land. This history led to economic and racial oppression, which is still experienced by many of their descendants today. Other Latinx groups (e.g., Guatemalans, Salvadorans, Mexicans) have migrated to the U.S. seeking better economic opportunities and family reunification. Alternatively, Cubans arrived into the U.S. seeking political asylum following the Cuban Revolution (1953–1959), which put Fidel Castro in power. Unlike other Latinxs, Cubans were able to benefit from a wide range of programs created by the U.S. federal government to guarantee their successful adjustment into the U.S. culture. As a result of their history, Cubans enjoy higher levels of education and socioeconomic status. Finally, Latinxs of Puerto Rican descent have a history of colonization resulting from political intervention by the U.S. in the island (Organista, 2007). Today's Puerto Rican community continues to be impacted by their history of double colonization, first by the Spaniards (1492) and then the U.S. government (1898). People of Puerto Rican descent fare worse than other ethnic groups on several indicators of well-being, including lower household incomes and homeownership, and are also more likely to live in poverty (Acosta-Belen & Santiago, 2006; Adames & Chavez-Dueñas, 2017; Brown & Patten, 2013). Puerto Ricans have also been reported to have an earlier age of onset of dementia, higher rates of cognitive impairment, and greater severity even after controlling for educational attainment (Livney et al., 2011).

The diverse histories of Latinxs in the U.S. highlight how the pan-ethnic label can be helpful in describing shared cultural characteristics which are important to consider when working with Latinx patients with dementia. However, knowledge about the complex U.S. history of each Latinx group is also necessary to help providers contextualize the sociopolitical realities that impact families and clients suffering from dementia. Such knowledge can also assist providers to dispel commonly held myths and stereotypes about this population such as all Latinxs are immigrants, all immigrants are undocumented, most Latinxs speak Spanish, Latinxs are a monolithic racial group, etc.

Immigration

Contrary to commonly held stereotypes, the majority of Latinxs in the U.S. (65% or 36 million individuals) are U.S.-born American citizens. The remaining 35% (19 million) are immigrants who have historically traveled to the U.S. for a variety of reasons including better economic opportunities, family reunification, or to escape violence and political turmoil in their countries of origin. However, access to economic opportunities available to Latinx immigrants has fluctuated depending on the economic circumstances and sociopolitical climate of the U.S. at the time of their arrival, as well as the need for cheap labor. To illustrate, during times of economic growth, when the demand for cheap labor is high, immigrants are welcomed and immigration laws become less stringent. However, during times of economic uncertainty, immigrants are vilified and scapegoated while immigration laws become stricter (Cornelius, 2009; Flores et al., 2008). As a result, the experience of many Latinx immigrants in the U.S. has been marked by exclusion, exploitation, and xenophobia (Chavez-Dueñas, Adames, Perez-Chavez, & Salas, 2019). Although there is a growing body of interdisciplinary scholarly work on Latinxs and health, the literature examining the role of immigration and all of its complexities on Latinx patients with dementia remains sparse. However, few available studies suggest that immigration plays an important role. In a study examining nativity status and sources of care among older Latinx immigrants of Mexican descent, Angel, Rote, Brown, Angel, and Markides (2014) posit that elderly Latinx immigrants were more likely to report cognitive impairment, mobility limitations, and difficulties with basic activities of daily living (ADLs) and instrumental activities of daily living (IADLs) compared to U.S.-born Mexicans. Interestingly, children of Latinx immigrants with dementia were less likely to seek support or assistance from other family members and social service agencies. These findings are important to consider when designing support services for caregivers of Latinx immigrants with dementia.

Unauthorized Immigrants Currently 8.5 million Latinxs live in the U.S. without permission of the U.S. government and are referred to as undocumented immigrants. This segment of the U.S. Latinx population is one of the most vulnerable groups that is susceptible to high levels of stress resulting from xenophobia, lack of access to resources, and discrimination. Moreover, due to their legal status, undocumented immigrants are also susceptible to exploitation and abuse from their employers who often deny them the most basic labor rights (e.g., breaks, access to water, restroom facilities) while at times exposing them to dangerous working conditions (Kronick & Hargis, 1998; National Council of La Raza, 1990, 2008; Suárez-Orozco & Suárez-Orozco, 2001; Velásquez, 1993). Overall, undocumented people experience a host of psychosocial stressors that require attention in the assessment, diagnosis, treatment, and outreach of individuals and families dealing with dementia (Chavez-Dueñas et al., 2019). For instance, undocumented immigrants often find it more difficult to secure good-paying jobs that offer benefits (Migration Policy Institute, 2014). As a result, the undocumented population has high rates of financial hardship with approximately 33% living in poverty and 66% earning less than twice the federal poverty line. Moreover, they often have low levels of formal education with only 25% obtaining a high school diploma or General Equivalency Diploman (GED; Migration Policy Institute, 2014). Additionally, with the exception of some emergency care, unauthorized immigrants are not eligible for most federally funded health insurance programs (e.g., Medicare, Medicaid, Affordable Care Act). Thus, it is not surprising that the share of uninsured individuals among this group (60%) is much higher than that of Latinx adults who are permanent legal residents or citizens (25%), and as compared to the general U.S. population (14%; Krogstad & Lopez, 2014; Migration Policy Institute, 2014). The lack of access to health insurance coupled with economic hardship serve as barriers to early detection of dementia. A delayed diagnosis prevents people from receiving treatment that may help slow the progression of cognitive decline. The circumstances that many Latinx immigrant families face leave them with less time to adjust and plan for the life changes and logistics associated with caring for someone with dementia (e.g., costs, medication management, palliative care). Unfortunately, undocumented patients and their families may not be able to afford the services they need, thus requiring healthcare providers to become creative in securing the necessary help. For instance, healthcare providers can help undocumented patients to connect with private medical and community organizations that may provide services at low or no cost.

Medical Repatriation Medical repatriation is the practice wherein hospitals and healthcare facilities send undocumented immigrant patients back to their country of origin without their consent (Donelson, 2015). Unfortunately, medical repatriation has received little attention in the dementia literature. Although unauthorized immigrants are ineligible for federally funded health insurance programs, hospitals are required to provide emergency medical care. Nonetheless, once medical staff deem patients medically stable, the staff are no longer obligated to provide health services (Center for Social Justice at Seton Hall Law School [CSJ] and the Health Justice Program at New York Lawyers for the Public Interest [NYLPI], 2012; Donelson, 2015). However, the Emergency Medical Treatment and Active Labor Act (EMTALA, 1986) mandates that hospitals create discharge plans, which include the transfer of patients to health facilities where continuity of care can be ensured. Lamentably, undocumented immigrants are unable to pay for continuity of care services; as a result, due to their status, centers and long-term facilities refuse to admit them as patients (Donelson, 2015). The lack of facilities that would admit undocumented immigrants, coupled with financial concerns for the level of care needed, often leaves hospitals with few choices regarding discharge planning. As a result, many medical facilities operate as de facto Immigration and Customs Enforcement (ICE) officials, deporting immigrants without their consent and, at times, through coercion (Donelson, 2015). While the exact statistics of undocumented immigrants impacted by medical repatriation is currently difficult to assess given the failure of legislation requiring such monitoring, there are a number of documented cases when undocumented Latinx immigrants have been sent back to Mexico, El Salvador, Guatemala, and Honduras (Center for Social Justice at Seton Hall Law School [CSJ] and the Health Justice Program at New York Lawyers for the Public Interest [NYLPI], 2012). The quote, "Apparently they see us as beasts of burden that can be dumped back over the border when we have outlived our usefulness" (Sontag, 2008, para. 5), captures the sense of frustration, betrayal, and no-win position that undocumented immigrants impacted by medical repatriation may experience. Sadly, given the disproportionate rates of dementia among Latinxs, along with the additional risk factors faced by immigrants, medical professionals will continue to grapple with the realities associated with providing care to people who are unable to afford health services, which, in turn, may lead to unethical practices such as medical repatriation.

Racial Diversity

The literature on Latinxs in the social science and medical fields often uses pan-ethnic terms (e.g., Hispanic, Latino, Latinx) to describe this diverse segment of the population. While pan-ethnic labels have political relevance and provide a sense of unity among individuals who have experienced a history of oppression, discrimination, and invisibility (Adames & Chavez-Dueñas, 2017; Chavez-Dueñas et al., 2014), they can also obscure within-group differences. Of particular relevance to Latinxs are differences in skin color and phenotype, which have been associated with a wide range of variables. Available studies examining how skin color and phenotype impacts Latinxs suggest that being darker and less European-looking can negatively affect their mental health (Montalvo, 2005; Montalvo & Codina, 2001; Ramos, Jaccard, & Guilamo-Ramos, 2003), educational attainment, and wages (Arce, Murgia, & Frisbie, 1987). Despite the diversity in skin color and phenotype that exists within this community, Latinxs are socialized to not identify themselves racially (Adames & Chavez-Dueñas, 2017; Chavez-Dueñas et al., 2014). In order not to succumb to commonly held stereotypes, it is important for health professionals to know that a "typical" Latinx phenotype does not exist. However, Latinxs may be phenotypically Black, Brown, White, and any range of skin color in between. Consideration that variations in skin color within Latinx families can impact family dynamics is important when developing programs and creating treatment protocols for individuals with dementia (Adames, Chavez-Dueñas, & Organista, 2016). Overall, it is imperative that health professionals not only incorporate the Latinx culture but also integrate the patient's racialized experiences into their assessment and treatment. For instance, practitioners working from traditional frameworks are not likely to consider the role that race and ethnicity plays in the lives of their Latinx patients with dementia. Moreover, when culture and race are incorporated, they are often used purely as descriptors. While a patient's background is important, it does not provide sufficient information on how the patient's racialized experience shapes their symptoms and response to care, which may be different among racially heterogeneous ethnic groups such as Latinxs.

Systemic Challenges and Their Impact on Dementia Care for Latinxs

Socioeconomic Status

One of the biggest challenges experienced by Latinx elders is poverty and lack of financial security. Overall, Latinx elders tend to have fewer financial resources such as pensions, housing assets, and retirement funds compared to their non-Latinx White counterparts. Approximately 25% of Latinx elders are reported to live below the poverty line, which represents more than double the rate of non-Latinx White elders (DeNavas, Proctor, & Smith, 2011). Additionally, Latinxs are half (39% vs. 72%) as likely as non-Latinx elders to have additional income from private retirement plans. Consequently, a significant percentage (40%) relies solely on social security benefits as a source of income (Angel & Angel, 2009). Given these grim statistics, Latinxs diagnosed with dementia or other chronic and progressive conditions are more likely to rely on the caregiving provided by their children, spouses, and extended family (Gassoumis, Wilber, Baker, & Torres-Gil, 2010). Latinx elders with dementia are also less likely to afford supportive services (e.g., home healthcare, respite care), putting the burden for all the necessary caregiving on their children and family members.

Xenophobia and the Rise in Anti-Latinx Sentiment

In addition to high levels of poverty, many Latinxs in the U.S. confront increasing degrees of xenophobia and a growing anti-Latinx sentiment (Chavez-Dueñas et al., 2019). While experiences of rejection and hostility against Latinxs are not new, several events have contributed to an increase in the anti-Latinx sentiment. First, the unprecedented changes in the U.S. demographics have made Latinxs more visible in every sphere of society, increasing fears about imminent changes in the social and cultural fabric of the population (Stacey, Carbone-López, & Rosenfeld, 2011). Second, the events that took place on September 11, 2001, resulted in an increase fear and rejection of people perceived as immigrants. In the wake of 9/11, rhetoric describing immigrants as "criminals" who may become a potential threat to the homeland security began to emerge. Although Latinxs were not involved in 9/11, they were deeply impacted by it. Following 9/11, a number of laws and policies opposing immigration, disproportionately targeting Latinx populations, increased dramatically. For instance, in 2005 there were approximately 300 bills related to immigration. This number increased to 1500 by 2009. During the same 9-year period, the number of legislations enacted increased by approximately 500%, from 38 to 222 new laws (National Conference of State Legislatures, 2011). Finally, the anti-Latinx and anti-immigrant rhetoric expressed by Donald J. Trump during the 2016 U.S. presidential election campaign took the anti-Latinx sentiment to new lows (Chavez-Dueñas et al., 2019; French et al., 2019). To illustrate, a hate crime report conducted by the Los Angeles (L.A.) County Commission on Human Relations indicates that hate crimes against Latinxs in L.A. increased by 69% in 2015 (Romero, 2016). These events and the resulting increase in rejection and hostility toward Latinxs have the potential to increase psychological distress and anxiety among individuals, families, and communities of Latinx descent. Among families and patients suffering with dementia, the current sociopolitical context may contribute to Latinxs being more reticent to seek assistance from health professionals.

Service Underutilization and Its Impact on Dementia Care for Latinxs

Studies suggest that Latinxs tend to underutilize formal caregiving services (Crist, Garcia-Smith, & Phillips, 2006) including long-term care facilities (e.g., nursing homes). This pattern of underutilization persists even among patients who have suffered a significant decline in their ability to function independently (Espino, Angel, Wood, Finely, & Ye, 2013). A complex set of factors contribute to the underutilization of mental health services by Latinxs including economic variables, access to health insurance, language barriers, lack of culturally congruent services, and racial/ethnic biases among healthcare providers. Lack of access to health insurance, particularly among low-income and undocumented Latinxs as discussed earlier in the chapter, is likely to delay the diagnosis and treatment of dementia

(Angel, Angel, McClellan, & Markides, 1996). Communication barriers are additional challenges experienced in the healthcare system by monolingual, Spanish-speaking patients (Fortinsky, 2014). Not speaking and understanding English often results in patients or their families not being able to describe the symptoms, signs, onset, and course of cognitive decline in detail, which are core elements of an accurate dementia diagnosis. As a result, communication difficulties often become challenges to treatment compliance. In addition to communication barriers, Latinxs may find that the health services they are offered are incongruent with their traditional cultural values (e.g., placing a family member in a long-term care facility). Thus, they may not follow the recommendations made by health professionals and may keep their family member in the home for longer periods of time, thus increasing their caregiver burden (Adames et al., 2014).

In addition to economic, language, and cultural barriers, racial and ethnic biases existing within the healthcare system are additional factors associated with service underutilization among Latinxs. First, research indicates that racial and ethnic minorities are more likely to receive lower-quality healthcare than non-Latinx Whites, even when access to healthcare and socioeconomic status are controlled for (Smedley, Stith, & Nelson, 2003). Second, healthcare providers have been found to interact less effectively with Patients of Color than with White patients (Cooper & Roter, 2003; Cooper et al., 2003), with racial and ethnic biases contributing to the disparity in healthcare quality received by Latinxs. For instance, scholars Tobin and colleagues (1987) and Van Ryn and Burke (2000) report that healthcare providers hold implicit biases based on patients' race, ethnicity, and other demographic variables (e.g., gender, sexual orientation). Implicit biases impact the ways healthcare professionals interpret signs and symptoms of different conditions as well as their clinical decision-making (Kunda & Sherman-Williams, 1993; Tobin et al., 1987). Moreover, implicit biases may impact how providers relate to patients and how much attention is paid to patient concerns. To illustrate, in a study conducted by Hinton, Chambers, Velasquez, Gonzalez, and Haan (2006), it is reported that while the racial and ethnic disparities in health service utilization between Latinxs and non-Latinx Whites may be attributed to Latinxs not seeking help, a significant percentage (80%) of caregivers reported that psychiatric symptoms related to dementia were not considered even when these were communicated to primary care physicians. In other words, Latinx caregivers reported that medical professionals neglected information related to the management of behavioral problems and as a result they did not provide recommendations for how to manage these symptoms. Thus, contrary to the belief that Latinxs do not seek support from medical professionals, evidence suggests that when they do, their needs are likely to be neglected (Hinton et al., 2006). These negative experiences with healthcare providers result in Latinxs rating their experiences with the medical system as less than satisfactory. In a study by Fortinsky (2014), Latinxs rated their experiences with health professional as fair to poor regardless of socioeconomic status.

Diagnosis and Dementia Symptom Management in the Latinx Patient

As described in the introduction section of this chapter, Latinxs are more likely to be diagnosed with dementia than their non-Latinx White counterparts. Researchers from the Institute of Aging project that the number of Latinxs living with Alzheimer's disease, the most common type of dementia, could show an alarming increase of 832% going from 379,000 in 2012 to 3.5 million in 2060 (Wu et al., 2016). Studies also suggest that, on an average, Latinxs have an onset of Alzheimer's that is approximately 7 years earlier compared to non-Latinx Whites (Alzheimer's Association, 2010). Despite what we know about the prevalence and symptom onset of dementia among Latinxs, there continues to be a dearth of literature examining the biopsychosocial factors that contribute to such statistics. Moreover, few studies have analyzed how systemic barriers, racial biases, and lack of culturally trained health professionals impact the early diagnosis and symptom management of dementia among Latinxs. This section provides a brief overview of the emerging literature on Latinxs, dementia diagnosis, and treatment.

Biopsychosocial Variables and Dementia Prevalence in Latinx Populations

The increased probability of developing dementia for Latinxs is arguably due to the combination of biomedical, environmental, and systemic variables that disproportionately impact Latinxs. Latinxs are more likely to be diagnosed with medical conditions such as hypertension, atherosclerosis, and diabetes mellitus that increase the risk of developing dementia. Interestingly, the disparity in age of onset among Latinxs has been arguably due to a higher prevalence of diabetes mellitus (Alzheimer's Association, 2010). Latinxs are also more likely to live in places (e.g., houses and neighborhoods) where they are exposed to environmental toxins (e.g., contaminated water, air pollution) that increase their risk of developing dementia (Santiago-Rivera, Adames, Chavez-Dueñas, & Benson-Florez, 2016). Finally, Latinxs are more likely to experience systemic barriers including low levels of academic attainment and high levels of poverty, which have been shown to decrease access to high-quality nutritious foods, thus further exacerbating their vulnerability for developing a dementia (Aggarwal, 2013; Killin, Starr, Shiue, & Russ, 2016). While currently no cure exists for dementia, early diagnosis and symptom management can improve the quality of life for both patients and their caregivers (National Hispanic Council on Aging, 2016). Unfortunately, Latinxs do not often benefit from an early diagnosis given that they are less likely to be screened for the condition due to lack of knowledge coupled with distrust in the medical system. Moreover, the percentage of Latinxs participating in dementia research is significantly lower than that of their non-Latinx White counterparts (Gallagher-Thompson et al., 2004; Mehta, Yin, Resendez, & Yaffe, 2005). It is thus imperative that health professionals develop ways to deliver culturally congruent information about dementia. Gallagher-Thompson et al. (2015) and Valle, Yamada, and Matiella (2006) use *fotonovelas*, stories conveyed through photographs and text in a comic book-like fashion, to increase awareness about the symptoms associated with dementia, to increase symptom management, and to teach Latinx caregivers how to cope with stress and depression.

Making a Timely Diagnosis of Dementia

A critical step in the management of dementia is making a prompt and accurate diagnosis. Nonetheless, a correct diagnosis depends on the health professional's ability to understand and accurately interpret the beliefs, values, and behaviors of patients as well as their barriers to early diagnosis (Karlawish et al., 2011). One of the most important facts that health professionals need to understand and consider is that many Latinxs lack basic knowledge about the common symptoms associated with dementia (Cox, 2007; Karlawish et al., 2011). For instance, Latinxs tend to attribute memory loss to the normal aging process, making it less likely that they would seek medical assistance during the early stages of dementia (National Hispanic Council on Aging, 2016; Neary & Mahoney, 2005). As a result, missed and delayed diagnoses are more common among aging Latinxs than non-Latinx Whites (National Hispanic Council on Aging, 2016). Latinxs also experience a longer delay between the time they recognize the signs and symptoms of the condition and the diagnosis of dementia (Connell, Roberts, McLaughlin, & Carpenter, 2009; Dilworth-Anderson, Hendrie, Manly, Khachaturian, & Fazio, 2008). Unfortunately, by the time they receive a diagnosis of dementia, many Latinx patients typically have already experienced a significant impairment in their overall cognitive functioning (Cox, 2007).

The shortage of bilingual Latinx healthcare professionals also contributes to the problem of early diagnosis of dementia (Wu et al., 2016). In 2013, only 4.5% of all practicing medical professionals were of Latinx descent; however, it is unknown how many providers in this group were also bilingual. Some obstacles to an early diagnosis of dementia can be improved by both increasing the number of outreach and recruitment programs that are culturally congruent and language specific and by developing culturally responsive training programs for Latinx-serving health professionals (e.g., primary care physicians, neuropsychologists, nurses). Programs with Latinx-specific foci may assist professionals in gaining knowledge on how to minimize their biases so they can develop *confianza* (trust) with their Latinx patients. Once *confianza* is established, Latinxs may be more likely to see medical settings as places where they can seek help. This assertion is supported by studies that have looked at the role of *confianza* which demonstrate that when Latinxs are encouraged by someone they already trust, they are more likely to engage in health screenings including for dementia (National Hispanic Council on Aging, 2016).

Dementia Symptom Management

After Latinxs receive a diagnosis of dementia, the management of symptoms becomes the main goal of treatment. However, effective symptom management requires that the patient follow the recommendations of health professionals. Among Latinxs, treatment compliance also requires that both the patient and the family understand the biomedical nature of dementia and how treatment can be useful. Thus, providing basic psychoeducation regarding the etiology and treatment of dementia is particularly important for Latinxs, given the lack of information and the persistence of misconceptions that exist about this condition within this community. Evidence indicates that Latinxs believe that dementia can be caused by high levels of stress, thinking too much, and taking things too seriously or that it may be simply the result of destiny (National Hispanic Council on Aging, 2016). It is this type of information and misinformation, combined with the systemic barriers discussed above, that likely contributes to Latinxs not taking advantage of the available treatment options that can help manage dementia-related symptoms and behaviors.

Due to the fact that many people in the Latinx community typically associate dementia with memory loss, they are often unaware of its neuropsychiatric symptoms such as apathy, depression, and irritability. Consequently, Latinx caregivers are more likely to attribute behavioral changes to other factors such as health conditions, interpersonal problems, personality, and stress (Ayalon & Areán, 2004; Hinton, Chambers, & Velásquez, 2009; Hinton, Franz, Yeo, & Levkoff, 2005). When asked what contributes to neuropsychiatric symptoms, less than 30% of Latinxs attribute these symptoms to dementia (Hinton et al., 2009). Unfortunately, when Latinxs recognize the behaviors associated with dementia, they often report that healthcare providers do not provide information on how to manage these symptoms (Hinton et al., 2009). Latinxs often end up having to find their own ways to manage these symptoms which tend to increase caregiving burden. The consequences of not addressing neuropsychiatric symptoms are exemplified by the Sacramento Area Latino Study on Aging (SALSA), one of the most well-known studies examining aging in the Latinx population. In the study, Haan and colleagues (2003) found high levels of behavioral symptoms among older participants of Mexican descent diagnosed with dementia. Despite the high prevalence of neuropsychiatric symptoms in this population, patients and their family members reported not receiving information on how to manage these symptoms. The patients were also not referred to specialists (e.g., psychologists) who could help with symptom management (Hinton, Haan, Gellar, & Mungas, 2003). As a result, the participants in the study were less likely to receive help from professional sources (Hinton, Chambers, Velásquez, Gonzalez, & Haan, 2006). Given that neuropsychiatric symptoms are associated with increased disability among cognitively impaired Latinx elderly (Hinton, Tomaszewski Farias, & Wegelin,

2008) and with symptoms of depression noted among caregivers (Hinton et al., 2003), it is essential that families are provided with ways to recognize these symptoms, so services can be secured.

The Latinx Culture, Context, and Caregiving

Caring for a family member with dementia is a particularly overwhelming task that can have serious adverse health consequences for the caregivers (Ory, Hoffman, Yee, Tennstedt, & Schulz, 1999; Zarit, Todd, & Zarit, 1986). Evidence suggests that caregiving is associated with a wide range of negative physical and mental health outcomes including depression and early mortality (Schulz, O'Brien, Bookwala, & Fleissner, 1995). While the literature on caregiving is extensive, most studies focus on non-Latinx White caregivers. At the same time, the study of the interplay between culture, context, and caregiving among Latinx populations remains sparse. This area requires further attention because the proportion of Latinxs who will be expected to become caregivers will grow exponentially. Evercare and the National Alliance for Caregiving (2008) reports that approximately eight million caregivers in the U.S. are of Latinx descent with 1.8 million of them providing care for a patient with Alzheimer's disease, as Latinxs are more likely to provide informal care to their family members than individuals from other racial/ethnic groups (Weiss, Gonzalez, Kabeto, & Langa, 2005). Research suggests that Latinx caregivers often live in multigenerational households where they are responsible for caring for not only a parent with a dementia but also children under the age of 18 (Evercare & The National Alliance for Caregiving, 2008).

Latinx Perceptions of Caregiver Burden

The few available studies on Latinx caregivers have produced mixed results. For instance, some studies suggest that Latinxs view caregiving as an honor and a responsibility, and as a result, these individuals are less likely to describe caregiving as a burden (Coon et al., 2004). Given their views on caregiving, Latinxs tend to perform the duties associated with the role even during challenging circumstances and with little to no formal support (Coon et al., 2004). The strong sense of responsibility for the care of their family contributes to Latinxs finding themselves in intensive caregiving situations where the patient requires constant monitoring and support (Evercare & The National Alliance for Caregiving, 2008). Other studies suggest that Latinxs do perceive caregiving as a burden with 63% of Latinxs caregivers reporting high levels of burden compared to 51% of non-Latinx caregivers (Evercare & The National Alliance for Caregiving, 2008). This group of Latinxs reports feeling anxious and in need of social support. When compared to their non-Latinx counterparts, Latinxs report being more bothered by the caregiving responsibilities such as feeding, dressing, hygiene, and the like (Cox & Monk, 1993; Valle, Cook-Gait, & Tazbaz, 1993). They also reported higher levels of somatic symptoms and stress. Overall, subjective experiences of caregiving burden impact mental health and predict depression among Latinxs (Hernandez & Bigatti, 2010). Interestingly, despite the challenges and burden experienced by Latinx caregivers, research also suggests that they see themselves as having the ability to respond to disruptive behaviors associated with dementia and have more control over negative thoughts related to the caregiving role (Depp et al., 2005).

Contextual and Cultural Beliefs in Caregiving

Systemic challenges related to race, gender, and immigration further complicate the experiences of caregiving among Latinxs. For instance, *marianismo* influences women's attempt to balance a perceived duty to the family (Arévalo-Flechas, Acton, Escamilla, Bonner, & Lewis, 2014). As a consequence, Latinx women are more likely to become caregivers while simultaneously juggling multiple responsibilities (Apesoa-Varano, Barker, & Hinton, 2012). In addition to the challenges they experience when caring for a person with dementia, Latinx caregivers also report difficulties associated with being a member of a racial and ethnic minority group including financial stress, health disparities, low levels of education, racism, and the like (Neary & Mahoney, 2005). Research indicates that immigration status also adversely affects caregivers' assertiveness. For instance, undocumented immigrants report feeling afraid when communicating with healthcare providers (Neary & Mahoney, 2005).

Cultural beliefs also have a central impact on the caregiving experience and decisions caregivers make about their family members with dementia. To illustrate, cultural beliefs influence a caregiver's decision to keep a family member with dementia in the home instead of placing them in a long-term care facility. Latinxs may view placement in a nursing home as abandoning their family responsibilities and resort to such measures as a last alternative (National Hispanic Council on Aging, 2016; Neary & Mahoney, 2005). Additionally, placing a family member in a nursing home can be overly stressful for patients as they are likely to miss aspects of their culture such as food, music, and language which may provide a sense of comfort (Neary & Mahoney, 2005).

The Mind May Forget but What If the Heart Remembers?

As described throughout this chapter, culture facilitates how patients and their caregivers understand, experience, and respond to medical conditions including dementia. Culture also facilitates and impacts how Latinx patients and their families make meaning of their conditions, respond to symptoms, and make decisions about their treatment options. In addition to the undeniable and pivotal role of culture on medical conditions, Latinxs are also impacted by context and sociopolitical factors that may limit access to resources, exacerbate their symptoms, or contribute to their disease's progression. Effective outreach, diagnosis, and treatment of Latinxs with dementia necessitate careful consideration of all of these factors. Without a planned integration of the patient and their family's cultural values, beliefs, and practices, along with the systemic barriers they experience, inferior quality of care is likely to be delivered (Adames & Chavez-Dueñas, 2017; Bosma, Apland, & Kazanjian, 2010). We further posit that even as patients' cognition begins to deteriorate (e.g., memory) as a result of dementia, a thoughtful integration of their culture and careful consideration for how lifelong experiences of oppression have impacted their life can produce healthcare that is truly humane. In order to craft and deliver comfort, dignity, and the best quality of life possible for Latinxs with dementia, healthcare providers must consider the role of cultural values, practices, and a history of oppression on the patient and the family. After all, while the patient's mind might forget, their heart may still remember images, odors, and sounds that are either threatening or comforting.

References

Acosta-Belen, E., & Santiago, C. E. (2006). *Puerto Ricans in the United States: A contemporary portrait.* Boulder, CO: Lynne Rienner Publications.

Adames, H. Y., & Chavez-Dueñas, N. Y. (2017). *Cultural foundations and interventions in Latino/a mental health: History, theory, and within group differences.* New York, NY: Routledge.

Adames, H. Y., Chavez-Dueñas, N. Y., Fuentes, M. A., Salas, S. P., & Perez-Chavez, J. G. (2014). Integration of Latino/a cultural values into palliative health care: A culture centered model. *Palliative & Supportive Care, 12*(2), 149–157. https://doi.org/10.1017/S147895151300028X

Adames, H. Y., Chavez-Dueñas, N. Y., & Organista, K. C. (2016). Skin color matters in Latino/a communities: Identifying, understanding, and addressing Mestizaje racial ideologies in clinical practice. *Professional Psychology: Research and Practice, 47*(1), 46–55. http://dx.doi.org/10.1037/pro0000062

Administration on Aging. (2015). *A statistical profile of Hispanic older Americans aged 65+.* Retrieved from http://www.aoa.acl.gov/Aging_Statistics/minority_aging/Facts-on-Hispanic-Elderly.aspx

Aggarwal, N. T. (2013). Diverse populations, health disparities and dementia. In *2013 NIH/ACL*

Alzheimer's Webinar Series. Retrieved from http://www.aoa.acl.gov/AoA_Programs/HPW/Alz_Grants/docs/2013_07_24-DiversityHealthDisparitiesDementia-slides.pdf

Alzheimer's Association. (2010). *Alzheimer's disease facts and figures: Includes a special report on race, ethnicity, and Alzheimer's disease*. Retrieved from http://www.alz.org/documents_custom/report_alzfactsfigures2010.pdf

Alzheimer's Association. (2016a). *Alzheimer's disease facts and figures*. Retrieved from www.alz.org/documents_custom/2016-facts-and-figures.pdf

Alzheimer's Association. (2016b). *Latinos and Alzheimer's*. Retrieved from http://www.alz.org/espanol/about/latinos_and_alzheimers.asp

Añez, L. M., Silva, M. A., Paris, M., & Bedregal, L. E. (2008). Engaging Latinos through the integration of cultural values and motivational interviewing principles. *Journal of Professional Psychology: Research and Practice, 39*, 155–159.

Angel, J. L., Angel, R. J., McClellan, J. L., & Markides, K. S. (1996). Nativity, declining health, and preferences in living arrangements among elderly Mexican Americans: Implications for long term care. *The Gerontologist, 36*, 464–473.

Angel, J. L., Rote, S. M., Brown, D. C., Angel, R. J., & Markides, K. S. (2014). Nativity status and sources of care assistance among elderly Mexican-origin adults. *Journal of Cross-Cultural Gerontology, 29*, 243–258. https://doi.org/10.1007/s10823-014-9234-9

Angel, R. J., & Angel, J. L. (2009). *Hispanic families at risk: The new economy, work, and the welfare state*. New York: Springer Sciences.

Apesoa-Varano, E. C., Barker, J. C., & Hinton, L. (2012). Mexican-American families and dementia: An exploration of "work" in response to dementia-related aggressive behavior. In *Aging, health, and longevity in the Mexican-origin population* (pp. 277–291). New York, NY: Springer.

Arce, C. H., Murgia, E., & Frisbie, W. P. (1987). Phenotype and life chances among Chicanos. *Hispanic Journal of Behavioral Sciences, 9*, 19–22.

Arévalo-Flechas, L. C., Acton, G., Escamilla, M. I., Bonner, P. N., & Lewis, S. L. (2014). Latino Alzheimer's caregivers: What is important to them? *Journal of Managerial Psychology, 29*(6), 661–684.

Atkinson, D. R., Morten, G., & Sue, D. W. (1989). *Counseling American minorities: A cross-cultural perspective*. Dubuque, IA: Brown.

Ayalon, L., & Areán, P. A. (2004). Knowledge of Alzheimer's disease in four ethnic groups of older adults. *International Journal of Geriatric Psychiatry, 19*(1), 51–57.

Bernal, G., Jiménez-Chafey, M. I., & Domenech Rodríguez, M. M. (2009). Cultural adaptation of treatments: A resource for considering culture in evidence-based practice. *Professional Psychology: Research and Practice, 40*(4), 361–368. https://doi.org/10.1037/a0016401

Berry, J. W. (1990). Psychology of acculturation. In R. W. Brislin (Ed.), *Applied cross-cultural psychology* (pp. 232–253). Newbury Park, CA: Sage.

Bosma, H., Apland, L., & Kazanjian, A. (2010). Cultural conceptualizations of hospice palliative care: More similarities than differences. *Journal of Palliative Medicine, 24*, 510–522.

Brown, A., & Patten, E. (2013). *Hispanics of Puerto Rican origin in the United States, 2011*. Retrieved from www.pewhispanic.org/2013/06/19/hispanics-of-puerto-rican-origin-in-the-united-states-2011/

Capielo Rosario, C., Adames, H. Y., Chavez-Dueñas, N. Y., & Rentería, R. (2019). Acculturation Profiles of Central Florida Puerto Ricans: Examining the influence of perceived ethnic-racial discrimination and neighborhood ethnic-racial composition. *Journal of Cross-Cultural Psychology, 50*(4), 556–576. https://doi.org/10.1177/0022022119835979

Carter, R. T. (1991). Cultural values: A review of empirical research and implications for counseling. *Journal of Counseling & Development, 70*(1), 164–173. https://doi.org/10.1002/j.1556-6676.1991.tb01579x

Center for Social Justice at Seton Hall Law School and the Health Justice Program at New York Lawyers for the Public Interest. (CSJ; NYLPI; 2012). *Discharge, deportation, and dangerous journeys: A study on the practice of medical repatriation*. Retrieved from https://law.shu.edu/ProgramsCenters/PublicIntGovServ/CSJ/upload/final-med-repat-report-2012.pdf

Centers for Disease Control and Prevention. (2015). *Hispanics' health in the United States*. Retrieved from http://www.cdc.gov/media/releases/2015/p0505-hispanic-health.html

Chandler, C. R. (1979). Traditionalism in a modern setting: A comparison of Anglo- and Mexican-American value orientations. *Human Organization, 38*(7), 153–159.

Chavez-Dueñas, N. Y., Adames, H. Y., & Organista, K. C. (2014). Skin-color prejudice and within-group racial discrimination: Historical and current impact on Latino/a populations. *Hispanic Journal of Behavioral Sciences, 36*(1), 3–26. https://doi.org/10.1177/0739986313511306

Chavez-Dueñas, N. Y., Adames, H. Y., Perez-Chavez, J. G., & Salas, S. P. (2019). Healing ethno-racial trauma in Latinx immigrant communities: Cultivating hope, resistance, and action. *American Psychologist, 74*(1), 49–62. http://dx.doi.org/10.1037/amp0000289

Connell, C. M., Roberts, J. S., McLaughlin, S. J., & Carpenter, B. D. (2009). Black and white adult family members' attitudes toward a dementia diagnosis. *Journal of the American Geriatrics Society, 57*(9), 1562–1568.

Coon, D. W., Rubert, M., Solano, N., Mausbach, B., Kraemer, H., Arguelles, T., … Gallagher-Thompson, D. (2004). Well-being, appraisal, and coping in Latina and Caucasian female dementia caregivers: Findings from the REACH study. *Aging & Mental Health, 8*(4), 330–345.

Cooper, L. A., & Roter, D. L. (2003). Patient-provider communication: The effect of race and ethnicity on process and outcomes of healthcare. In B. D. Smedley, A. Stith, & A. R. Nelson (Eds.), *Unequal treatment: Confronting racial and ethnic disparities in health care* (pp. 552–593). Washington, DC: The National Academic Press.

Cooper, L. A., Roter, D. L., Johnson, R. L., Ford, D. E., Steinwachs, D. M., & Powe, N. R. (2003). Patient-centered communication, ratings of care, and concordance of patient and physician race. *Annals of Internal Medicine, 11*, 907–915.

Cornelius, W. A. (2009). Ambivalent reception: Mass public responses to the "New" Latino immigration to the United States. In M. M. Suárez-Orozco & M. M. Paez (Eds.), *Latinos remaking America* (pp. 165–189). Berkeley, CA: University of California Press.

Cox, C., & Monk, A. (1993). Hispanic culture and family care of Alzheimer's patients. *Health & Social Work, 18*(2), 92–100.

Cox, C. B. (2007). *Dementia and social work practice: Research and interventions*. New York, NY: Springer.

Crist, J. D., Garcia-Smith, D., & Phillips, L. R. (2006). Accommodating the stranger en casa: How Mexican American elders and caregivers decide to use formal care. *Research and Theory for Nursing Practice: An International Journal, 20*(2), 109–126. https://doi.org/10.1891/088971806780641791

Crist, J. D., McEwen, M. M., Herrera, A. P., Kim, S. S., Pasvogel, A., & Hepworth, J. T. (2009). Caregiving burden, acculturation, familism, and Mexican American elders' use of home care services. *Research and Theory for Nursing Practice: An International Journal, 23*(3), 165–166. https://doi.org/10.1891/1541-6577.23.3.165

Cuéllar, I., Arnold, B., & Maldonado, R. (1995). Acculturation rating scale for Mexican Americans-II: A revision of the original ARSMA scale. *Hispanic Journal of Behavioral Sciences, 17*, 275–304.

De La Cancela, V. (1986). A critical analysis of Puerto Rican machismo: Implications for clinical practice. *Psychotherapy Theory, Research, & Practice, 23*(2), 291–296. https://doi.org/10.1037/h0085611

DeNavas, C., Proctor, B. D., & Smith, J. C. (2011). *Income, poverty, and health insurance coverage in the United States: 2010*. Washington, DC: United States Census Bureau.

Depp, C., Sorocco, K., Kasl-Godley, J., Thompson, L., Rabinowitz, Y., & Gallagher-Thompson, D. (2005). Caregiver self-efficacy, ethnicity, and kinship differences in dementia caregivers. *The American Journal of Geriatric Psychiatry, 13*(9), 787–794.

Dilworth-Anderson, P., Hendrie, H. C., Manly, J. J., Khachaturian, A. S., & Fazio, S. (2008). Diagnosis and assessment of Alzheimer's disease in diverse populations. *Alzheimer's & Dementia: The Journal of the Alzheimer's Association, 4*(4), 305–309. https://doi.org/10.1016/j.jalz.2008.03.001

Donelson, K. (2015). Medical repatriation: The dangerous intersection of health care law and immigration. *Journal of Health Care Law and Policy, 18*(2), 347–369.

Emergency Medical Treatment and Active Labor Act (EMTALA; 1986). Retrieved from https://www.cms.gov/Regulations-and-Guidance/Legislation/EMTALA/

Espino, D. V., Angel, J. L., Wood, R. C., Finely, M. R., & Ye, Y. (2013). Characteristics of Mexican American elders admitted to nursing facilities in the United States: Data from the Hispanic established populations for epidemiologic studies of the elderly (EPESE) study. *Journal of the American Directors Association, 14*(3), 226.e221–226.e224.

Evercare and the National Alliance for Caregiving. (2008). *Evercare study of Hispanic family caregiving in the U.S.: Findings from a national study*. Retrieved from http://www.caregiving.org/data/Hispanic_Caregiver_Study_web_ENG_FINAL_11_04_08.pdf

Falicov, C. J. (1989). *Latino families in therapy* (2nd ed.). New York, NY: Guilford Press.

Ferri, C. P., Prince, M. J., Brayne, C., Brodaty, H., Fratiglioni, L., Ganguli, M., … Alzheimer's Disease International. (2005). Global prevalence of dementia: A Delphi consensus study. *Lancet, 366*(9503), 2112–2117. https://doi.org/10.1016/S0140-6736(05)67889-0

Flores, E., Tschann, J. M., Dimas, J. M., Bachen, E. A., Pasch, L. A., & de Groat, C. L. (2008). Perceived discrimination, perceived stress, and mental and physical health among Mexican-origin adults. *Hispanic Journal of Behavioral Sciences, 30*(4), 401–424. https://doi.org/10.1177/0739986308323056

Fortinsky, R. H. (2014). *Challenges in supporting people with dementia from Latino backgrounds and their family careers*. Workshop presented at Making Research Count Practice Development. King's College, London.

French, B. H., Lewis, J. A., Mosely, D., Adames, H. Y., Chaves-Dueñas, N. Y., Chen, G. A., & Neville, H. A. (2019). Toward a psychological framework of radical healing in Communities of Color. The Counseling Psychologist. https://doi.org/10.1177/0011000019843506

Gallagher-Thompson, D., Singer, L. S., Depp, C., Mausbach, B. T., Cardenas, V., & Coon, D. W. (2004). Effective recruitment strategies for Latino and Caucasian dementia family caregivers in intervention research. *The American Journal of Geriatric Psychiatry, 12*(5), 484–490.

Gallagher-Thompson, D., Tzuang, M., Hinton, L., Alvarez, P., Rengifo, J., Valverde, I., … Thompson, L. W. (2015). Effectiveness of a fotonovela for reducing depression and stress in Latino dementia family caregivers. *Alzheimer Disease and Associated Disorders, 29*(2), 146.

Gassoumis, Z. D., Wilber, K. H., Baker, L. A., & Torres-Gil, F. M. (2010). Who are the Latino baby boomers? Demographic and economic characteristics of a hidden population. *Journal of Aging & Social Policy, 22*(1), 53–68.

Gil, R. M., & Vasquez, C. I. (1996). *The Maria paradox*. New York, NY: Perigee.

Haan, M. N., Mungas, D. M., Gonzalez, H. M., Ortiz, T. A., Acharya, A., & Jagust, W. J. (2003). Prevalence of dementia in older Latinos: The influence of type 2 diabetes mellitus, stroke and genetic factors. *Journal of the American Geriatrics Society, 51*, 169–177.

Helms, J. E. (1990). *Black and white racial identity: Theory, research, and practice*. Westport, CT: Greenwood Press.

Hernandez, A. M., & Bigatti, S. M. (2010). Depression among older Mexican American caregivers. *Cultural Diversity and Ethnic Minority Psychology, 16*(1), 50–58. https://doi.org/10.1037/a0015867

Hinton, L., Chambers, D., & Velásquez, A. (2009). Making sense of behavioral disturbances in persons with dementia: Latino family caregiver attributions of neuropsychiatric inventory domains. *Alzheimer's Disease and Associated Disorders, 23*(4), 401–405. https://doi.org/10.1097/WAD.0b013e3181a6bc21

Hinton, L., Chambers, D., Velásquez, A., Gonzalez, H., & Haan, M. (2006). Dementia neuropsychiatric symptom severity, help-seeking patterns, and unmet needs in the Sacramento Area Latino Study on Aging (SALSA). *Clinical Gerontologist, 29*(4), 1–16.

Hinton, L., Franz, C. E., Yeo, G., & Levkoff, S. E. (2005). Concepts of dementia in a multiethnic sample of caregivers. *Journal of the American Geriatrics Society, 53*(8), 1405–1410.

Hinton, L., Haan, M., Gellar, S., & Mungas, D. (2003). Neuropsychiatric symptoms in Latino elders with dementia or cognitive impairment without dementia and factors that modify their association with caregiver depression. *The Gerontologist, 43*(5), 669–677.

Hinton, L., Tomaszewski Farias, S., & Wegelin, J. (2008). Neuropsychiatric symptoms are associated with disability in cognitively impaired Latino elderly with and without dementia: Results from the Sacramento Area Latino Study on Aging. *International Journal of Geriatric Psychiatry, 23*(1), 102–108. https://doi.org/10.1002/gps.1952

Inclan, J. (1985). Variations in value orientations in mental health work with Puerto Ricans. *Psychotherapy: Theory, Research, Practice, Training, 22*(2S), 324–334. https://doi.org/10.1037/h0085511

John, R., Resendiz, R., & de Vargas, L. W. (1997). Beyond familism?: Familism as explicit motive for eldercare among Mexican American caregivers. *Journal of Cross-Cultural Gerontology, 12*, 145–162.

Kalibatseva, Z., & Leong, F. L. (2014). A critical review of culturally sensitive treatments for depression: Recommendations for intervention and research. *Psychological Services, 11*(4), 433–450. https://doi.org/10.1037/10036047

Karlawish, J., Barg, F. K., Augsburger, D., Beaver, J., Ferguson, A., & Nunez, J. (2011). What Latino Puerto Ricans and non-Latinos say when they talk about Alzheimer's disease. *Alzheimer's & Dementia: The Journal of the Alzheimer's Association, 7*(2), 161–170. https://doi.org/10.1016/j.jalz.2010.03.015

Killin, L. O., Starr, J. M., Shiue, I. J., & Russ, T. C. (2016). Environmental risk factors for dementia: A systematic review. *BMC Geriatrics, 16*(1), 175. https://doi.org/10.1186/s12877-016-0342-y

Kluckhohn, F. R., & Strodtbeck, F. L. (1961). *Variations in value orientations*. Evanston, IL: Row, Peterson.

Kohatsu, E. L., Concepcion, W. R., & Perez, P. (2010). Incorporating levels of acculturation in counseling practice. In J. G. Ponterotto, J. M. Casas, L. A. Suzuki, & C. M. Alexander (Eds.), *Handbook of multicultural counseling* (2nd ed., pp. 343–456). Thousand Oaks, CA: Sage.

Krogstad, J. M., & Lopez, M. H. (2014). *Hispanic immigrants more likely to lack health insurance than U.S.-born*. Retrieved from http://www.pewresearch.org/fact-tank/2014/09/26/higher-share-of-hispanic-immigrants-than-u-s-born-lack-health-insurance/

Kronick, R. F., & Hargis, C. H. (1998). *Dropouts: Who drops out and why—And the recommended action*. Springfield, IL: Charles C. Thomas Publishing.

Kunda, Z., & Sherman-Williams, B. (1993). Stereotypes and the construal of individuating information. *Personality and Social Psychology Bulletin, 19*, 90–99.

Livney, M. G., Clark, C. M., Karlawish, J. H., Cartmell, S., Negrón, M., Nuñez, J., ... Arnold, S. E. (2011). Ethnoracial differences in the clinical characteristics of Alzheimer's disease at initial presentation at an urban Alzheimer's disease center. *The American Journal of Geriatric Psychiatry, 19*(5), 430–439. https://doi.org/10.1097/JGP.0b013e3181f7d881

Lopez, G., & Patten, E. (2015). *The impact of slowing immigration: Foreign-born share falls among 14 largest U.S. Hispanic origin groups*. Washington, DC: Pew Research Center. Retrieved from www.pewhispanic.org/2015/09/15/the-impact-of-slowing-immigration-foreign-born-share-falls-among-14-largest-us-hispanic-origin-groups/

Losada, A., Shurgot, G. R., Knight, B. G., Márquez, M., Montorio, I., Izal, M., & Ruiz, M. A. (2006). Cross-cultural study comparing the association of familism with burden and depressive symptoms in two samples of Hispanic dementia caregivers. *Aging & Mental Health, 10*(1), 69–76.

Magaña, S., Schwartz, S., Rubert, M., & Szapocznik, J. (2006). Hispanic caregivers of adults with mental retardation: The importance of family functioning. *American Journal on Mental Retardation, 111*, 250–262.

Manly, J. J., & Mayeux, R. (2004). Ethnic differences in dementia and Alzheimer's disease. In N. B. Anderson, R. A. Bulatao, & B. Cohen (Eds.), *Critical perspectives on racial and ethnic differences in health in late life* (pp. 95–141). Washington, DC: The National Academic Press.

Mehta, K. M., Yin, M., Resendez, C., & Yaffe, K. (2005). Ethnic differences in acetylcholinesterase inhibitor use for Alzheimer disease. *Neurology, 65*(1), 159–162.

Mendez-Luck, C. A., Applewhite, S. R., Lara, V. E., & Toyowaka, N. (2016). The concept of familism in the lived experiences of Mexican-origin caregivers. *Journal of Marriage and Family, 78*, 813–829. https://doi.org/10.1111/jomf.12300

Migration Policy Institute. (2014). *Profile of the unauthorized population: United States*. Retrieved from http://www.migrationpolicy.org/data/unauthorized-immigrant-population/state/US

Miller, W. R., & Rollnick, S. (2013). *Motivational interviewing: Helping people change*. New York, NY: Guilford Press.

Montalvo, F. F. (2005). Surviving race: Skin color and the socialization and acculturation of Latinas. *Journal of Ethnic and Cultural Diversity in Social Work, 13*(3), 25–43.

Montalvo, F. F., & Codina, G. E. (2001). Skin color and Latinos in the United States. *Ethnicities, 1*(3), 321–341. https://doi.org/10.1177/146879680100100303

National Conference of State Legislatures. (2011). *Immigrant policy project: 2010 Immigration-related laws and resolutions in the States*. Retrieved from www.ncsl.org/research/immigration/2010-immigration-related-laws-and-resolutions-in-t.aspx

National Council of La Raza. (1990). *Immigration reform*. Retrieved from http://www.nclr.org/content/topics/detail/500

National Council of La Raza. (2008). *Five facts about undocumented workers in the United States*. Retrieved from http://publications.nclr.org/bitstream/handle/123456789/1002/FS_FiveFacts_1.pdf?sequence=1&isAllowed=y

National Hispanic Council on Aging. (2016). *Status of Hispanic older adults: Insights from the field report*. Retrieved from http://www.nhcoa.org/wp-content/uploads/2016/09/2016-NHCOA-Status-of-Hispanic-Older-Adults-report-.pdf

Neary, S. R., & Mahoney, D. F. (2005). Dementia caregiving: The experience of Hispanic/Latino caregivers. *Journal of Transcultural Nursing, 16*(2), 163–170. https://doi.org/10.1177/1043659604273547

Organista, K. C. (2007). *Solving Latino psychosocial and health problems: Theory, practice, and populations*. Hoboken, NJ: Wiley.

Ory, M. G., Hoffman, R. R., Yee, J. L., Tennstedt, S., & Schulz, R. (1999). Prevalence and impact of caregiving: A detailed comparison between dementia and nondementia caregivers. *The Gerontologist, 39*(2), 177–186.

Phinney, J. (1989). Stages of ethnic identity development in minority group adolescents. *Journal of Early Adolescence, 9*, 34–49.

Ramos, B., Jaccard, J., & Guilamo-Ramos, V. (2003). Dual ethnicity and depressive symptoms: Implications of being black and Latino/a in the United States. *Hispanic Journal of Behavioral Sciences, 25*(2), 147–173. https://doi.org/10.1177/0739986303025002002

Rodríguez, J. J. L., Ferri, C. P., Acosta, D., Guerra, M., Huang, Y., Jacob, K. S., … Prince, M. J. (2008). Prevalence of dementia in Latin America, China, and India: A population-based cross-sectional survey. *Lancet, 372*, 464–474.

Rodríguez, J. J. L., Valhuerdi, A., Sanchez, I. I., Reyna, C., Guerra, M. A., Copeland, J. R., … Prince, M. J. (2008). The prevalence, correlates and impact of dementia in Cuba: A 10/66 group population-based survey. *Neuroepidemiology, 31*, 243–251. https://doi.org/10.1159/000165362

Romero, A. J., Cuéllar, I., & Roberts, R. E. (2000). Ethnocultural variables and attitudes toward cultural socialization of children. *Journal of Community Psychology, 28*(1), 79–89. https://doi.org/10.1002/(SICI)1520-6629(200001)28:1<79::AID-JCOP8>3.0.CO;2-N

Romero, D. (2016). *In the era of trump, anti-Latino hate crimes jumped 69% in L.A.* Retrieved from http://www.laweekly.com/news/in-the-era-of-trump-anti-latino-hate-crimes-jumped-69-in-la-7443401

Rozario, P. A., & DeRienzis, D. (2008). Familism beliefs and psychological distress among African American women caregivers. *The Gerontologist, 48*(6), 772–780. https://doi.org/10.1093/geront/48.6.772

Saetermoe, C. L., Beneli, I., & Busch, R. M. (1999). Perceptions of adulthood among Anglo and Latino parents. *Current Psychology, 18*(2), 171–184. https://doi.org/10.1007/s12144-999-1026-y

Santiago-Rivera, A. L., Adames, H. Y., Chavez-Dueñas, N. Y., & Benson-Florez, G. (2016). The impact of racism on communities of color: Historical contexts and contemporary issues. In A. N. Alvarez, C. T. H. Liang, & H. A. Neville (Eds.), *The cost of racism for people of color: Contextualizng Expereinces of discrimination* (pp. 229–245). Washington, DC: APA Books. https://doi.org/10.1037/14852-011

Santiago-Rivera, A. L., Arredondo, P., & Gallardo-Cooper, M. (2002). *Counseling Latinos and la familia: A practical guide*. Thousand Oaks, CA: Sage Publications.

Scharlach, A. E., Kellam, R., Ong, N., Baskin, A., Goldstein, C., & Fox, P. J. (2006). Cultural attitudes and caregiver service use: Lessons from focus groups with racially and ethnically diverse family caregivers. *Journal of Gerontological Social Work, 47*(1–2), 133–156. https://doi.org/10.1300/J083v47n01_09

Schulz, R., O'Brien, A. T., Bookwala, J., & Fleissner, K. (1995). Psychiatric and physical morbidity effects of dementia caregiving: Prevalence, correlates, and causes. *The Gerontologist, 35*(6), 771–791.

Shurgot, G. R., & Knight, B. G. (2005). Influence on neuroticism, ethnicity, familism, and social support on perceived burden in dementia caregivers: Pilot test of the transactional stress and social support model. *Journal of Gerontology: Psychological Sciences, 60*(6), P331–P334. https://doi.org/10.1093/geronb/60.6.P331

Smedley, B. D., Stith, A., & Nelson, A. R. (2003). *Unequal treatment: Confronting racial and ethnic disparities in health care*. Washington, DC: The National Academic Press.

Sontag, D. (2008). *Departed from U.S. in a coma, returned to U.S. to be saved*. Retrieved from http://www.nytimes.com/2008/11/09/world/americas/09iht-deport.1.17653904.html

Sousa, R. M., Ferri, C. P., Acosta, D., Guerra, M., Huang, Y., Jacob, K. S., … Prince, M. (2010). The contribution of chronic diseases to the prevalence of dependence among older people in Latin America, India, and China: A 10/66 dementia research group population-based survey. *BioMed Central Geriatrics, 10*(53). https://doi.org/10.1186/1471-2318-10-53

Stacey, M., Carbone-López, K., & Rosenfeld, R. (2011). Demographic change and ethnically motivated crime: The impact of immigration on anti-Hispanic hate crime in the United States. *Journal of Contemporary Criminal Justice, 27*, 278–298.

Suárez-Orozco, C., & Suárez-Orozco, M. (2001). *Children of immigration*. Cambridge, MA: Harvard University Press.

Szapocznik, J., Scopetta, M. A., & King, O. E. (1978). Theory and practice in matching treatment to the special characteristics and problems of Cuban immigrants. *Journal of Community Psychology, 6*(2), 112–122. https://doi.org/10.1002/1520-6629(197804)6:2<112::AID-JCOP2290060203>3.0.CO:2-R

Talamantes, M. A., Trejo, L., Jiminez, D., & Gallagher-Thompson, D. (2006). Working with Mexican American families. In G. Yeo & D. Gallagher-Thompson (Eds.), *Ethnicity and the dementias* (2nd ed., pp. 327–340). New York, NY: Routledge.

Tobin, J. N., Wassertheil-Smoller, S., Wexler, J. P., Steingart, R. M., Budner, N., Lense, L., & Wachspress, J. (1987). Sex bias in considering coronary bypass surgery. *Annals of Internal Medicine, 107*(1), 19–25. https://doi.org/10.7326/0003-4819-107-1-19

U.S. Census Bureau. (2004). *U.S. interim projections by age, sex, race, and Hispanic origin*. Retrieved from http://www.census.gov/ipc/www.usinterimproj/

U.S. Census Bureau. (2016). *Profile America facts for features: Hispanic heritage month 2016*. Retrieved from http://www.census.gov/content/dam/Census/newsroom/facts-for-features/2016/cb16-ff16.pdf

Valle, R., Cook-Gait, H., & Tazbaz, D. (1993). *The cross-cultural Alzheimer/dementia caregiver comparison study*. Paper presented at 46th scientific meeting of the Gerontological Society of America. New Orleans, LA.

Valle, R., Yamada, A. M., & Matiella, A. C. (2006). Fotonovelas: A health literacy tool for educating Latino older adults about dementia. *Clinical Gerontologist, 30*(1), 71–88.

Van Ryn, M., & Burke, J. (2000). The effect of patient race and socio-economic status on physicians' perceptions of patients. *Social Science & Medicine, 50*(6), 813–828. https://doi.org/10.1016/S0277-9536(99)00338-X

Velásquez, L. C. (1993). *Migrant adults' perceptions of schooling, learning, and education*. Doctoral dissertation College of Education, University of Tennessee, Knoxville, TN.

Villarruel, F. A., Carlo, G., Grau, J. M., Azmitia, M., Cabrera, N. J., & Chahin, T. J. (2009). *Handbook of U.S. Latino psychology: Development and community-based perspectives*. Thousand Oaks, CA: Sage Publications.

Wallace, S. P., & Facio, E. L. (1987). Moving beyond familism: Potential contributions of gerontological theory to studies of Chicano/Latino aging. *Journal of Aging Studies, 10*(1), 337–354.

Weiss, C. O., Gonzalez, H. M., Kabeto, M. U., & Langa, K. M. (2005). Differences in amount of informal care received by non-Hispanic whites and Latinos in a nationally representative sample of older Americans. *Journal of the American Geriatrics Society, 53*(1), 146–151.

Wu, S., Vega, W. A., Resendez, J., & Jin, H. (2016). *Latinos & Alzheimer's disease: New numbers behind the crisis*. Retrieved from https://roybal.usc.edu/wp-content/uploads/2016/10/Latinos-and-AD_USC_UsA2-Impact-Report.pdf

Youn, G., Knight, B. G., Jeong, H. S., & Benton, D. (1999). Differences in familism values and caregiving outcomes among Korean, Korean American, and white American dementia caregivers. *Psychology and Aging, 14*(3), 355–364. https://doi.org/10.1037/0882-7974.14.3.355

Zarit, S. H., Todd, P. A., & Zarit, J. M. (1986). Subjective burden of husbands and wives as caregivers: A longitudinal study. *The Gerontologist, 26*(3), 260–266.

Meta-Analysis of Dementia Rates in Central America, South America, and the Caribbean

3

Carlos A. Rodriguez

Abstract

There has been a rapid increase in epidemiological surveys on the prevalence of dementia in the countries of Central America, South America, and the Caribbean. However, no past review of these studies has provided an overall quantitative estimate of the prevalence of dementia or its major subtypes (e.g., Alzheimer's disease and vascular dementia) in the populations of these regions. Further, the majority of past reviews have described the frequency of this disorder without regard to sources of variance between studies. This chapter provides estimates of the prevalence of dementia and two of its major subtypes based on meta-analysis. Investigation of the sources of heterogeneity in prevalence between studies is reported. These estimates are then discussed in terms of how they compare to the prevalence of dementia and its subtypes in the available studies of Latinos in the United States of America (USA). Knowledge of the occurrence of dementia and its subtypes is necessary to inform public health policies to prepare for the rapid changes that are expected in the population age structures of these countries.

C. A. Rodriguez (✉)
Spectrum Health, Grand Rapids, MI, USA
e-mail: usc.c.rod@gmail.com

Introduction

Population aging in Central America, South America, and the Caribbean is occurring at an unprecedented rate (United Nations Department of Economic and Social Affairs, 2007). Of the age-related concerns facing countries in these regions, the cost of dementia care is estimated to be increasingly high (Allegri et al., 2007). Similarly, in the United States of America (USA), Latinos represent one of the fastest growing segments of the older adult population (U.S. Administration on Aging, 2011). Further, Latinos with dementia in the USA have been observed to have higher levels of cognitive impairment at the first presentation to the initial evaluation and report significantly longer periods of cognitive decline by the time they are referred to diagnostic services compared to non-Hispanic Whites (Cooper, Tandy, Balamurali, & Livingston, 2010). Timely diagnosis and treatment of Alzheimer's disease (AD) and other dementias have the potential to offset immense costs associated with institutionalized care, increasing savings for the countries of these regions (Weimer & Sager, 2009). Knowing the prevalence of dementia and its various subtypes among Latinos throughout the USA, Central America, South America, and the Caribbean can help prioritize and plan healthcare programs and policies aimed at targeting undetected cases of dementia among Latinos.

H. Y. Adames, Y. N. Tazeau (eds.), *Caring for Latinxs with Dementia in a Globalized World*,
https://doi.org/10.1007/978-1-0716-0132-7_3

The purpose of the current meta-analysis is to provide an update of the prevalence of dementia and two of its major subtypes in Central America, South America, and the Caribbean, and compare this to rates among Latinos in the USA. The populations from the 22 countries reviewed are composed of heterogeneous groups of individuals with varying ethnic and racial backgrounds. The goal of providing pooled estimates for regions of such widely diverse groups is to provide a starting point for determining the needs of a targeted population. Later, if differences emerge between or within regions or subgroups, clues as to the genetic, social, and environmental contributions to a disease process can be gleaned from the epidemiological findings. For example, comparing the prevalence of dementia and AD between Nigerian Africans and African Americans – two groups with the same ethnic background but living in starkly different environments – Hendrie et al. (1995) found the rates of dementia and AD to be lower among Nigerian Africans than African Americans, pointing to possible environmental or gene–environmental interactions. Similarly, determining the prevalence of dementia and its subtypes among Latinos in the USA – in the context of the larger group of countries with whom they have current or historical ties – can further the aim of determining the needs of this targeted population.

Past Reviews of Dementia Prevalence

In earlier reviews of the literature on the prevalence of dementia in Central America, South America, and the Caribbean, studies focused mainly on describing the instruments used to diagnose dementia, the availability of dementia care services in various countries throughout the region, and the need to establish better epidemiological strategies in various counties (Arizaga, Mangone, Allegri, & Ollari, 1999; Mangone & Arizaga, 1999). The few surveys conducted at the time on the prevalence of dementia were not described in enough detail in terms of their methodology to draw comparisons between studies or provide summary estimates.

For example, Ferri et al. (2005) developed a consensus panel of experts to generate estimates of the prevalence of dementia in all the World Health Organization (WHO)-defined regions. This was based on a systematic review of the world literature of population-based studies from 1980 to 2004 as indexed on MEDLINE and from the United Nations population projections. However, only three studies from Latin American countries on the prevalence of dementia were identified by the systematic review: One was from Brazil (Herrera, Caramelli, Silveira, & Nitrini, 2002), another from a study of incidence of dementia in Brazil (Nitrini et al., 2004), and one from Colombia (Roselli et al., 2000), which defined dementia based only on performance on the Mini-Mental State Examination (MMSE; Folstein, Folstein, & McHugh, 1975). The consensus panel estimated the prevalence of dementia in Latin America and the Caribbean in adults 60 years and older at 4.6%.

Kalaria et al. (2008) reported estimates of prevalence based on 11 studies for seven countries: Cuba (8.2% for 60 and older), Argentina (11.5% for 65 and older), Brazil (5.3% for 65 and older), Chile (4.3%), Colombia (1.8% for 65 and older; 3.4% for 75 and older), Peru (6.7%), and Venezuela (8.0% for 55 and older; 10.3% for 65 and older). Estimates of the prevalence of subtypes of dementia, AD, and vascular dementia (VaD) were available for three countries: Cuba (60 and older: 5.1% for AD and 1.9% for VaD), Brazil (65 and older: 2.7% for AD and 0.9% for VaD), and Venezuela (55 and older: 4.0% for AD and 2.1% for VaD). However, the methodologies used to define, sample, and assess dementia in the populations of these studies were not specified, other than to list the version of the Diagnostic and Statistical Manual of Mental Disorders (DSM; APA, 2000) used to define cases of dementia; no specification as to how AD and VaD cases were made was reported.

Nitrini et al. (2009) conducted a collaborative study on the prevalence of dementia in the Latin America countries and focused their review on studies of large, general populations that had data available to stratify dementia prevalence into 5-year age periods starting at age 65, and by gender and levels of education (i.e., illiterate vs. literate). They searched the MEDLINE and LILACS databases and contacted authors of population surveys to ask if they would be willing to collaborate. Final data were presented from eight studies in six countries. The unadjusted pooled estimate of the prevalence of dementia was calculated as 7.1%; age-standardized to the world population, it was 5.97%. The high rate of dementia found in the Latin American countries stood in contrast to the estimations of Ferri et al. (2005) that had concluded the prevalence of dementia in developing regions was lower than that in developed regions (Nitrini et al., 2009). Prevalence varied widely between countries, from 2% in Brazil to 13% in Venezuela. Higher rates of dementia prevalence were found in the 65 to 69 age ranges compared to a systematic review of European studies (Lobo et al., 2000). There were slightly higher rates for women than men, and rates were twice as high in illiterate participants than literate participants. Details regarding criteria used for estimating the prevalence of AD and VaD were not provided. Further, the authors presented only the ranges of the proportion of dementia cases caused by AD (49.9% in Venezuela to 84.5% in Chile) and VaD (8.7% in Peru to 26.5% in Venezuela) in those 65 and older.

In 2009, Alzheimer's Disease International developed a World Alzheimer Report based on a systematic review of the world literature and meta-analysis in regions when possible (the same findings were also presented by Prince et al., 2013). Strict criteria for inclusion and exclusion were applied, focusing on population-based studies of the prevalence of dementia among adults 60 and older (according to the DSM-IV or International Classification of Diseases-10 edition [ICD-10] or similar criteria), for which fieldwork started on or after January 1, 1980, and using the PUBMED/MEDLINE databases through March 2009. Prince et al. (2013) included 11 studies for their meta-analysis of the Latin American region. The estimated standardized prevalence, based on the Western European population, for Latin America was 8.5%. By individual region in 2010, the estimation of dementia prevalence in adults 60 and older for the Caribbean was 6.5% (calculated not by meta-analysis, but reliance on individual studies from Cuba and the Dominican Republic), Andean Latin America (5.6%), Central Latin America (6.1%), Southern Latin America (7.0%), and Tropical Latin America (5.5%). Dementia prevalence doubled for every 5.5-year increment in age in Latin America. The prevalence in men was lower than that of women, with a trend for the discrepancy to increase with age, though this was not statistically significant. They noted significant heterogeneity in estimates from Latin America. However, investigation of the sources of this heterogeneity was achieved only in the sample of studies from Western Europe where methodological factors (e.g., higher prevalence in studies that incorrectly applied two-phase methodology; higher prevalence were reflected in studies using an informant interview), year of study (higher prevalence was noted in studies carried out in the 1990s compared to earlier), and country accounted for portions of the overdispersion (i.e., the extent to which values of a variable differ from a fixed value such as the mean).

A systematic review with meta-analysis examined studies conducted in Brazil (Fagundes, Silva, Silva Thees, & Pereira, 2011). The study identified articles in MEDLINE, EMBASE, LILACS, SciELO, and a Brazilian thesis database, and found studies by reviewing the references sections. The authors focused on population-based studies using cognitive tests to assess dementia among individuals 60 years of age and older and excluded articles dealing with test validation or those that sampled participants with cognitive impairment due to sources other than dementia. They developed a uniform quality scoring system to code each of their eleven articles. Studies reported rates between 5.1% and 9.0%, and the authors made observations that rates increased with age, were higher in lower socioeconomic populations, were higher in those with less years of education, and were higher in women. Study quality was evaluated based on eight criteria, each receiving a score of zero or one for random sample or whole population, unbiased sampling frame, adequate sample size (>300 subjects), standard measurements, outcomes measured by unbiased assessors, adequate response rate (>70%), and subjects described as refusing treatment, confidence intervals, and subgroup analysis. Studies receiving a total score of 6 to 8 were considered high quality, while studies receiving a 5 or below were considered low quality. Only five studies were ranked as high quality, and in the analyses of sources of heterogeneity among the studies, study quality predicted heterogeneity, with lower-quality studies overestimating the prevalence of dementia. Many of the studies in their review diagnosed dementia based only on a cutoff score of a single screening measure, thus making it difficult to draw comparisons with studies using more rigorous criteria for case definition.

Prevalence of Dementia Among Latinos in the USA

Systematic reviews of the prevalence of dementia among Latinos in the USA are lacking. Narrative reviews of ethnic minorities covering dementia epidemiology, treatment, and access are more readily found. Harwood and Ownby (2000) identified two articles reporting the prevalence of dementia among Hispanic Americans in the USA (Perkins et al., 1997; Gurland et al., 1999). Perkins sampled from a population of municipal retirees in Houston, Texas, and found an age-adjusted (60 years–80 years) prevalence of dementia of 4.75%, using the National Institute of Neurological and Communicative Disorders and Stroke and the Alzheimer's Disease and Related Disorders Association criteria (NINCDS-ADRDA; McKhann et al., 1984). Gurland et al. (1999) randomly sampled from a population of Medicare beneficiaries in Northern Manhattan, as well as nursing home residents gathered from informants at health and social service sites. The cases were defined based on algorithms combining psychometric tests, laboratory, and clinical findings. Hispanic Americans had a prevalence of dementia of 7.5% for 65–79 year olds, 27.9% for 75–84 year olds, and 62.9% for 85 years and older. AD accounted for 88.3% of all dementia cases.

Three other studies provided information on convenience or clinical samples regarding rates of dementia and/or the relative frequency of dementia subtypes. In a community outreach sample in California's East San Fernando Valley (a region with a high concentration of elderly Hispanic Americans, mostly of Mexican descent), Fitten, Ortiz, and Ponton (2001) recruited 100 community-dwelling Hispanic Americans. Based on the DSM-IV criteria, they identified 65 cases of dementia, of which 38.5% had AD (based on the National Institute of Neurological and Communicative Disorders and Stroke–Alzheimer's Disease and Related Disorders Association (NINCDS–ADRDA) criteria) and 38.5% had VaD (based on the National Institute of Neurological Disorders and Stroke–Association International pour la Recherche et L'Enseignement en Neurosciences criteria (NINDS–AIREN); Roman et al., 1993). Hou, Yaffe, Perez-Stable, and Miller (2006) assessed cases of dementia in the registry of the Alzheimer's Disease Research Centers of California between 1992 and 2002. Dementia was based on the DSM-III-R criteria, AD was defined by the NINCDS–ADRDA criteria, and VaD was defined by the State of California Alzheimer's Disease Diagnostic and Treatment Centers criteria (ADDTC; Román, Erkinjuntti, Wallin, Pantoni, & Chui, 2002). AD accounted for 82.5% of their cases of their dementia among Hispanic Americans and 10.5% of their cases of VaD. Carrion-Baralt, Suarez-Perez, del Rio, Moore, and Silverman (2010) reported results on a population of Puerto Rican veterans aged 65 and older who used the Veterans Affairs Caribbean Healthcare System, in San Juan, Puerto Rico, between October 2005 and March 2007. Based on the ICD-10 criteria, the prevalence of dementia in the sample was 12.69%, and the proportion of dementia cases diagnosed as VaD was 15.14%.

The specific prevalence of AD in this sample was not reported.

Each of the studies reviewed in this section used samples drawn from a mix of clinical, service-user populations, including a random sample from a Medicare registry. Morgan et al. (2008) provided data suggesting Hispanic Americans are more likely to be unfamiliar with the dynamics of the Medicare system than non-Hispanic Whites, and therefore populations reliant on this registry are arguably unrepresentative of the larger general population of Hispanic Americans living in New York. Attempts to generalize rates of dementia and its subtypes from these studies are wrought with difficulties, and therefore less biased samples need to be relied on to provide accurate rates of occurrence of this disorder. Other studies reporting on the prevalence of dementia and its subtypes in Hispanic Americans using population-based samples are included in the current systematic review.

Variation in Rates of Dementia and Its Subtypes

From the studies described above, a picture of variable rates of dementia emerges from populations in Central America, South America, the Caribbean, and the Latinos in the USA. Many studies worldwide, though not all, have observed variable rates between men and women and according to level of education (Sharp & Gatz, 2011). In addition, the two systematic reviews with meta-analyses reporting on Latin America and the Caribbean postulated that study quality and methodological features account for a portion of this variation between prevalence studies (Fagundes et al. 2011; Prince et al., 2013). These characteristics included sample size, accurate use of multiphase methodology, and incorporation of an informant interview. Fagundes et al. (2011) established a relationship between study quality and rates of dementia within samples of Brazilian studies. Prince et al. (2013) reported heterogeneity between Latin American and Caribbean studies, but did not assess the source of heterogeneity in this region due to an insufficient number of available studies. A study of sources of variation in prevalence rates worldwide provided evidence of variable rates in AD based on use of laboratory studies, use of both urban and rural populations, nonuse of neuroimaging, as well as lack of adjustment for false negatives (Corrada, Brookmeyer, & Kawas, 1995). In a rigorous systematic review and meta-analysis of studies directly comparing prevalence and incidence of dementia between geographical locations at various scales of geography, Russ, Batty, Hearnshaw, Fenton, and Starr (2012) reported differences between urban and rural sites, particularly for AD, and especially when early life rural living was taken into account.

Another source of variation in studies may be the criteria used to define cases of dementia. For example, the DSM-IV was used as one way of defining cases by the 10/66 Dementia Research Group (Llibre Rodriguez et al., 2008) in their collaborative studies of the prevalence of dementia in six sites throughout Latin America. They also defined cases based on an algorithm derived from multiple measures that they had piloted in older adults in 25 centers in India, China and Southeast Asia, Latin America and the Caribbean, and Africa (Prince, Acosta, Chiu, Scazufca, & Varghese, 2003). Dementia prevalence of 10/66 was higher than that of DSM-IV dementia and more consistent across sites. The majority of the cases defined by the 10/66 algorithm, but not confirmed by the DSM-IV, were supported by high levels of associated disability, suggesting the DSM-IV may not be operationalized equally in different regions and may be insensitive to milder stages of the disorder. Similarly, rate of VaD diagnosed was found to vary according to the criteria used (Román et al., 2002), with higher rates found using the DSM-IV, moderate rates found using the ADDTC, and lower rates identified by using the NINDS–AIREN criteria.

The current meta-analysis aimed to establish an updated prevalence of dementia in Central America, South America, the Caribbean, and among Latinos in the USA. Only one previous study (Kalaria et al., 2008) incorporated a systematic review of prevalence of dementia subtypes in Latin American and the Caribbean, but they did not describe criteria used to define cases or investigate sources of heterogeneity, nor were the methods of sampling detailed. No systematic review of Latinos in the USA has yet provided data on the rates of dementia or its subtypes. Given that past systematic reviews of the prevalence of dementia have discovered a large amount of variance between studies, the current meta-analysis will also investigate various sources of expected heterogeneity in prevalence based on past literature (Corrada et al., 1995; Fagundes et al., 2011; Prince et al., 2013). This includes variation by age, sex, education, overall quality of study, case definition criteria, appropriate use of multiphase methodology, use of a multi-domain cognitive battery, use of informant interview, use of laboratory testing, use of neuroimaging, and geography (country and rurality). Specifically, prevalence of dementia was hypothesized to increase with age. Women were expected to have a higher prevalence of dementia. Higher prevalence of dementia was expected to be found among those with lower education. Lower-quality studies were expected to have higher estimates of dementia prevalence. Dementia prevalence was expected to be higher among studies that did not use blood laboratory testing, did not use neuroimaging, and did not use a multi-domain cognitive battery and studies conducted in rural sites. Lower prevalence was expected in studies that did not rely on an informant interview.

Methods

The current review targeted a specific set of studies with the primary goal of identifying articles with enough uniformity to provide summary estimates of the prevalence of dementia and dementia subtypes using meta-analysis. When sufficient data were not reported for a particular meta-analysis, results from the studies were described narratively.

Search Strategies

Studies were identified for the current review based on two methods. First, the author searched the PUBMED/MEDLINE, PsycINFO, and SciELO databases of all published studies on the prevalence of dementia and dementia subtypes in the USA, Central America, South America, and the Caribbean up until February 2013. The search was limited to the three main languages spoken in the populations of Latinos in these regions with the intention of having all Portuguese articles translated when necessary. The following search terms were entered in English, Spanish, and Portuguese: (dementia or Alzheimer's or vascular dementia or multi-infarct dementia or frontotemporal dementia or mixed dementia or frontotemporal lobar degeneration or primary progressive aphasia or Pick's disease or (dementia and Parkinson's) or Lewy body dementia or dementia subtype or dementia etiology) and (Hispanic and the USA) or (Hispanic and America) or (Latino and the USA) or (Latino and America) or Hispanic or Latino or Argentina or Belize or Bolivia or Brazil or Chile or Colombia or Costa Rica or Cuba or Dominican Republic or Ecuador or El Salvador or Guatemala or Honduras or Jamaica or Mexico or Nicaragua or Panama or Paraguay or Peru or Trinidad or Tobago or Uruguay or Venezuela or Latin America or Caribbean) and (prevalence or epidemiology or frequency or rate).

This search strategy was supplemented by examination of the references sections of identified papers and past relevant reviews. Inclusion criteria were as follows:

1. Population-based studies of dementia or AD or other subtypes or causes of dementia in general populations.
2. Studies that have target populations of participants 60 years of age or older; this included studies that ascertained dementia in populations of older age groups (e.g., 65 years of age and older, 70 years of age and older)
3. Diagnosis of dementia according to the DSM-IV, ICD-10, or Cambridge Examination for Mental Disorders of the Elderly

(CAMDEX; Roth et al., 1986), NINCDS–ADRDA, or similar criteria.
4. The fieldwork for the study must have started on or after January 1, 1980.

The exclusion criteria related to method of sampling and case definition were as follows:

1. Studies provided prevalence data from the follow-up phase of a population cohort (such as is seen in incidence studies).
2. Studies of unrepresentative service-user populations (e.g., nursing home, residential care populations, veterans sampled from hospitals, studies relying mainly on registries of private or public insurance networks, etc.).
3. Studies that defined dementia solely based on cognitive impairment or only on a cut point of a screening measure.
4. Studies that defined dementia based only on history of seeking treatment for dementia.
5. Studies that targeted subtypes of dementia without assessing for total rate of dementia; this exclusion criteria was implemented so that relative frequencies of dementia subtypes could be established.
6. Studies targeting young-onset dementia (less than 60 years; e.g., Huntington's disease).

Data Extraction

The methods and characteristics of the populations were extracted from studies or were obtained through related articles, for example, presentation of methods separate from reporting of prevalence data in large collaborative studies. In studies using more than one diagnostic criteria and providing prevalence data separately according to these definitions, the prevalence of the most widely used and most recent version of the criteria across studies were extracted, typically the DSM-IV or ICD-10, in order to reduce heterogeneity between studies.

The final selection of studies was evaluated for quality using an adapted version of the scoring system used by Prince et al. (2013) to facilitate comparisons. The adaptation consisted of assigning zero points in a quality category if insufficient information was available to designate a score. The total score could range from 0 (poor quality) to 11 (high quality):

(a) With 500 participants, a study could estimate a true prevalence of 6% (± 2.1%) for adults 65 years and older; with 1500 participants, precision increases to (± 1.2%); and with a sample size of 3000, precision increases to ±0.8% (Alzheimer's Disease International, 2009). For the current review, if no information was available about sample size, the study received 0 points; < 500, 0.5 point; 500–1499, 1 point; 1500–2999, 1.5 points; and ≥ 3000, 2 points. NB: The type of sampling methodology was not included as part of the quality index as the inclusion/exclusion criteria removed studies based on nonrandom, convenience, or clinical sampling from the final pool.
(b) For study design: A two-phase study with no sampling of screen negatives or no information available was given 0 points; a multiphase study with sampling of screen negatives but no weighting back was given 1 point; and a one-phase study or multiphase study with appropriate sampling of screen negatives and weighting back was given 2 points.
(c) For response proportion: If no information was available, 0 points; < 60%, 1 point; 60%–79%, 2 points; and ≥ 80%, 3 points.
(d) For diagnostic assessment: 1 point each for using a multiple domain cognitive test battery (two or more domains; screening tests such as the MMSE would not count toward this quality category), formal disability assessment, informant interview, and clinical interview.

Analysis

Prevalence data were extracted from studies by obtaining either the numerator and denominator (i.e., sample size), or prevalence and denominator, or prevalence and 95% confidence intervals. When otherwise unavailable, 95% confidence intervals were calculated using Clopper and Pearson's (1934). Exact method (Newcombe, 1998) was used for rates below 4%; in all other cases, the score method (Wilson, 1912) was used.

Estimates of prevalence were standardized for age using the direct method (Lilienfeld & Stolley, 1994) based on two standard populations: (a) the Latin American and Caribbean population and (b) the Western European population. The Western European population was chosen as a standard population in order to facilitate comparisons with previous international reviews (e.g., Ferri et al., 2005; Prince et al., 2013). The crude prevalence estimates, that is, number of cases/sample size or sample size of subgroup, are presented in tables, along with the standardized prevalence estimates when appropriate.

Comparisons between males and females within each country in the prevalence of dementia were made using χ^2 test using the SPSS 16 program (SPSS, Inc., Chicago, IL). In order to compare the prevalence of dementia by education, studies providing prevalence data between participants with no formal years of education and those with 1 or more years of formal education were reviewed and relevant data extracted. Education was dichotomized because of the variable ways education was measured, reported, and analyzed among the studies. Those with no years of education were labeled illiterate, and those with 1 year or more were coded as literate, recognizing the common way illiteracy was defined among the studies. This classification is far from ideal, considering the more standard definition of illiteracy as the percentage of the population over 15 who cannot both read and write a comprehensible, short simple statement on their everyday life (UNESCO, 2006). As with sex, education was evaluated with a χ^2 test. Random effects meta-analyses were used to compare associations between sex and prevalence of dementia across countries, as well as to compare the relationship between education and prevalence of dementia across countries, using a α-value of 0.05.

In each of the meta-analyses and meta-regressions, prevalence estimates were transformed to logits to improve their statistical accuracy. After analyses, these estimates were then transformed back to prevalence for ease of interpretation. Two methods of quantifying the heterogeneity among studies were used. As a first step in determining whether the produced estimates were derived from studies with substantial variation among them, Cochran's Q was calculated (1954). Cochran's Q tests the null hypothesis that the variance in the estimates of prevalence from among the studies is not significantly greater than expected from sampling error alone, using a α-value of 0.05.

While Cochran's Q is a useful method of testing for heterogeneity, the statistic has been shown to have poor power when there are few studies and overestimates the level of heterogeneity between studies when there are many (Hardy & Thompson, 1998). In order to give a more appropriate idea of the impact of heterogeneity existing between studies, Higgins and Thompson (2002) developed the I^2 statistic to gauge the impact of heterogeneity. I^2 can generally be interpreted as the proportion of variance in point estimates due to heterogeneity, with mild heterogeneity generally considered to be less than 30% and substantial heterogeneity indicated by more than 50%. Estimates of Higgins I^2 greater than 30% are noted in the text.

If there was moderate or substantial heterogeneity, a random effects meta-analysis was performed, evaluating significance based on a α-value less than 0.05. Meta-analyses and meta-regressions were used to evaluate the source of heterogeneity among studies, using the following binary predictors: case definition criteria (e.g., DSM-IV or other), use of a multi-domain cognitive battery, use of proper multiphase methodology, use of routine blood laboratory testing, and use of neuroimaging, rurality, and country. Meta-regressions were inspected visually using a funnel plot of the logit event rate by the continuous predictor of interest. Meta-regressions were further evaluated using a regression test for funnel plot asymmetry.

Results

Evidence Base

The literature searches retrieved 1450 studies (including duplicates). All study abstracts in Portuguese were also available in English and

Spanish because of the translation policies of the SciELO. A total of 881 studies remained after duplicates had been removed. Two studies did not have an abstract available for review and could not be examined further (Codner, Vergara, Fredes, & Guzman, 1984; Morales-Virgen, 1997). After comparing the titles and abstracts against inclusion criteria, 42 studies were identified for detailed examination. Copies of the full, published versions of 40 of these articles were available for review against inclusion and exclusion criteria. An additional five articles were added after reviewing the references sections of 42 of the identified articles and from examining the references of past reviews, increasing the pool of studies for detailed review to 47. Forty-four were available in English and Spanish. Three studies did not have fully published articles available for review (Albala et al., 1997; Ketzoian et al., 1997; Teixeira et al., 2009). Teixeira et al. (2009) could not be considered further as it had no information available regarding sampling methods. However, sufficient study characteristics and prevalence data were available from additional sources for the other two studies and were subsequently added to the final list of included studies. Study characteristics and prevalence data by Albala et al. (1997) are presented as cited by Nitrini et al. (2009). Study characteristics and prevalence data by Ketzoian et al. (1997) are presented as cited in Arizaga et al. (1999) and Nitrini et al. (2009), from a description of the pilot phase of the study (Seccion Neuroepidemiologia, Instituto de Neurologia, 1992) and from an online presentation of results of the study (Seccion Neuroepidemiologia, Instituto de Neurologia, 2004).

Thirteen of the remaining 44 articles were excluded due to sampling methods, for example, reliance on lists from hospital or nursing home registries. Seven articles were excluded because they either reported results using a follow-up wave of an original population cohort study (e.g., incidence studies) or were reporting on risk factors associated with dementia in a population cohort study previously presented in an earlier article. In either case, the current review reports the prevalence data from the original population-based studies. Four studies were excluded since they either defined dementia using only a cognitive screening test or were using the term dementia to describe organic brain syndrome broadly defined. Two studies were excluded because upon further review only presentation of the methodology and results of pilot screening of a longitudinal study were provided. A study from Colombia (Diaz-Cabezas, Ruano-Restrepo, Chacon-Cardona, & Vera-Gonzalez, 2006) was excluded because they reported the results from evaluations of adults 50 years and older, with no estimates of dementia in either the 60 years and older or 65 years and older age groups.

Eighteen articles were included in the final selection for the current systematic review. In one study, Llibre Rodriguez et al. (2008) reported on the results from a population-based cohort study coordinated through the 10/66 Dementia Research Group in seven separate sites in Latin America and the Caribbean, allowing the current review to incorporate results from a total of 23 studies. The countries with the best coverage were Brazil and Cuba, with five and four studies, respectively. Next were Mexico, Peru, and the USA, with three studies each. Venezuela had two studies, whereas Chile, the Dominican Republic, and Uruguay had only one study each.

Five studies cited earlier publications for details of their methodology; the current review obtained information from these earlier publications when necessary. Study characteristics for Molero et al. (2007) were obtained from Maestre et al. (2002). Additional study characteristics for Plassman et al. (2007) were obtained from Langa et al. (2005). Supplementary study characteristics for Scazufca et al. (2008) and Llibre Rodriguez et al. (2008) were obtained from Prince et al. (2007).

Quality of Studies

The main characteristics of the studies included in this systematic review are provided in Table 3.1 Using the scoring system described earlier to evaluate the quality of studies, studies ranged in quality between 4 and 10.5. The average quality

Table 3.1 Study characteristics

Reference	Setting (rural vs. urban)	Source (*N* of phase)	Assessment/diagnostic criteria	Total *N* (response %) [attrition[a]]	Dementia *N*	Quality score
Albala et al. (1997)	Concepcion, Chile (urban)	Census, random sample (2 phases)	MMSE, PFAQ, CDR, lab. eval., brain CT, DSM-III-R ICD-10, informant interview	2213 (NA) [NA]	97	5.5
Ketzoian et al. (1997)	Cerro and Casabo, Montevideo, Uruguay (urban)	Census, whole population of 2 districts (2 phases)	MMSE, Npsych. eval., neurological eval., ICD-9	2731 (93.1%) [NA]	85	7.5
Llibre et al. (1999)	Mariano, Havana, Cuba (urban)	Census and medical registries, cluster random sampling (1 phase)	MMSE, HIS, GDS, clinical interview, DSM-III-R NINCDS-ADRDA	779 (NA) [1 phase]	64	4
Herrera et al. (2002)	Catanduva, Brazil (urban)	Census, random sample of whole population (2 phases)	MMSE, PFAQ, HIS, CDR, Neurological eval., Npsych. eval., lab. eval., brain CT consensus using DSM-III-R, NINCDS-ADRDA, NINDS- AIREN, McKeith et al.'s (1996) criteria, Lund and Manchester criteria (1994)	1656 (98.5%) [6%]	118	8.5
Haan et al. (2003)	Sacramento Valley, California, USA (Rural and urban)	Census, whole population of 5 counties phone/ address list with surnames (3 phase)	3MSE, SEVLT, SENAS, IADL, lab. eval., IQCODE, consensus requiring ≥2 impaired cognitive domains and decline from premorbid functioning and clinical impairment of independent functioning, NINCDS–ADRDA, ADDTC criteria (1992), MRI	1789 (85%) [7.4]	69	10.5
Demirovic et al. (2003)[b]	Miami Dade County, Florida, USA (urban)	Census, probability sample, 6 ethnically homogenous areas (2 phases)	SPMSQ, MMSE, DAFS Neuro. eval, Npsych. eval., lab. eval., MRI, consensus using NINCDS–ADRDA	616 (68.1%) (NA)	80	8
Molero, Pino-Ramirez, and Maestre (2007)	Santa Lucia, Maracaibo, Venezuela (urban)	Door-to-door survey, whole population of 1 district (3 phase)	SPMSQ, IADL, DQ, BDS, Neurological eval., Npsych. eval., lab. eval., brain MRI, consensus: DSM-IV, NINCDS–ADRDA, NINDS–AIREN, ADDTC criteria (1992); McKeith et al.'s (1996) criteria; Neary et al. (1998)	3457 (94.5%) [29.4%]	196	10

(continued)

Table 3.1 (continued)

Reference	Setting (rural vs. urban)	Source (*N* of phase)	Assessment/diagnostic criteria	Total *N* (response %) [attrition[a]]	Dementia *N*	Quality score
Plassman et al. (2007)	42 states of USA (Rural and urban)	Census, stratified random subsample of nationally representative study that oversampled Hispanics (1 phase)	TICS, IQCODE, CDR, BDS, MHIS, DSRS, neurological eval., Npsych. eval., lab. eval., consensus using DSM-III-R, DSM-IV, NINCDS–ADRDA, NINDS–AIREN, Lund and Manchester criteria (1994) McKeith et al.'s (1996)	155 (54.2%) [1 phase]	23	6
Magalhaes et al. (2008)	Lagoa Pequena, Santo Estevao, Bahia, Brazil (rural)	Door-to-door survey, whole population of 1 district (1 phase)	CAMDEX, CAMCOG, lab. Eval., neurological eval.	466 (91.7%) [1 phase]	231	9
Llibre Guerra, Guerra Hernandez, and Perera Minet. (2008)	Mariano, Havana, Cuba (urban)	Census and medical registry, cluster random sampling (2 phases)	MMSE, HIS, GDS, clinical interview, informant interview, NINCDS ADRDA, DSM-IV	300 (NA) [1 phase]	36	4
Scazufca et al. (2008)	Butanta, Sao Paulo, Brazil (urban)	66 census sectors, whole population (1 phase)	CSI-D, GMS, neurological eval., CERAD 10-word list learning with delayed recall, CERAD animal naming verbal fluency, HAS-DDS, NEUROEX, BPSD, NPI-Q, lab. eval. DSM-IV	2072 (91.4%) [1 phase]	105	10.5
Llibre Rodriguez et al. (2008)[c]	[Lisa and Luyano and Mariano and Playa and Plaza], Havana and [Milanes] Matanzas, Cuba (urban)	Census, door-to-door survey of all catchment areas (1 phase)	CSI-D, GMS, neurological eval., CERAD 10-word list learning with delayed recall, CERAD animal naming verbal fluency, HAS-DDS, NEUROEX, BPSD, NPI-Q, lab. eval. 10/66 dementia algorithm, DSM-IV	2944 (94%) [1 phase]	189 [DSM-IV] 318 [10/66]	10.5
	Santo Domingo, Dominican Republic (urban)			2011 (95%) [1 phase]	109 [DSM-IV] 236 [10/66]	10.5
	Lima, Peru (urban)			1381 (80%) [1 phase]	43 [DSM-IV] 129 [10/66]	10

(continued)

Table 3.1 (continued)

Reference	Setting (rural vs. urban)	Source (*N* of phase)	Assessment/diagnostic criteria	Total *N* (response %) [attrition[a]]	Dementia *N*	Quality score
Llibre Rodriguez et al. (2008)[c]	Mexico City, Mexico (urban)	Census, door-to- survey of all catchment areas (1 phase)	CSI-D, GMS, neurological eval., CERAD 10-word list learning with delayed recall, CERAD animal naming verbal fluency, HAS-DDS, NEUROEX, BPSD, NPI-Q, lab. eval., 10/66 dementia algorithm, DSM-IV	1002 (84%) [1 phase]	42 [DSM-IV] 87 [10/66]	10
	Morelos, Mexico (rural)			1000 (86%)	22 [DSM-IV] 85 [10/66]	10
	Cañete, Peru (rural)			552 (88%) [1 phase]	3 [DSM-IV] 36 [10/66]	10
	Caracas, Venezuela (urban)			1904 (80%) [1 phase]	37 [DSM-IV] 109 [10/66]	10.5
Bottino et al. (2008)	Sao Paulo, Brazil (urban)	Census, 3 districts, each with 30 sectors cluster random sampling (2 phases)	MMSE, FOME, IQCODE, B-ADL, neurological eval., brain CT or MRI, lab. eval., CAMDEX, CAMCOG, DSM-IV	1563 (70%) [34.4%]	107	9.5
Custodio, Garcia, Montesinos, Escobar, and Bendezu (2008)	Lima, Peru (urban)	Census, cluster random sampling (3 phases)	MMSE, CDT, PFAQ, neurological eval., Npsych. eval., HIS, CDR, lab. eval., brain CT, DSM-IV, NINCDS– ADRDA, NINDS–AIREN, McKeith et al.'s (1996) criteria, Neary et al.'s (1998) criteria, informant interview	1531 (83.3%) [8.3%]	103	8.5
Llibre Rodriguez et al. (2008)	Playa, Havana, Cuba (urban)	Census, health registries, door-to-door survey (2 phases)	MMSE, CDR, CERAD 10 word learning tasks, IADL, brain CT, lab. eval., clinical interview, semi-structured informant interview, consensus using DSM-IV, NINCDS-ADRDA, NINDS- AIREN	18351 (96.4%) [5.5%]	1499	10
Mejia-Arango and Gutierrez (2011)	Mexico (rural and urban)	Census, nationally representative sample (2 phases)	CCCE (cut points based on ≥2 failed domains as validated in clinical sample diagnosed with DSM-IV) history of cognitive decline IQCODE, IADL	7166 (NA) [4.4%]	357	8

(continued)

Table 3.1 (continued)

Reference	Setting (rural vs. urban)	Source (*N* of phase)	Assessment/diagnostic criteria	Total *N* (response %) [attrition[a]]	Dementia *N*	Quality score
Lopes, Ferrioli, Nakano, Litvoc, and Bottino (2012)	Ribeirao Preto, Sao Paulo, Brazil (urban)	Census, cluster sampling (2 phases)	MMSE, FOME, IQCODE, B-ADL, CAMDEX, CAMCOG, neurological eval., lab. eval., brain CT, DSM-IV	1145 (62.7%) [40%]	68	9

3MSE Modified Mini-Mental State Examination, *ADDTC* State of California Alzheimer's Disease Diagnostic and Treatment Centers criteria (1992), *B-ADL* Bayer Activities of Daily Living Scale, *BDS* Blessed Dementia Scale, *BPSD* Behavioral and Psychological Symptoms of Dementia, *CAMCOG* Cambridge Cognitive Examination, *CAMDEX* Cambridge Examination for Mental Disorders in the Elderly, *CCCE* Cross Cultural Cognitive Examination, *CDR* Clinical Dementia Rating, *CERAD* Consortium to Establish a Registry for Alzheimer's Disease, *CDT* Clock Drawing Test, *CSI-D* Community Screening Interview for Dementia, *CT* computerized tomography, *DAFS* Direct Assessment of Functional Status, *DQ* Dementia Questionnaire, *DSM* Diagnostic and Statistical Manual of Mental Disorders, *DSRS* Dementia Severity Rating Scale, *eval* evaluation, *FOME* Fuld Object Memory Evaluation, *GDS* Global Deterioration Scale, *GMS* Geriatric Mental State, *HAS-DDS* History and Aetiology Schedule Dementia Diagnosis and Subtype, *HIS* Hachinski Ischemic Scale, *IADL* Instrumental Activities of Daily Living, *ICD* International Classification of Diseases, *IQCODE* Informant Questionnaire on Cognitive Decline in the Elderly, *lab* laboratory, *MHIS* Modified Hachinski Ischemic Score, *MMSE* Mini-Mental State Examination, *MRI* magnetic resonance imaging, *N* number of individuals, *NA* not available, *NEUROEX* a brief, structured neurological assessment with objective quantifiable measures of lateralizing signs, parkinsonism, ataxia, apraxia, primitive "release" reflexes, *NINDS–AIREN* National Institute of Neurological Disorders and Stroke – Association International pour la Recherche et L'Enseignement en Neurosciences, *NINCDS–ADRDA* National Institute of Neurological and Communicative Disorders and Stroke – Alzheimer's Disease and Related Disorders Association, *NPI-Q* Neuropsychiatric Inventory, *Npsych* neuropsychological, *PFAQ* Pfeffer Functional Activities Questionnaire, *SENAS* Spanish and English Neuropsychological Assessment Scales, *SEVLT* Spanish and English Verbal Learning Test, *SPMSQ* Short Portable Mental Status Questionnaire, *TICS* Telephone Interview for Cognitive Status, Total N number of subjects before attrition

[a]Attrition between screening and final assessment phases

[b]Only data for Cuban-American women were available

[c]10/66 multisite study; source and assessment/diagnostic criteria are the same for all sites

of studies from Brazil was 9.6, with a standard deviation (SD) of 0.76. Cuba averaged an overall score of 7.13 (SD = 3.61); Mexico had an average score of 9.33 (SD = 1.15); Peru had an average score of 9.5 (SD = 0.87); the USA had an average score of 8.17 (SD = 2.25); and Venezuela had an average score of 10.25 (0.35). Chile, Uruguay, and the Dominican Republic had quality scores of 5.5, 7.5, and 10.5, respectively.

One source of lost points on the scoring system was the inappropriate use of multiphase methodology. Eleven studies utilized one phase methodology in their designs. Two studies used two-phase methodology, yet not enough information was available to decide whether they sampled screen negatives and weighted back appropriately. Of the remaining 10 studies, six did not appropriately use multiphase methodology.

Sample size was generally adequate. Two studies had a sample size greater than 3000; nine studies had a sample size between 1500 and 3000; nine studies had a sample size between 1500 and 500; and three studies had a sample size less than 500. Thirteen of the studies reported a response rate greater than 80%; three studies reported response proportions between 60% and 80%; three studies reported response proportions less than 60%; and four studies either did not report response proportions or the information was unavailable. Most (90%) of the studies assessed participants with a multiple domain cognitive battery. Of the 23 studies, 21 explicitly reported using an informant interview.

While not scored within the quality scoring system, four of the 23 studies reported evaluating participants with routine blood laboratory tests to screen for the presence of other underlying causes of cognitive impairment. Of the 23 studies, 21 explicitly reported using an informant interview. Fifteen studies used the DSM-IV criteria in their diagnoses of dementia, two used the DSM-III-R, one used the ICD-10, one used the ICD-9, and one used the CAMDEX. Two studies defined dementia as an impairment in two or more cognitive domains (as assessed through formal testing), plus a history of decline and impairment in activities of daily living as reported by an informant. One study diagnosed dementia based on a consensus of clinicians using a neurological evaluation, a multi-domain cognitive battery, impairment in activities of daily living, routine blood panel, and a brain MRI (Demirovic et al., 2003). The seven studies reported in the Llibre Rodriguez et al. (2008) article provided estimates of dementia prevalence defined by two definitions, the DSM-IV and dementia as defined by the 10/66 Dementia Research Group algorithm (Prince et al., 2003). All prevalence data included in the current review rely on their estimates of dementia defined by the DSM-IV criteria in order to reduce variance in estimates across the larger pool of studies reviewed.

Prevalence of Dementia

Seventeen studies throughout Latin America and the Caribbean and one study reporting only on Cuban American women in the USA provided data on participants 65 years and older. Demirovic et al. (2003) reported only on the females in the Cuban American sample because their estimates of Cuban American males were unstable due to the relatively low specificity in their screening measure, the Short Portable Mental Status Questionnaire (SPMSQ; Pfeiffer, 1975), for this subgroup. Cuban American females had a prevalence of dementia of 12.9% (95% CI, 10.4–15.4), after adjusting for the sensitivity and specificity of the SPMSQ.

A meta-analysis of the pooled prevalence of dementia across the 17 studies reporting from Central America, South America, and the Caribbean, representing 2910 cases of dementia, estimated the pooled prevalence of dementia as 7.15% (95 CI%, 6.9–7.41). However, there was a high degree of heterogeneity among the studies, $Q(16) = 382.3$, $p < 0.05$, and Higgins $I^2 = 95.81$. An overall random effects meta-analysis taking into account this overdispersion estimated the prevalence of dementia as 5.34% (95% CI, 4.32–6.6). Sixteen of these countries provided data for 5-year age bands (see Table 3.2). The unadjusted pooled prevalence of dementia was 5.04% (95% CI, 4.08–6.2). Excluding the Ketzoian et al. (1997) study because sample size per age group was unavailable, the total age-adjusted prevalence – using Central America, South America, and the Caribbean as the standard population – was 5.99%; for the age-adjusted to the Western European standard population, it was 6.96%.

Comparing age-standardized rates of dementia among adults 65 years and older between countries, Cuba had the highest prevalence of dementia (8.11%), followed in order from the highest to lowest were Brazil (7.43%), Peru (6.36%), Chile (4.91%), the Dominican Republic (4.61%), Uruguay (2.7%), Mexico (2.68), and Venezuela (2.29%). A random effects model comparing prevalence of dementia in four age groups (65–69, 70–74, 75–79, and 80 years and older) revealed significant differences by age, $p < 0.05$. Comparison of sequential age groups in three random effects models was each statistically significant; generally, the rate of dementia increased exponentially, doubling about each 5-year increment in age. In Peru, the prevalence of dementia increased drastically from 8.17 (6.28–10.44) in ages 80–84 to 45.05 (35.24–55.27) in age 85 years and older. This sudden exponential increase is, in part, due to the sole inclusion of data reported by Custodio et al. (2008) for the 85 years and older age group, whereas two additional studies from the 10/66 Dementia Research Group (Llibre Rodriguez et al., 2008) were included in the estimates for

Table 3.2 Prevalence of dementia (%) and 95% confidence intervals by age groups and country

	Age group (years)								
Country	65–69 [N]	70–74 [N]	75–79 [N]	80–84 [N]	85+ [N]	Crude prevalence for ≥65 years [N]	Standardized[a] prevalence for ≥65 years [N]	Standardized[b] prevalence for ≥65 years [N]	Overall quality of studies Mean (SD) [N]
Brazil	2.05	4	8.55	14.34	27.9	6.81	7.43	8.57	9.34 (0.85)
	(1.49–2.76)	(3.14–5.08)	(6.98–10.42)	(11.72–17.42)	(23.53–32.73)	(6.18–7.5)			
	[4]	[4]	[4]	[4]	[4]	[4]	[4]	[4]	[4]
Chile	1.25	2.39	5.48	11.93	16.67	4.38	4.91	5.73	5.5
	(0.6–2.28)	(1.35–3.92)	(3.51–8.1)	(8.15–16.66)	(10.48–24.57)	(3.57–5.33)			
	[1]	[1]	[1]	[1]	[1]	[1]	[1]	[1]	[1]
Cuba	2.99	4.34	6.86	13.19	23.05	8	8.11	7.88	8.17 (3.62)
	(2.58–3.44)	(3.81–4.95)	(6.16–7.62)	(12.09–14.38)	(21.42–24.75)	(7.65–8.36)			
	[3]	[3]	[3]	[3]	[3]	[3]	[3]	[3]	[3]
Dominican Republic	1.31	3.79	4.76	11.17[c]	NA	5.38	4.61	4.96	10.5
	(0.53–2.69)	(2.36–5.88)	(3.09–7.35)	(8.88–14.11)		(4.47–6.44)			
	[1]	[1]	[1]	[1]	[1]	[1]	[1]	[1]	[1]
Mexico	0.70	1.11	2.62	8.16[c]	NA	2.91	2.68	2.97	10(0)
	(0.2–0.19)	(0.49–2.47)	(1.3–4.57)	(6.01–11.1)		(2.21–2.73)			
	[2]	[2]	[2]	[2]	[2]	[2]	[2]	[2]	[2]
Peru	1.16	1.37	4.29	8.17	45.05	5.67	6.36	7.76	9.5 (0.87)
	[3]	[3]	[3]	[3]	[1]	[3]	[3]	[3]	[3]
Venezuela	0.68	0.64	2.37	7.1[c]	NA	2	2.29	2.54	10.25 (0.35)
	(0.27–1.16)	(0.13–0.89)	(1.02–4.58)	(4.78–10.82)		(1.41–2.67)			
	[1]	[1]	[1]	[1]	[1]	[1]	[1]	[1]	[1]

(continued)

Table 3.2 (continued)

Country	Age group (years) 65–69 [N]	70–74 [N]	75–79 [N]	80–84 [N]	85+ [N]	Crude prevalence for ≥65 years [N]	Standardized[a] prevalence for ≥65 years [N]	Standardized[b] prevalence for ≥65 years [N]	Overall quality of studies Mean (SD) [N]
Uruguay	0.88	0.67	2.94	5.88	11.41	3.11	2.7	3.17	7.5
	(0.38–1.72)	(0.22–1.57)	(1.61–4.88)	(3.72–8.78)	(6.79–17.67)	(2.5–3.85)			
	[1]	[1]	[1]	[1]	[1]	[1]	[1]	[1]	[1]
Pooled data[d]	1.82[e]	3.05[e]	5.82[e]	11.43[e]	25.16	5.04	5.99	6.96	9.03 (1.90)
	(1.36–2.42)	(2.25–4.11)	(4.66–7.25)	(9.59–13.57)	(18.62–33.06)	(4.08–6.20)			
	[15]	[15]	[15]	[15]	[7]	[16]	[15]	[15]	[16]

Note. N number of studies, *NA* not available, *SD* standard deviation. Only data for those 65 years and older are shown
[a]Direct standardization for age using Latin America and the Carribean as the standard population
[b]Direct standardization for age using Western Europe as the standard population
[c]Prevalence in individuals 80 years of age and over
[d]Meta-analyzed estimates of prevalence of dementia using random effects models
[e]Meta-analyses revealed significant differences between successive age groups (65–69, 70–74, 75–79, 80+), $p < 0.05$

the earlier age bands. Peru is also notable for having a stable prevalence rate in their 65–69 and 70–75 age groups, thereafter increasing exponentially; similar exceptions to the typical 5-year doubling rate were Cuba, Venezuela, and Uruguay. Across the seven Latin American sites in their collaborative studies, Llibre Rodriguez et al. (2008) observed a doubling with every 7.5-year increment in age, using both of their definitions of dementia.

The prevalence of dementia was estimated in variable age groups by five other studies. Four of these studies provided prevalence data by multiple age groups (see Table 3.2). One study conducted in Cuba (Llibre et al., 1999) reported a prevalence of 8.23% (95% CI, 6.38–10.37) among a sample of adults 60 years of age and older. This estimate was pooled using a fixed effects meta-analysis with the three studies in Table 3.3 that provided estimates on adults 60 years of age and older, representing 719 cases of dementia. The results revealed a high degree of heterogeneity among the studies, Q (3) = 780.25, $p < 0.05$, and Higgins I^2 = 99.62. An estimate of the pooled prevalence of these studies using a random effects model was 10.59 (95% CI, 2.64–3.41). Across the four studies in Table 3.3, a general trend of increasing prevalence with older age groups is noted.

The effect of sex on the prevalence of dementia was examined using χ^2 tests for each group of studies (or study) within each country for 18 studies that provided relevant data. The results are shown in Table 3.4. Females in Mexico, Peru, and Venezuela had a significantly higher prevalence of dementia compared to males within their same country. In addition, a random effects model was used to compare the prevalence of dementia between males and females across all 18 studies regardless of country. The meta-analysis revealed no significant differences between females (6.1%; 95% CI, 4.59–8.07) and males (5%; 95% CI, 3.48–7.13) in the prevalence of dementia, p = 0.39.

Eighteen studies in the current review examined the association between prevalence of dementia and sex using χ^2 tests and/or logistic regression. Significant differences were found in one Brazilian study (Herrera et al., 2002), two Cuban studies (Llibre et al., 1999; Llibre Guerra et al., 2008), and one Peruvian study (Custodio et al., 2008). In pooling their estimates from six studies (rural Peru was not assessed since it had too few cases of dementia), the 10/66 Dementia Research Group reported a significant association between sex and the prevalence of dementia (Llibre Rodriguez et al., 2008).

No studies directly tested the interaction of age and sex associated with the prevalence of dementia. A study of Latinos in the USA, the majority of which were Mexican American, that stratified dementia cases by age and sex showed the difference in prevalence of dementia between females and males was not statistically significant in any age group. A similar pattern was observed in the data presented in one study of Venezuela (Molero et al., 2007), one study in Brazil (Scazufca et al., 2008), and one study of Cuba (Llibre Rodriguez et al., 2009). The 10/66 Dementia Research Group studies showed this same pattern among their Cuban, Dominican Republic, Mexican, and Venezuelan sites. Their rural Peru study could not be evaluated due to the few number of cases identified. However, in their urban Peru study, the prevalence of dementia was not statistically different between females and males up until age 80, at which point the data suggest the prevalence of females may increase more sharply than males. In a study conducted in Mexico, Mejia-Arango and Gutierrez (2011) made the opposite observation: prevalence of dementia was higher in females until age 80, above which men showed a higher prevalence than women.

The effect of education on the prevalence of dementia was assessed using χ^2 tests for each group of studies within each country for nine studies that provided relevant data. The results are shown in Table 3.5. Participants in Brazil, Cuba, Peru, and Venezuela, classified as illiterate, had a significantly higher prevalence of dementia than literate participants within their own country. Additionally, a random effects model found a significant difference between the prevalence of

Table 3.3 Prevalence of dementia (%) and 95% confidence intervals by variable age groups and country

Country					Crude prevalence for ≥60 years	Standardized prevalence[a] for ≥60 years	Standardized prevalence[b] for ≥60 years	Quality of study
	Age group (years)							
	60–69	70–79	80–89	90+				
Brazil[c]	44	52.7	53.42	86.7	49.67	48.67	49.86	9
	(37.67–50.53)	(44.69–60.68)	(42.1–64.41)	(62.12–96.26)	(45.13–54.22)			
	60–69	70–79	80+					
Mexico[d]	1.1	1.9	2.2		5.21	1.5	1.41	8
	(0.8–1.3)	(1.6–2.3)	(1.9–2.7)		(4.71–5.77)			
	60–69	70–79	80–84	85+				
USA[e]	0.92	4.69	9.68	19.72	3.88	7.45	4.81	10.5
	(0.4–1.81)	(3.36–6.51)	(5.62–16.16)	(12.54–29.7)	(3.03–4.89)			
	65–74	75–84	85+					
Venezuela[f]	5.52	19.75	44		13.27[†]	6.49[†]	11.25[†]	10
	(4.07–7.29)	(16.17–23.91)	(35.61–52.75)		(11.5–15.1)			

[a]Direct standardization for age using Latin America and the Carribean as the standard population
[b]Direct standardization for age using Western Europe as the standard population
[†]Prevalence in individuals 65 years of age and older
[c]Magalhaes et al. (2008)
[d]Mejia-Arango and Gutierrez (2011); caution is due in comparing these data as the rates appear to represent proportion of the total sample rather than per age group
[e]Haan et al. (2003)
[f]Molero et al. (2007); only data for 65 years and older are shown

Table 3.4 Prevalence of dementia (%) and 95% confidence intervals by sex and country

Country	Female		Male		N (studies)	Overall quality of studies mean (SD)
	Prevalence (%)	95% CI	Prevalence (%)	95% CI		
Brazil	9.5	(8.69–10.47)	9.03	(7.98–10.20)	5	9.3 (0.76)
Cuba	8.06	(7.64–8.63)	7.72	(7.15–8.34)	3	8.17 (3.62)
Dominican Republic	5.28	(4.2–6.62)	5.7	(4.2–7.7)	1	10.5
Mexico	5.13	(4.55–5.78)*	4.19	(3.6–4.86)	3	9.33 (1.16)
Peru	4.62	(3.8–5.61)*	2.81	(2.01–3.83)	3	9.5 (0.87)
Venezuela	6.01	(5.19–6.94)*	4.21	(3.31–5.35)	2	10.25 (0.35)
USA	3.8	(2.69–5.10)	4.05	(2.85–5.73)	1	10.5
Pooled data[a]	6.1	(4.59–8.07)[b]	5	(3.48–7.13)	18	9.25 (1.82)

*$p < 0.05$, using χ^2 to compare female to male subjects with dementia within country
[a]Meta-analyzed estimates of prevalence of dementia
[b]Not statistically significant as estimated by random effects meta-analysis

Table 3.5 Prevalence of dementia (%) and 95% confidence intervals by education and country

Country	Illiterate		Literate		N (Studies)	Overall quality of studies mean (SD)
	Prevalence (%)	95% CI	Prevalence (%)	95% CI		
Brazil	18.14	(15.89–20.62)*	1.07	(0.77–1.47)	4	9.25 (0.87)
Chile	5.03	(3.7–6.81)	4.03	(3.13–5.18)	1	5.5
Cuba	34.48	(30.08–39.03)*	7.99	(7.63–8.37)	2	10.25 (0.35)
Peru	15.24	(11.44–20.02)*	4.91	(3.85–6.24)	1	8.5
Venezuela	23.43	(18.89–28.67)*	9.96	(8.3–11.92)	1	10
Pooled data[a]	19.72	(12.54–29.62)[b]	8.1	(5.12–12.58)	9	8.94 (1.47)

Note. Categorization as illiterate was defined as having zero or no formal years of schooling
*$p < 0.05$, using χ^2 to compare literate with illiterate subjects with dementia
[a]Meta-analyzed estimates of prevalence of dementia
[b]Significant difference between illiterate and literate participants as estimated by meta-analysis

dementia between illiterate (19.72%; 95% CI, 12.54–29.62) and literate (8.1%; 95% CI, 5.12–12.58) participants across all nine studies regardless of country, $p = 0.05$. Of the 23 studies included in the current review, 19 reported results of testing an association between education and the prevalence of dementia using χ^2 tests and/or logistic regression and various means of defining education (e.g., years of formal education, 0 vs. 1–4 years vs. 4 years or more). Rural Peru was not evaluated due to insufficient number of cases identified. Significant differences were found in twelve studies: four in Brazil (Bottino et al., 2008; Herrera et al., 2002; Lopes et al., 2012; Scazufca et al., 2008), three in Cuba (Llibre Guerra et al., 2008; Libre Rodriguez et al., 2008, 2009), two in Mexico (Llibre Rodriguez et al., 2008; Mejia-Arango & Gutierrez, 2011), two in Venezuela (Llibre Rodriguez et al., 2008; Molero et al., 2007), one in Peru (Custodio et al., 2008), and one in the USA (Haan et al., 2003). Each of these studies reported finding a statistically significant association between education and the prevalence of dementia, where more education was associated with a greater prevalence of dementia.

Prevalence of Subtypes of Dementia in Adults 65 Years and Older

Six studies from four countries (Brazil, Cuba, Peru, and Venezuela) provided frequencies of subtypes of dementia in adults 65 years of age and older, presenting 967 cases of AD and 441 cases of VaD. Their results are presented in Table 3.6. Results of other subtypes were too few to include in the current meta-analyses, and not all studies reported frequencies of these other subtypes. A fixed effects meta-analysis was used to pool the estimates of the prevalence of AD among studies. Results revealed a high degree of heterogeneity among the studies, $Q = 64.98$, $p < 0.05$, and Higgins $I^2 = 92.3$. A random effects model estimated the pooled prevalence of AD among the studies as 4.1 (95% CI, 3.01–5.56). Venezuela had a higher prevalence of AD than Brazil, $p < 0.05$, and was also higher than Peru, $p < 0.05$. Another random effects model estimated the pooled relative frequency of AD among all dementia cases in these studies as 49.53% (95% CI, 43.14–55.93). No statistically significant differences were found between the countries in the proportion of dementia cases accounted for by AD, $p = 0.68$.

A fixed effects meta-analysis was used to pool the estimates of the prevalence of VaD among these studies. Results also revealed a high degree of heterogeneity among the studies, $Q = 45.78$, $p < 0.05$, and Higgins $I^2 = 89.07$. A random effects model estimated the pooled prevalence of VaD among the studies as 1.57

Table 3.6 Prevalence of Alzheimer's disease and vascular dementia in adults 65 years and older by study and country

	Alzheimer's disease		Vascular dementia		Quality of study
Country reference	Prevalence in 65+ (%) [95% CI]	Proportion of dementia cases in 65+ (%) [95% CI]	Prevalence in 65+ (%) [95% CI]	Proportion of dementia cases (%) [95% CI]	
Brazil					
Herrera et al. (2002)	3.92	55.08	0.66	9.32	8.5
	[3.04–4.98]	[46.09–63.76]	[0.37–1.19]	[5.29–15.92]	
Scazufca et al. (2008)	1.64	32.38	1.64	32.38	10.5
	[1.14–2.29]	[24.19–41.82]	[1.14–2.29]	[24.19–41.82]	
Cuba					
Llibre Rodriguez et al. (2008)	3.79	46.43	1.8	22.08	10
	[3.53–4.08]	[43.92–48.96]	[1.62–2.01]	[20.05–24.25]	
Llibre Guerra et al. (2008)	7.7	63.89	2.67	22.22	4
	[5.16–11.24]	[43.92–48.96]	[1.16–5.19]	[11.72–38.08]	
Peru					
Custodio et al. (2008)	3.79	56.31	0.59	8.74	8.5
	[2.89–4.87]	[46.68–65.49]	[0.27–1.11]	[4.67–15.78]	
Venezuela					
Molero et al. (2007)	6.67	50.28	3.52	26.52	10
	[5.47–8.12]	[43.06–57.48]	[2.61–4.64]	[20.62–33.39]	
Pooled data[a]	4.1[b]	49.53	1.57	19.67	8.58[c] (2.4)
	[3.01–5.56]	[43.14–55.93]	[1.03–2.38]	[14.29–26.45]	

[a]Meta-analyzed estimates using random effects models from six studies

[b]Significant difference between the prevalence of AD and the prevalence of VaD as analyzed in a random effects meta-analysis

[c]Overall quality of studies used in meta-analyses; mean (standard deviation)

(95% CI, 1.03–2.38). Venezuela had a higher prevalence of VaD than Brazil ($p < 0.05$), Cuba, ($p < 0.05$), and Peru ($p < 0.05$). Cuba had a higher prevalence of VaD than Peru, $p < 0.05$. A random effects model estimated the pooled proportion of dementia cases accounted for by VaD among the studies as 19.67% (95% CI, 14.29–26.45). Cuba had a significantly higher proportion of dementia cases accounted for by VaD than Peru, $p < 0.05$.

Using a random effects model to compare the prevalence of AD to the prevalence of VaD in these studies, the results revealed that the prevalence of AD was significantly greater than the prevalence of VaD, $p < 0.05$. Similarly, the proportion of cases of dementia accounted for by AD was significantly higher than the proportion accounted for by VaD, $p < 0.05$.

Prevalence of Subtypes of Dementia in Adults 60 Years and Older

Four studies from three countries (Brazil, Cuba, and the USA) provided frequencies of subtypes of dementia in adults 60 years of age and older, representing 165 cases of AD and 59 cases of VaD. Their results are presented in Table 3.7. A fixed effects meta-analysis was used to pool the estimates of the prevalence of AD among studies. Results revealed a high degree of heterogeneity among the studies, $Q = 27.26$, $p < 0.05$, and Higgins $I^2 = 88.99$. A random effects model estimated the pooled prevalence of AD among the studies as 3.21 (95% CI, 2.02–5.08). Brazil ($p < 0.05$) and Cuba ($p < 0.05$) individually had a higher prevalence of AD than the study conducted in the USA, of which a majority of the

Table 3.7 Prevalence of Alzheimer's disease and vascular dementia in adults 60 years and older by study and country

	Alzheimer's disease		Vascular dementia		Quality of study
Country reference	Prevalence in 65+ (%) [95% CI]	Proportion of dementia cases (%) [95% CI]	Prevalence in 65+ (%) [95% CI]	Proportion of dementia cases (%) [95% CI]	
Brazil					
Bottino et al. (2008)	4.33	59.81	1.15	15.89	9.5
	[3.41–5.5]	[50.34–68.61]	[0.67–1.84]	[10.16–23.98]	
Lopes et al. 2012)	2.62	44.12	1.14	19.12	9
	[1.77–3.72]	[32.95–55.92]	[0.61–1.93]	[11.53–30.01]	
Cuba					
Llibre et al. (1999)	5.13	62.5	1.93	23.44	4
	[3.79–6.92]	[50.25–73.33]	[1.08–3.16]	[14.75–35.13]	
USA					
Haan et al. (2003)	1.74	56.52	0.79	20.29	10.5
	[1.19–2.47]	[44.79–67.57]	[0.47–1.32]	[12.49–31.22]	
Pooled data[a]	3.93	55.69	1.36	19.02	7.5[b] (3.04)
	(2.75–5.59)	[44.8–66.06]	[0.97–1.91]	[14.5–24.53]	
Pooled data[c]	3.21	53.07	1.19	19.31	8.25[b] (2.9)
	(2.02–5.08)	[43.65–62.28]	[0.83–1.7]	[15.26–24.13]	

Note. Categorization as illiterate was defined as having zero or no formal years of schooling
*$p < 0.05$, using χ^2 to compare literate with illiterate subjects with dementia
[a]Meta-analyzed estimates for Brazil and Cuba using random effects models
[b]Overall quality of studies used in meta-analyses; mean (standard deviation)
[c]Meta-analyzed estimates for Brazil, Cuba, and the USA using random effects mode

sample were Mexican American. Another random effects model estimated the pooled relative frequency of AD among all dementia cases in these studies as 53.07% (95% CI, 43.65–62.28). Cuba had a significantly higher proportion of dementia cases accounted for by AD than the study conducted in the USA, $p = 0.05$.

A fixed effects meta-analysis was used to pool the estimates of the prevalence of VaD among these studies. Results revealed a moderate degree of heterogeneity among the studies, $Q = 45.78$, $p = 0.11$, and Higgins $I^2 = 49.81$. A random effects model estimated the pooled prevalence of VaD among the studies as 1.19 (95% CI, 0.83–1.7). Cuba had a higher prevalence of VaD than the study in the USA ($p < 0.05$). A random effects model estimated the pooled proportion of dementia cases accounted for, by VaD, among the studies as 19.31 (95% CI, 15.26–24.13). No statistically significant differences between the countries were found in the proportion of dementia cases accounted for by VaD, $p = 0.54$.

Using a random effects model to compare the prevalence of AD to the prevalence of VaD across these studies, the results revealed that the prevalence of AD was significantly greater than the prevalence of VaD, $p < 0.05$. Similarly, the proportion of cases of dementia accounted for by AD was significantly higher than the proportion accounted for by VaD across the studies, $p < 0.05$.

Analysis of Sources of Heterogeneity in the Prevalence of Dementia

An investigation of the sources of heterogeneity among studies estimating the prevalence of dementia among adults 65 years and older using random effects modeling was conducted utilizing indicators of study quality. Meta-regression was used to inspect heterogeneity, modeling the prevalence of dementia by quality of the studies using the criteria described previously. The random effects meta-regression indicated that study quality did not account for a statistically significant portion of heterogeneity either through visual inspection of a funnel plot of the logit event rate of dementia by study quality or by formal testing of funnel plot asymmetry ($z = -1.13$, $p = 0.26$).

Investigating study quality in more detail, a meta-analysis found no statistically significant difference in the prevalence of dementia between 14 studies using the DSM-IV criteria of dementia (5.54%; 95% CI, 4.40–6.95) and three studies that did not (4.66%; 95% CI, 2.83–7.58), $p = 0.54$. Nor was there a statistically significant difference when comparing studies using the DSM-IV or ICD-10 criteria (5.44%; 95% CI, 4.35–6.79) versus other criteria of dementia (4.8%; 95% 2.05–10.82), $p = 0.78$.

In a further investigation of the sources of variation among the studies, a comparison of 15 studies using a multi-domain cognitive battery (5.14; 95% CI, 4.09–6.43) and two studies that did not (7.26%; 95% CI, 2.62–18.56), $p = 0.51$, revealed no statistically significant differences. No statistically significant differences were found comparing six studies using multiphase methodology that did not accurately sample screen negatives and weigh these data back into their estimates of the prevalence of dementia (6.55; 95% CI, 4.68–9.08) with the 11 other studies (4.68; 95% CI, 3.516.22), $p = 0.14$. Comparing 15 studies that utilized routine blood laboratory testing in their dementia workup (5.3%; 95% CI, 4.28–6.54) to two studies that did not (6.17; 95% CI, 1.57–21.33) revealed no statistically significant differences, $p = 0.83$. The prevalence of dementia estimated in five studies that incorporated neuroimaging (brain MRI or CT) into their diagnoses was compared to 12 studies that did not. Studies using neuroimaging (8.59%; 95% CI: 6.51–11.27) were found to have higher estimates of the prevalence of dementia than those that did not (4.24; 95% CI: 3.19–5.62), $p < 0.05$.

Two additional sources of variation were investigated relating to geographical differences. The prevalence of dementia among adults 65 years and older in 17 studies categorized as being undertaken in urban settings was compared to two rural studies. The prevalence of dementia was lower in studies at rural sites (1.01%; 95% CI, 0.17–5.67) versus studies conducted in urban locations (5.93%; 95% CI, 4.81–7.28), $p < 0.05$.

Finally, a random effects model was used to compare the prevalence of dementia among adults 65 years and older between countries. This model was statistically significant, $p < 0.05$. This is partly the reason why prevalence of dementia is presented by country in Tables 3.2 to 3.7.

Analysis of Sources of Heterogeneity in the Prevalence of Dementia Subtypes

An investigation of the sources of heterogeneity among studies estimating the prevalence of subtypes of dementia among adults 65 years and older using random effects modeling was conducted. Meta-regression was used to inspect heterogeneity, modeling the prevalence of AD by quality of the studies using the criteria described previously. The random effects meta-regression indicated a slight trend for study quality to account for some heterogeneity among studies, based on visual inspection of a funnel plot of the logit event rate of AD by study quality and by formal testing of funnel plot asymmetry ($z = -1.78$, $p = 0.08$). This suggests one lower-quality study (Llibre Guerra et al., 2008; quality score = 4) may have slightly overestimated the prevalence of AD.

Investigating study quality in more detail, a meta-analysis found a significant difference in the prevalence of AD between 5 studies using the NINCDS–ADRDA criteria (McKhann et al., 1984) of AD (4.81%; 95% CI, 3.67–6.28) and one study that did not (1.64%; 95% CI, 1.17–2.29), $p = 0.05$. No significant differences in the prevalence of AD were found between four studies using the NINDS–AIREN criteria (Roman et al., 1993) for VaD ($p = 0.79$) and two studies that did not. There were no significant differences in the prevalence of AD between four studies that took the results of neuroimaging into their diagnoses and two that did not, $p = 0.4$.

The sources of heterogeneity among studies in the prevalence of VaD were evaluated similarly to the analyses in AD. Meta-regression was used to inspect heterogeneity, modeling the prevalence of VaD by quality of the studies using the criteria described previously. The random effects meta-regression did not account for a substantial amount of heterogeneity among studies, based on visual inspection of a funnel plot of the logit event rate of VaD by study quality and by formal testing of funnel plot asymmetry ($z = -0.11$, $p = 0.91$).

Investigating study quality in more detail, a meta-analysis found no significant differences in the prevalence of VaD between four studies using the NINDS–AIREN criteria (Roman et al., 1993) for VaD and two studies that did not, $p = 0.06$. Comparing the five studies that used the DSM-IV (1.83%; 95% CI, 1.21–2.77) to the one study that did not (0.66%; 95% CI, 0.37–1.2) revealed a significant difference in the prevalence of VaD, $p < 0.05$. No significant differences in the prevalence of VaD were found between five studies using the NINCDS–ADRDA criteria (McKhann et al., 1984) of AD ($p = 0.79$) and one study that did not, $p = 0.83$. There were no significant differences in the prevalence of VaD between four studies that took the results of neuroimaging into their diagnoses and two that did not, $p = 0.38$.

Discussion

The current review identified 18 articles representing 23 studies of the prevalence of dementia in Central America, South America, the Caribbean, and among Latinos in the USA that met strict inclusion and exclusion criteria. Twenty studies from Central America, South America, and the Caribbean represented nine countries: Brazil, Chile, Cuba, the Dominican Republic, Mexico, Peru, Uruguay, and Venezuela. Brazil and Cuba had the most studies that qualified for the review, whereas Chile, the Dominican Republic, and Uruguay each had only one study that qualified. As expected, a statistically significant amount of variability in dementia prevalence was discovered between studies. The pooled estimate of our meta-analysis with 17 studies that takes heterogeneity into account by using a random effects model was 5.32% for Central America, South America, and the Caribbean, overall. In order to compare our study with other reviews of dementia prevalence among adults 65 years and older that do not take similar approaches to accounting for study variability, the pooled estimate from the studies reviewed in this paper is presented using a fixed effects model. Though recognized as imperfect, such estimates are still useful indicators of the public health burden of a disorder.

The unadjusted pooled estimate was 7.2%. This figure is similar to that reported by Nitrini et al. (2009) among six countries in Latin America and the Caribbean (7.1%; unadjusted) from populations of adults 65 years and older. The age-adjusted prevalence of dementia standardized to the Western European population structure was 8.1% (estimated in a fixed effects meta-analysis available from 15 studies), which appears to be higher than the age-standardized 6.4% reported for 11 population-based cohorts aged 65 and older in eight European countries (Lobo et al., 2000). The current review's estimate of dementia is also higher than that found in a global systematic review of dementia prevalence from 1994 to 2000 among all continents, reporting a range of 4.2% to 7.2% in those over 65 years (Lopes, Hototian, Reis, Elkis, & Bottino, 2007).

The meta-analysis was able to replicate the association between education and dementia, especially in prevalence studies (Sharp & Gatz, 2011). Using pooled estimates of dementia across studies, participants with no formal years of education had more than double the prevalence of dementia (19.72%) than participants with one or more years of education (8.1%). This was similarly found in a previous review of dementia prevalence studies in Latin American and the Caribbean (Nitrini et al., 2009). We were also able to replicate the finding of an association between age and the prevalence of dementia with our meta-analysis of 15 studies in Latin America and the Caribbean. Prevalence increased exponentially with age, as a whole appearing to double with every 5-year increment in age, as seen in other population surveys throughout the world (Jorm, Korten, & Henderson, 1987; Lobo et al., 2000; Prince et al., 2013). This effect appeared to be moderated by country; however, in Cuba, Peru, Uruguay, and Venezuela, the rate of increase remained stable in the earliest age groups (65–69 to 70–75), subsequently increasing exponentially in accordance with the typical 5-year doubling rate.

In regard to differences between men and women, there was no significant difference when pooling prevalence across countries by sex. Many other studies have noted an interaction with sex and age, wherein women have a higher prevalence of dementia, especially after age 80 (Lobo et al., 2000; Prince et al., 2007). Within country comparisons in Mexico, Peru, and Venezuela, we found that women had a higher prevalence of dementia than men (Lobo et al., 2000). These were the countries with the lowest percentage of older adults of the ones included in our study, with the exception of the Dominican Republic (United Nations Department of Economic and Social Affairs, 2007). This may suggest this difference is due to survival differences between men and women. As with all risk factor associations, incidence studies are needed to truly establish a link between sex and dementia.

To our knowledge, this is the first study to provide estimates based on meta-analysis on the frequency of dementia subtypes in Central America, South America, and the Caribbean. Due to the range in the number of dementia subtypes for which studies provided data, we were only able to analyze data on the prevalence and relative frequency of AD and VaD. In many of the studies providing data on dementia subtypes in Central America, South America, and the Caribbean, dementia subtypes by age, sex, and education were not available. Furthermore, as with the prevalence of total dementia, there was a great degree of heterogeneity found between studies. Nevertheless, we present pooled estimates in order to guide public health policy in this region. Among adults 65 years and older, the prevalence of AD was 4.1%, representing 49.5% of dementia cases across six studies from four countries (Brazil, Cuba, Peru, and Venezuela). VaD had a prevalence of 1.6%, representing 19.67% of dementia cases. AD was the most frequent subtype of dementia in all studies. These rates are lower than those provided by Plassman et al. (2007) who reported an AD prevalence of 26.8%

pooled across their sample of non-Hispanic White, non-Hispanic Black, and Hispanic adults in the USA. AD accounted for 74.4% of all dementia cases in that study. The prevalence of VaD was 5.6%, accounting for 15.6% of all dementia cases. This difference in rates is not surprising, however, given that the sample in Plassman et al. (2007) was composed of adults 71 years of age and older. Indeed, it is very difficult to compare prevalence rates between studies using different age brackets, especially because of the noted pattern that AD increases in the proportion of dementia cases accounted for with age (Jorm et al., 1987). The rates of AD and VaD in Central America, South America, and the Caribbean do seem to be similar to those found in European population studies of adults 65 years and older (Lobo et al., 2000).

Prevalence of Dementia Among Latinos in the USA

How do Latinos in the USA compare to populations in Central America, South America, and the Caribbean? Few studies were available for review on the prevalence of dementia among Latinos in the USA that met the strict criteria for this review (Demirovic et al., 2003; Haan et al., 2003; Plassman et al., 2007). Plassman et al. (2007) used a nationally representative sample to estimate the prevalence of dementia in the USA. However, only a total of 84 USA Latinos (country of origin or descent was unspecified) were sampled. The prevalence reported for the USA Latinos aged 71 and older was 27.4% (unadjusted), which is much higher than the pooled estimate of 10.8% in adults 70 years and older in our 15 Central American, South American, and Caribbean studies. The low number of USA Latinos in their sample makes it difficult to draw conclusions about this difference.

A better comparison can be made with the other two identified USA studies that qualified for this review. Demirovic et al. (2003) sampled from a community-based population of Cuban Americans in Miami Dade, Florida. Due to the low specificity of their screening measure for Cuban American males, only the prevalence of Cuban American females was available (12.9%). Cuban American women had a higher prevalence of dementia than Cuban women age 65 and older (8%; unadjusted pooled estimate) reviewed in our sample. In a sample comprised of a majority of Mexican Americans in Sacramento County, California, Haan et al. (2003) reported a prevalence of dementia of 4.8%. This is similar to the estimate of prevalence reported for adults 60 years and older in Mexico (5.21%; Mejia-Arango & Gutierrez, 2011). However, if one compares the prevalence of dementia using those cases that were fully evaluated in their sample (3.9%) to the study in Mexico, we see a lower prevalence among Mexican Americans.

This suggests a difference between these two comparisons of USA Latinos and the populations of their respective countries of origin or ancestry. This pattern of differences remains even when only comparing Mexican American women (3.8%) to women in Mexico (5.8%). Nearly half of all Mexican Americans in the Haan et al. (2003) sample were born in the USA, whereas the Cuban Americans in the Demirovic et al. (2003) study were all born in Cuba. These differences in nativity and the historically different patterns of migration between Mexicans and Cubans to the USA may underlie the differences in contrasts with their respective countries of origin or ancestry. In a further analysis of their sample of Mexican Americans in Sacramento County, California, Haan, Al-Hazzouri, and Aiello (2011) compared Mexican Americans born in Mexico to those born in the USA and found that both nativity and migration predicted lifetime trajectories of socioeconomic status (SES) and modified the relationship between lifetimes SES and late life cognitive health. Immigrants were more likely to remain poor over the life course, whereas those whose parents or grandparents were born in the USA were more likely to improve in SES. Those with higher SES in childhood and adulthood experienced fewer declines in global cognition and verbal memory than the less advantaged. Perhaps Cuban Americans in the Demirovic et al. (2003) study experienced a greater lifetime accumulation of disadvantage compared to the half of Mexican American born in the USA who comprised the sample in Haan et al. (2003). The age of migration to the USA was not reported in the Demirovic et al. (2003) study; thus, it is not possible to determine the accumulation of lifetime advantage or disadvantage. Nonetheless, the possible role of socio-environmental factors such as SES, nativity, and migration factors in the prevalence of dementia warrants further attention.

Variation in Dementia Prevalence

In generating models for the estimates of the prevalence, great variation was noted among studies in the prevalence of dementia and dementia subtypes. The sources of this variation were investigated. Two geographical indicators were found to predict variation in dementia: rurality and country, supporting findings by Russ et al. (2012) in their systematic review of the geographical variation of dementia. For example, in the current review, among adults 60 years and older, the prevalence of AD was higher in Brazil and Cuba than in Mexican Americans in the USA. Further, AD accounted for a greater proportion of dementia cases in Cuba than in Mexican Americans in the USA.

Using the quality scoring system modeled after the Alzheimer's Disease International World Alzheimer Report (2009), it was observed that many of the studies in the current review were of good quality. It is perhaps not surprising that variation between studies in terms of the global quality index in the prevalence of dementia was not found. When examined further, no significant differences were found between 14 studies using the DSM-IV criteria to define dementia and those that did not. This is in contrast to the work shown by Llibre Rodriguez et al. (2008) that examined differences between using the DSM-IV criteria versus their own algorithm-based definition of dementia. The three studies not using the DSM-IV criteria based their definitions on the ICD-9 and ICD-10 or incorporated the NINCDS–ADRDA criteria for dementia, each of which is generally similar. Alternatively, the current study did find that use of the NINCDS–ADRDA criteria to define AD was associated with a greater prevalence of AD than a study using only the DSM-IV, suggesting underestimation by the DSM-IV criteria. In this study, no differences were found in the diagnosis of VaD between studies using the NINDS–AIREN criteria and two studies that did not. This may be due to the reliance of these two other studies on other standard criteria for VaD. Llibre Guerra et al. (2008) used a combination of the criteria for multi-infarct dementia in the DSM-IV and scores on the Hachinski Ischemic Scale (Hachinski, 1983). Scazufca et al. (2008) relied on the DSM-IV criteria for multi-infarct dementia and on information derived from the History and Aetiology Schedule Dementia Diagnosis and Subtype interview to differentially diagnose VaD. Another possibility is the sample size of studies not using the NINDS–AIREN criteria which may explain failure to reach statistical significance.

Other aspects of methodology such as the incorrect use of multiphase screening or lack of use of routine laboratory testing in identifying other possible sources of cognitive symptoms were not found to predict variation in dementia prevalence. Corrada et al. (1995) found evidence for the importance of such factors in predicting heterogeneity in prevalence rates of AD. Only two studies did not use routine laboratory testing, so perhaps the sample size of this subgroup was too small to find statistical significance. Similarly, no differences emerged between studies that used a multi-domain cognitive battery and two studies that did not.

In comparison, there were only five studies that incorporated neuroimaging into their results, and they produced higher estimates of the prevalence of dementia than those that did not. The opposite pattern was found by Corrada et al. (1995) where higher rates (although of AD not dementia) were found among studies not using brain CT than those that did. Studies that used neuroimaging in the current review tended to have more comprehensive assessment approaches, and perhaps that covaried with prevalence rates. Surprisingly, no significant differences were found between studies using neuroimaging to diagnose AD or VaD and those that did not. Again, all of these comparisons were limited by sample size and may have prevented the statistical tests from reaching significance.

Limitations

The current systematic review did not investigate publication bias due to the large variability in scores across studies. Conceptually, prevalence studies are not likely to be biased in in terms of which get published according to estimated prevalence rates. A greater threat to the current review is the selection bias of studies found in mainstream journals, which are more likely to be indexed in PUBMED/MEDLINE and PSYCINFO. This was partially addressed by searching the SciELO database, which has a strong collection of papers from Central American, South American, and Caribbean countries.

The choice of using a dichotomous measure of education may have limited our incorporation of findings from several studies into our meta-analysis (Cohen, 1983). An attempt was made to narratively review individual findings regarding the association between education and dementia in order to supplement the statistical tests. Furthermore, our classification of participants with no years of education as illiterate does not conform to the standard definition provided by the UNESCO (2006). We did so only following the pattern established by many of the individual studies in the current review. Another issue of definition regards our classification of studies into rural vs. urban sites. This was mainly done by reliance of studies' self-descriptions of settings. This problem was recognized in Russ et al. (2012), where various definitions of rurality were used throughout the studies in their meta-analyses. A universal standard for the definition for both education and rurality remains elusive.

AD and VaD were categorized as mutually exclusive entities in the current review. Yet, it is widely recognized that it is very difficult, even for recognized experts, to make distinctions between these two diagnostic categories, especially in large-scale epidemiological studies (Chui et al., 2000). Variation in the way studies reported prevalence or frequency of cases prevented further distinctions along the continuum between AD, VaD, and mixed dementias. Future epidemiological surveys would provide a great service to the scientific and clinical literature by reporting rates of other subtypes of dementia, even if none are identified in their samples.

Few studies were identified that provided data on Hispanic Americans in the USA or any of the 14 other Central American, South American, and Caribbean countries not represented in the current review. By developing strict criteria for inclusion, we necessarily excluded studies that provide insightful clues as to the occurrence of dementia in a particular setting. This was done so as to provide a concise way of communicating the public health burden of dementia across a wide variety of settings. Public health policies should incorporate knowledge of the sources of variation highlighted in this review into their assessments of population needs for services. It was not possible to recognize all possible sources of variability in dementia prevalence. As more population studies in the region accumulate, it may facilitate the coding of more study characteristics that influence rates of reported dementia.

Conclusion

Dementia among countries in Central America, South American, and the Caribbean is a major public health problem. The estimates of prevalence in the countries that have adequate population-based surveys of dementia and dementia subtypes suggest rates that stand in contrast to earlier estimates of the region (Ferri et al., 2005). National programs would do well to prepare for the expected increase in the scope of this problem, knowing that populations in these countries are growing older and becoming more likely to encounter issues of age-related cognitive decline. Like in other population-based studies throughout the world, AD accounts for the most amount of dementia cases. The population of Latinos in the USA is similarly expected to age rapidly, and current estimates suggest dementia is as much of a problem as for other groups in the nation. The complex interaction of nativity, immigration histories, and genetic risk may place certain immigrant groups at higher risk for dementia than others. Ongoing monitoring of these issues can help identify areas of need for these populations and suggest priorities in variables to target in the search of risk and protective factors of dementia.

References

Albala, C., Quiroga, P., Klaassen, G., Rioseco, P., Perez, H., & Calvo, C. (1997). Prevalence of dementia and cognitive impairment in Chile. (Abstract). *World Congress of Gerontology*, Adelaide, Australia, 483.

Allegri, R. F., Butman, J., Arizaga, R. L., Machnicki, G., Serrano, C., Taragano, F. E., … Lon, L. (2007). Economic impact of dementia in developing countries: An evaluation of costs of Alzheimer-type dementia in Argentina. *International Psychogeriatrics, 19*(4), 705–718.

Alzheimer's Disease International. (2009). *World Alzheimer Report 2009*. London: Alzheimer's Disease International.

American Psychiatric Association. (2000). *Diagnostic and Statistical Manual of Mental Disorders* (4th ed.). Washington, DC: American Psychiatric Association.

Arizaga, E., Mangone, C. A., Allegri, R. F., & Ollari, J. A. (1999). Vascular dementia: The Latin American perspective. *Alzheimer's Disease & Associated Disorders, 13*(3), S201–S205.

Bottino, C. M., Azevedo, D., Jr., Tatsch, M., Hototian, S. R., Moscoso, M. A., Folquitto, J., … Litvoc, J. (2008). Estimate of dementia prevalence in a community sample from Sao Paulo, Brazil. *Dementia and Geriatric Cognitive Disorders, 26*(4), 291–299.

Carrion-Baralt, J. R., Suarez-Perez, E., del Rio, R., Moore, K., & Silverman, J. (2010). Prevalence of dementia in Puerto Rican veterans is higher than in mainland U.S. veterans. *Journal of the American Geriatrics Society, 58*(4), 798–799.

Chui, H. C., Mack, W., Jackson, J. E., Mungas, D., Reed, B. R., Tinklenberg, J., … Jagust, W. J. (2000). Clinical criteria for the diagnosis of vascular dementia: a multicenter study of comparability and interrater reliability. *Archives of Neurology, 57*(2), 191–196.

Clopper, C. J., & Pearson, E. S. (1934). The use of confidence or fiducial limits illustrated in the case of the binomial. *Biometrica, 26*(4), 404–413.

Cochran, W. G. (1954). The combination of estimates from different experiments. *Biometrics, 10*(1), 101–129.

Codner, S., Vergara, F., Fredes, A., & Guzman, N. (1984). Prevalence of neuropsychiatric disorders in an elderly population. *Revista Médica de Chile, 112*(10), 998–1001.

Cohen, J. (1983). The cost of dichotomization. *Applied Psychological Measurement, 7*(3), 249–253.

Cooper, C., Tandy, A. R., Balamurali, T. B. S., & Livingston, G. (2010). A systematic review and meta-analysis of ethnic differences in use of dementia treatment, care, and research. *The American Journal of Geriatric Psychiatry, 18*(3), 193–203.

Corrada, M., Brookmeyer, R., & Kawas, C. (1995). Sources of variability in prevalence rates of Alzheimer's disease. *International Journal of Epidemiology, 24*(5), 1000–1005.

Custodio, N., Garcia, A., Montesinos, R., Escobar, J., & Bendezu, L. (2008). Prevalencia de demencia en una poblacion urbana de Lima-Peru: Estudio puerta a puerta. *Anales de la Facultad de Medicina, 69*(4), 233–238.

Demirovic, J., Prineas, R., Loewenstein, D., Bean, J., Duara, R., Sevush, S., & Szapocznik, J. (2003). Prevalence of dementia in three ethnic groups: The South Florida program on aging and health. *Annals of Epidemiology, 13*(6), 472–478.

Diaz-Cabezas, R., Ruano-Restrepo, M. I., Chacon-Cardona, J. A., & Vera-Gonzales, A. (2006). Neuroepidemiology profile of the central zone of the department of Caldas (Colombia), years 2004-2005. *Revista de Neurologia, 43*(11), 646–652.

Fagundes, S. D., Silva, M. T., Silva Thees, M. R., & Periera, M. G. (2011). Prevalence of dementia among elderly Brazilians: A systematic review. *São Paulo Medical Journal, 129*(1), 46–50.

Ferri, C. P., Prince, M., Brayne, C., Brodaty, H., Fratiglioni, H., Ganguli, M., … Scazufca, M. (2005). Global prevalence of dementia: A Delphi consensus study. *Lancet, 366*(9503), 2112–2117.

Fitten, L. J., Ortiz, F., & Ponton, M. (2001). Frequency of Alzheimer's disease and other dementias in a community-outreach sample of Hispanics. *Journal of the American Geriatrics Society, 49*(10), 1301–1308.

Folstein, M. F., Folstein, S. E., & McHugh, P. R. (1975). Mini-mental state: A practical method for grading the cognitive state of patients for the clinician. *Journal of Psychiatric Research, 12*(3), 189–198.

Gurland, B. J., Wilder, D. E., Lantigua, R., Stern, Y., Chen, J., Killeffer, E. H., & Mayeux, R. (1999). Rates of dementia in three ethnoracial groups. *International Journal of Geriatric Psychiatry, 14*(6), 481–493.

Haan, M. N., Mungas, D. M., Gonzalez, H. M., Ortiz, T. A., Acharya, A., & Jagust, W. J. (2003). Prevalence of dementia in older Latinos: The influence of type 2 diabetes mellitus, stroke and genetic factors. *Journal of the American Geriatrics Society, 51*(2), 169–177.

Haan, M. N., Zeki Al-Hazzouri, A., & Aiello, A. E. (2011). Life-span socioeconomic trajectory, nativity, and cognitive aging in Mexican Americans: the Sacramento Area Latino Study on Aging. *The Journals of Gerontology. Series B, Psychological Sciences and Social Sciences, 66*(suppl_1), i102–i110.

Hachinski, J. C. (1983). Differential diagnosis of Alzheimer's dementia: Multi-infarct dementia. In B. Reisberg (Ed.), *Alzheimer's Disease* (pp. 188–192). New York, NY: Free Press.

Hardy, R. J., & Thompson, S. G. (1998). Detecting and describing heterogeneity in meta-analysis. *Statistics in Medicine, 17*(8), 841–856.

Harwood, D. G., & Ownby, R. L. (2000). Ethnicity and dementia. *Current Psychiatry Reports, 2*(1), 40–45.

Hendrie, H. C., Osuntokun, B. O., Hall, K. S., Ogunniyi, A. O., Hui, S. L., Unverzagt, F. W., ... Musik, B. S. (1995). Prevalence of Alzheimer's disease and dementia in two communities: Nigerian Africans and African Americans. *American Journal of Psychiatry, 152*(10), 1485–1492.

Herrera, E., Jr., Caramelli, P., Silveira, A. S., & Nitrini, R. (2002). Epidemiologic survey of dementia in a community dwelling Brazilian population. *Alzheimer Disease & Associated Disorders, 16*(2), 103–108.

Higgins, J. P., & Thompson, S. G. (2002). Quantifying heterogeneity in a meta-analysis. *Statistics in Medicine, 21*(11), 1539–1558.

Hou, C. E., Yaffe, K., Perez-Stable, E. J., & Miller, B. L. (2006). Frequency of dementia etiologies in four ethnic groups. *Dementia and Geriatric Cognitive Disorders, 22*(1), 42–47.

Jorm, A. F., Korten, A. E., & Hendersen, A. S. (1987). The prevalence of dementia: A quantitative integration of the literature. *Acta Psychiatrica Scandinavica, 76*(5), 465–479.

Kalaria, R. N., Maestre, G. E., Arizaga, R., Friedland, R. P., Galasko, D., Hall, K., ... Antuono, P. (2008). Alzheimer's disease and vascular dementia in developing countries: Prevalence, management, and risk factors. *Lancet Neurology, 7*(9), 812–826.

Ketzoian, C., Romero, S., Dieguez, E., Coirolo, G., Rega, I., & Caseres, R. (1997). Estudio de la prevalencia de las principales enfermedades neurologicas en una poblacion del Uruguay. *La Prensa Medica Uruguaya, 17*, 9–26.

Langa, K. M., Plassman, B. L., Wallace, R. B., Herzog, A. R., Heeringa, S. G., Ofstedal, M. B., ... Willis, R. J. (2005). The aging, demographics, and memory study: Study design and methods. *Neuroepidemiology, 25*(4), 181–191.

Lilienfeld, D. E., & Stolley, P. D. (1994). *Foundations of epidemiology*. New York, NY: Oxford University Press.

Llibre Guerra, J. C., Guerra Hernandez, M. A., & Perera Miniet, E. (2008). Comportamiento del sindrome demencial y la enfermedad del Alzheimer. *Revista Habanera de Ciencas Medicas, 2*(1), 1–13.

Llibre, J. J., Guerra, M. A., Perez-Cruz, H., Bayarre, H., Fernandez-Ramirez, S., Gonzalez-Rodriguez, M., & Samper, J. A. (1999). Sindrome demencial y factores de riesgo en adultos mayores de 60 anos residentes en la Habana. *Revista de Neurologia, 29*(10), 908–942.

Llibre Rodriguez, J. J., Fernandez, Y., Marcheco, B., Contreras, N., Lopez, A. M., Otero, M., ... Bayarre, H. (2009). Prevalence of dementia and Alzheimer's disease in a Havana municipality: A community-based study among elderly residents. *MEDICC Review, 11*(2), 29–35.

Llibre Rodriguez, J. J., Ferri, C. P., Acosta, D., Guerra, M., Huang, Y., Jacob, K. S., ... Prince, M. (2008). Prevalence of dementia in Latin America, India, and China: A population-based cross-sectional survey. *Lancet, 372*(9637), 464–474.

Lobo, A., Launer, L. J., Fratiglioni, L., Andersen, K., Di Carlo, A., Breteler, M. M., ... Hofman, A. (2000). Prevalence of dementia and major subtypes in Europe: A collaborative study of population-based cohorts. *Neurology, 54*(Suppl. 5), S4–S9.

Lopes, M. A., Ferrioli, E., Nakano, E. Y., Litvoc, J., & Bottino, C. M. (2012). High prevalence of dementia in a community-based survey of older people from Brazil: Association with intellectual activity rather than education. *Journal of Alzheimer's Disease, 32*(2), 307–316.

Lopes, M. A., Hototian, S. R., Reis, G. C., Elkis, H., & Bottino, C. M. C. (2007). Systematic review of dementia prevalence 1994 to 2000. *Dementia and Neuropsychologia, 1*(3), 230–240.

Lund & Manchester Groups. (1994). Consensus statement: Clinical and neuropathological criteria for frontotemporal dementia: The Lund and Manchester groups. *Journal of Neurology, Neurosurgery, and Psychiatry, 57*(4), 416–418.

Maestre, G. E., Pino-Ramirez, G., Molero, A. E., Silva, E. R., Zambrano, R., Falque, L., ... Sulbaran, T. A. (2002). The Maracaibo Aging Study: Population and Methodological Issues. *Neuroepidemiology, 21*(4), 194–201.

Magalhaes, M. O., Peixoto, J. M., Frank, M. H., Gomes, I., Rodrigues, B. M., Menezes, C., ... Melo, A. (2008). Risk factors for dementia in a rural area of northeastern Brazil. *Arquivos de Neuro-Psiquiatria, 66*(2), 157–162.

Mangone, C. A., & Arizaga, R. L. (1999). Dementia in Argentina and other Latin-American countries: An overview. *Neuroepidemiology, 18*(5), 231–235.

McKeith, I. G., Galasko, D., Kosaka, K., Perry, E. K., Dicskon, D. W., Hansen, L. A., ... Perry, R. H. (1996). Consensus guidelines for the clinical and pathologic diagnosis of dementia with Lewy bodies (DLB): Report of the consortium on DLB international workshop. *Neurology, 47*(5), 1113–1124.

McKhann, G., Drachman, D., Folstein, M., Katzman, R., Price, D., & Stadlan, E. M. (1984). Clinical diagnosis of Alzheimer's disease: Report of the NINCDS-ADRDA work group under the auspices of department of health and human services task force on Alzheimer's disease. *Neurology, 34*(7), 939–944.

Mejia-Arango, S., & Gutierrez, L. M. (2011). Prevalence and incidence rates of dementia and cognitive impairment no dementia in the Mexican population: Data from the Mexican health and aging study. *Journal of Aging and Health, 23*(7), 1050–1074.

Molero, A. E., Pino-Ramirez, G., & Maestre, G. E. (2007). High prevalence of dementia in a Caribbean population. *Neuroepidemiology, 29*(1–2), 107–112.

Morales-Virgen, J. J. (1997). The epidemiology of dementias. *Gaceta Médica de México, 133*(2), 159–161.

Morgan, R. O., Teal, C. R., Hasche, J. C., Petersen, L. A., Byrnne, L. M., Paterniti, D. A., & Vernig, B. A. (2008). Does poorer familiarity with Medicare translate into worse access to health care? *Journal of the American Geriatrics Society, 56*(11), 2053–2060.

Neary, D., Snowden, J. S., Gustafson, L., Passant, U., Stuss, D., Black, S., Benson, D. F. (1998). Frontotemporal lobar degeneration: A consensus on clinical diagnostic criteria. *Neurology, 51*(6), 1546–1554.

Newcombe, R. (1998). Two-sided confidence intervals for the single proportion: Comparison of seven methods. *Statistics in Medicine, 17*(8), 857–872.

Nitrini, R., Bottino, C. M., Albala, C., Custodio Capuñay, N. S., Ketzoian, C., Llibre Rodriguez, J. J., … Caramelli, P. (2009). Prevalence of dementia in Latin America: A collaborative study of population-based cohorts. *International Psychogeriatrics, 21*(4), 622–630.

Nitrini, R., Caramelli, P., Herrera, E., Jr., Bahia, V. S., Caixeta, L. F., Radanovic, M., … Takahasi, D. Y. (2004). Incidence of dementia in a community-dwelling Brazilian population. *Alzheimer Disease & Associated Disorders, 18*(4), 241–246.

Perkins, P., Annegers, J. F., Doody, R. S., Cooke, N., Aday, L., & Vernon, S. W. (1997). Incidence and prevalence of dementia in a multiethnic cohort of municipal retirees. *Neurology, 49*(1), 44–50.

Pfeiffer, E. (1975). Short portable mental state questionnaire for assessment of organic brain deficit in elderly patients. *Journal of the American Geriatric Society, 23*(10), 433–441.

Plassman, B. L., Langa, K. M., Fisher, G. G., Heeringa, S. G., Weir, D. R., Ofstedal, M. B., … Wallace, R. B. (2007). Prevalence of dementia in the United States: The aging, demographics, and memory study. *Neuroepidemiology, 29*(1–2), 125–132.

Prince, M., Acosta, D., Chiu, H., Scazufca, M., & Varghese, M. (2003). Dementia diagnosis in developing countries: A cross-cultural validation study. *Lancet, 361*(9361), 909–917.

Prince, M., Bryce, R., Albanese, E., Wimo, A., Ribeiro, W., & Ferri, C. P. (2013). The global prevalence of dementia: A systematic review and metaanalysis. *Alzheimer's & Dementia, 9*(1), 63–75.

Prince, M., Ferri, C. P., Acosta, D., Albanese, E., Arizaga, R., Dewey, M., … Uwakwe, R. (2007). The protocols for the 10/66 dementia research group population-based research programme. *BioMed Central Public Health, 20*(7), 165.

Román, G. C., Erkinjuntti, T., Wallin, A., Pantoni, L., & Chui, H. C. (2002). Subcortical ischaemic vascular dementia. *The Lancet Neurology, 1*(7), 426–436.

Roman, G. C., Tatemichi, T. K., Erkinjuntti, T., Cummings, J. L., Masdeu, J. C., Garcia, J. H., … Schienberg, P. (1993). Vascular dementia: Diagnostic criteria for research studies. Report of the NINDS-AIREN international workshop. *Neurology, 43*(2), 250–260.

Roselli, D., Ardila, A., Pradilla, G., Morillo, L., Bautista, L., Rey, O., & Camacho, M. (2000). The Mini-Mental State Examination as a selected diagnostic test for dementia: A Colombian population study. *Revista de Neurologia, 30*(5), 428–432.

Roth, M., Tym, E., Mountjoy, C. Q., Huppert, F. A., Hendrie, H., Verma, S., & Goddard, R. (1986). CAMDEX: A standardised instrument for the diagnosis of mental disorder in the elderly with special reference to the early detection of dementia. *British Journal of Psychiatry, 149*, 698–709.

Russ, T. C., Batty, G. D., Hearnshaw, G. F., Fenton, C., & Starr, J. M. (2012). Geographical variation in dementia: Systematic review with meta-analysis. *International Review of Epidemiology, 41*(4), 1012–1032.

Scazufca, M., Menezes, P. R., Vallada, H. P., Crepaldi, A. L., Pasto-Valero, M., Coutinho, L. M., … Almeida, O. P. (2008). High prevalence of dementia among older adults from poor socioeconomic backgrounds in Sao Paulo, Brazil. *International Psychogeriatrics, 20*(2), 394–405.

Seccion Neuroepidemiologia, Instituto de Neurologia. (1992). Estudio piloto de la prevalencia de las principals enfermedades neurologicas en una poblacion de Uruguay. *Revista Medica de Uruguay, 8*(3), 191–205.

Seccion Neuroepidemiologia, Instituto de Neurologia. (2004). *Epidemiología y factores de riesgo de los síndromes demenciales en el Uruguay* [Powerpoint slides]. Retrieved from http://www2.bago.com.bo/sbn/eventos/html/pres_neurosur/arch_pdf/epid_demencia.pdf

Sharp, E. S., & Gatz, M. (2011). Relationship between dementia and education: An updated systematic review. *Alzheimer's Disease and Associated Disorders, 25*(4), 289–304.

Teixeira, A. L., Caramelli, P., Barbosa, M. T., Santos, A. P., Pellizzaro, M., Beato, R. G., ... Sakurai, E. (2009). Prevalence of major psychiatric disorders in a cohort of oldest old in Brazil: The Pietà study. *Alzheimer's & Dementia, 5*(4), 392.

U.S. Administration on Aging. (2011). *65+ minority population comparison using census 2000 and census 2010* [Data file]. Retrieved from http://www.aoa.gov/Aging_Statistics/Census_Population/census2010/Index.aspx

UNESCO Institute for Statistics. (2006). *Global education digest*. Retrieved from http://www.uis.unesco.org/glossary/

United Nations Department of Economic and Social Affairs. (2007). *World Population Ageing 2007*. New York, NY: United Nations Publications.

Weimer, D. L., & Sager, M. A. (2009). Early identification and treatment of Alzheimer's disease: Social and fiscal outcomes. *Alzheimer's & Dementia: The Journal of the Alzheimer's Association, 5*(3), 215–226.

Wilson, E. B. (1912). Probable inference, the law of succession, and statistical inference. *Journal of the American Statistical Association, 22*(158), 209–212.

Part II

Evaluation, Assessment, and Diagnosis

4 Treating Medical Comorbidities Associated with Dementia Among Latinos

Samuel C. Gable and Maureen K. O'Connor

Abstract

Latinos in the United States (U.S.) experience greater prevalence and worse course for certain health conditions associated with dementia when compared to the general population. Close scrutiny of these discrepancies is crucial in order to improve the early and accurate detection of neurodegenerative processes, as well as intervene in potentially treatable causes of cognitive decline. Such scrutiny involves the examination of biological and sociocultural contributors of medical comorbidities among different Latino subgroups, with the largest U.S.-residing populations being Mexican Americans and Caribbean Latinos. This chapter will review the medical comorbidities most often associated with dementia and cognitive decline as they pertain to Latinos in the U.S., including type 2 diabetes mellitus (T2DM), thyroid disease, infectious diseases (i.e., HIV, hepatitis C), obstructive sleep apnea (OSA), depression, alcohol abuse, and nutritional deficiencies (i.e., vitamin B12, folate). Following this section, the four major dementias are reviewed with an emphasis on the prevalence, genetics, pathology, and clinical presentation for Latino populations: Alzheimer's disease (AD), vascular dementia (VaD), frontotemporal lobar degeneration (FTLD), and Lewy body dementias (LBD), including dementia with Lewy bodies (DLB) and Parkinson's disease dementia (PDD).

S. C. Gable (✉)
University of Massachusetts Boston,
Boston, MA, USA
e-mail: samgable@gmail.com

M. K. O'Connor
Bedford Veterans Hospital, Center for Translational Cognitive Neuroscience, Boston University School of Medicine, Boston, MA, USA
e-mail: maureen.oconnor@med.va.gov

Introduction

Medical and health practitioners who work with elderly Latinos in the United States (U.S.) will be aware of the immense diversity that must be considered for this population when assessing for cognitive impairment and dementia. Sociodemographic variables such as the education histories of Latin American countries, for example, can introduce crucial (though difficult to account for) variability in performance during assessment, such as that of neuropsychological testing (Rivera Mindt, Byrd, Saez, & Manly, 2010). Medical comorbidities that accompany or masquerade as neurodegenerative conditions are also quite variable and complex, depending on the particular patient history and demographics. For instance, Latinos who immigrated as adults are consistently found to hold superior health status and reduced mortality compared to U.S.-born

H. Y. Adames, Y. N. Tazeau (eds.), *Caring for Latinxs with Dementia in a Globalized World*,
https://doi.org/10.1007/978-1-0716-0132-7_4

Latinos and non-Hispanic Whites (Crimmins, Kim, Alley, Karlamangla, & Seeman, 2007). This "immigrant health paradox" is only partially explained by factors such as the "salmon effect," in which the most unwell return to their countries of origin and erroneously deflate the estimates of morbidity and mortality among those who remain (Palloni & Arias, 2004; Vega, Rodriguez, & Gruskin, 2009). However, despite this apparent health boon, Latinos in the U.S. experience greater prevalence and worse health outcomes than the general U.S. population for some conditions that are understood to deletiriously impact neurological health.

In this chapter, we review comorbidities as they pertain to cognitive impairment and dementia among Latinos. Medical comorbidities to be discussed include: type 2 diabetes mellitus, thyroid disease, infectious diseases (i.e., HIV, hepatitis C), and obstructive sleep apnea. In addition, the influence of psychiatric comorbidities, in particular depression and alcohol abuse, is reviewed, as well as nutritional deficiencies (i.e., vitamin B12, folate). Following this review, the four major neurodegenerative diseases are briefly described including: Alzheimer's disease (AD), vascular dementia (VaD), frontotemporal lobar degeneration (FTLD), and Lewy body dementias (LBD), including dementia with Lewy bodies (DLB) and Parkinson's disease dementia (PDD). Special attention to the neurodegenerative diseases' prevalence and course among Latinos is provided when this information is available.

Comorbidities in Dementia

In clinical practice, prior to making a diagnosis of dementia, it is important to first consider the often co-occurring medical and psychiatric comorbidities that may precipitate, exacerbate, or, in some cases, mask neurodegenerative processes. These contributing factors may be amenable to treatment that stabilizes and possibly reverses the process of decline. For the comorbidities to be discussed in this chapter, Latinos have higher prevalence rates and worse health outcomes when compared to the general U.S. populations, which is often attributed to poorly understood biological and socioeconomic factors (Vega et al., 2009).

Regarding the biological factors, the incidence of dementia and medical comorbidities can vary between Latino subpopulations due in part to genotypic differences between regions. Most of the literature on Latinos has investigated Latinos of either Mexican or Caribbean heritage. Mexican genetic heritage is primarily European and Amerindian (e.g., Mayan) (Campos, Edland, & Peavy, 2013; Fitten, Ortiz, & Pontón, 2001) as well as African, but to a lesser degree (Chavez-Dueñas, Adames, & Organista, 2014). Mexican American communities are most concentrated in the western U.S. region and make up 63.36% of total Latinos in the USA and 11.14% of the total U.S. population (U.S. Census Bureau, 2015); therefore, studies of Latinos from states such as California will have most likely recruited a largely Mexican or Mexican American sample. The second Latino sub-ethnicity most represented in the scientific literature are Caribbeans, such as Puerto Ricans, Dominican Republicans, and Cubans, who share a combination of European, Amerindian, and African ancestry (Chavez-Dueñas et al., 2014). Caribbean Latinos live primarily along the eastern coast and account for 16.55% of the Latino U.S. population and 2.9% of the total population (U.S. Census Bureau, 2015). As the incidence of comorbidities and dementia can vary among these and less represented populations of Latinos (e.g., Central or South Americans), we provide more specific ethnic designations within the broader "Latino" category when this information is available.

Socioeconomic status (SES) variables are also influential in the prevalence of dementia and medical comorbidities. Institutional barriers confronting many Latino communities due to poverty and language contribute to the prevalence and often worse health outcomes of these conditions. These barriers include reduced access to adequate and language-appropriate healthcare, lack of health insurance, and undocumented immigrant status with accompanying concerns for safety when interacting with institutions (Centers for Disease Control and Prevention [CDC], 2016; Livney et al., 2011; Vega et al., 2009). Immigrants often live in urban, economically depressed communities that may expose them to higher levels of stress, environmental toxins, traumatic experiences, worsened sleep quality, and limited access to high-quality nutrition (Horowitz, Colson, Herbert, & Lancaster, 2004; Loredo et al., 2010; Vega et al., 2009). Finally, the SES variable of education should be considered when examining research relating to cognitive performance and dementia among Latinos. For example, elderly Latinos who are referred to neuropsychological testing practices hold a wide range of educational attainment, spanning zero years of schooling to the graduate level. Low levels of education and illiteracy are implicated in not only artificially depressed neuropsychological test performance, but also higher risk for developing dementia (Heaton, Ryan, & Grant, 2009).

Type 2 Diabetes Mellitus (T2DM)

Type 2 diabetes mellitus (T2DM) is a major comorbidity of dementia affecting Latinos. Latinos experience a higher prevalence of T2DM than non-Hispanic Whites (Pabon-Nau, Cohen, Meigs, & Grant, 2010). Latinos with T2DM also have worse health outcomes, including diabetes-related complications (e.g., nephropathy, retinopathy, and amputations) and mortality (Kirk et al., 2008; Lora et al., 2009; Wu et al., 2003). Rates of T2DM vary by particular country of origin. For example, in a review of Latino health characteristics from the 2000–2005 National Health Interview Survey, diabetes prevalence was found to be highest among Puerto Ricans as well as U.S.-born Mexican Americans as compared to Mexicans. Foreign-born Cubans, Dominicans, Central Americans, and South Americans were found to have half the prevalence of diabetes compared to Mexicans (Pabon-Nau et al., 2010).

T2DM as well as its precursor, prediabetes, are associated with acute changes in cognition due to glycemic fluctuations that are transient and typically resolve once glucose is controlled (Claessens, Geijselaers, & Sep, 2016; Kodl & Seaquist, 2008). Both hypo- and hyperglycemic fluctuations are found to alter cognitive and emotional functioning. Acute hypoglycemia in diabetic patients (types 1 and 2) has been found to impact overall cognitive functioning, and deficits are typically noted in areas of attention, reaction time, processing speed, working memory, executive functioning, and memory (McNay & Cotero, 2010; Warren & Frier, 2005). Complex tasks seem to be especially affected, such as driving, as are hippocampal-mediated tasks due to the sensitivity of this region to glucose fluctuations. Cognitive dysfunction associated with hyperglycemia can include mental slowing, attention, and working memory (Cox et al., 2005; Sommerfield, Deary, & Frier, 2004; Waclawski, 2012). A study by Cox et al. (2005) found that mental slowing associated with hyperglycemia was highly variable, occurring in approximately half of type 1 and type 2 diabetic patients.

Regarding emotional functioning, acute hypo- and hyperglycemia in diabetics have been associated with increased depression and anxiety (Lustman & Clouse, 2005; Warren & Frier, 2005). Depression in diabetes is associated with poor treatment adherence, thus perpetuating the effects of glycemic instability. Acute hypoglycemia may masquerade as psychiatric and medical conditions such as psychosis, delirium, traumatic brain injury, and stroke (Griswold, Del Regno, & Berger, 2015; Luber et al., 1996; Sahoo, Mehra, & Grover, 2016). Regarding the latter, imaging abnormalities due to acute hypoglycemia have been found to resemble ischemic stroke in a few cases (Yong, Morris, Shuler, Smith, & Wardlaw, 2012).

Hypothyroidism

Thyroid dysfunction, and particularly hypothyroidism, is associated with cognitive and neuropsychiatric difficulties. These symptoms can be alleviated with thyroid hormone replacement therapy (Resta et al., 2012). Prevalence statistics for thyroid disease among Latinos is scarce. One study comparing hypothyroidism incidence with bone mineral density and fractures among post-menopausal Puerto Rican women found a high prevalence (24.2%) of hypothyroidism (González-Rodríguez, Felici-Giovanini, & Haddock, 2013). Latinos experience high rates of certain health conditions that are associated with thyroid dysfunction, including T2DM and hepatitis C (Rodríguez-Torres, Ríos-Bedoya, Ortiz-Lasanta, Marxuach-Cuétara, & Jiménez-Rivera, 2008; Wang, 2013). Congenital hypothyroidism has also been found to be more common in Latino, Asian, and Pacific Islander newborns compared to other ethnicities (Hinton et al., 2010). Regarding protective factors, Latinos do not appear to be at higher risk for autoimmune forms of thyroid disease (e.g., Grave's, Hashimoto's) or thyroid cancer compared to non-Hispanic Whites in some research (Magreni, Bann, Schubart, & Goldenberg, 2015; McLeod, Caturegli, Cooper, Matos, & Hutfless, 2014).

The onset and course of cognitive symptoms due to thyroid dysfunction are often insidious and subtle. Thyroid dysfunction is associated with impairments in global cognition, attention and concentration, psychomotor speed, executive function, visuospatial abilities, new learning, and memory (Bégin, Langlois, Lorrain, & Cunnane, 2008; Dugbartey, 1998). Cognitive changes may occur in the context of subclinical hypothyroidism among older adults (Resta et al., 2012). The impact of hypothyroidism and its subclinical manifestation also appear to increase the risk for dementia (especially vascular dementia), although the mechanisms remain unclear (Etgen, Sander, Bickel, & Förstl, 2011; Gan & Pearce, 2012).

Depression, apathy, and anxiety often accompany thyroid dysfunction. As with mood symptoms from other etiologies, their presence may interfere with performance on cognitive testing with the effect of inflated estimates of neurological impairment (Davis & Tremont, 2007; Samuels, 2008). More severe psychotic symptoms can be observed in hypothyroidism, including hallucinations, paranoid ideation, and delirium (Heinrich & Grahm, 2003).

Infectious Diseases

HIV-associated neurocognitive disorder (HAND) is part of the potential sequelae of human immunodeficiency virus (HIV) infection. HAND diagnoses range in severity, from "asymptomatic" (i.e., evidence of mild cognitive impairment without functional decline) to HIV-associated dementia (HAD). Early detection and treatment of HIV is important in order to preserve cognitive health. With the advent of highly active antiretroviral therapy (HAART), HIV treatment has been shown to prevent, stabilize, and sometimes improve cognitive impairment associated with HAND (Cohen et al., 2001; Cysique, Maruff, & Brew, 2006; McCutchan et al., 2007; Parsons, Braaten, Hall, & Robertson, 2006).

Latinos in the U.S. experience an elevated prevalence of HIV infection. In 2015, Latinos accounted for 24% of new HIV diagnoses despite representing 17% of the population (Centers for Disease Control and Prevention [CDC], 2017). Diagnosis among Latino gay and bisexual men increased 13% between 2010 and 2014. The rate of infections among heterosexual Latina women in 2015 was 5% greater than males and three times that of White women. Among individuals over the age of 50, Latinos accounted for 16% of all HIV diagnoses, which was the third highest subpopulation for this age cohort, and below that of African Americans and Whites (CDC, 2016). HIV-related mortality is also higher among Latinos residing in the U.S. compared to Whites, with the greatest mortality rate found among Puerto Ricans and the lowest among Mexicans (Clark, Surendera Babu, Harris, & Hardnett, 2015).

The onset of cognitive symptoms in HAND is typically insidious, though rapid decompensation can occur in the event of opportunistic infections. Early symptoms of HAND can include confusion, slowed mental processing and psychomotor speed, reductions in attention/concentration, and memory impairment (Hardy & Vance, 2009). This frontal-subcortical pattern of impairment can be reflected on medical diagnostic imaging of the brain, which indicates pathological changes of the basal ganglia (often cited are the caudate and putamen), frontal cortex, and white matter pathways (Woods, Moore, Weber, & Grant, 2009). Depression, fatigue, irritability, and apathy often accompany HAND. More severe neurobehavioral symptoms of delirium, mania, auditory and visual hallucinations, and paranoia are also reported (Singer & Thames, 2016).

Hepatitis C (HCV)

Latinos in the U.S. are at higher risk for hepatitis C (HCV) infection than the general population, though this is divisible by particular groups. Puerto Rican men in the U.S. have the highest prevalence rate at 11.6%, whereas South American men in the U.S. were found to be below the U.S. national average at 0.4% (Franciscus, 2014; Kuniholm et al., 2014). Compared to other ethnic groups in the U.S., Latinos tend to experience a more aggressive HCV disease course. They are more likely to become infected at a younger age, develop chronic liver cirrhosis, have comorbid HIV infection, and are less likely to be tested, initiate treatment, or be referred to specialists (Blessman, 2008; Cummins, Erlyana, Fisher, & Reynolds, 2015).

Neuropsychiatric and cognitive symptoms are relatively common in HCV and are reported in approximately half of patients (Senzolo et al., 2011). These include fatigue, depression, anxiety, and cognitive dysfunction especially in the areas of attention and memory. Cognitive symptoms with HCV infection may be reported irrespective of liver complications, substance use and other medical/psychiatric comorbidities, or medical diagnostic imaging evidence of neurological injury (Huckans et al., 2009; Monaco et al., 2015). Inflammation is a current area of investigation as a potential mechanism by which HCV infection could bring about these symptoms (Irwin & Terrault, 2008; Senzolo et al., 2011).

Obstructive Sleep Apnea (OSA)

Obstructive sleep apnea (OSA) is a condition in which breathing is obstructed repeatedly during sleep. With consistent use of a positive airway pressure (PAP) machine to treat OSA, the neurocognitive changes associated with this condition may be ameliorated or avoided (Macey, 2012). Although few high-quality studies have explored the prevalence of OSA among Latinos, results have indicated that Latinos experience a high prevalence of OSA and that it often goes undiagnosed in this population (Ramos et al., 2015; Redline et al., 2014). Sleep-disordered breathing, including OSA, is a potential causal mechanism for the metabolic and vascular morbidities with high prevalence among Latinos, including diabetes, obesity, and hypertension (Redline et al., 2014; Sands-Lincoln et al., 2013).

The cognitive symptoms associated with OSA resemble those of sleep deprivation, with deficits in selective and sustained attention, vigilance, processing speed, working memory, and executive functioning (Alhola & Polo-Kantola, 2007; Stranks & Crowe, 2016; Wen, Wang, Liu, Yan, & Xin, 2012). Individuals with OSA also experience a higher prevalence of depression (Ejaz, Khawaja, Bhatia, & Hurwitz, 2011). Repeated hypoxemia caused by OSA is theorized to result in changes to synaptic plasticity and microvasculature affecting gray and white matter (Lin et al., 2017; Stranks & Crowe, 2016).

Depression

Depression has been called a "pseudodementia" as it can accompany reversible cognitive symptoms that overlap with those seen in neurodegenerative diseases, and which are not fully accounted for by apathy or poor effort as observed in neuropsychological testing settings (Ganguli, 2009; Kang et al., 2014). The rates of depression in the elderly increase with compounding medical morbidity (Taylor, 2014). The elderly are also likely to experience life events that are precipitants to depression, such as the loss of loved ones or loss of mobility through chronic illness. Depression is an influential covariate in studies that explore the incidence of dementia among Latinos (Clark et al., 2005; Fitten et al., 2001; González, Haan, & Hinton, 2001; Livney et al., 2011; Russell & Taylor, 2009). Poverty, low English-language acquisition, and lower levels of acculturation have been implicated in the increased prevalence of depression among Latinos in these studies. Latino elders who live alone were also found to experience higher levels of depression compared to other ethnic minority peers (Russell & Taylor, 2009). Cognitive impairments in individuals with depression have typically been found in the areas of attention/concentration, psychomotor and processing speed, executive functioning, and memory (Kang et al., 2014). Depression may also occur in the context of a neurodegenerative process, such as Alzheimer's disease, vascular dementia, Parkinson's disease, Huntington's disease, or HAND (Lezak, Howieson, Bigler, & Tranel, 2012).

Alcohol Abuse

Chronic alcohol abuse is a known risk factor for dementia. The cognitive deficits attributed to alcohol have the potential to improve during abstinence, especially within the first year (Bernardin, Maheut-Bosser, & Paille, 2014). Rates of alcohol and other substance use among Latinos, overall, are at or below the U.S. national average (Alvarez, Jason, Olson, Ferrari, & Davis, 2007; Substance Abuse and Mental Health Services Administration [SAMHSA], 2013). Latinos in the U.S. have lower rates of alcohol consumption than non-Hispanic Whites but also a higher likelihood of problem drinking for those who do drink. Higher levels of acculturation and being born in the U.S. are correlated with greater alcohol consumption. Rates among the U.S. Latino subgroups are variable, with Puerto Rican men and women having the highest rates of drinking and the lowest rates found among Cuban men and Mexican women (Ríos-Bedoya & Freile-Salinas, 2014; SAMHSA, 2013). Importantly, a history of alcohol consumption among Latino patients presenting with cognitive complaints should be considered within the context of the higher prevalence of chronic liver disease and cirrhosis for this population (Vega et al., 2009).

Cognitive deficits associated with chronic alcohol abuse have been reported in the areas of attention, working memory, executive functioning (i.e., mental flexibility, problem-solving, and reasoning), learning and recollection in episodic memory, and visuospatial organization and construction (Bernardin et al., 2014). Chronic alcohol abuse is associated with brain white matter changes as well as gray matter atrophy in the prefrontal cortex, parietal lobes, limbic structures (e.g., mammillary bodies and hippocampus), and cerebellar vermis (Oscar-Berman & Marinković, 2007; Shanmugarajah et al., 2016). Permanent cognitive impairment is seen in the context of prolonged, excessive drinking in combination with insufficient intake of vitamin B1 (thiamine). Wernicke–Korsakoff syndrome (WKS) can result from alcohol-related thiamine deficiency, and it is the combination of the often co-occurring conditions of Wernicke encephalopathy and Korsakoff's syndrome (Korsakoff's dementia). Patients with WKS may present with ataxia, abnormal or uncontrolled eye movements (e.g., nystagmus), confusion, psychosis, executive dysfunction, and memory impairment (Martin, Singleton, & Hiller-Sturmhöfel, 2003; Oscar-Berman & Marinković, 2007). The deficits in WKS are associated with neuronal loss of thiamine-dependent neurons in limbic structures (i.e., thalamus, hypothalamus, and mammillary bodies), frontal lobes, and the cerebellum. Damage to thalami and mammillary bodies is thought to primarily account for the amnestic profile in WKS and Korsakoff's syndrome, although hippocampal damage has been detected in these patients as well (Kopelman, 2015).

Nutritional Deficiencies

Vitamin B12 and folate (vitamin B9) are the most common nutritional deficiencies that result in potentially reversible cognitive impairment and are associated with greater risk for vascular disease, dementia, and mood symptoms in the elderly (Doets et al., 2013; Garrod et al., 2008; Ramos et al., 2005). Low levels of either vitamin B12 or folate in the elderly can be treated with supplementation. In their analysis of the U.S. national survey data exploring nutritional status among the elderly, Hinds, Johnson, Webb, and Graham (2011) found that the "other Hispanic" group, that is, non-Mexican Americans, had the highest prevalence of vitamin B12 deficiency compared to other ethnicities. Mexican Americans had one of the lowest levels of B12 deficiency, and folate deficiencies were low or nonexistent for all ethnic groups. Kwan, Bermudez, and Tucker (2002) also found a high prevalence of vitamin B12 deficiency within their sample of Caribbean Latino elderly residing in Massachusetts, but not in the non-Hispanic White cohort to which they were compared.

Immigration history is likely a relevant variable when considering the influence of nutritional causes for cognitive impairment. In industrialized nations, vitamin B12 deficiency is less common due to the abundance of fortified and animal-sourced food. Even so, the elderly in these nations remain at risk for B12 deficiency due to age-related worsening of vitamin absorption from dietary sources (Allen, 2008, 2009). In developing countries, including much of Latin America, higher prevalence of low or deficient B12 levels have been found among children and breastfeeding mothers (Allen, 2008). This is a cause for concern considering the role of vitamin B12 in neurological development (de Benoist, 2008). Folate deficiency demonstrates a reversed global trend from vitamin B12, in that it is mainly sourced from leafy greens and grain products and therefore is more plentiful in the diets of lower-income countries (Allen, 2009).

For both vitamin B12 and folate, deficient levels are associated with cognitive impairment including dementia and psychiatric morbidity (Moretti et al., 2004; Reynolds, 2002). Patients with deficient B12 levels were found to have deficits in attention, mental speed, working memory, mental flexibility, and memory difficulties in both encoding and retrieval. Psychiatric symptoms include depression, delirium, and acute psychotic states. Deficient folate has been linked to impairments in the areas of visuomotor function, reasoning and abstraction, and memory. Deficient folate is more associated with depression than vitamin B12 (Moretti et al., 2004).

Deficiencies in both folate and vitamin B12 are found when homocysteine levels are elevated, which is an inflammatory marker associated with cerebrovascular disease, cognitive decline, and dementia (de Benoist, 2008; Miller et al., 2003). Using data from the Sacramento Area Latino Study on Aging (SALSA), Haan et al. (2007) and Miller et al. (2003) found elevated homocysteine to be a risk factor for cognitive impairment and dementia among this elderly, primarily Mexican American cohort. Increasing plasma vitamin B12 levels was put forth as a potential mechanism for protecting against cognitive impairment.

Dementia Syndromes

Once the above contributory factors have been considered and ruled out or treated, the clinician will be more prepared to determine the presence of a neurodegenerative disease. The four most common neurodegenerative dementias are presented below. This section begins with an overview of Alzheimer's disease (AD) and vascular dementia (VaD), after which a brief discussion is presented on the contributions of metabolic and vascular comorbidities in these two conditions. Following this discussion, frontotemporal lobar degeneration (FTLD) and Lewy body dementias (LBD) including Lewy body dementia (LBD) and Parkinson's disease dementia (PDD) are presented.

According to the latest criteria from the National Institute on Aging and the Alzheimer's Association (NIA-AA; McKhann et al., 2011), all-cause dementia ranges from mild to severe stages and encompasses cognitive and behavioral/neuropsychiatric symptoms. Dementia symptoms must interfere with functional ability, represent a decline from previous levels of functioning, be inexplicable by delirium or psychiatric comorbidities alone, and entail cognitive and behavioral decline in two or more domains (e.g., memory, language) as detected by history taking, mental status exam, or neuropsychological testing.

Alzheimer's Disease (AD)

Prevalence AD is the most common cause of dementia in Western countries (Rizzi, Rosset, & Roriz-Cruz, 2014). It accounts for 60–80% of cases of dementia in the U.S., which includes "mixed dementia" types in which AD pathology co-occurs with that of other dementias (e.g., cerebrovascular disease, Lewy body pathology; Alzheimer's Association, 2016). There is a slightly greater incidence of AD among women, which is only partially explained by increased longevity in women (Viña & Lloret, 2010). Prevalence of AD among Latinos is also higher than among non-Hispanic Whites, which is thought to be due to the contribution of greater metabolic and vascular burden, as well as yet unknown genetic contributors (Chin, Negash, & Hamilton, 2011; Haan et al., 2003; Tang et al., 2001).

The age of onset has been found to be earlier among Latinos of Mexican American, Central American, and Caribbean ancestry for Alzheimer's disease (AD) and vascular dementia (VaD) (Clark et al., 2005; Fitten et al., 2014; Gurland et al., 1999; Livney et al., 2011; Tang et al., 2001). Depending on the variables considered in these studies, the differences in age of onset in these studies was accounted for by medical comorbidities (e.g., vascular disease, diabetes mellitus type 2, hyperlipidemia), APOE e4 genotype, or depression. Although Gurland et al. (1999) and Livney et al. (2011) hypothesized that factors such as lower educational attainment and poverty may be contributory, this was not supported in the other abovementioned studies.

Genetics Regarding genetic risk factors, approximately 5% of early-onset cases, that is, before the age of 65, are autosomal dominant due to mutations on the following genes: presenilin 1 and 2 and the amyloid precursor protein gene (Farrer et al., 1997; Mendez, 2012). However, these mutations are not considered to account for the approximate 4 years earlier onset for Latinos overall. The fourth genetic risk for AD regards certain apolipoprotein E (APOE) genotypes that are found to incur increased risk for the development of AD. The APOE gene is involved in cholesterol transport. It has three common alleles (e2, e3, and e4), the combination of which can modulate one's risk for developing AD. The APOE e4/e4 genotype typically confers the greatest risk, followed by e3/e4 and e2/e4. Protective genotypes are e2/e2 and e2/e3 (Farrer et al., 1997; Tang et al., 1998).

The risk linked to these APOE genotypes, as well as their distribution in the population, varies by racial/ethnic group. APOE e4/e4 appears to be most widely distributed among Caucasians, followed by African-Americans and Latinos. In the meta-analysis by Farrer et al. (1997) that pooled clinical and autopsy results of 5930 patients and 8607 controls, frequencies of this genotype were found to be 14.8% among Caucasian cases, 12.3% among African American cases, 2.7% among Latino cases, and 8.9% among Japanese cases. Among non-AD controls, the rates were 1.8%, 2.1%, 1.9%, and 0.8%, respectively. The actual risk or protective influence conferred by APOE genotypes also varied by group. Latinos in this largely Mexican American sample had the lowest-conferred odds from the e4/e4 genotype (OR = 2.2). These results have been replicated elsewhere and have also been found among Caribbean Latinos in the U.S. (Campos et al., 2013; Tang et al., 1998).

Pathology AD pathology is characterized by a cascade of a neurochemical changes and inflammation, followed by and coincident with the pathological aggregation of amyloid-β (Aβ) protein and neuronal loss (Wenk, 2003). An early neurochemical change is the reduction in choline acetyltransferase and acetylcholinesterase enzyme activity, as well as the attenuated ability to break down neuritic plaques via reductions in corticotropin-releasing factor and somatostatin. Finally, the increased stimulation of glutamate receptors in this process contributes to neuronal death. The typical pattern of atrophy in AD targets the medial temporal (i.e., hippocampus and entorhinal cortex) and post-temporoparietal cortices most significantly, spreading to the frontal lobes and other areas of the neocortex in later stages (Miller-Thomas et al., 2016; Rabinovici et al., 2007). The visual cortex, basal ganglia, thalami, cerebellum, and brain stem are relatively spared in this typical pattern of atrophy (Rabinovici et al., 2007).

Clinical presentation AD is above all a clinical diagnosis, which can be supported by biomarker evidence to increase confidence (Jack et al., 2010; McKhann et al., 2011). It is characterized as having an insidious onset and gradual progression. The typical (or "amnestic") cognitive profile for AD pathology follows a clinical course of initial impairment in episodic memory, which is often noted by the patient or close relations. Early signs may include increased difficulty with remembering names, poor retention of learned information such as recent news stories, or becoming lost in formerly familiar locations. Cognitive symptoms then progress to other cognitive domains including complex attention (e.g., working memory), semantic memory, problem-solving and reasoning, visuospatial ability, and language. As with dementia generally, the pattern of impairment eventually becomes global. AD patients often retain insight into their condition while in the mild or even moderate stages, which understandably may result in depression or anxiety symptoms. In later stages, psychiatric comorbidities such as depression can be the result of structural and neurochemical changes due to the dementia process, and lack of insight is common.

Latinos have been found to differ in their severity of cognitive and psychiatric symptoms in AD when compared to other ethnicities. In a study comparing clinical characteristics in AD between ethnicities in the U.S., Livney et al. (2011) found that the largely Caribbean Latino sample had more severe cognitive symptoms and depression at initial presentation compared to the other cohorts. The authors contextualized these findings within the context of lower average years of education and lower socioeconomic status among the Latino sample, which partially mediated this discrepancy. Furthermore, Salazar, Royall, and Palmer (2015) found elevated rates of neuropsychiatric symptoms among their sample of 1079 elderly Mexican Americans with dementia. The neuropsychiatric profiles suggested that the Mexican American sample were cared for in family homes until dementia severity was severe enough to necessitate medical intervention.

Vascular Dementia (VaD)

Prevalence Vascular dementia (VaD), which may be grouped within the broader category of vascular cognitive impairment (VCI), is considered the second most common dementia after AD and is a condition of deteriorated cognitive functioning due to cerebrovascular disease (Gorelick et al., 2011; O'Brien & Thomas, 2015). VaD is more common in men compared to women (Rizzi et al., 2014). Contributors to VaD and VCI are multiple and include infarcts (e.g., strategic, multiple, or watershed), small vessel ischemic disease, cerebral amyloid angiopathy (CAA), and familial causes such as cerebral autosomal dominant arteriopathy with subcortical infarcts and leukoencephalopathy (CADASIL). The shared presentation of cerebrovascular disease with other degenerative pathology, such as amyloid-β accumulation, is more common than pure vascular dementia. Pure VaD cases are estimated to account for only approximately 10% of cases (O'Brien & Thomas, 2015).

Not surprisingly, risk factors for vascular dementia align with those for vascular disease more generally, including hypertension, hyperlipidemia, diabetes, obesity, atrial fibrillation, raised homocysteine levels, and smoking (O'Brien & Thomas, 2015). As noted earlier in this chapter, Latinos tend to experience a high prevalence of these conditions as well, which suggests a similarly high cerebrovascular burden (Haan et al., 2003; Pabon-Nau et al., 2010; Vega et al., 2009). Notably, it appears that the incidence of these conditions at midlife is especially associated with eventual cognitive decline in old age (Barnes & Yaffe, 2011; Lista, Dubois, & Hampel, 2015). Even with timely intervention of these conditions at midlife, such as through medications (e.g., statins) or lifestyle changes (e.g., smoking cessation, dietary changes), some detectable cognitive impairment that has been incurred may still be incurred and irreversable in old age.

Genetics The genetic bases for VaD and VCI remain poorly understood in the scientific literature. The Notch 3 mutation on chromosome 19 that leads to CADASIL is the most extensively studied contributor, though this accounts for a small minority (~4.6 per 100,000) of mostly early-onset VaD cases (Joutel et al., 1996; Moreton, Razvi, Davidson, & Muir, 2014). The few genetic contributors that have been detected are often nonspecific to VaD, such as the APOE e4 genotype that is influential in AD, as well as the C677T polymorphism of the MTHFR gene related to homocysteine metabolism (Iadecola, 2013; O'Brien & Thomas, 2015).

Pathology The many etiologies of VaD and VCI correspond to similarly heterogeneous pathological and cognitive profiles. Individuals presenting with VCI typically have multiple sites of cerebrovascular damage (Gorelick et al., 2011). VCI pathology includes cortical and subcortical infarcts, ischemic lesions, and microhemorrhages (Gorelick et al., 2011; O'Brien & Thomas, 2015).

AD pathology exists as a separate risk factor for vascular-related cognitive impairment and dementia. Amyloid-β acts as a vasoconstrictor which, when its aggregation has become pathological, can interfere with cerebral microcirculation and cerebrovascular autoregulation, as well as cause cerebral hemorrhages and infarcts (Gorelick et al., 2011). Relatedly, AD and VaD share common risk factors that are harmful to the cerebrovascular system, such as hypertension, diabetes, obesity, and smoking. In CAA, which accompanies Alzheimer's disease in 80% to 100% of cases (and approximately 10–30% of unselected brains), the resulting hemorrhage is more likely to target areas seen in AD, including the temporal and parietal lobes.

Clinical presentation The particular source of cerebrovascular disease will determine the pattern of cognitive deficits, as well as their onset and progression (Deramecourt et al., 2012; O'Brien et al., 2003). When the etiology in VaD or VCI is due to infarcts, the initial impairments are often focal to the particular lesion. For example, a left fronto-parietal stroke may incur difficulties with language such as in Wernicke's aphasia. Common cognitive signs of infarct-caused VaD and VCI include attentional difficulties, disinhibition, aphasia, hemiparesis, and sensory impairment. Cognitive decline caused by multiple infarcts typically follows a stepwise progression, meaning the patient experiences periods of stabilization before continued decline as the result of new cerebrovascular incidents. However, it is important to note that this stepwise course is not always followed, with many cases displaying a more insidious and slowly progressive course mimicking AD. An insidious onset of cognitive symptoms with a progressive decline is often associated with subcortical ischemic vascular disease (O'Brien et al., 2003), although it may also be seen with cortical infarcts. Subcortical dementia is characterized by cognitive slowing and executive dysfunction, including impaired working memory and sustained attention (O'Brien & Thomas, 2015). Additional signs include motor abnormalities such as gait disturbance (frontal or magnetic gait), incontinence, and pseudobulbar affect. In the presence of a single infarct, the impairment may stabilize assuming further infarcts are avoided. This patient may still meet criteria for VaD, as the impairment need not be progressive.

Medical Comorbidities Contributing to AD and VaD

As mentioned previously, Latinos have a higher prevalence of metabolic and vascular morbidity, including T2DM and, more variably, hypertension. Both of these conditions are considered risk factors in the incidence of AD and VaD. A brief outline of the mechanisms by which these conditions are found to contribute to dementia incidence is now provided.

The presence of T2DM increases the risk of cerebrovascular disease, including white matter disease of the brain and infarcts, which in turn results in a higher likelihood of VaD (Bordier, Doucet, Boudet, & Bauduceau, 2014; Fitten et al., 2001). The relationship with AD pathology is somewhat more controversial considering the lack of conclusive evidence from autopsy studies (Luchsinger, 2012; Starr & Convit, 2007). T2DM sets in motion different mechanisms that are theorized to prevent the clearing of amyloid-β (Aβ) and tau, two proteins that are linked to AD and other neurodegenerative diseases. Excessive levels of insulin in the brain from hyperinsulinemia is thought to result in the excess deposits of these proteins. Chronic hyperinsulinemia in T2DM may also, over time, result in a reduction of insulin crossing the blood-brain barrier, thus depriving insulin to the receptors in the hippocampus and entorhinal cortex and resulting in Alzheimer's-like cognitive impairment. Furthermore, this process may result in the reduction of insulin-degrading enzyme (IDE), which is a mechanism for clearing AB. Lipid lipoprotein-related proteins (LRP) are also reduced in T2DM with the effect of impeding the clearance of AB (Bordier et al., 2014; Claessens et al., 2016; Luchsinger, 2012).

Hypertension, on the other hand, is a highly prevalent risk factor for dementia and stroke both in the U.S. and globally (Forouzanfar et al., 2017; Nwankwo, Yoon, Burt, & Gu, 2013). U.S. Latinos are typically reported to have hypertension rates on par with non-Hispanic Whites, though U.S. Caribbean Latinos have been found to have elevated blood pressure relative than Whites or other U.S. Latino populations (Fei et al., 2017; Pabon-Nau et al., 2010). Low socioeconomic status and degree of acculturation to the U.S. society, in addition to metabolic comorbidities, are variables that increase the risk profile of cardiovascular morbidity such as hypertension among Latinos (Daviglus et al., 2012; Vega et al., 2009).

Hypertension is considered to be a risk factor for VaD and AD (Obisesan, 2009; Tzourio, 2007). Chronic hypertension is most consistently implicated in frontal and subcortical symptoms of late-life cognitive impairment and dementia, including deficits in attention, slowed speed of processing, and executive dysfunction (Matthews & Richter, 2003; O'Brien et al., 2003; Obisesan, 2009). Hypertension is thought to cause cognitive decline through microvascular changes, affecting both white and gray matter in the brain, due to reductions in blood flow and subsequent deprivation of oxygen and nutrients over time. Such circulatory changes are considered to be part of normal aging but may be accelerated with chronic hypertension (Gąsecki, Kwarciany, Nyka, & Narkiewicz, 2013). White matter changes are typically detected as white matter hyperintensities (WMH) on medical diagnostic imaging, with the frontal lobes being particularly vulnerable even over relatively short time frames (i.e., 1 year; Raz et al., 2012). Gray matter regions that appear particularly sensitive to hypertension in combination with advanced age include the prefrontal cortex, the hippocampus, and the inferior temporal and parietal cortices (Gąsecki et al., 2013).

Frontotemporal Degeneration (FTD)

Prevalence Frontotemporal degeneration (FTD) is a progressive dementia with two major subdivisions, which are behavioral or frontal variant FTD (bvFTD), and three language-based variants that are conceptualized within the category of primary progressive aphasias (PPA): non-fluent or agrammatic, semantic, and logopenic. These clinical presentations share pathological similarities, despite diverging in their clinical presentation. Prevalence statistics for FTD in Latinos were not found at the time of this review. Men and women appear to be equivalently at risk for developing FTD (Onyike & Diehl-Schmid,

2013). FTD is characterized by presenile onset, with 75–80% of cases occurring before the age of 65 (Rabinovici & Miller, 2010).

Genetics Family history and autosomal dominant cases in FTD are relatively common (McKhann, 2001). Although research is ongoing as to the genetic etiology, mutations on chromosomes 1, 3, 9, 16, and 17 and TDP-43 have been noted (McKhann et al., 2011; Neary et al., 1998; Rabinovici & Miller, 2010). The chromosome 17 mutation, which regards the microtubule-associated protein tau (MAPT) of the chromosome, is accompanied by earlier onset of symptoms and a tendency toward more symmetric temporal lobe atrophy. Additionally, the progranulin (GRN) gene mutations on chromosome 17 have been implicated with both social-behavioral and language impairments (e.g., anomia, word comprehension difficulty) in FTD and are implicated in Alzheimer's disease among other dementias (Gorno-Tempini et al., 2011). The chromosome 9 mutation is implicated in inherited behavioral variant frontotemporal dementia, which may be combined with motor neuron disease, and increases the likelihood of psychotic features.

Pathology A shared pathology has been found underlying both behavioral variant and PPA syndromes. Pathology is categorized as either frontal lobe degeneration type, which comprises a nonspecific histology, or Pick type, in which astrocytic gliosis is exhibited (Neary et al., 1998). In the latter presentation, ballooned cells and inclusion bodies may or may not be present, although these were integral in the initial conceptualization of the disease (i.e., Pick disease; McKhann et al., 2001). FTD variants (behavioral, non-fluent, semantic, and logopenic) typically exhibit the tau-positive, ubiquitin/TDP-43-positive FTD-type pathology, but amyloid-type pathology may also be found (Gorno-Tempini et al., 2011; Rabinovici & Miller, 2010). Logopenic PPA is more typically found to have both AD- and tau-type pathologies.

The pattern of degeneration in FTD is predominantly localized in the frontal or temporal lobes (Gorno-Tempini et al., 2011; Rascovsky et al., 2011). In imaging-confirmed behavioral variant FTD, the pathology affects the frontal lobes, is bilateral, and is typically symmetric. In non-fluent PPA, this pattern is asymmetric and affects the left frontotemporal lobes. More specifically, atrophy will be found within the posterior fronto-insular region (e.g., inferior frontal gyrus, insula, premotor, and supplemental areas). In the semantic PPA variant, atrophy is often more prevalent in the left hemisphere, though damage can present bilaterally with the ventral and lateral anterior temporal neocortex lobes most prominently affected. The bilateral nature of the lesions in the semantic variant have been found to correspond to cognitive deficits in both verbal and visual domains. Finally, in logopenic PPA, the atrophy is localized in the left temporoparietal junction of the posterior temporal, supramarginal, and angular gyri.

Clinical presentation The clinical profiles outlined in the latest consensus criteria (Gorno-Tempini et al., 2011; Rascovsky et al., 2011) for the FTD subtypes are described below. For precise diagnostic criteria, the reader is directed to these publications as the description herein is meant to provide only a general overview. As with other dementias, these profiles describe the most prominent symptoms typically at early to intermediate stages of the disease process.

Behavioral or frontal variant FTD is distinguished by an alteration is social or personal conduct which is prominent in the early stages of the disease when clinical signs are first present. Individuals may exhibit inappropriate or impulsive behaviors (e.g., swearing, angry outbursts, stealing) and display avolitional and apathetic tendencies (Rascovsky et al., 2011). Perseverative behaviors are also common (e.g., repetitive movements, stereotypy of speech), as is hyperorality and overeating. Neuropsychologically, patients present with a dysexecutive pattern of impairment in areas of attention, reasoning, and problem-solving, with relative sparing of memory, language, and visuospatial functioning. Memory impairments that do occur on neuropsychological testing are most often attributed to frontal dysfunction and are characterized by difficulty with encoding and retrieval. Patients who exhibit a more disinhibited and active presentation, rather than inactive and avolitional, may demonstrate less executive impairment, although they will likely be impaired in measures of selective attention (Neary et al., 1998).

The cognitive profile in non-fluent or agrammatic PPA is characterized as agrammatism and apraxia of speech (Gorno-Tempini et al., 2011). Speech is effortful, halting, and with notable word-finding difficulties. Additional impairment may occur for comprehension when the content has complex syntax structure. Phonological and grammatical errors are often present in speech, and difficulties with reading and writing may occur. Word comprehension and object knowledge are intact. Impairment in other cognitive domains is absent, though behavioral changes may occur in later stages. In semantic variant PPA, the primary cognitive deficits are in word comprehension and confrontation naming. Patients may also exhibit impaired object knowledge, especially for infrequent or unfamiliar items, and surface dyslexia or dysgraphia. Repetition and speech production are intact. In logopenic PPA, the hallmark clinical feature is impaired single-word retrieval and repetition. Patients may also exhibit phonemic paraphasias, that is, substitutions with a nonword that sounds similar.

Lewy Body Dementias

Prevalence Lewy body dementias, under which are grouped dementia with Lewy bodies (DLB) and Parkinson's disease dementia (PDD), represent a neurodegenerative disease class characterized by abnormal Lewy body formation (McKeith et al., 2017; Walker, Possin, Boeve, & Aarsland, 2015). The pathological process in Lewy body dementias is the pathological aggregation and accumulation of alpha-synuclein protein within neurons. The outcome of this process are Lewy bodies, also called inclusion bodies, which can be found either cortically or subcortically, including in the brain stem. DLB and PDD are increasingly considered synonymous synucleinopathies with overlapping clinical and pathological criteria (McKeith et al., 2017). Similar to FTD, prevalence statistics of LBD for Latinos were not available at the time of this review. Men have an increased risk for developing LBD (Savica et al., 2014). The age of onset in DLB is typically 50 years and older, and in PDD the typical range begins after 75 years of age (Vasconcellos & Pereira, 2015).

Genetics Lewy body dementias are typically sporadic, although an autosomal dominant pattern can occur in rare cases with mutations in SNCA and LRRK2 (Walker et al., 2015). A study by Nervi et al. (2011) found DLB and core clinical features to aggregate in families in their sibling study of the U.S. Caribbean Latino families. The genetic bases for Lewy body dementias overall are poorly understood, though several genes are implicated (McKeith et al., 2017; Vasconcellos & Pereira, 2015). These include mutations of the alpha-synuclein (SNCA) and glucocerebrosidase (GBA) genes, the latter of which is somewhat overrepresented in these patients. Biomarkers typically associated with other dementias are recommended as part of the diagnostic process in Lewy body dementias as well, in particular amyloid-β and tau. This is both because of the commonly shared pathology with Lewy body dementias and other dementias, and also due to the high occurrence of misdiagnoses (e.g., with Alzheimer's disease or frontotemporal dementias; McKeith et al., 2005, 2017; Walker et al., 2015).

Pathology The current international criteria (McKeith et al., 2017) outlines the available biomarkers tests to assist in the diagnosis of DLB and PDD. Direct biomarkers of Lewy bodies are not available, but several procedures can yield highly predictive results. These biomarkers include reduced dopamine active transporter (DAT) uptake in the basal ganglia or reduced uptake on single-photon emission computed tomography (SPECT) or positron emission tomographic (PET) scan, reduced uptake on metaiodobenzylguanidine (MIBG) myocardial scintigraphy, and findings congruent with rapid eye movement (REM) sleep behavior disorder on polysomnography. Supportive biomarkers include absent medial temporal lobe atrophy on structural imaging such as computerized axial tomography (CAT) or magnetic resonance imaging (MRI), which will help to rule out the typical differential of Alzheimer's disease; occipital hypometabolism and preserved posterior or midcingulate metabolism on SPECT and/or PET scan, which relate to visual cortex pathology and will help to rule out tauopathy; and electroencephalogram (EEG) results that indicate a pre-alpha-dominant frequency.

Clinical presentation Clinically, DLB and PDD share many clinical features, with the main distinction being made in temporal course: in PDD, parkinsonism precedes cognitive symptoms by at least a year (Walker et al., 2015). Due to this overlap, we will include the latest international consensus criteria for DLB (McKeith et al., 2017), though the reader is reminded that PDD is largely represented in this description.

The essential clinical features in the diagnosis of DLB is cognitive impairment, for which the typical neuropsychological profile shows prominent attentional, dysexecutive, and visual processing deficits. Memory and object naming are likely to be spared earlier in the disease process. If this essential clinical feature has been met, patients are diagnosed with probable DLB if they meet two or more core clinical features, whether or not biomarkers are present, or if they have at least one clinical feature with one or more indicative biomarker. Possible DLB is diagnosed in the presence of either one of the core clinical features only, or one or more indicative biomarkers in the absence of core clinical features.

The core clinical features include fluctuation in attention and arousal. This resembles delirium and can include episodes of disorientation, variable or lapsing attention, incoherent speech, or altered consciousness. Patients may be drowsy or lethargic during the day. Visual hallucinations are the second core clinical feature, which occur in up to 80% of patients (McKeith et al., 2017). These are well formed and typically feature animals, or people of short stature. Patients may not find the hallucinations to be disturbing, and in the earlier stages they are more able to distinguish them from reality. The next core feature is spontaneous parkinsonism. The final core clinical feature is REM sleep behavior disorder, in which atonia is lacking during the REM sleep phase causing the patient to "act out" dreams. This is usually ascertained by soliciting the account of a bed partner or questioning the patient if upon waking they have found that their position in the bed changes notably, bed linens or nearby objects are in disarray.

Severe neuroleptic sensitivity to antipsychotic medication is now considered a supportive clinical feature since the 2005 criteria was published (McKeith et al., 2017). This category includes symptoms that are commonly occurring in DLB but lack specificity. The motor and cognitive symptoms of patients with Lewy body disease can be exacerbated when prescribed antipsychotics, such as for the commonly occurring visual hallucinations (Aarsland et al., 2005). More recently, though, D2 receptor blocking antipsychotics have become less prescribed, which has reduced the prevalence of this clinical sign (Walker et al., 2015). Other supportive clinical features include alterations and sleep patterns (hypersomnia), hyposmia (i.e., reduced olfactory ability), and episodes of nonresponsiveness resembling syncope.

Summary Latinos are a diverse, large minority population within the U.S. with unique trends in dementia and medical comorbidities. In particular, a higher rate of metabolic and vascular disorders places Latinos at a greater risk for cerebrovascular burden, which in turn is likely to influence the prevalence of Alzheimer's disease and vascular dementia in particular. Continued research is necessary to further understand the specific genetic contributors of populations originating from Latin America in regard to dementia prevalence and course, as well as medical comorbidities. Important differences in social/cultural and biological variables exist between particular subgroups of Latinos, with the major divisions considered in the literature being between Latinos of Mexican or Caribbean origin. Socioeconomic variables such as poverty, access to quality healthcare, educational history, immigration history, English-language acquisition, and acculturation to the U.S. society are also important considerations in clinical practice when working with elderly Latinos in the context of neurodegenerative diseases and medical comorbidities.

References

Aarsland, D., Perry, R., Larsen, J. P., McKeith, I. G., O'Brien, J. T., Perry, E. K., … Ballard, C. G. (2005). Neuroleptic sensitivity in Parkinson's disease and parkinsonian dementias. *The Journal of Clinical Psychiatry, 66*(5), 633–637.

Alhola, P., & Polo-Kantola, P. (2007). Sleep deprivation: Impact on cognitive performance. *Neuropsychiatric Disease and Treatment, 3*(5), 553–567.

Allen, L. H. (2008). Causes of vitamin B12 and folate deficiency. *Food and Nutrition Bulletin, 29*(2 Suppl), S20–S34; discussion S35–37. https://doi.org/10.1177/15648265080292S105

Allen, L. H. (2009). How common is vitamin B-12 deficiency? *The American Journal of Clinical Nutrition, 89*(2), 693S–696S. https://doi.org/10.3945/ajcn.2008.26947A

Alvarez, J., Jason, L. A., Olson, B. D., Ferrari, J. R., & Davis, M. I. (2007). Substance abuse prevalence and treatment among Latinos and Latinas. *Journal of Ethnicity in Substance Abuse, 6*(2), 115–141.

Alzheimer's Association. (2016). 2016 Alzheimer's disease facts and figures. *Alzheimer's & Dementia: The Journal of the Alzheimer's Association, 12*(4), 459–509.

Barnes, D. E., & Yaffe, K. (2011). The projected effect of risk factor reduction on Alzheimer's disease prevalence. *The Lancet. Neurology, 10*(9), 819–828. https://doi.org/10.1016/S1474-4422(11)70072-2

Bégin, M. E., Langlois, M. F., Lorrain, D., & Cunnane, S. C. (2008). Thyroid function and cognition during aging. *Current Gerontology and Geriatrics Research, 2008*, 1–11. https://doi.org/10.1155/2008/474868

Bernardin, F., Maheut-Bosser, A., & Paille, F. (2014). Cognitive impairments in alcohol- dependent subjects. *Frontiers in Psychiatry, 5*. https://doi.org/10.3389/fpsyt.2014.00078

Blessman, D. J. (2008). Chronic hepatitis C in the Hispanic/Latino population living in the United States: A literature review. *Gastroenterology Nursing: The Official Journal of the Society of Gastroenterology Nurses and Associates, 31*(1), 17–25. https://doi.org/10.1097/01.SGA.0000310931.64854.5f

Bordier, L., Doucet, J., Boudet, J., & Bauduceau, B. (2014). Update on cognitive decline and dementia in elderly patients with diabetes. *Diabetes & Metabolism, 40*(5), 331–337. https://doi.org/10.1016/j.diabet.2014.02.002

Campos, M., Edland, S. D., & Peavy, G. M. (2013). Exploratory study of apolipoprotein E ε4 genotype and risk of Alzheimer's disease in Mexican Hispanics. *Journal of the American Geriatrics Society, 61*(6), 1038–1040. https://doi.org/10.1111/jgs.12292

Centers for Disease Control and Prevention [CDC]. (2016). *HIV among people aged 50 and over.* Retrieved from https://www.cdc.gov/hiv/group/age/olderamericans/index.html

Centers for Disease Control and Prevention [CDC]. (2017). *HIV among Latinos*. Retrieved from https://www.cdc.gov/hiv/group/racialethnic/hispaniclatinos/index.html

Chavez-Dueñas, N. Y., Adames, H. Y., & Organista, K. C. (2014). Skin-color prejudice and within-group racial discrimination: Historical and current impact on Latino/a populations. *Hispanic Journal of Behavioral Sciences, 36*(1), 3–26. https://doi.org/10.1177/0739986313511306

Chin, A. L., Negash, S., & Hamilton, R. (2011). Diversity and disparity in dementia: The impact of ethnoracial differences in Alzheimer's disease. *Alzheimer Disease & Associated Disorders, 25*(3), 187–195. https://doi.org/10.1097/WAD.0b013e318211c6c9

Claessens, D., Geijselaers, S. L. C., & Sep, S. J. S. (2016). Glycaemia, cognition, and type 2 diabetes. *MaRBLe, 2*(0). Retrieved from http://openjournals.maastrichtuniversity.nl/Marble/article/view/296

Clark, C. M., DeCarli, C., Mungas, D., Chui, H. I., Higdon, R., Nuñez, J., … van Belle, G. (2005). Earlier onset of Alzheimer disease symptoms in latino individuals compared with anglo individuals. *Archives of Neurology, 62*(5), 774–778. https://doi.org/10.1001/archneur.62.5.774

Clark, H., Surendera Babu, A., Harris, S., & Hardnett, F. (2015). HIV-related mortality among adults (≥18 years) of various Hispanic or Latino subgroups—United States, 2006–2010. *Journal of Racial and Ethnic Health Disparities, 2*(1), 53–61. https://doi.org/10.1007/s40615-014-0047-x

Cohen, R. A., Boland, R., Paul, R., Tashima, K. T., Schoenbaum, E. E., Celentano, D. D., … Carpenter, C. C. (2001). Neurocognitive performance enhanced by highly active antiretroviral therapy in HIV-infected women. *AIDS (London, England), 15*(3), 341–345.

Cox, D. J., Kovatchev, B. P., Gonder-Frederick, L. A., Summers, K. H., McCall, A., Grimm, K. J., & Clarke, W. L. (2005). Relationships between hyperglycemia and cognitive performance among adults with type 1 and type 2 diabetes. *Diabetes Care, 28*(1), 71–77.

Crimmins, E. M., Kim, J. K., Alley, D. E., Karlamangla, A., & Seeman, T. (2007). Hispanic paradox in biological risk profiles. *American Journal of Public Health, 97*(7), 1305–1310. https://doi.org/10.2105/AJPH.2006.091892

Cummins, C. A., Erlyana, E., Fisher, D. G., & Reynolds, G. L. (2015). Hepatitis C infection among Hispanics in California. *Journal of Addictive Diseases, 34*(4), 263–273. https://doi.org/10.1080/10550887.2015.1074500

Cysique, L. A. J., Maruff, P., & Brew, B. J. (2006). Variable benefit in neuropsychological function in HIV-infected HAART-treated patients. *Neurology, 66*(9), 1447–1450. https://doi.org/10.1212/01.wnl.0000210477.63851.d3

Daviglus, M. L., Talavera, G. A., Avilés-Santa, M. L., Allison, M., Cai, J., Criqui, M. H., … Stamler, J. (2012). Prevalence of major cardiovascular risk factors and cardiovascular diseases among Hispanic/Latino individuals of diverse backgrounds in the United States. *JAMA, 308*(17), 1775–1784. https://doi.org/10.1001/jama.2012.14517

Davis, J. D., & Tremont, G. (2007). Neuropsychiatric aspects of hypothyroidism and treatment reversibility. *Minerva Endocrinologica, 32*(1), 49–65.

de Benoist, B. (2008). Conclusions of a WHO technical consultation on folate and vitamin B12 deficiencies. *Food and Nutrition Bulletin, 29*(2 Suppl), S238–S244. https://doi.org/10.1177/15648265080292S129

Deramecourt, V., Slade, J. Y., Oakley, A. E., Perry, R. H., Ince, P. G., Maurage, C.-A., & Kalaria, R. N. (2012). Staging and natural history of cerebrovascular pathology in dementia. *Neurology, 78*(14), 1043–1050. https://doi.org/10.1212/WNL.0b013e31824e8e7f

Doets, E. L., van Wijngaarden, J. P., Szczecińska, A., Dullemeijer, C., Souverein, O. W., Dhonukshe-Rutten, R. A. M., … de Groot, L. C. P. G. M. (2013). Vitamin B12 intake and status and cognitive function in elderly people. *Epidemiologic Reviews, 35*, 2–21. https://doi.org/10.1093/epirev/mxs003

Dugbartey, A. T. (1998). Neurocognitive aspects of hypothyroidism. *Archives of Internal Medicine, 158*(13), 1413–1418. https://doi.org/10.1001/archinte.158.13.1413

Ejaz, S. M., Khawaja, I. S., Bhatia, S., & Hurwitz, T. D. (2011). Obstructive sleep apnea and depression. *Innovations in Clinical Neuroscience, 8*(8), 17–25.

Etgen, T., Sander, D., Bickel, H., & Förstl, H. (2011). Mild cognitive impairment and dementia: The importance of modifiable risk factors. *Deutsches Arzteblatt International, 108*(44), 743–750. https://doi.org/10.3238/arztebl.2011.0743

Farrer, L. A., Cupples, L. A., Haines, J. L., Hyman, B., Kukull, W. A., Mayeux, R., … van Duijn, C. M. (1997). Effects of age, sex, and ethnicity on the association between apolipoprotein E genotype and Alzheimer disease. A meta-analysis. APOE and Alzheimer Disease Meta Analysis Consortium. *JAMA, 278*(16), 1349–1356.

Fei, K., Rodriguez-Lopez, J. S., Ramos, M., Islam, N., Trinh-Shevrin, C., Yi, S. S., … Thorpe, L. E. (2017). Racial and ethnic subgroup disparities in hypertension prevalence, New York City health and nutrition examination survey, 2013–2014. *Preventing Chronic Disease, 14*. https://doi.org/10.5888/pcd14.160478

Fitten, L. J., Ortiz, F., Fairbanks, L., Bartzokis, G., Lu, P., Klein, E., … Ringman, J. (2014). Younger age of dementia diagnosis in a Hispanic population in southern California. *International Journal of Geriatric Psychiatry, 29*(6), 586–593. https://doi.org/10.1002/gps.4040

Fitten, L. J., Ortiz, F., & Pontón, M. (2001). Frequency of Alzheimer's disease and other dementias in a community outreach sample of Hispanics. *Journal of the American Geriatrics Society, 49*(10), 1301–1308.

Forouzanfar, M. H., Liu, P., Roth, G. A., Ng, M., Biryukov, S., Marczak, L., … Murray, C. J. L. (2017). Global burden of hypertension and systolic blood pressure of at least 110 to 115 mm Hg, 1990–2015. *JAMA, 317*(2), 165–182. https://doi.org/10.1001/jama.2016.19043

Franciscus, A. (2014). *Hepatitis C and U.S. Hispanics*. Retrieved from https://npin.cdc.gov/publication/hcsp-fact-sheet-hepatitis-c-and-us-hispanics

Gan, E. H., & Pearce, S. H. S. (2012). Clinical review: The thyroid in mind: Cognitive function and low thyrotropin in older people. *The Journal of Clinical Endocrinology and Metabolism, 97*(10), 3438–3449. https://doi.org/10.1210/jc.2012-2284

Ganguli, M. (2009). Depression, cognitive impairment and dementia: Why should clinicians care about the web of causation? *Indian Journal of Psychiatry, 51*(Suppl 1), S29–S34.

Garrod, M. G., Green, R., Allen, L. H., Mungas, D. M., Jagust, W. J., Haan, M. N., & Miller, J. W. (2008). Fraction of total plasma vitamin B12 bound to transcobalamin correlates with cognitive function in elderly Latinos with depressive symptoms. *Clinical Chemistry, 54*(7), 1210–1217. https://doi.org/10.1373/clinchem.2007.102632

Gąsecki, D., Kwarciany, M., Nyka, W., & Narkiewicz, K. (2013). Hypertension, brain damage and cognitive decline. *Current Hypertension Reports, 15*(6), 547–558. https://doi.org/10.1007/s11906-013-0398-4

González, H. M., Haan, M. N., & Hinton, L. (2001). Acculturation and the prevalence of depression in older Mexican Americans: Baseline results of the Sacramento Area Latino Study on Aging. *Journal of the American Geriatrics Society, 49*(7), 948–953.

González-Rodríguez, L. A., Felici-Giovanini, M. E., & Haddock, L. (2013). Thyroid dysfunction in an adult female population: A population-based study of Latin American Vertebral Osteoporosis Study (LAVOS) - Puerto Rico site. *Puerto Rico Health Sciences Journal, 32*(2), 57–62.

Gorelick, P. B., Scuteri, A., Black, S. E., Decarli, C., Greenberg, S. M., Iadecola, C., … American Heart Association Stroke Council, Council on Epidemiology and Prevention, Council on Cardiovascular Nursing, Council on Cardiovascular Radiology and Intervention, and Council on Cardiovascular Surgery and Anesthesia. (2011). Vascular contributions to cognitive impairment and dementia: A statement for healthcare professionals from the american heart association/american stroke association. *Stroke, 42*(9), 2672–2713. https://doi.org/10.1161/STR.0b013e3182299496

Gorno-Tempini, M. L., Hillis, A. E., Weintraub, S., Kertesz, A., Mendez, M., Cappa, S. F., … Grossman, M. (2011). Classification of primary progressive aphasia and its variants. *Neurology, 76*(11), 1006–1014. https://doi.org/10.1212/WNL.0b013e31821103e6

Griswold, K. S., Del Regno, P. A., & Berger, R. C. (2015). Recognition and differential diagnosis of psychosis in primary care. *American Family Physician, 91*(12), 856–863.

Gurland, B. J., Wilder, D. E., Lantigua, R., Stern, Y., Chen, J., Killeffer, E. H., & Mayeux, R. (1999). Rates of dementia in three ethnoracial groups. *International Journal of Geriatric Psychiatry, 14*(6), 481–493.

Haan, M. N., Miller, J. W., Aiello, A. E., Whitmer, R. A., Jagust, W. J., Mungas, D. M., … Green, R. (2007). Homocysteine, B vitamins, and the incidence of dementia and cognitive impairment: Results from the Sacramento Area Latino Study on Aging. *The American Journal of Clinical Nutrition, 85*(2), 511–517.

Haan, M. N., Mungas, D. M., Gonzalez, H. M., Ortiz, T. A., Acharya, A., & Jagust, W. J. (2003). Prevalence of dementia in older latinos: The influence of type 2 diabetes mellitus, stroke and genetic factors. *Journal of the American Geriatrics Society, 51*(2), 169–177.

Hardy, D. J., & Vance, D. E. (2009). The neuropsychology of HIV/AIDS in older adults. *Neuropsychology Review, 19*(2), 263–272. https://doi.org/10.1007/s11065-009-9087-0

Heaton, R. K., Ryan, L., & Grant, I. (2009). Demographic influences and use of demographically corrected norms in neuropsychological assessment. In I. Grant & K. M. Adams (Eds.), *Neuropsychological assessment of neuropsychiatric and neuromedical disorders* (3rd ed., pp. 127–157). New York, NY: Oxford University Press.

Heinrich, T. W., & Grahm, G. (2003). Hypothyroidism presenting as psychosis: Myxedema madness revisited. *Primary Care Companion to The Journal of Clinical Psychiatry, 5*(6), 260–266.

Hinds, H. E., Johnson, A. A., Webb, M. C., & Graham, A. P. (2011). Iron, folate, and vitamin B12 status in the elderly by gender and ethnicity. *Journal of the National Medical Association, 103*(9–10), 870–877.

Hinton, C. F., Harris, K. B., Borgfeld, L., Drummond-Borg, M., Eaton, R., Lorey, F., … Pass, K. A. (2010). Trends in incidence rates of congenital hypothyroidism related to select demographic factors: Data from the United States, California, Massachusetts, New York, and Texas. *Pediatrics, 125*(Supplement 2), S37–S47. https://doi.org/10.1542/peds.2009-1975D

Horowitz, C. R., Colson, K. A., Herbert, P. L., & Lancaster, K. (2004). Barriers to buying healthy foods for people with diabetes: Evidence of environmental disparities. *American Journal of Public Health, 94*(9), 1549–1554. https://doi.org/10.2105/ajph.94.9.1549

Huckans, M., Seelye, A., Parcel, T., Mull, L., Woodhouse, J., Bjornson, D., … Hauser, P. (2009). The cognitive effects of hepatitis C in the presence and absence of a history of substance use disorder. *Journal of the International Neuropsychological Society, 15*(01), 69. https://doi.org/10.1017/S1355617708090085

Iadecola, C. (2013). The pathobiology of vascular dementia. *Neuron, 80*(4), 844–866. https://doi.org/10.1016/j.neuron.2013.10.008

Irwin, J., & Terrault, N. (2008). Cognitive impairment in hepatitis C patients on antiviral therapy. *Gastroenterology & Hepatology, 4*(1), 65–67.

Jack, C. R., Knopman, D. S., Jagust, W. J., Shaw, L. M., Aisen, P. S., Weiner, M. W., … Trojanowski, J. Q. (2010). Hypothetical model of dynamic biomarkers of the Alzheimer's pathological cascade. *The Lancet. Neurology, 9*(1), 119–128. https://doi.org/10.1016/S1474-4422(09)70299-6

Joutel, A., Corpechot, C., Ducros, A., Vahedi, K., Chabriat, H., Mouton, P., … Tournier-Lasserve, E. (1996). Notch3 mutations in CADASIL, a hereditary adult-onset condition causing stroke and dementia. *Nature, 383*(6602), 707–710. https://doi.org/10.1038/383707a0

Kang, H., Zhao, F., You, L., Giorgetta, C., Venkatesh, D., Sarkhel, S., & Prakash, R. (2014). Pseudo- dementia: A neuropsychological review. *Annals of Indian Academy of Neurology, 17*(2), 147–154. https://doi.org/10.4103/0972-2327.132613

Kirk, J. K., Passmore, L. V., Bell, R. A., Narayan, K. M. V., D'Agostino, R. B., Arcury, T. A., & Quandt, S. A. (2008). Disparities in A1C levels between Hispanic and non-Hispanic white adults with diabetes: A meta-analysis. *Diabetes Care, 31*(2), 240–246. https://doi.org/10.2337/dc07-0382

Kodl, C. T., & Seaquist, E. R. (2008). Cognitive dysfunction and diabetes mellitus. *Endocrine Reviews, 29*(4), 494–511. https://doi.org/10.1210/er.2007-0034

Kopelman, M. D. (2015). What does a comparison of the alcoholic Korsakoff syndrome and thalamic infarction tell us about thalamic amnesia? *Neuroscience & Biobehavioral Reviews, 54*, 46–56. https://doi.org/10.1016/j.neubiorev.2014.08.014

Kuniholm, M. H., Jung, M., Everhart, J. E., Cotler, S., Heiss, G., McQuillan, G., … Ho, G. Y. F. (2014). Prevalence of hepatitis C virus infection in US Hispanic/Latino adults: Results from the NHANES 2007-2010 and HCHS/SOL studies. *The Journal of Infectious Diseases, 209*(10), 1585–1590. https://doi.org/10.1093/infdis/jit672

Kwan, L. L., Bermudez, O. I., & Tucker, K. L. (2002). Low vitamin B-12 intake and status are more prevalent in Hispanic older adults of Caribbean origin than in neighborhood- matched non-Hispanic whites. *The Journal of Nutrition, 132*(7), 2059–2064.

Lezak, M. D., Howieson, D. B., Bigler, E. D., & Tranel, D. (2012). *Neuropsychological assessment* (5th ed.). New York, NY: Oxford University Press.

Lin, K.-H., Lin, H.-Z., Lin, Y.-P., Meng, F.-C., Lin, F., Wu, H.-T., & Lin, G.-M. (2017). Obstructive sleep apnea and retinal microvascular characteristics: A brief review. *Neuropsychiatry, 7*(1), 12–21. https://doi.org/10.4172/Neuropsychiatry.1000173

Lista, S., Dubois, B., & Hampel, H. (2015). Paths to Alzheimer's disease prevention: From modifiable risk factors to biomarker enrichment strategies. *The Journal of Nutrition, Health & Aging, 19*(2), 154–163. https://doi.org/10.1007/s12603-014-0515-3

Livney, M. G., Clark, C. M., Karlawish, J. H., Cartmell, S., Negrón, M., Nuñez, J., … Arnold, S. E. (2011). Ethnoracial differences in the clinical characteristics of Alzheimer's disease at initial presentation at an urban Alzheimer's disease center. *The American Journal of Geriatric Psychiatry, 19*(5), 430–439. https://doi.org/10.1097/JGP.0b013e3181f7d881

Lora, C. M., Daviglus, M. L., Kusek, J. W., Porter, A., Ricardo, A. C., Go, A. S., & Lash, J. P. (2009). Chronic kidney disease in United States Hispanics: A growing public health problem. *Ethnicity & Disease, 19*(4), 466–472.

Loredo, J. S., Soler, X., Bardwell, W., Ancoli-Israel, S., Dimsdale, J. E., & Palinkas, L. A. (2010). Sleep health in U.S. Hispanic population. *Sleep, 33*(7), 996–997.

Luber, S. D., Brady, W. J., Brand, A., Young, J., Guertler, A. T., & Kefer, M. (1996). Acute hypoglycemia masquerading as head trauma: A report of four cases. *The American Journal of Emergency Medicine, 14*(6), 543–547. https://doi.org/10.1016/S0735-6757(96)90094-7

Luchsinger, J. A. (2012). Type 2 diabetes and cognitive impairment: Linking mechanisms. *Journal of Alzheimer's Disease: JAD, 30*(Suppl 2), S185–S198. https://doi.org/10.3233/JAD-2012-111433

Lustman, P. J., & Clouse, R. E. (2005). Depression in diabetic patients: The relationship between mood and glycemic control. *Journal of Diabetes and its Complications, 19*(2), 113–122. https://doi.org/10.1016/j.jdiacomp.2004.01.002

Macey, P. M. (2012). Is brain injury in obstructive sleep apnea reversible? *Sleep, 35*(1), 9–10. https://doi.org/10.5665/sleep.1572

Magreni, A., Bann, D. V., Schubart, J. R., & Goldenberg, D. (2015). The effects of race and ethnicity on thyroid cancer incidence. *JAMA Otolaryngology – Head & Neck Surgery, 141*(4), 319–323. https://doi.org/10.1001/jamaoto.2014.3740

Martin, P. R., Singleton, C. K., & Hiller-Sturmhöfel, S. (2003). The role of thiamine deficiency in alcoholic brain disease. *Alcohol Research & Health: The Journal of the National Institute on Alcohol Abuse and Alcoholism, 27*, 134–142.

Matthews, K. D., & Richter, R. W. (2003). Binswanger's disease: Its association with hypertension and obstructive sleep apnea. *The Journal of the Oklahoma State Medical Association, 96*(6), 265–268; quiz 269–270.

McCutchan, J. A., Wu, J. W., Robertson, K., Koletar, S. L., Ellis, R. J., Cohn, S., … Williams, P. L. (2007). HIV suppression by HAART preserves cognitive function in advanced, immune-reconstituted AIDS patients. *AIDS, 21*(9), 1109–1117. https://doi.org/10.1097/QAD.0b013e3280ef6acd

McKeith, I. G., Boeve, B. F., Dickson, D. W., Halliday, G., Taylor, J.-P., Weintraub, D., … Kosaka, K. (2017). Diagnosis and management of dementia with Lewy bodies: Fourth consensus report of the DLB Consortium. *Neurology, 89*(1), 88–100. https://doi.org/10.1212/WNL.0000000000004058

McKeith, I. G., Dickson, D. W., Lowe, J., Emre, M., O'Brien, J. T., Feldman, H., … Consortium on DLB. (2005). Diagnosis and management of dementia with Lewy bodies: Third report of the DLB Consortium. *Neurology, 65*(12), 1863–1872. https://doi.org/10.1212/01.wnl.0000187889.17253.b1

McKhann, G. M., Albert, M. S., Grossman, M., Miller, B., Dickson, D., Trojanowski, J. Q., & Work Group on Frontotemporal Dementia and Pick's Disease. (2001). Clinical and pathological diagnosis of frontotemporal dementia: Report of the work group on Frontotemporal Dementia and Pick's Disease. *Archives of Neurology, 58*(11), 1803–1809.

McKhann, G. M., Knopman, D. S., Chertkow, H., Hyman, B. T., Jack, C. R., Kawas, C. H., … Phelps, C. H. (2011). The diagnosis of dementia due to Alzheimer's disease: Recommendations from the National Institute on Aging-Alzheimer's Association workgroups on diagnostic guidelines for Alzheimer's disease. *Alzheimer's & Dementia: The Journal of the Alzheimer's Association, 7*(3), 263–269. https://doi.org/10.1016/j.jalz.2011.03.005

McLeod, D. S. A., Caturegli, P., Cooper, D. S., Matos, P. G., & Hutfless, S. (2014). Variation in rates of autoimmune thyroid disease by race/ethnicity in US military personnel. *JAMA, 311*(15), 1563–1565. https://doi.org/10.1001/jama.2013.285606

McNay, E. C., & Cotero, V. E. (2010). Mini-review: Impact of recurrent hypoglycemia on cognitive and brain function. *Physiology & Behavior, 100*(3), 234–238. https://doi.org/10.1016/j.physbeh.2010.01.004

Mendez, M. F. (2012). Early-onset Alzheimer's disease: Nonamnestic subtypes and type 2 AD. *Archives of Medical Research, 43*(8), 677–685. https://doi.org/10.1016/j.arcmed.2012.11.009

Miller, J. W., Green, R., Ramos, M. I., Allen, L. H., Mungas, D. M., Jagust, W. J., & Haan, M. N. (2003). Homocysteine and cognitive function in the Sacramento Area Latino Study on Aging. *The American Journal of Clinical Nutrition, 78*(3), 441–447.

Miller-Thomas, M. M., Sipe, A. L., Benzinger, T. L. S., McConathy, J., Connolly, S., & Schwetye, K. E. (2016). Multimodality review of amyloid-related diseases of the central nervous system. *Radiographics, 36*(4), 1147–1163. https://doi.org/10.1148/rg.2016150172

Monaco, S., Mariotto, S., Ferrari, S., Calabrese, M., Zanusso, G., Gajofatto, A., … Dammacco, F. (2015). Hepatitis C virus-associated neurocognitive and neuropsychiatric disorders: Advances in 2015. *World Journal of Gastroenterology, 21*(42), 11974–11983. https://doi.org/10.3748/wjg.v21.i42.11974

Moreton, F. C., Razvi, S. S. M., Davidson, R., & Muir, K. W. (2014). Changing clinical patterns and increasing prevalence in CADASIL. *Acta Neurologica Scandinavica, 130*(3), 197–203. https://doi.org/10.1111/ane.12266

Moretti, R., Torre, P., Antonello, R. M., Cattaruzza, T., Cazzato, G., & Bava, A. (2004). Vitamin B12 and folate depletion in cognition: A review. *Neurology India, 52*(3), 310–318.

Neary, D., Snowden, J. S., Gustafson, L., Passant, U., Stuss, D., Black, S., ... Benson, D. F. (1998). Frontotemporal lobar degeneration: A consensus on clinical diagnostic criteria. *Neurology, 51*(6), 1546–1554.

Nervi, A., Reitz, C., Tang, M.-X., Santana, V., Piriz, A., Reyes, D., ... Mayeux, R. (2011). Familial aggregation of dementia with Lewy bodies. *Archives of Neurology, 68*(1), 90–93. https://doi.org/10.1001/archneurol.2010.319

Nwankwo, T., Yoon, S. S., Burt, V., & Gu, Q. (2013). Hypertension among adults in the United States: National Health and Nutrition Examination Survey, 2011-2012. *NCHS Data Brief, 133*, 1–8.

O'Brien, J. T., Erkinjuntti, T., Reisberg, B., Roman, G., Sawada, T., Pantoni, L., ... DeKosky, S. T. (2003). Vascular cognitive impairment. *The Lancet. Neurology, 2*(2), 89–98.

O'Brien, J. T., & Thomas, A. (2015). Vascular dementia. *The Lancet, 386*(10004), 1698–1706. https://doi.org/10.1016/S0140-6736(15)00463-8

Obisesan, T. O. (2009). Hypertension and cognitive function. *Clinics in Geriatric Medicine, 25*(2), 259–288. https://doi.org/10.1016/j.cger.2009.03.002

Onyike, C. U., & Diehl-Schmid, J. (2013). The epidemiology of frontotemporal dementia. *International Review of Psychiatry, 25*(2), 130–137. https://doi.org/10.3109/09540261.2013.776523

Oscar-Berman, M., & Marinković, K. (2007). Alcohol: Effects on neurobehavioral functions and the brain. *Neuropsychology Review, 17*(3), 239–257. https://doi.org/10.1007/s11065-007-9038-6

Pabon-Nau, L. P., Cohen, A., Meigs, J. B., & Grant, R. W. (2010). Hypertension and diabetes prevalence among U.S. Hispanics by country of origin: The National Health Interview Survey 2000–2005. *Journal of General Internal Medicine, 25*(8), 847–852. https://doi.org/10.1007/s11606-010-1335-8

Palloni, A., & Arias, E. (2004). Paradox lost: Explaining the hispanic adult mortality advantage. *Demography, 41*(3), 385–415. https://doi.org/10.1353/dem.2004.0024

Parsons, T. D., Braaten, A. J., Hall, C. D., & Robertson, K. R. (2006). Better quality of life with neuropsychological improvement on HAART. *Health and Quality of Life Outcomes, 4*, 11. https://doi.org/10.1186/1477-7525-4-11

Rabinovici, G. D., & Miller, B. L. (2010). Frontotemporal lobar degeneration: Epidemiology, pathophysiology, diagnosis and management. *CNS Drugs, 24*(5), 375–398. https://doi.org/10.2165/11533100-000000000-00000

Rabinovici, G. D., Seeley, W. W., Kim, E. J., Gorno-Tempini, M. L., Rascovsky, K., Pagliaro, T. A., ... Rosen, H. J. (2007). Distinct MRI atrophy patterns in autopsy-proven Alzheimer's disease and frontotemporal lobar degeneration. *American Journal of Alzheimer's Disease and Other Dementias, 22*(6), 474–488. https://doi.org/10.1177/1533317507308779

Ramos, A. R., Tarraf, W., Rundek, T., Redline, S., Wohlgemuth, W. K., Loredo, J. S., ... González, H. M. (2015). Obstructive sleep apnea and neurocognitive function in a Hispanic/Latino population. *Neurology, 84*(4), 391–398. https://doi.org/10.1212/WNL.0000000000001181

Ramos, M. I., Allen, L. H., Mungas, D. M., Jagust, W. J., Haan, M. N., Green, R., & Miller, J. W. (2005). Low folate status is associated with impaired cognitive function and dementia in the Sacramento Area Latino Study on Aging. *The American Journal of Clinical Nutrition, 82*(6), 1346–1352.

Rascovsky, K., Hodges, J. R., Knopman, D., Mendez, M. F., Kramer, J. H., Neuhaus, J., ... Miller, B. L. (2011). Sensitivity of revised diagnostic criteria for the behavioural variant of frontotemporal dementia. *Brain: A Journal of Neurology, 134*(Pt 9), 2456–2477. https://doi.org/10.1093/brain/awr179

Raz, N., Yang, Y. Q., Rodrigue, K. M., Kennedy, K. M., Lindenberger, U., & Ghisletta, P. (2012). White matter deterioration in 15 months: Latent growth curve models in healthy adults. *Neurobiology of Aging, 33*(2), 429.e1–429.e5. https://doi.org/10.1016/j.neurobiolaging.2010.11.018

Redline, S., Sotres-Alvarez, D., Loredo, J., Hall, M., Patel, S. R., Ramos, A., ... Daviglus, M. L. (2014). Sleep-disordered breathing in Hispanic/Latino individuals of diverse backgrounds. The Hispanic Community Health Study/Study of Latinos. *American Journal of Respiratory and Critical Care Medicine, 189*(3), 335–344. https://doi.org/10.1164/rccm.201309-1735OC

Resta, F., Triggiani, V., Barile, G., Benigno, M., Suppressa, P., Giagulli, V. A., ... Sabbà, C. (2012). Subclinical hypothyroidism and cognitive dysfunction in the elderly. *Endocrine, Metabolic & Immune Disorders Drug Targets, 12*(3), 260–267.

Reynolds, E. H. (2002). Folic acid, ageing, depression, and dementia. *BMJ: British Medical Journal, 324*(7352), 1512–1515.

Ríos-Bedoya, C. F., & Freile-Salinas, D. (2014). Incidence of alcohol use disorders among Hispanic subgroups in the USA. *Alcohol and Alcoholism (Oxford, Oxfordshire), 49*(5), 549–556. https://doi.org/10.1093/alcalc/agu032

Rivera Mindt, M., Byrd, D., Saez, P., & Manly, J. (2010). Increasing culturally competent neuropsychological services for ethnic minority populations: A call to action. *The Clinical Neuropsychologist, 24*(3), 429–453. https://doi.org/10.1080/13854040903058960

Rizzi, L., Rosset, I., & Roriz-Cruz, M. (2014). Global epidemiology of dementia: Alzheimer's and vascular types. *BioMed Research International, 2014*. https://doi.org/10.1155/2014/908915

Rodríguez-Torres, M., Ríos-Bedoya, C. F., Ortiz-Lasanta, G., Marxuach-Cuétara, A. M., & Jiménez-Rivera, J. (2008). Thyroid dysfunction (TD) among chronic hepatitis C patients with mild and severe hepatic fibrosis. *Annals of Hepatology, 7*(1), 72–77.

Russell, D., & Taylor, J. (2009). Living alone and depressive symptoms: The influence of gender, physical disability, and social support among Hispanic and non-Hispanic older adults. *The Journals of Gerontology Series B: Psychological Sciences and Social Sciences, 64B*(1), 95–104. https://doi.org/10.1093/geronb/gbn002

Sahoo, S., Mehra, A., & Grover, S. (2016). Acute hyperglycemia associated with psychotic symptoms in a patient with Type 1 Diabetes Mellitus: A case report. *Innovations in Clinical Neuroscience, 13*(11–12), 25–27.

Salazar, R., Royall, D. R., & Palmer, R. F. (2015). Neuropsychiatric symptoms in community- dwelling Mexican-Americans: Results from the Hispanic Established Population for Epidemiological Study of the Elderly (HEPESE) study. *International Journal of Geriatric Psychiatry, 30*, 300–307. https://doi.org/10.1002/gps.4141

Samuels, M. H. (2008). Cognitive function in untreated hypothyroidism and hyperthyroidism. *Current Opinion in Endocrinology, Diabetes, and Obesity, 15*(5), 429–433. https://doi.org/10.1097/MED.0b013e32830eb84c

Sands-Lincoln, M., Grandner, M., Whinnery, J., Keenan, B. T., Jackson, N., & Gurubhagavatula, I. (2013). The association between obstructive sleep apnea and hypertension by race/ethnicity in a nationally representative sample. *The Journal of Clinical Hypertension, 15*(8), 593–599. https://doi.org/10.1111/jch.12144

Savica, R., Grossardt, B. R., Bower, J. H., Boeve, B. F., Ahlskog, J. E., & Rocca, W. A. (2014). Incidence of dementia with Lewy bodies and Parkinson disease dementia. *JAMA Neurology, 70*(11), 1396–1402. https://doi.org/10.1001/jamaneurol.2013.3579

Senzolo, M., Schiff, S., D'Aloiso, C. M., Crivellin, C., Cholongitas, E., Burra, P., & Montagnese, S. (2011). Neuropsychological alterations in hepatitis C infection: The role of inflammation. *World Journal of Gastroenterology, 17*(29), 3369. https://doi.org/10.3748/wjg.v17.i29.3369

Shanmugarajah, P. D., Hoggard, N., Currie, S., Aeschlimann, D. P., Aeschlimann, P. C., Gleeson, D. C., … Hadjivassiliou, M. (2016). Alcohol-related cerebellar degeneration: Not all down to toxicity? *Cerebellum & Ataxias, 3*, 17. https://doi.org/10.1186/s40673-016-0055-1

Singer, E. J., & Thames, A. D. (2016). Neurobehavioral manifestations of human immunodeficiency virus/AIDS. *Neurologic Clinics, 34*(1), 33–53.

Sommerfield, A. J., Deary, I. J., & Frier, B. M. (2004). Acute hyperglycemia alters mood state and impairs cognitive performance in people with type 2 diabetes. *Diabetes Care, 27*(10), 2335–2340.

Starr, V. L., & Convit, A. (2007). Diabetes, sugar-coated but harmful to the brain. *Current Opinion in Pharmacology, 7*(6), 638–642. https://doi.org/10.1016/j.coph.2007.10.007

Stranks, E. K., & Crowe, S. F. (2016). The cognitive effects of obstructive sleep apnea: An updated meta-analysis. *Archives of Clinical Neuropsychology, 31*(2), 186–193. https://doi.org/10.1093/arclin/acv087

Substance Abuse and Mental Health Services Administration [SAMHSA]. (2013). *Hispanic subgroups differ in rates of substance use treatment need and receipt*. National survey on drug use and Health (NSDUH) Report, October 24, 2013. Retrieved from http://www.samhsa.gov/data/sites/default/files/spot128-hispanic-treatment-2013/spot128-hispanic-treatment-2013.pdf

Tang, M. X., Cross, P., Andrews, H., Jacobs, D. M., Small, S., Bell, K., … Mayeux, R. (2001). Incidence of AD in African-Americans, Caribbean Hispanics, and Caucasians in northern Manhattan. *Neurology, 56*(1), 49–56.

Tang, M. X., Stern, Y., Marder, K., Bell, K., Gurland, B., Lantigua, R., … Mayeux, R. (1998). The APOE-epsilon4 allele and the risk of Alzheimer disease among African Americans, whites, and Hispanics. *JAMA, 279*(10), 751–755.

Taylor, W. D. (2014). Depression in the elderly. *New England Journal of Medicine, 371*(13), 1228–1236. https://doi.org/10.1056/NEJMcp1402180

Tzourio, C. (2007). Hypertension, cognitive decline, and dementia: an epidemiological perspective. *Dialogues in Clinical Neuroscience, 9*(1), 61–70.

U.S. Census Bureau. (2015). *American Community Survey B03001 1-Year estimates: Hispanic or Latino origin by specific origin*. Retrieved September 22, 2016.

Vasconcellos, L. F. R., & Pereira, J. S. (2015). Parkinson's disease dementia: Diagnostic criteria and risk factor review. *Journal of Clinical and Experimental Neuropsychology, 37*(9), 988–993. https://doi.org/10.1080/13803395.2015.1073227

Vega, W. A., Rodriguez, M. A., & Gruskin, E. (2009). Health disparities in the Latino population. *Epidemiologic Reviews, 31*, 99–112. https://doi.org/10.1093/epirev/mxp008

Viña, J., & Lloret, A. (2010). Why women have more Alzheimer's disease than men: Gender and mitochondrial toxicity of amyloid-beta peptide. *Journal of Alzheimer's Disease: JAD, 20*(Suppl 2), S527–S533. https://doi.org/10.3233/JAD-2010-100501

Waclawski, E. R. (2012). Hyperglycaemia and cognitive function – acute and chronic effects and work. *Occupational Medicine (Oxford, England), 62*(4), 236–237. https://doi.org/10.1093/occmed/kqs017

Walker, Z., Possin, K. L., Boeve, B. F., & Aarsland, D. (2015). Lewy body dementias. *The Lancet, 386*(10004), 1683–1697. https://doi.org/10.1016/S0140-6736(15)00462-6

Wang, C. (2013). The relationship between type 2 diabetes mellitus and related thyroid diseases. *Journal of Diabetes Research, 2013*. https://doi.org/10.1155/2013/390534

Warren, R. E., & Frier, B. M. (2005). Hypoglycaemia and cognitive function. *Diabetes, Obesity and Metabolism, 7*(5), 493–503. https://doi.org/10.1111/j.1463-1326.2004.00421.x

Wen, X., Wang, N., Liu, J., Yan, Z., & Xin, Z. (2012). Detection of cognitive impairment in p atients with obstructive sleep apnea hypopnea syndrome using mismatch negativity. *Neural Regeneration Research, 7*(20), 1591–1598. https://doi.org/10.3969/j.issn.1673-5374.2012.20.010

Wenk, G. L. (2003). Neuropathologic changes in Alzheimer's disease. *The Journal of Clinical Psychiatry, 64*(Suppl 9), 7–10.

Woods, S. P., Moore, D. J., Weber, E., & Grant, I. (2009). Cognitive neuropsychology of HIV- associated neurocognitive disorders. *Neuropsychology Review, 19*(2), 152–168. https://doi.org/10.1007/s11065-009-9102-5

Wu, J. H., Haan, M. N., Liang, J., Ghosh, D., Gonzalez, H. M., & Herman, W. H. (2003). Diabetes as a predictor of change in functional status among older Mexican Americans: A population-based cohort study. *Diabetes Care, 26*(2), 314–319.

Yong, A. W., Morris, Z., Shuler, K., Smith, C., & Wardlaw, J. (2012). Acute symptomatic hypoglycaemia mimicking ischaemic stroke on imaging: A systemic review. *BMC Neurology, 12*, 139. https://doi.org/10.1186/1471-2377-12-139

Functional Assessment in Latinos with Dementia: A Review of Tools and Cultural Considerations

5

Philip Sayegh and Catherine V. Piersol

Abstract

As the American elderly population continues to grow, so will the number of Latinos living with dementia. A critical aspect of dementia evaluation involves functional assessment, referring to the evaluation of the ability to carry out basic activities of daily living (ADLs) and instrumental activities of daily living (IADLs). Functional assessment bears relevance in numerous fashions, including safety and care planning, diagnostic conclusions, treatment planning, and recommendations. A number of questionnaire- and performance-based measures exist to assess the capacity to execute ADLs and IADLs, each possessing their own unique strengths and limitations. However, a number of cultural and linguistic influences can impact the validity of results from these measures, ranging from language barriers to cultural beliefs about illness to shame and stigma, which can alter reporting styles on questionnaires or execution of tasks on performance-based measures. Moreover, the psychometric properties of these measures, as well as translated versions of these measures, need to be considered yet are often overlooked, calling into question the suitability of their use and findings among Latinos. This chapter reviews key issues pertaining to the functional assessment of Latinos with dementia, including the general strengths and limitations of available measures, key cultural and linguistic considerations, and clinical and research recommendations to help improve the accuracy of the functional assessment of Latinos with dementia.

P. Sayegh (✉)
University of California, Los Angeles,
Los Angeles, CA, USA
e-mail: psayegh@psych.ucla.edu

C. V. Piersol
Thomas Jefferson University, Philadelphia, PA, USA
e-mail: catherine.v.piersol@jefferson.edu

Introduction

The U.S. Census Bureau, Population Division (2016) estimated that as of July 1, 2015, there were approximately 57 million Latino individuals living in the United States of America (USA), comprising about 17.3% of the total population. This estimate is projected to more than double by 2060, as the predicted number of Latinos in the USA will be approximately 119 million, representing 28.6% of the population at that time (Colby & Ortman, 2015). It is anticipated that as the population of Latinos residing in the USA continues to burgeon and age, so will the prevalence of dementia among Latinos given that older age is a prominent risk factor for dementia (e.g., Zarit & Zarit, 2007) and that Latino Americans have been shown to be about 1.5 times as likely to have dementia than their non-Latino White counterparts (e.g.,

H. Y. Adames, Y. N. Tazeau (eds.), *Caring for Latinxs with Dementia in a Globalized World*,
https://doi.org/10.1007/978-1-0716-0132-7_5

Alzheimer's Association, 2016; Gurland et al., 1999). Therefore, a deeper and empirically informed understanding of dementia assessment procedures among this large and diverse ethnic group is sorely needed.

A critical aspect of the clinical evaluation of dementia pertains to the assessment of functional abilities, or functional assessment. Functional assessment involves the evaluation of a person's ability to perform tasks that are required for living. Healthcare professionals working with individuals with dementia must often measure a wide range of functional abilities, including basic activities of daily living (ADLs), such as oral hygiene, dressing, and grooming (Chisholm, Toto, Raina, Holm, & Rogers, 2014), and instrumental activities of daily living (IADLs), including money management (e.g., Loewenstein et al., 1989), telling time, using a calendar, and managing medications (e.g., Cullum et al., 2001; Loewenstein et al., 1989). Clearly, accurate functional assessment is needed to measure functional capacity that can directly assist in safety and care planning for individuals with dementia. In addition, a decline in the ability to independently execute important everyday activities (ranging from grooming and dressing to managing medications and finances) is related to cognitive deficits and is one of the key diagnostic criteria for major neurocognitive disorder, often referred to as dementia, as noted in the *Diagnostic and Statistical Manual of Mental Disorders* (5th ed.; *DSM-5*; American Psychiatric Association, 2013). Nonetheless, the literature on best practices and cultural and linguistic considerations for the functional assessment of Latinos with dementia is limited. Research has demonstrated that information on patients' functional abilities (obtained from an informant-report questionnaire) significantly predicted a clinical diagnosis of dementia to relatively equivalent degrees across Latino and non-Latino White outpatients (Sayegh & Knight, 2013a) and highlighted the key role of functional assessment in the dementia diagnostic process in Latinos. Therefore, a richer understanding of cultural and linguistic influences on the functional assessment of Latinos with dementia is warranted to assist with improving diagnostic accuracy and assessment outcomes (e.g., treatment plans and recommendations, referrals for services and to other providers, emotional reactions of patients and relatives, and stigma).

The purpose of this chapter is to provide an overview of and describe the strengths and limitations of both questionnaire- and performance-based measures of functional assessment and present key cultural and linguistic considerations that can influence the assessment process and results among Latinos. We conclude by providing recommendations regarding best practices and areas of future research regarding the functional assessment of Latinos with dementia.

Questionnaire-Based Measures of Functional Abilities

Brief Overview

Some of the most commonly used tools in the functional assessment of dementia include questionnaires that measure ADLs, IADLs, or a combination of the two that can be filled out by patients themselves, family members, caregivers, certain healthcare professionals (e.g., nursing staff), and/or other knowledgeable informants. One commonly used questionnaire that measures only basic ADLs is the *Katz Index of Independence in Activities of Daily Living* (*Katz ADL*; Katz, Downs, Cash, & Grotz, 1970). The *Katz ADL* asks informants to rate patients as either *independent* (1 point) or *dependent* (0 point) regarding their ability to carry out six ADLs (i.e., bathing, dressing, toileting, transferring, fecal and urinary continence, and feeding), with lower scores suggestive of limited or reduced functional independence. Similarly, the *Barthel Index of Activities of Daily Living* (Mahoney & Barthel, 1965) measures 10 basic ADLs (e.g., bathing, grooming, dressing, feeding, and toilet use) using an ordinal scale ranging from 0 to 1 or 0 to 3 (depending on the ADL being assessed), with lower scores representative of higher levels of functional dependence.

Other questionnaires of functional capacity measure only IADLs, such as the *Lawton Instrumental Activities of Daily Living Scale* (*Lawton IADL*; Lawton & Brody, 1969), which assesses eight domains of function. When the scale was first developed, women were scored on all eight areas of function (e.g., medication and finance management and ability to use the telephone), whereas for men, the areas of food preparation, housekeeping, and laundering were excluded. At present, clinicians are encouraged to assess all eight areas and note whether any domains are not applicable for each individual client (Ward, Jagger, & Harper, 1998). Clients are scored according to their highest level of functioning in each category, resulting in a summary score that can range from 0 (*low function, dependent*) to either 8 or the highest number of items administered (*high function, independent*). The *Functional Activities Questionnaire* (*FAQ*; Pfeffer, Kurosaki, Harrah, Chance, & Filos, 1982) is another frequently used scale that measures functional capacity pertaining to IADLs. Although healthcare professionals may complete this questionnaire, their responses to each item are based on information provided directly by a caregiver or other reliable informant. The *FAQ* assesses a total of 10 IADLs (e.g., traveling out of the neighborhood, preparing balanced meals, and medication and financial management) with response options ranging from 0 (*normal*) to 3 (*dependent*) as well as *not applicable* (e.g., *never did*). Higher scores are suggestive of reduced functional capacity regarding IADLs.

Many questionnaire-based measures assess both ADLs and IADLs. For example, the *Bayer Activities of Daily Living Scale* (*B-ADL*; Hindmarch, Lehfeld, de Jong, & Erzigkeit, 1998) consists of 25 items answered by a caregiver or other knowledgeable informant. The first items assess patients' ability to manage the ADLs and more complex IADLs and capacity for self-care, whereas other items evaluate specific tasks. The last set of items assesses cognitive abilities that are important for successfully executing ADLs. Respondents rate each ability on a 1 (*never*) to 10 (*always*) scale, with lower scores indicative of higher levels of functional disability and dependence. Similarly, the *Blessed Dementia Scale* (Blessed, Tomlinson, & Roth, 1968) is an informant-report clinical rating scale that measures both ADLs and IADLs with 22 items that measure changes in performance of everyday activities (8 items), self-care habits (3 items), and changes in personality, interests, and drives (11 items). Ratings are based on behavior over the preceding 6 months. In addition, the *Bristol Activities of Daily Living Scale* (Bucks, Ashworth, Wilcock, & Siegfried, 1996) is a 20-item questionnaire designed to measure the ability of individuals with dementia to carry out ADLs and IADLs, such as dressing, preparing food, and using transport. Informants make severity judgments that range from 0 (*independence – no help required*) to 3 (*dependence – unable even with supervision*), producing a total score range of 0–60. Respondents can choose to score an item as *not applicable* if the person with dementia never engaged in that activity before the onset of dementia.

The choice of whether to use questionnaires of functional capacity that tap only ADLs, IADLs, or both depends on various factors, such as the clinical presentation of the patient and referral question. For example, in cases of more advanced dementia with a referral question regarding ability for self-care, a measure of more basic ADLs may be more useful and informative. In contrast, among patients in the earlier stages of dementia referred for diagnostic clarification, it may be most appropriate to measure more complex activities such as IADLs to increase the chance of detecting any clinically meaningful functional decline that bears relevance in terms of diagnosis and treatment.

Strengths

There are a number of strengths to the use of questionnaire-based measures of functional abilities during the assessment of individuals with dementia, contributing to their frequent utilization in clinical evaluations. Not only are they usually free or inexpensive, short, and relatively easy and quick for respondents to complete; they also require minimal training, time, or effort on the part of the healthcare provider in terms of administration, scoring, or interpretation. These measures also have an additional advantage over another commonly used source of information, that of neuropsychological test performance. In terms of assessing and diagnosing dementia, the questionnaire-based measures of functional abilities are comparatively less affected by patients' premorbid ability, education level, and proficiency in the culture's dominant language as they can be for neuropsychological test performance (Jorm, 2004). Moreover, Malmstrom et al. (2009) stated that informant-report questionnaires reduce the potential for bias due to differences in education, culture, and language as they assess patients against their prior cognitive functioning instead of using norms that may lack validity for Latinos and individuals from other cultural and ethnic minority groups. They added that these measures have also been shown to perform at least as well at screening for dementia or other cognitive decline as conventional cognitive screening tests, such as the *Mini-Mental State Examination* (*MMSE*; Folstein, Folstein, & McHugh, 1975). Similarly, Erzigkeit et al. (2001) found that the *B-ADL* was as effective as or even superior to the *MMSE* in terms of identifying individuals with clinically manifest dementia symptoms. Finally, Malmstrom and colleagues also noted that informant-based questionnaires benefit from face validity because cognitive abilities are evaluated with regard to the ability to carry out ADLs and IADLs. In short, these measures are getting directly at what healthcare providers are trying to measure, rather than abilities that are associated with executing ADLs and IADLs, such as cognitive abilities (often assessed through neuropsychological assessment).

Limitations

Despite their numerous strengths, questionnaire-based measures of patients with dementia's functional abilities suffer from a number of limitations as well. For example, they generally display weaker face validity, replicability, and sensitivity to change than performance-based measures (Guralnik & Simonsick, 1993), which are discussed later in this chapter. They are also subject to more subjectivity and bias than standardized performance-based measures that require direct observation of the execution of ADLs and IADLs either in the home or healthcare setting (Harvey, Velligan, & Bellack, 2007). In addition, there may be environmental challenges to using questionnaire-based measures in that some hospitals, skilled nursing facilities, and other healthcare settings do not have bathing facilities or toilets readily available, making assessment in these particular areas difficult. Furthermore, some nursing staff members help patients with dementia with ADLs simply to save time even when patients are capable of executing them autonomously (Wallace & Shelkey, 2007).

In addition, when completing the questionnaires themselves, individuals with dementia may lack insight into their true functional capacity and tend to overestimate their functional capacity (e.g., Saper, 2003) or, in cases of more advanced dementia, may lack the cognitive skills or mental stamina necessary to accurately comprehend, respond to, and complete the questionnaires (Sheehan, 2012). Informant reports of patients' functional abilities may be contaminated by other factors, such as the emotional state of the informant and the quality of the relationship between the patient and informant (Jorm, 2004). A whole host of informant-related characteristics can clearly impact responses, such as level of education, reading and health literacy, or even denial or minimization of functional decline in patients associated with difficulty accepting or coping with the decline. Despite these limitations, Jorm suggested that using informant-based reports in combination with other sources of clinically relevant information such as neuropsychological test performance can increase the accuracy of the screening and diagnosing of dementia, as the informant reports provide information that is complementary to the other sources. More simply put, using both methods provides more clinically relevant information than using one method alone. These suggestions provide added importance for

Latinos for whom the use of neuropsychological test performance, as the primary basis for diagnosis, may be less valid (Sayegh, 2015). This is not to say that questionnaire-based measures of patients' functional abilities do not suffer from cultural and linguistic limitations, which is discussed in further detail in the following sections.

Cultural and Linguistic Considerations

Cultural values and beliefs about illness and disease can shape the way Latinos assign meaning to dementia symptoms (Dilworth-Anderson & Gibson, 2002), thereby potentially influencing symptom reporting. Accordingly, there are a number of important considerations pertaining to the potential influence of culture and language on questionnaire-based measures of functional abilities among Latinos with dementia. For instance, willingness to report problems with self-care may be particularly strongly connected to fear of disclosing one's dependence on others by older adults from certain groups such as those from Latino backgrounds (Talamantes & Sanchez-Reilly, 2010). Relatedly, culturally influenced values and beliefs about dementia, such as perceptions of shame and stigma regarding dementia, may limit symptom reporting by both the patients themselves and their informants (Sayegh & Knight, 2013b). Shame and stigma in the setting of family-centered, collectivistic cultural values, often associated with individuals from Latino backgrounds, may serve as significant barriers to reporting on symptoms (such as a decline in functional abilities) because families may elect to keep such possibly embarrassing dementia-related symptoms as private as possible (Gallagher-Thompson, Solano, Coon, & Areán, 2003). On the other hand, some studies have reported that Latino caregivers of individuals with dementia may be more sensitive to noncognitive changes, such as functional decline as opposed to cognitive decline, as compared to their non-Latino counterparts (e.g., Ortiz, Fitten, Cummings, Hwang, & Fonseca, 2006), which could result in an emphasis on these kinds of symptoms.

Aspects of acculturation, such as English language proficiency and level of immersion within the dominant culture, may also influence symptom reporting. In their recently proposed *Sociocultural Health Belief Model*, Sayegh and Knight (2013b) posited that acculturation directly influences culturally influenced beliefs and accurate (i.e., biomedical) knowledge about dementia. For example, Latinos with lower acculturation levels may perceive and/or report on patients' dementia symptoms differently than more acculturated Latinos or those from the dominant culture, which could clearly impact diagnostic validity. Taken together, these findings highlight the need for healthcare professionals who conduct functional assessments among Latinos with dementia to carefully reflect on and consider the potentially influential roles of various cultural and linguistic factors on the assessment process and results.

The level and quality of educational attainment are also key considerations when conducting functional assessments with questionnaire-based measures administered to older adults and their informants. This consideration is especially important when assessing older Latinos who have historically had less of a chance to progress in formal school settings. Care should be taken to consider the educational level and quality of patients who may not be familiar with the type of questions that are asked in many functional assessment questionnaires. Educational quality is not always tantamount to educational quantity and can be influenced by numerous factors, such as student–teacher ratio, teacher quality, and availability of specialized classes and resources (Rohit et al., 2007). While information regarding quantity of education (e.g., number of completed years of formal schooling and highest degree completed) generally can be easily and quickly assessed, *quality* of education is more difficult to quantify. However, there are different methods of assessing quality of education that can be evaluated objectively (e.g., using published school rankings), directly from patients (e.g., via self-report questionnaires of educational quality, such as the one created by Baird, 2007), or by way of validated proxy measures (e.g., the *Wide Range Achievement Test* [4th Edition; Wilkinson & Robertson, 2006]: *Word Reading* subtest; Sayegh, Arentoft, Thaler, Dean, & Thames, 2014). However, proxy measures that rely strongly on English language reading abilities may not be applicable to Latinos who are not monolingual English speakers, for example.

Another important yet often overlooked consideration is the psychometric properties of many questionnaire-based measures that have not been specifically normed on or assessed among Latino populations. The measurement properties of questionnaires being administered to diverse groups need to be evaluated prior to administration and interpretation. Researchers have long challenged the cross-cultural equivalence of standard scales and their items for use among diverse populations (e.g., Hui & Triandis, 1985). Specifically, they have expressed doubt about the ability to make meaningful comparisons across cultural groups when using measures that were not normed on the groups to whom they are being administered (Chapleski, Lamphere, Kaczynski, Lichtenberg, & Dwyer, 1997).

Although very few studies, to our knowledge, have examined the cross-cultural psychometric properties of questionnaires of patient's functional abilities, one study by Erzigkeit et al. (2001) found a one-factor structure for the *B-ADL* for separate samples of individuals with dementia of varying severity in three European countries (the United Kingdom, Germany, and Spain) despite mean differences in scores. Specifically, all *B-ADL* items loaded on a factor they termed *dementia severity*, and they were not related to age, sex, education, or country. Their findings provided evidence for the invariance of the one-factor structure of the *B-ADL* across the three European countries and suggested that the mean differences in scores were likely meaningful and not due to measurement artifact. In addition, Sayegh and Knight (2014) examined the measurement invariance of the *FAQ* across a sample of English-speaking Latino and non-Latino White outpatients and informants by conducting a multigroup confirmatory factor analysis. Although the patients on whom informants reported were assessed in English, the languages in which informants provided information for *FAQ* completion were not indicated. Similar to the Erzigkeit and colleagues study, results revealed that the *FAQ* had the same one-factor structure of *dementia severity* across both ethnic groups, indicating that this scale is likely functioning similarly across groups and is measuring the same latent construct of *dementia severity* pertaining to IADLs. However, the authors did not find evidence for the scalar invariance of the scale, which suggests that ethnic group comparisons of the mean scores from this scale's *dementia severity* factor cannot be meaningfully made. The authors posited that language barriers among some Hispanic informants or the observed significant differences in levels of education across groups may have made it more challenging for them to understand and convey patients' functional capacity. These studies highlight the significance of carefully evaluating the psychometric properties of questionnaire-based measures when applying them to diverse groups lest erroneous conclusions and recommendations be made.

Lastly, there have been many successful attempts at translating and validating various questionnaire-based measures of functional abilities. It is critical that healthcare providers assessing Latinos with dementia use measures that have undergone careful translation and are accurately modified and tested for the Spanish (or other) language with the target population (Erkut, Alarcón, Coll, Tropp, & García, 1999). Doing so provides support for the psychometric equivalence of measures across diverse groups, thereby increasing the likelihood of the accuracy of functional assessment findings. A number of questionnaire-based measures have been translated and validated in different languages. For example, Folquitto et al. (2007) assessed the suitability of a version of the *B-ADL* translated into Brazilian Portuguese (Mapi Research Institute, 1999) in a Brazilian outpatient sample for discerning between healthy older adults and patients with dementia, as well as between mild and moderate dementia cases. They reported that the Brazilian version of the *B-ADL* demonstrated good applicability and high internal consistency (Cronbach's alpha = 0.981) in their sample in that it successfully discriminated patients with mild to moderate dementia from healthy participants and distinguished participants based on various levels of cognitive impairment. The *FAQ* has also been translated for use among Brazilian older adults. The *FAQ* was translated and back-translated by Sanchez, Correa, and Lourenço (2011) and was subsequently evaluated for its psychometric properties for use in Brazil by the authors of the Brazilian *FAQ* themselves as well as more recently by additional researchers (e.g., Assis, de Paula, Assis, de Moraes, & Malloy-Diniz, 2014; Dutra, Ribeiro, Pinheiro, Melo, & Carvalho, 2015). The overall results of these studies suggested that the adapted version of the questionnaire is largely a reliable and valid measure and is appropriate for functional assessment in Brazilian older adults.

As an example involving the Spanish language, the *Activities of Daily Living Questionnaire* (*ADLQ*; Johnson, Barion, Rademaker, Rehkemper, & Weintraub, 2004), an informant-report measure of functional abilities, was translated into Spanish and assessed for its psychometric properties (Gleichgerrcht, Camino, Roca, Torralva, & Manes, 2009). The *ADLQ* was adapted to Spanish through two translations from English to Spanish based on the original (i.e., English language) questionnaire, followed by two back-translations from Spanish to English. The results revealed that the Spanish version of the *ADLQ* (*ADLQ-SV*) had strong psychometric properties (e.g., strong internal consistency [Cronbach's alpha = 0.88] and concurrent validity) for assessing functional capacity among individuals with dementia and can be appropriately administered to Spanish-speaking caregivers. Healthcare providers are encouraged to consult and seek out information on the availability, suitability, and psychometric properties of translated questionnaires of functional abilities when assessing Latinos with dementia.

Performance-Based Measures of Functional Abilities

Brief Overview

Performance-based measures evaluate an individual's ability to execute both basic ADLs and IADLs. In a recent systematic review of assessments of task performance, Wesson, Clemson, Brodaty, and Reppermund (2016) reviewed 21 performance-based instruments that "used direct observation of complex task performance using everyday objects" (p. 337). They included single-domain, multi-domain, and global instruments in their review. Single-domain measures include the *Hopkins Medication Schedule* (Carlson, Fried, Xue, Tekwe, & Brandt, 2005) for medication management and the *Financial Capacity Instrument* (*FCI*; Marson et al., 2000) for financial management. Multi-domain instruments, in which two or more functional domains are assessed, include the *Direct Assessment of Functional Skills* (*DAFS*; Loewenstein et al., 1989; McDougall et al., 2010), the *Cognitive Performance Test* (*CPT*; Burns, Mortimer, & Merchak, 1994), the *Performance Assessment of Self-care Skills* (*PASS*; Chisholm et al., 2014; Rodakowski et al., 2014; Rogers & Holm, 1989), the *Texas Functional Living Scale* (*TFLS*; Cullum et al., 2001), the *Independent Living Scales* (*ILS*; Loeb, 1996), and the *Executive Function Performance Test* (*EFPT*; Baum et al., 2008). Global instruments use specific activities to assess performance skills (e.g., problem-solving, paying attention, following directions) that can then be generalized to performance on other tasks. These instruments include the *Assessment of Motor and Process Skills* (*AMPS*; Fisher & Jones, 2010), the *Large Allen Cognitive Level Screen, version 5* (*ACLS-5*; Allen et al., 2007), and the *Naturalistic Action Test* (*NAT*; Schwartz, Segal, Veramonti, Ferraro, & Buxbaum, 2002).

Performance-based measures are typically administered either in unfamiliar, artificial settings (e.g., clinic or skilled nursing facility) or in familiar, natural environments (e.g., personal residence or family's residence). Each context requires a potentially different set-up, materials, and instructions depending on the location. The benefit of administering performance-based measures in familiar, natural environments is the opportunity to use clients' personal items, such as their toothbrush, kitchen utensils, and medication(s), when administering an assessment in the home. For example, the *PASS* has both a clinic and home version. As with the *EFPT*, the home version of the *PASS* is designed to use the client's kitchen, medication(s), and household supplies. In doing so, clients are familiar with the items and the structure of the environment, thereby increasing the ecological validity, that is, real-world meaningfulness, of results. When a performance-based instrument is implemented as an outcome measure, the setting in which the performance-based assessment is administered should be documented and replicated at baseline (initial evaluation) and post-treatment (discharge evaluation).

Strengths

Performance-based assessment is likely less biased than indirect, cognitive testing. As previously noted, clients with dementia may have impaired insight, leading to the tendency for overestimation of their ability (e.g., Saper, 2003), whereas informants have been shown to both overestimate and underestimate the functional capacity of the person for whom they are caring (e.g., Doble, Fisk, & Rockwood, 1999) for a variety of factors. Moreover, the psychometric properties of self- and informant-report IADL measures have often been found to have poor to moderate quality, with the recommendation for additional examination (Sikkes, Lange-de Klerk, Pijnenburg, & Scheltens, 2009), and there is minimal research on the cross-cultural measurement equivalence of such measures (Sayegh & Knight, 2014). Thus, the strength of using performance-based assessment to evaluate functional status in persons with dementia centers around the ability of the professional to directly observe and interpret functional performance without relying on self- or informant-report, thereby allowing for identification of subtle changes in performance to which questionnaires may not be sensitive. For people with dementia, performance-based measures identify preserved capacities, or what the person "can do" under directed optimal conditions (Glass, 1998). This objective information can be shared with the family and caregivers and interpreted appropriately so that caregivers can learn to both help manage as needed and strive to optimize the client's performance in daily activities.

Limitations

As discussed above, performance-based assessment is often implemented in a contrived clinic environment, and performance ability is measured at a single point in time (Moore, Palmer, Patterson, & Jeste, 2007). Accordingly, assessment findings may be influenced by fluctuations in effort, engagement, or mental status as well as the inability to use routine environmental cues or materials or compensatory strategies (e.g., Marson & Hebert, 2006). For these reasons, it has been argued that certain informant-report questionnaires are equally capable of detecting IADL decline in persons with mild cognitive impairment relative to performance-based measures (e.g., Gold, 2012).

In addition, performance-based measures often require advance preparation by health professionals and a relatively great deal of time to administer. The administration process can total 30 or 45 minutes, which is typically much more time-consuming than self- or informant-based scales that can often be completed in 15 to 20 minutes. In a fast-paced hospital, clinic, or other healthcare facility or practice, extended time to complete a performance-based measure may not always be feasible. Furthermore, the space, materials, and equipment necessary to adhere to standardized administration procedures may limit their use within the healthcare setting environment.

Finally, training and cost may place limits on the use of performance-based measures. Competency in the use of certain performance-based measures may require either informal or formal training (with certification), and measures range from free to costly. For example, the *EFPT* is a public-domain instrument, and therefore, the User Manual, assessment forms, and additional resources are available for free download (http://www.ot.wustl.edu/about/resources/executive-function-performance-test-efpt-308). Occupational therapists can independently learn to administer and score the five sub-tasks of this functional assessment measure. Similarly, a variety of health professionals can independently learn to administer the *TFLS*, though this tool must be purchased (http://www.pearsonclinical.com/products/100000222/texas-functional-living-scale-tfls.quick.html). In contrast, the *AMPS* requires occupational therapists to attend a 5-day intensive training course and complete rater calibration prior to administering the tool. Training courses are held in the USA and abroad, with cost ranging from U.S. $900 to $1500 (https://www.innovativeotsolutions.com/content/amps/courses/).

Cultural and Linguistic Considerations

Culture is shaped by norms, values, and behavior patterns that guide thought and actions within the environment (e.g., Krefting, 1991). Cultural beliefs and norms may emerge through the performance of daily activities and established habits and routines (Bonder, Martin, & Miracle, 2004). As such, assessing functional ability in Latino clients with dementia may best be achieved through performance-based measures that permit the consideration of culture-bound information and some customization while maintaining standardization. In contrast, administering functional assessment measures that employ unfamiliar language and assess irrelevant culture-bound daily activities may limit optimal performance and the validity and meaningfulness of results in many Latino clients with dementia (Paul, 1995).

There are performance-based functional assessment measures that promote cultural sensitivity while adhering to standardization. The *AMPS* assessment protocol requires clients to plan and complete tasks of their choice using culturally familiar objects. Similarly, of the 26 subtasks of the *PASS*, the occupational therapist is able to choose to administer only those that are relevant to the client, thereby increasing the validity of results by allowing for adaptation of administration based on consideration of possible cultural influences on performance. Moreover, for both of these instruments, the instructions to clients can be adjusted for language and content as deemed appropriate. These examples demonstrate how cultural and linguistic factors may sometimes be considered by allowing for culturally appropriate modifications when administering certain performance-based assessments. However, the standardization of some measures, such as the *CPT* and *DAFS*, limits the integration of culturally relevant objects and cultural and linguistic adaptation into the assessment procedure and interpretation process.

Healthcare professionals must make every effort to consider cultural and linguistic factors when assessing and interpreting clients' functional abilities. Among people with dementia, information about cultural values and routines are often understood by asking questions of family members and other informal caregivers. It is the responsibility of the healthcare professional to consider and interpret if the cultural factors and linguistic features of an assessment are playing a modifying role in the performance of the client. Ultimately, the objective of administering performance-based assessment measures is to determine clients' optimal level of functioning within a safe environment.

Recommendations

Research on the functional assessment of Latinos with dementia remains rather sparse. Nonetheless, several important conclusions can be gleaned from the extant literature. First, healthcare professionals must strive to increase their multicultural competency, thereby allowing them to make better-informed decisions regarding the functional assessment process among Latinos with dementia, including which measure(s) to use and how to most appropriately interpret findings in the context of each individual client's unique background. When selecting and interpreting results from functional assessment measures, whether questionnaire- or performance-based, clinicians should educate themselves on the measures' reliability and validity, and in particular whether their psychometric properties have been examined both within and across Latino populations. For example, measures that have been normed on inpatients in an urban hospital in New York City may not be applicable to patients hailing from a border community in rural South Texas, despite that both speak Spanish. Even within Latinos of similar English language proficiency, older adults who are more recent immigrants may construe items differently than their counterparts who have resided in the USA for longer (Talamantes & Sanchez-Reilly, 2010).

Moreover, as part of multicultural competency, we as healthcare professionals should be aware that we interpret patients' understandings of their symptoms from our own ethnic and cultural backgrounds (Kleinman, 1996). This may in part explain the finding from Zayas, Torres, and Cabassa (2009) that non-Latino White clinicians rated Latino psychiatric patients' functional capacity and symptom severity as significantly worse than did the Latino clinicians. Probably based on sociocultural biases, healthcare professionals are likely influenced by dissimilar factors and create attributions of pathology differently, in addition to evaluating its magnitude in very disparate fashions (e.g., Malgady & Costantino, 1998).

Second, given the unique and shared strengths, limitations, and cultural considerations presented in this chapter that pertain to both questionnaire- and performance-based measures of functional abilities, it is recommended that both be used in combination if possible, especially when assessing Latinos, and in line with the recommendation provided by Jorm (2004). Questionnaire-based measures are short, free or inexpensive, easy to administer and interpret, and face-valid, but are limited in part because they depend on the subjective responses of patients and informants. Performance-based measures allow for a more direct and objective assessment of functional abilities but are more difficult to conduct in the healthcare setting and may not always directly reflect functional capacity in the home (e.g., Guralnik, Reuben, Buchner, & Ferrucci, 1995). Performance-based assessments administered in the home (e.g., Allen et al., 2007; Baum et al., 2008; Rogers & Holm, 1989) are available; however, conditions for providing home healthcare services must be met (Vance, 2016).

As discussed, both types of measures require careful consideration of cultural and linguistic factors that could influence performance and interpretation of results. Combining information obtained from both self-report and performance-based measures has been demonstrated to augment prognostic value, particularly among high-functioning older adults (Reuben et al., 2004). Although it is not often practical or feasible to administer both types of measures, the choice of which measure or measures to administer largely depends on factors such as training of and time constraints of the healthcare providers, coupled with the necessity for the most reliable and valid clinical information (Talamantes & Sanchez-Reilly, 2010). An interprofessional, team-based approach to functional assessment can promote the use of both types of measures, where, for example, the psychologist and occupational therapist administer questionnaire- and performance-based assessments, respectively. In this scenario, results can be shared in order to develop the most effective treatment plan to achieve the best clinical outcomes.

Third, when healthcare providers and patients do not speak the same language, they are encouraged to either provide a timely referral to another qualified provider with proficiency in the patient's language or seek a professional translator or interpreter. There may be important yet detrimental clinical outcomes due to deficient interpretation, such as missing pieces of clinical information obtained from patients and informants, as well as information conveyed to the patients and informants from the healthcare provider, and inadequate or incomplete patient and informant psychoeducation (Talamantes & Sanchez-Reilly, 2010). Other important problems associated with insufficient interpretation include the administration of needless testing or, on the other hand, failing to recognize important signs that would lead providers to conduct or order tests that are actually indicated or needed (Woloshin, Bickell, Schwartz, Gany, & Welch, 1995).

Finally, more research is sorely needed aimed at both developing new culturally and linguistically appropriate questionnaire- and performance-based measures of functional abilities and examining the cross-cultural psychometric properties of such existing measures among Latinos and other diverse groups. Such research would assist in helping healthcare professionals determine whether observed variations in functional capacity across individuals from diverse ethnic groups reflect real differences or simply measurement error from the tools used in the assessment (Talamantes & Sanchez-Reilly, 2010), which may likely be related to cultural and linguistic factors. Ideally, measures should be created concurrently in both English and Spanish, described as the "decentering technique" (Marin & Marin, 1991). This technique facilitates the incorporation of meaningful constructs for that particular culture rather than simply utilizing the translated constructs formerly developed that may be less culturally relevant (Talamantes & Sanchez-Reilly, 2010). Until research has accomplished these goals, healthcare professionals are encouraged to consider our recommendations in the interest of best serving Latino patients with dementia and their caregivers.

References

Allen, C. K., Austin, S. L., David, S. K., Earhart, C. A., McCraith, D. B., & Riska-Williams, L. (2007). *Manual for the Allen Cognitive Level Screen-5 (ACLS -5) and Large Allen Cognitive Level Screen-5 (LACLS-5).* Camarillo, CA: ACLS and LACLS Committee.

Alzheimer's Association. (2016). *2016 Alzheimer's disease facts and figures*. Retrieved May 25, 2017 from https://www.alz.org/documents_custom/2016-facts-and-figures.pdf.

American Psychiatric Association. (2013). *Diagnostic and statistical manual of mental disorders* (5th ed.). Washington, DC: Author.

Assis, L. D. O., de Paula, J. J., Assis, M. G., de Moraes, E. N., & Malloy-Diniz, L. F. (2014). Psychometric properties of the Brazilian version of Pfeffer's functional activities questionnaire. *Frontiers in Aging Neuroscience, 6*, 255.

Baird, R. A. E. (2007). *The effects of educational quality on the cognitive performance of minority and Caucasian HIV+ subjects* (Doctoral dissertation). Retrieved October 28, 2016 from ProQuest, UMI Dissertations Publishing. (3283618).

Baum, C. M., Connor, L. T., Morrison, T., Hahn, M., Dromerick, A. W., & Edwards, D. F. (2008). Reliability, validity, and clinical utility of the executive function performance test: A measure of executive function in a sample of people with stroke. *American Journal of Occupational Therapy, 62*, 446–455.

Blessed, G., Tomlinson, B. E., & Roth, M. (1968). The association between quantitative measures of dementia and of senile change in the cerebral grey matter of elderly subjects. *British Journal of Psychiatry, 114*, 797–781.

Bucks, R. S., Ashworth, D. L., Wilcock, G. K., & Siegfried, K. (1996). Assessment of activities of daily living in dementia: Development of the Bristol activities of daily living scale. *Age and Ageing, 25*, 113–120.

Burns, T., Mortimer, J. A., & Merchak, P. (1994). Cognitive performance test: A new approach to functional assessment in Alzheimer's disease. *Journal of Geriatric Psychiatry and Neurology, 7*, 46–54.

Bonder, B. R., Martin, L., Miracle, A. W. (2004). Culture emergent in occupation. *American Journal of Occupational Therapy, 58*(2), 159–168.

Carlson, M. C., Fried, L. P., Xue, Q. L., Tekwe, C., & Brandt, J. (2005). Validation of the Hopkins Medication Schedule to identify difficulties in taking medications. *The Journals of Gerontology, Series A: Biological Sciences and Medical Sciences, 60*, 217–223.

Chapleski, E. E., Lamphere, J. K., Kaczynski, R., Lichtenberg, P. A., & Dwyer, J. W. (1997). Structure of a depression measure among American Indian elders: Confirmatory factor analysis of the CES-D scale. *Research on Aging, 19*, 462–485.

Chisholm, D., Toto, P., Raina, K., Holm, M., & Rogers, J. (2014). Evaluating capacity to live independently and safely in the community: Performance assessment of self-care skills. *British Journal of Occupational Therapy, 77*, 59–63.

Colby, S. L., & Ortman, J. M. (2015). *Projections of the size and composition of the U.S. population: 2014 to* 2060. *Population estimates and projections* (Current Population Reports, P25–1143). Washington, DC: U.S. Census Bureau.

Cullum, C. M., Saine, K., Chan, L. D., Martin-Cook, K., Gray, K., & Weiner, M. F. (2001). Performance-based instrument to assess functional capacity in dementia: The Texas Functional Living Scale. *Neuropsychiatry, Neuropsychology, and Behavioral Neurology, 14*, 103–108.

Dilworth-Anderson, P., & Gibson, B. E. (2002). The cultural influence of values, norms, meanings, and perceptions in understanding dementia in ethnic minorities. *Alzheimer Disease & Associated Disorders, 16*, S56–S63.

Doble, S. E., Fisk, J. D., & Rockwood, K. (1999). Assessing the ADL functioning of persons with Alzheimer's disease: Comparison of family informants' ratings and performance-based assessment findings. *International Psychogeriatrics, 1*, 399–409.

Dutra, M. C., Ribeiro, R. D. S., Pinheiro, S. B., Melo, G. F. D., & Carvalho, G. D. A. (2015). Accuracy and reliability of the Pfeffer Questionnaire for the Brazilian elderly population. *Dementia e Neuropsychologia, 9*, 176–183.

Erkut, S., Alarcón, O., Coll, C. G., Tropp, L. R., & García, H. A. V. (1999). The dual-focus approach to creating bilingual measures. *Journal of Cross-Cultural Psychology, 30*, 206–218.

Erzigkeit, H., Lehfeld, H., Peña-Casanova, J., Bieber, F., Yekrangi-Hartmann, C., Rupp, M., ... Hindmarch, I. (2001). The Bayer-Activities of Daily Living Scale (B-ADL): Results from a validation study in three European countries. *Dementia and Geriatric Cognitive Disorders, 12*, 348–358.

Fisher, A. G., & Jones, K. (2010). *Assessment of Motor and Process Skills, Vol. 1: Development, standardization, and administration manual* (revised). Fort Collins, CO: Three Star Press.

Folquitto, J. C., Bustamante, S. E., Barros, S. B., Azevedo, D., Lopes, M. A., Hototian, S. R., ... Bottino, C. (2007). The Bayer: Activities of Daily Living Scale (B-ADL) in the differentiation between mild to moderate dementia and normal aging. *Revista Brasileira de Psiquiatria, 29*, 350–353.

Folstein, M. F., Folstein, S. E., & McHugh, P. R. (1975). "Mini-Mental State:" A practical method for grading the cognitive state of patients for the clinician. *Journal of Psychiatric Research, 12*, 189–198.

Gallagher-Thompson, D., Solano, N., Coon, D., & Areán, P. (2003). Recruitment and retention of Latino dementia family caregivers in intervention research: Issues to face, lessons to learn. *The Gerontologist, 43*, 45–51.

Glass, T. A. (1998). Conjugating the "tenses" of function: Discordance among hypothetical, experimental, and enacted function in older adults. *The Gerontologist, 38*, 101–112.

Gleichgerrcht, E., Camino, J., Roca, M., Torralva, T., & Manes, F. (2009). Assessment of functional impairment in dementia with the Spanish version of the Activities of Daily Living Questionnaire. *Dementia and Geriatric Cognitive Disorders, 28*, 380–388.

Gold, D. A. (2012). An examination of instrumental activities of daily living assessment in older adults and mild cognitive impairment. *Journal of Clinical and Experimental Neuropsychology, 34*, 11–34.

Guralnik, J. M., Reuben, D. B., Buchner, D. M., & Ferrucci, L. (1995). Performance measures of physical function in comprehensive geriatric assessment. In L. Z. Rubenstein, D. Wieland, & R. Bernabel (Eds.), *Geriatric assessment technology: The state of the art* (pp. 59–74). New York: Springer Publishing Co.

Guralnik, J. M., & Simonsick, E. M. (1993). Physical disability in older Americans. *Journal of Gerontology, 48*(Special Issue), 3–10.

Gurland, B. J., Wilder, D. E., Lantigua, R., Stern, Y., Chen, J., Killeffer, E. H., & Mayeux, R. (1999). Rates of dementia in three ethnoracial groups. *International Journal of Geriatric Psychiatry, 14*, 481–493.

Harvey, P. D., Velligan, D. I., & Bellack, A. S. (2007). Performance-based measures of functional skills: Usefulness in clinical treatment studies. *Schizophrenia Bulletin, 33*, 1138–1148.

Hindmarch, I., Lehfeld, H., de Jong, P., & Erzigkeit, H. (1998). The Bayer Activities of Daily Living (B-ADL) scale. *Dementia and Geriatric Cognitive Disorders, 9*, 20–26.

Hui, C. H., & Triandis, H. C. (1985). Measurement in cross-cultural psychology: A review and comparison of strategies. *Journal of Cross-Cultural Psychology, 16*, 131–152.

Johnson, N., Barion, A., Rademaker, A., Rehkemper, G., & Weintraub, S. (2004). The Activities of Daily Living Questionnaire: A validation study in patients with dementia. *Alzheimer Disease & Associated Disorders, 18*, 223–230.

Jorm, A. F. (2004). The Informant Questionnaire on Cognitive Decline in the Elderly (IQCODE): A review. *International Psychogeriatrics, 16*, 275–293.

Katz, S., Downs, T. D., Cash, H. R., & Grotz, R. C. (1970). Progress in development of the index of ADL. *The Gerontologist, 10*(1 Part 1), 20–30.

Kleinman, A. (1996). How is culture important for DSM-IV? In J. E. Mezzich, A. Kleinman, H. Fabrega Jr., & D. L. Parron (Eds.), *Culture and psychiatric diagnosis: A DSM-IV perspective* (pp. 15–25). Washington, DC: American Psychiatric Press.

Krefting, L. (1991). The culture concept in the everyday practice of occupational and physical therapy. *Physical & Occupational Therapy in Pediatrics, 11*, 1–16. https://doi.org/10.1080/J006v11n04_01

Lawton, M. P., & Brody, E. M. (1969). Physical self-maintenance scale (functional assessment). *Gerontologist, 9*, 179–186.

Loeb, P. A. (1996). *ILS: Independent Living Scales manual*. San Antonio, TX: Psychological Corp.

Loewenstein, D. A., Amigo, E., Duara, R., Guterman, A., Hurwitz, D., Berkowitz, N., ... Eisdorfer, C. (1989). A new scale for the assessment of functional status in Alzheimer's disease and related disorders. *Journal of Gerontology, 44*, P114–P121.

Mahoney, F. I., & Barthel, D. W. (1965). Functional evaluation: The Barthel index. *Maryland State Medical Journal, 14*, 61–65.

Malgady, R. G., & Costantino, G. (1998). Symptom severity in bilingual Hispanics as a function of clinician ethnicity and language of interview. *Psychological Assessment, 10*, 120–127.

Malmstrom, T. K., Miller, D. K., Coats, M. A., Jackson, P., Miller, J. P., & Morris, J. C. (2009). Informant-based dementia screening in a population-based sample of African Americans. *Alzheimer's Disease and Associated Disorders, 23*, 117–123.

Mapi Research Institute. (1999). *Cultural adaptation of the Bayer Activities of Daily Living Scale (B-ADL) into Brazilian Portuguese*. Report (pp. 1–19). Lyon, France: Mapi Research Institute.

Marin, G., & Marin, B. V. (1991). *Research with Hispanic populations. Applied social research methods series* (Vol. 23). Newbury Park, CA: Sage Publications, Inc.

Marson, D., & Hebert, K. (2006). Functional assessment. In D. Attix & K. Welsh-Bohmer (Eds.), *Geriatric neuropsychology assessment and intervention* (pp. 158–189). New York, NY: The Guilford Press.

Marson, D. C., Sawrie, S. M., Snyder, S., McInturff, B., Stalvey, T., Boothe, A., … Harrell, L. E. (2000). Assessing financial capacity in patients with Alzheimer disease: A conceptual model and prototype instrument. *Archives of Neurology, 57*, 877–884.

McDougall, G. J., Becker, H., Pituch, K., Acee, T. W., Vaughan, P. W., & Delville, C. L. (2010). The SeniorWISE study: Improving everyday memory in older adults. *Archives of Psychiatric Nursing, 24*, 291–306.

Moore, D. J., Palmer, B. W., Patterson, T. L., & Jeste, D. V. (2007). A review of performance-based measures of functional living skills. *Journal of Psychiatric Research, 41*, 97–118.

Ortiz, F., Fitten, L. J., Cummings, J. L., Hwang, S., & Fonseca, M. (2006). Neuropsychiatric and behavioral symptoms in a community sample of Hispanics with Alzheimer's disease. *American Journal of Alzheimer's Disease and Other Dementias, 21*, 263–273.

Pfeffer, R. I., Kurosaki, T. T., Harrah, C. H., Chance, J. M., & Filos, S. (1982). Measurement of functional activities in older adults in the community. *Journal of Gerontology, 37*, 323–329.

Paul, S. (1995) Culture and its influence on occupational therapy evaluation. *Canadian Journal of Occupational Therapy, 62*, 154–161. https://doi.org/10.1177/000841749506200307

Reuben, D. B., Seeman, T. E., Keeler, E., Hayes, R. P., Bowman, L., Sewall, A., … Guralnik, J. M. (2004). Refining the categorization of physical functional status: The added value of combining self-reported and performance-based measures. *The Journals of Gerontology, Series A: Biological Sciences and Medical Sciences, 59*, M1056–M1061.

Rodakowski, J., Skidmore, E. R., Reynolds, C. F., Dew, M. A., Butters, M. A., Holm, M. B., … Rogers, J. C. (2014). Can performance on daily activities discriminate between older adults with normal cognitive function and those with mild cognitive impairment? *Journal of the American Geriatrics Society, 62*, 1347–1352.

Rogers, J. C., & Holm, M. B. (1989). *Performance assessment of self-care skills*. Unpublished performance test. University of Pittsburgh.

Rohit, M., Levine, A., Hinkin, C., Abramyan, S., Saxton, E., Valdes-Sueiras, M., & Singer, E. (2007). Education correction using years in school or reading grade-level equivalent? Comparing the accuracy of two methods in diagnosing HIV-associated neurocognitive impairment. *Journal of the International Neuropsychological Society, 13*, 462–470.

Sanchez, M. A. S., Correa, P. C. R., & Lourenço, R. A. (2011). Cross-cultural adaptation of the "Functional Actives Questionnaire – FAQ" for use in Brazil. *Dementia e Neuropsychologia, 5*, 322–327.

Saper, J. (2003). The dimensions of insight in people with dementia. *Aging & Mental Health, 7*, 113–122.

Sayegh, P. (2015). Cross-cultural issues in the neuropsychological assessment of dementia. In F. R. Ferraro (Ed.), *Minority and cross-cultural aspects of neuropsychological assessment: Enduring and emerging trends* (2nd ed., pp. 54–71). Hove, England: Psychology Press.

Sayegh, P., Arentoft, A., Thaler, N. S., Dean, A. C., & Thames, A. D. (2014). Quality of education predicts performance on the Wide Range Achievement Test – 4th Edition Word Reading subtest. *Archives of Clinical Neuropsychology, 29*, 731–736.

Sayegh, P., & Knight, B. G. (2013a). Assessment and diagnosis of dementia in Hispanic and non-Hispanic White outpatients. *The Gerontologist, 53*, 760–769.

Sayegh, P., & Knight, B. G. (2013b). Cross-cultural differences in dementia: The Sociocultural Health Belief Model. *International Psychogeriatrics, 25*, 517–530.

Sayegh, P., & Knight, B. G. (2014). Functional assessment and neuropsychiatric inventory questionnaires: Measurement invariance across Hispanics and non-Hispanic Whites. *The Gerontologist, 54*, 375–386.

Schwartz, M. F., Segal, M., Veramonti, T., Ferraro, M., & Buxbaum, L. J. (2002). The Naturalistic Action Test: A standardised assessment for everyday action impairment. *Neuropsychological Rehabilitation, 12*, 311–339.

Sheehan, B. (2012). Assessment scales in dementia. *Therapeutic Advances in Neurological Disorders, 5*, 349–358.

Sikkes, S. A. M., de Lange-de Klerk, E. S. M., Pijnenburg, Y. A. L., & Scheltens, P. (2009). A systematic review of Instrumental Activities of Daily Living scales in dementia: Room for improvement. *Journal of Neurology, Neurosurgery & Psychiatry, 80*, 7–12.

Talamantes, M., & Sanchez-Reilly, S. (2010). Health and health care of Hispanic/Latino American older adults. In V. S. Periyakoil (Ed.), *eCampus-geriatrics*. Stanford, CA: Stanford School of Medicine. Retrieved from https://geriatrics.stanford.edu/ethnomed/latino.html.

U.S. Census Bureau, Population Division. (2016). *FFF: Hispanic heritage month 2016*. Retrieved from https://www.census.gov/newsroom/facts-for-features/2016/cb16-ff16.html.

Vance, K. (2016). *Home health care: A guide for occupational therapy practice*. Bethesda, MD: American Occupational Therapy Association.

Wallace, M., & Shelkey, M. (2007). Katz index of independence in activities of daily living (ADL). *Urologic Nursing, 27*, 93–94.

Ward, G., Jagger, C., & Harper, W. (1998). A review of instrumental ADL assessments for use with elderly people. *Reviews in Clinical Gerontology, 8*, 65–71.

Wesson, J., Clemson, L., Brodaty, H., & Reppermund, S. (2016). Estimating functional cognition in older adults using observational assessments of task performance in complex everyday activities: A systematic review and evaluation of measurement properties. *Neuroscience & Biobehavioral Reviews, 68*, 335–360.

Wilkinson, G. S., & Robertson, G. J. (2006). *Wide range achievement test – Fourth edition: Professional manual*. Lutz, FL: Psychological Assessment Resources.

Woloshin, S., Bickell, N. A., Schwartz, L. M., Gany, F., & Welch, H. G. (1995). Language barriers in medicine in the United States. *Journal of the American Medical Association, 273*, 724–728.

Zayas, L. H., Torres, L. R., & Cabassa, L. J. (2009). Diagnostic, symptom, and functional assessments of Hispanic outpatients in community mental health practice. *Community Mental Health Journal, 45*, 97–105.

Zarit, S. H., & Zarit, J. M. (2007). *Mental disorders in older adults* (2nd ed., pp. 40–77). New York, NY: The Guilford Press.

Latinos with Dementia and Depression: Contemporary Issues and Assessment Challenges

6

Azara Santiago-Rivera, Gregory Benson-Flórez, and Kathleen D. Warman

Abstract

Latino older adults may be at greater risk for dementia and depression. The barriers to adequate healthcare clearly affect early diagnosis, treatment, and management of depression and dementia, a situation which becomes even more complicated when there is a presence of other comorbid health conditions. The relationship between depression and dementia, including the challenges associated with assessment and diagnosis for the Latino aging population, is highlighted in this chapter. Along with the focus on the Latino with dementia, a discussion is also provided regarding the role of the caregiver of the person with dementia and the challenges that Latino caregivers face. This chapter concludes with recommendations for further research such as conducting longitudinal studies that examine the impact of age, education, language, acculturation, and daily functioning on the progression of cognitive decline and dementia. The authors also comment on the need for greater attention to Latino older adults' emotional functioning, in general, and that of Latinos with dementia, in particular, as well as the needs of their caregivers.

A. Santiago-Rivera (✉)
Merrimack College, North Andover, MA, USA
e-mail: santiagoal@merrrimack.edu

G. Benson-Flórez · K. D. Warman
The Chicago School of Professional Psychology, Chicago, IL, USA
e-mail: gbenson-florez@thechicagoschool.edu; kwarman@ego.thechicagoschool.edu

Introduction

It is well recognized that the number of older adults is steadily increasing in the United States (U.S.). By the year 2050, it is estimated that the overall aging population (65 years of age and older) will be 83 million (Ortman, Velkoff, & Hogan, 2014). Ortman et al. (2014) also note that the aging population is clearly becoming much more ethnically and racially diverse. Demographic data indicate that the groups with the most growth since 2012 continue to be Asians, American Indians and Alaska Natives, African Americans, Latinos, Native Hawaiians, and other Pacific Islanders, while the rate of growth for non-Hispanic Whites is much smaller. Moreover, Latinos are among the fastest-growing segment of this population. For instance, in 2012 the aging Latino population, 65 and older, was at 7%, and demographic projections indicate that by 2060 it is expected to grow to approximately 22% (U.S. Department of Human Health Services, 2017).

With the unprecedented growth in the Latino aging population, more attention is directed at capturing their overall health status. For example, Carson et al. (2011) found that Latino older adults had a higher incidence of lifetime risk for

H. Y. Adames, Y. N. Tazeau (eds.), *Caring for Latinxs with Dementia in a Globalized World*,
https://doi.org/10.1007/978-1-0716-0132-7_6

developing hypertension compared to Whites, while other authors such as Angel, Kahlert, and Whitfield (2006) observed that Latinos have higher rates of diabetes which often contribute to health complications and disabilities. In addition, concerns about mental illness such as depression and its relationship to both cognitive decline and diabetes among Latino older adults are at the forefront (Katon et al., 2012). Together, these health problems of hypertension and diabetes will have a significant impact on healthcare, families, and communities.

The primary purpose of this chapter is to discuss the interplay between depression and dementia in older Latino adults. Specifically, this chapter begins by describing depression in older adults and by outlining several health factors that may exacerbate mood disorders, such as diabetes which is a common illness among the Latino population. In addition, we explain the challenges associated with accurately diagnosing depression in the Latino population, as well as provide a brief description of commonly used assessment measures. We also discuss caring for a family member with depression and dementia and describe some of the challenges that their caregivers face. Finally, we provide recommendations that emphasize a culturally competent treatment approach including a familiarity of the psychosocial factors and diverse worldview perspectives that exist between and within older adult Latino subgroups.

Depression and Older Latino Adults

Depression in older adults is considered common; however, it is not a normal part of the aging process. Often referred to as "depression in late life," depression has been linked to cognitive and functional impairment, chronic medical problems (e.g., diabetes, cardiovascular disease, hyperthyroidism, hypothyroidism), and stressful life events such as the loss of a loved one, reduced financial resources, and lack of social support (APA, 2014; Bowen & Ruch, 2015; Fiske, Wetherell, & Gatz, 2009). More importantly, Fiske et al. (2009) highlight the complex interplay of physical illness, psychosocial risk factors, neurological disorders, and possible predispositions (e.g., genetic) that make diagnosis and treatment extremely challenging. Furthermore, older adults may present symptoms of depression differently such as complaints about pain, loss of appetite, fatigue, sleeplessness, memory difficulties, and hopelessness rather than depressed mood (Kok & Reynolds, 2017).

With respect to Latino older adults, some risk factors may be similar to those of other ethnic and racial groups such as low socioeconomic status; however, there are unique factors that may exacerbate depression such as the acculturation process (Chavez-Korell, Benson-Florez, Rendon, & Farías, 2014) and the long-term effects of traumatic life events associated with the migration experience (Weisman et al., 2005). Along these lines, Liang, Xu, Quiñones, Bennet, and Ye (2011) examined depressive symptoms, over time, among a large sample of African Americans, Latinos, and White middle-age and older adults and found not only that Latinos were more likely to report elevated depressive symptoms but also that these symptoms were persistently high over time.

Impact of Acculturation

The role of acculturation and its impact on psychological and physical well-being have been widely studied (Berry, 2003). Acculturation is the multidimensional process of an individual or a group of people in adapting to a new culture and its values. Acculturation can be conceptualized as a bidirectional interaction between the individual and the environment in which changes in values, behaviors, and beliefs can occur over time (Berry, 2003). This adaptation process, especially for recent immigrants, may lead to considerable strain known as *acculturative stress* (Berry & Annis, 1974). The process may significantly vary among immigrants and generations which may result in acculturation "gaps" within families (Szapocznik et al., 1986). These gaps may create family conflicts and amplify stressors associated with the adaptation process. In fact, González,

Haan, and Hinton (2001) found that depressive symptoms were highest among the least acculturated older adults. They speculated that the language barrier, as well as lack of knowledge about mainstream U.S. culture, greatly influences psychological well-being.

For Latinos, acculturation is also a process of attitudinal and behavioral changes as a result of coming into contact with a new cultural environment (Marín, 1992, 1993). The possibility that acculturation toward mainstream U.S. culture may affect cognitive functioning in older Latino adults is supported by Farias, Mungas, Hinton, and Haan's (2011) study in which they found that acculturation toward American culture and being tested in English were significantly related to lower global cognitive functioning in a sample of 639 Latinos, age 60 and older. Equally important, they also found that older age was associated with a faster rate of cognitive decline over time.

Barriers to Medical Care Faced by Older Latinos

Older Latino adults face numerous challenges to affordable and quality healthcare services. The National Hispanic Council on Aging (2012) reported that approximately 5% of Latino older adults were uninsured compared to 1% of the general older population in this country. Moreover, for those who have chronic health conditions and are seeking healthcare, Latino older adults more often report having difficulty not only making appointments but also waiting longer to be seen by providers (Shirey & Summer, 2003). Recent data from the Kaiser Family Foundation (2013) revealed that Latinos with healthcare insurance were nearly 50% more likely to receive medical care and roughly twice as likely to get preventive care. Unfortunately, a proposed repeal in 2017 to the Affordable Care Act (Sorrell, 2012) by the current governmental administration may have a detrimental impact on access to healthcare. This may accentuate the barriers in place for Latino older adults seeking Medicare and/or Medicaid and potentially limit lifesaving health services (e.g., health screenings, annual physicals).

Not only do these external barriers prevent access, but personal beliefs about illness can also result in some Latinos not seeking healthcare services. For instance, Ortiz and Fitten (2000) found that Latino older adults who had cognitive impairment reported that a significant barrier for them was the belief that memory loss was a normal part of aging, thus delaying diagnosis and treatment. Other significant barriers consistently reported in the literature (e.g., Farias et al., 2011) include low socioeconomic status and limited or no English language proficiency.

Dementia and Depression

Trujillo et al. (2016) describe dementia as a "neurodegenerative condition characterized by the deterioration in cognitive abilities, memory, and daily functioning (e.g., bathing, eating, and dressing)" (p.1–2), and it is not uncommon for individuals with dementia to exhibit behaviors such as mood and personality changes, aggression, and wandering. Likewise, the severity of dementia and its progression make it difficult to treat. Some studies have shown that Latinos may be at greater risk for dementia due to certain biological factors and health conditions (Chin, Negash, & Hamilton, 2011), and other studies have concluded that the rates of dementia may be higher for Latinos in comparison with other ethnic groups (Sloan & Wang, 2005; Valle, Garrett, & Velaquez, 2013; Weisman et al., 2005).

An added dimension to consider for individuals who have dementia is that they are often diagnosed with depression (Hansen & Cabassa, 2012; Rapp et al., 2011). Specific to Latinos, Hinton, Haan, Geller, and Mungas (2003) found that in their sample of older adults diagnosed with dementia, at least 30% showed significantly higher scores for depression, irritability, anxiety, aggression, apathy, and disinhibition compared to those who were cognitively impaired and were not demented. Furthermore, there is evidence to support the claim that depression is often present before cognitive decline starts to occur, especially among ethnically and racially diverse older adults (Simões et al., 2016; Zahodne, Stern, & Manly, 2014).

Challenges of Screening for Depression and Dementia

As previously noted earlier, depression in older adults has been associated with dementia and other chronic physical conditions making diagnosis and treatment complex. One of the greater challenges is accurately screening for both dementia and depression in Latino adults. A variety of factors have been identified as contributing to the problem of an inaccurate diagnosis such as low levels of formal education, literacy levels, socioeconomic status, limited English language proficiency, and unfamiliarity with the assessment process (Early et al., 2013). Another complicating factor is that Latinos may distrust the medical establishment due to past experiences with discrimination and lack of culturally sensitive treatment approaches. Thus, Latino older adults may refuse to undergo any type of physical exam, as well as screening for dementia and/or for psychological conditions such as depression (Jimenez, Cook, Bartels, & Alegría, 2013). As a result, opportunities for early detection and the management of illness may be at stake.

One of the major challenges in accurately measuring depression and cognitive functioning centers on whether or not existing assessment tools are culturally and linguistically sensitive, as well as reliable and valid for use with Latinos. For example, cultural and language differences in the conceptualization of illness, meaning, and expression of depressive symptoms (Aguilar-Gaxiola, Kramer, Resendez, & Magaña, 2008; Canino & Alegría, 2009; Santiago-Rivera, Benson-Flórez, Santos, & Lopez, 2015) have been emphasized as contributing to misdiagnosis. In support of this perspective, Lewis-Fernández, Das, Alfonso, Weissman, and Olfson (2005) provide a case illustration in which an older adult female of Mexican heritage described her symptoms of depression as suffering from *nervios* [nerves], feeling *sofocada* [out of breath], and having headaches and heart palpitations. Because of the somatic focus of the symptoms, the initial assessment centered on potential cardiac and respiratory health conditions rather than a possible mood problem. Similarly, certain words in the Spanish language may more accurately describe depression. For instance, Noguera et al. (2009) set out to explore which Spanish language terms better described depressed mood in a sample of Latino adults receiving cancer treatment and found that they preferred to use the term *desanimado* [discouraged] rather than the term *deprimido* which is considered the literal translation of "depressed" in English. These findings suggest that accurately screening for depression must take into consideration the specific meaning and connotation of words in the Spanish language. In addition, accurately diagnosing dementia and other forms of cognitive decline can be problematic for older Latino adults because lower level of education and limited English language proficiency are confounding variables that influence neurological tests results (Ortiz, Fitten, Cummings, Hwang, & Fonseca, 2006; Weissberger, Salmon, Bondi, & Gollan, 2013).

Measures of Depression

Despite these challenges, a number of assessments commonly used to screen for depression in Latinos are the (1) Beck Depression Inventory-II (BDI; Beck, Ward, Mendelson, Mock, & Erbaugh, 1961), (2) Patient Health Questionnaire-9 (PHQ-9; Spitzer, Kroenke, & Williams, 1999), and (3) the Center for Epidemiologic Studies Depression Scale (CES-D; Radloff, 1977). More importantly, assessments such as the Geriatric Depression Scale (GDS; Yesavage et al., 1983) have received considerable attention as reliable screening tools for diagnosing depression in older adults who are experiencing cognitive decline. All of these assessment tools have published psychometric properties available for both the Spanish and English versions.

The Beck Depression Inventory-II (BDI-II; Beck, Steer, & Brown, 1996) is a well-established and widely used screening instrument for depression. It is a 21-item, self-report measure assessing a variety of symptoms occurring in the 2 weeks prior to the actual assessment. Each item is rated on a scale from 0 (no depressive symptoms) to 3 (severe depressive symptoms), and the total range of scores is 0–63. With respect to interpreting the scores, 0–13 indicates minimal depression, 14–19 indicates mild depression, 20–28 indicates moderate depression, and 29–63 indicates severe depression. There is considerable evidence indicating that the BDI-II has sound psychometric properties for both the English and Spanish versions. Specifically, Wiebe and Penley (2005) found that the English version had a reliability coefficient of 0.89 for the English version and a 0.91 coefficient for the Spanish version in a sample of college students. Moreover, results of an earlier study of older Mexican Americans and Whites, with an average age of 70, showed that the English version had an internal consistency coefficient of 0.80 (Gatewood-Colwell, Kaczmarek, & Ames, 1989). Novy, Stanley, Averill, and Daza (2001) reported internal consistency coefficients of 0.94 for each of the English and Spanish versions in a diverse sample of bilingual Latinos from various countries of origin (i.e., Mexico, South America, Central America, Cuba, Puerto Rico), ranging in age from 18 to 75. In essence, there is sufficient evidence demonstrating the BDI-II's utility in accurately screening for depression among Latinos, in general. In another study, Segal, Coolidge, Cahill, and O'Riley (2008) examined the psychometric properties of the BDI-II in a sample of younger (17–29 years of age) and older (55–90 years of age) adults. For purposes of this discussion, we note that their findings concerning the older adults showed good internal reliability (i.e., how well all the items consistently measure the same general construct of depression) on the BDI-II. Specifically, the internal reliability alpha was 0.86 for the older adult group. Moreover, this study showed strong, significant, and positive correlations with the Center for Epidemiologic Studies Depression Scale (CES-D; Radloff, 1977; $r = 0.69$), and the depression subscale of the Coolidge Axis II Inventory (CATI: Coolidge, 2004; Coolidge & Merwin, 1992; $r = 0.66$) demonstrated solid convergent validity (i.e., how the inventory correlates with other psychometrically sound measures of depression). In addition, the researchers note that there were no significant cross-sectional differences of age and gender on BDI-II scores; however, they acknowledge the lack of adequate ethnic diversity in their sample, particularly in the older adult group. As such, the results indicating no ethnic differences must be interpreted with caution. Nonetheless, a strength of this study is that it focused on a nonclinical community sample of older adults. However, and in summary, more research is needed to explore the use of the BDI-II with Latino older adults who are experiencing cognitive decline.

A popular measure that has been widely used with ethnic and racial minorities is the Patient Health Questionnaire-9 (PHQ-9; Spitzer et al., 1999). Overall, there is substantial evidence showing that it has good internal reliability with an alpha coefficient ranging from 0.86 to 0.89 (Kroenke, Spitzer, & Williams, 2001). The PHQ-9 is designed to measure the extent to which a variety of symptoms associated with depression are present such as anhedonia (lack of pleasure), difficulties sleeping and concentrating, changes in appetite, suicidal thoughts, and fatigue. Each item is rated on a 4-point scale from 0 (not at all) to 3 (nearly every day); the total range of scores is 0–27. With respect to the interpretation of scores, 0–4 indicates none or minimal symptoms present, 5–9 indicates mild depression, 10–14 indicates moderate depression, 15–19 indicates moderately severe depression, and 20–27 indicates severe depression (Kroenke & Spitzer, 2002).

Donlan and Lee (2010) used the Spanish version of the PHQ-9 (Diez-Quevado, Rangil, Sanchez-Planell, Kroenke, & Spitzer, 2001) with a sample of Mexican migrant farmworkers and found that the measure had good internal consistency reliability ($\alpha = 0.92$) and composite reliability ($\alpha = 0.88$) (i.e., the extent to which the items, as a whole, consistently measure the construct of depression). More recently, the factor structure of the same Spanish version of the PHQ-9 revealed that the internal consistency is good for the full scale ($\alpha = 0.89$) which is consistent with other empirical findings (Familiar et al., 2015).

Finally, there is evidence to support that the PHQ-9 is a reliable measure of depression for older Latino adults, ranging in age from 65 to 97 (Chavez-Korell et al., 2014).

As noted earlier, the CES-D is another measure that is commonly used in clinical, nonclinical, and community samples, as well as adolescents and adults ranging in ages (e.g., 15–96). The utility of the CES-D 10-item scale has been examined as a reliable measure of depression in older adults (e.g., Andresen, Malmgren, Carter, & Patrick, 1994). With respect to its usefulness in accurately screening for depression among Latino older adults, the 10-item measure has been shown to be a more sensitive screening tool for depression among a Puerto Rican sample of older adults with a specificity rate of 70% (Robison, Gruman, Gaztambide, & Blanks, 2002). In a more recent study consisting of a large ($N = 15{,}487$) and diverse sample of Latinos, consisting of Dominicans, Central Americans, Cubans, Mexican Americans, Puerto Ricans, and South Americans, González et al. (2016) found that the CES-D 10-item measure showed acceptable internal consistency (Cronbach's alpha = 0.80 to 0.86). Although the ages ranged from 18 to 76 in this large sample, ideally there was a representative subsample of older adults given that the data are derived from the Hispanic Community Health Study/Study of Latinos (HCHS/SOL) and for which households were randomly selected from various communities in four distinct geographic areas (i.e., Bronx, New York; San Diego, California; Chicago, Illinois; Miami, Florida).

The Geriatric Depression Scale (GDS; Yesavage et al., 1983) is another measure used to screen for depression in older adults, especially for those with mild to moderate dementia. An advantage of using this measure is that it consists of a forced choice format of yes or no responses, thus making it easier for older adults with cognitive impairment to respond. There are a 30-item and a 15-item versions of the GDS, and their validity and reliability have been supported through clinical and research endeavors (Kørner et al., 2006; Lopez, Quan, & Carvajal, 2010). There has been growing interest in using the GDS, especially the 15-item measure, with the Latino older adult population. While the results of earlier studies are mixed for the Spanish version of the GDS, Reuland et al. (2009) found sensitivities ranging from 76% to 89% and specificities ranging from 64% to 98% indicating that the measure appears to be valid and reliable. Lucas-Carrasco (2012) reported acceptable internal consistency of the GDS (Cronbach's $\alpha = 0.81$). In sum, the Spanish version of the GDS has significant clinical relevance in reliably assessing depression in older adults who have dementia.

Neuropsychological Assessment and Dementia

It is well recognized that neuropsychological assessments play a vital role in the diagnosis of cognitive deterioration in older adults because of their greater breadth and depth as compared to screening devices alone. Appropriately assessing the initial stages of dementia and its progression are key in developing effective strategies to manage the disease. As mentioned earlier, cultural and language differences may influence the performance on cognitive measures, rendering an inaccurate diagnosis (Siedlecki et al., 2010). This is particularly relevant for older Latino adults who are often predominantly Spanish language speakers and are more likely to be less acculturated to mainstream U.S. culture (Ortiz et al., 2006).

In an effort to address concerns about the utility of diagnostic measures for cognitive impairment among Latino older adults, the Spanish and English Neuropsychological Assessment Scales (SENAS; Mungas, Reed, Crane, Haan, & González, 2004; Mungas, Reed, Marshall, & González, 2000) was developed. SENAS is designed to assess cognitive abilities in older adults, and it consists of 16 scales measuring such areas as verbal and nonverbal conceptual thinking, object naming, verbal and visual attention span, and verbal comprehension. Mungas et al. (2004) reported its sound psychometric properties, as well as noted that the various scales in both the Spanish and English versions correspond well, thus reducing measurement bias.

Similarly, the Consortium to Establish a Registry for Alzheimer's Disease (CERAD) Neuropsychological Battery (Morris et al., 1989) consists of measures that assess specific aspects of cognitive functioning such as verbal fluency, naming pictures, and word recognition and recall. The Spanish version of the battery is a promising measure that can assist in assessing cognitive deficits related to neurodegenerative diseases (Fillenbaum, Kuchibhatla, Henderson, Clark, & Taussig, 2007). Although this chapter is not intended to describe neuropsychological tests in detail, it is important to acknowledge the progress made in developing such measures to improve the assessment of cognitive decline in Latino older adults.

Medications

There continue to be significant racial and ethnic disparities in the use of medications to treat dementia. In particular, studies have consistently found that African American and Latinos are less likely to be prescribed anti-dementia medication compared to non-Hispanic Whites for a variety of reasons such as lack of access to appropriate healthcare and prescription drug insurance coverage, as well as cultural bias in testing for dementia (Zukerman et al., 2008). Xiong, Filshtein, Beckett, and Hinton (2015) investigated the prevalence of antipsychotic drug use (i.e., medications to treat behavioral problems such as aggression and anxiety) among community-dwelling African American, Latinos, and non-Hispanic Whites diagnosed with dementia and neuropsychiatric symptoms. They found that Latinos were more likely to be taking antipsychotic medication compared to non-Hispanic Whites, especially among those with severe dementia. Even more striking, Latinos reported higher neuropsychiatric symptoms than African Americans and non-Hispanic Whites, indicating that the higher the symptoms, the greater the likelihood of psychotic medication use. It is important to note that this type of medication is often prescribed to manage problematic behaviors such as agitation and hallucinations, but has known side effects such as drowsiness, dizziness, low blood pressure, and vomiting (NIMH, 2016). These side effects can be particularly troublesome for an older adult who may already have other serious health conditions.

Caring for a Family Member with Dementia

As noted earlier in this chapter, dementia is a progressive disease, and the individual's ability to independently perform basic daily activities diminishes over time, requiring regular care by someone. Demographic trends indicate that the Latino caregiver is most often an adult female, is younger in age, and has more children at home compared to non-Latino Whites (Gallagher-Thompson, Solano, Coon, & Areán, 2003). According to Koerner, Shirai, and Pedroza (2013), the driving forces for what they refer to as a "pattern of care provision" among Latinos are cultural beliefs and values about family loyalty, described as *familismo*, as well as language and economic barriers that limit access to services, and a different understanding of the etiology of dementia and other mental illnesses such as depression. As a result of these forces, a Latino family member is more likely to care for an aging family member. Although the role of caregiver may be, to an extent, a rewarding one, there are a variety of challenges that often create stress and strain within the family.

The daily care of someone with dementia alone or in combination with other health issues can be considerably stressful and overwhelming. For example, mental health problems such as high anxiety and depression and low life satisfaction have been linked to the stress experienced by Latino caregivers (Losada et al., 2015; Weisman et al., 2005). In an effort to tease apart possible contributing factors to caregiver depression, Hinton et al. (2003) investigated the extent to which the intensity of neuropsychiatric symptoms commonly exhibited by individuals with dementia (i.e., depression, irritability, anxiety, aggression, apathy, inappropriate elation, disinhibition, hallucinations, delusions, aberrant motor activity) was related to depression in a sample of Latino caregivers who were primarily women and found that the more intense the patient's symptoms, the greater the depressive symptoms in the caregiver. More importantly, they found that caregivers who were caring for a spouse had higher depression scores. Other contributing factors to "caregiver burden" include reduced social support, poor coping strategies and quality of life, as well as physical health problems (Herrera, Lee, Nanyonjo, Laufman, & Torres-Vigil, 2009; Trujillo et al., 2016).

If a caregiver is experiencing physical and mental health issues, how are they able to offer quality care to an individual who has dementia and who might be depressed as well? This is not an insignificant question. In a recent study in which a number of focus groups were conducted consisting of Spanish-speaking caregivers, Turner et al. (2015) identified important themes that provide insight on how they coped and managed the behavior problems of a person with dementia. The participants reported that accepting the individual's illness was critical and that expressing love and being patient and flexible made the caregiver's role manageable. Conversely, some of the themes that were of major concern were not having the support of other family members, unresponsive providers, and the inability to maintain a much-needed income. As noted by Turner et al. (2015), the caregiver plays a vital role in the lives of individuals with dementia; therefore, it is especially important to provide them with supportive interventions such as training in effective coping strategies, psychoeducation, and, in some instances, treatment for depression if necessary.

Some progress has been made in developing interventions that are culturally and linguistically appropriate. One such approach is the use of *fotonovelas* (i.e., a story with pictures) that depict ways to use coping strategies that lead to effective stress management (Gallagher-Thompson et al., 2015). Another example showing promising results are the positive effects of both Cognitive-Behavioral Therapy (CBT) and Acceptance and Commitment Therapy (ACT) for reducing depressive symptoms and anxiety among caregivers (Losada et al., 2015).

Future Directions

As discussed in this chapter, depression, especially late in life, is related to significant cognitive decline. This is particularly relevant given research findings suggesting that Latinos are significantly more likely than non-Hispanic Whites to experience increases in depressive symptoms over time (Liang et al., 2011). Furthermore, dementia is associated with higher rates of depression among older adults (Snowden et al., 2015).

Since studies show that depression is common in older Latino adults who have dementia and other health conditions such as diabetes, an integrated behavioral health framework of care, especially in community health facilities, could be beneficial. Snowden et al. (2015) make a similar recommendation by stating that "… community-based interventions for treatment of dementia will need to be prepared to address depression and may need to include high levels of collaboration between nurse managers and physicians in order to significantly improve depressive symptoms" (p. 902). Likewise, another consideration is the need to continue conducting longitudinal studies that would help identify factors that may contribute to cognitive decline and dementia. It has been suggested (Snowden et al., 2015) that looking at such factors as age, education, language, acculturation, and daily functioning overtime would not only further our understanding of its progression but also help to improve overall healthcare services. Similarly, more research is needed to further the understanding of depression in older adults, of how it is manifested, and of ways to treat it. This includes investigating psychosocial and behavioral treatment approaches other than the standard use of medications to help alleviate symptoms. Along this line of reasoning, more research is needed to identify which neuropsychological tests are better suited to predict dementia in Latinos. Such efforts include examining which assessments are good predictors of cognitive decline such as for Alzheimer's disease (e.g., Weissberger et al., 2013), as well as furthering our understanding of the validity of tests that have been translated from English to Spanish (Siedlecki et al., 2010). Such attempts are commendable, and it is critically important to continue such efforts.

In closing, the older Latino adult population will continue to grow, demonstrating the need for culturally competent healthcare providers. Knowledge of the American Psychological Association's (2014) guidelines for psychological practice with older adults along with its guidelines for Multicultural Competency in Geropsychology (APA, 2009) and consideration of the impact of specific psychosocial issues in the treatment of Latino older adults such as language needs, acculturative stress, and *familismo* are essential to effective treatment. In addition, familiarity with the assessments for depression and dementia described in this chapter will help reduce cultural bias, misdiagnosis, and ineffective treatment of older Latino adults. Finally, continuing to develop interventions for caregivers that are culturally appropriate and that focus on increasing social support, reducing stress, and treating health conditions is critically important.

References

Aguilar-Gaxiola, S. A., Kramer, E. J., Resendez, C., & Magaña, C. G. (2008). *Depression in Latinos: Assessment, treatment, and prevention*. New York, NY: Springer.

American Psychological Association. (2009). *Multicultural competency in Geropsychology*. Washington, D.C.: American Psychological Association.

American Psychological Association. (2014). Guidelines for psychological practice with older adults. *American Psychologist, 69*(1), 34–65. http://dx.doi.org.ezproxy.lib.uwm.edu/10.1037/a0035063

Andresen, E. M., Malmgren, J. A., Carter, W. B., & Patrick, D. L. (1994). Screening for depression in well older adults: Evaluation of a short form of the CES-D (Center for Epidemiologic Studies Depression Scale). *American Journal of Preventive Medicine, 10*, 77–84. https://ezproxy.lib.uwm.edu/login?url=https://search-proquest-com.ezproxy.lib.uwm.edu/docview/618509615?accountid=15078

Angel, J. L., Kahlert, R. C., & Whitfield, K. E. (2006). *Hispanic health and aging in a new Century*. Second Conference on Aging in the Americas (SCAIA), Lyndon B. Johnson School of Public Affairs, University of Texas at Austin, National Alliance for Hispanic Health, Washington DC.

Beck, A. T., Steer, R. A., & Brown, G. K. (1996). *Manual for the revised Beck depression inventory*. San Antonio, TX: Psychological Corporation.

Beck, A. T., Ward, C. H., Mendelson, M., Mock, J., & Erbaugh, J. (1961). An inventory for measuring depression. *Archives of General Psychiatry, 4*, 561–571. http://dx.doi.org.ezproxy.lib.uwm.edu/10.1001/archpsyc.1961.01710120031004

Berry, J. W. (2003). Conceptual approaches to understanding acculturation. In K. M. Chun, P. B. Organista, & G. Marín (Eds.), *Acculturation: Advances in theory, measurement, and applied research* (pp. 17–38). Washington, DC: American Psychological Association.

Berry, J. W., & Annis, R. C. (1974). Acculturative stress: The role of ecology, culture and differentiation. *Journal of Cross-Cultural Psychology, 5*(4), 382–406. http://dx.doi.org.ezproxy.lib.uwm.edu/10.1177/002202217400500402

Bowen, M. E., & Ruch, A. (2015). Depressive symptoms and disability risk among older white and Latino adults by nativity status. *Journal of Aging and Health, 27*, 1286–1305. http://dx.doi.org.ezproxy.lib.uwm.edu/10.1177/0898264315580121

Canino, G., & Alegría, M. (2009). Understanding psychopathology among the adult and child Latino population from the United States and Puerto Rico. Handbook of US Latino psychology: Developmental and community-based perspectives, 31.

Carson, A. P., Howard, G., Burke, G. L., Shea, S., Levitan, E. B., & Munter, P. (2011). Ethnic differences in hypertension incidence among middle-ages and older U.S. adults: The Multi-ethnic study of atherosclerosis. *Hypertension, 57*, 1100–1107. https://doi.org/10.1161/HYPERTENSIONAHA.110.168005

Chavez-Korell, S., Benson-Florez, G., Rendon, D. A., & Farías, R. (2014). Examining the relationship between physical functioning, ethnic identity, acculturation, *familismo,* and depressive symptoms for Latino older adults. *The Counseling Psychologist, 42*, 255–277. http://dx.doi.org.ezproxy.lib.uwm.edu/10.1177/0011000013477906

Chin, A. L., Negash, S., & Hamilton, R. (2011). Diversity and disparity in dementia: The impact of ethnoracial differences in Alzheimer disease. *Alzheimer Disease and Associated Disorders, 25*, 187–195. http://dx.doi.org.ezproxy.lib.uwm.edu/10.1097/WAD.0b013e318211c6c9

Coolidge, F. L. (2004). *Coolidge Axis II inventory manual*. Colorado Springs, CO: Author.

Coolidge, F. L., & Merwin, M. M. (1992). Reliability and validity of the Coolidge Axis II inventory: A new inventory for the assessment of personality disorders. *Journal of Personality Assessment, 59*, 223–238. http://dx.doi.org.ezproxy.lib.uwm.edu/10.1207/s15327752jpa5902_1

Diez-Quevado, C., Rangil, T., Sanchez-Planell, L., Kroenke, K., & Spitzer, R. L. (2001). Validation and utility of the patient health questionnaire in diagnosing mental disorders. In 1003 general hospital Spanish inpatients. *Psychosomatic Medicine, 63*, 679–686. http://dx.doi.org.ezproxy.lib.uwm.edu/10.1097/00006842-200107000-00021

Donlan, W., & Lee, J. (2010). Screening for depression among indigenous Mexican migrant Farmworkers using the patient health questionnaire-9. *Psychological Reports, 106*, 419–432. http://dx.doi.org.ezproxy.lib.uwm.edu/10.2466/PR0.106.2.419-432

Early, D. R., Widaman, K. F., Harvey, D., Beckett, L., Park, L. Q., Farias, S. T., … Mungas, D. (2013). Demographic predictors of cognitive change in ethnically diverse older persons. *Psychology and Aging, 28*(3), 633–645. http://dx.doi.org.ezproxy.lib.uwm.edu/10.1037/a0031645

Familiar, I., Ortiz-Panozo, E., Hall, B., Vieitez, I., Romieu, I., Lopez-Ridaura, R., & Lajous, M. (2015). Factor structure of the spanish version of the patient health questionnaire-9 in Mexican women. *International Journal of Methods in Psychiatric Research, 24*(1), 74–82. http://dx.doi.org.ezproxy.lib.uwm.edu/10.1002/mpr.1461

Farias, S. T., Mungas, D., Hinton, L., & Haan, M. (2011). Demographic, neuropsychological, and functional predictors of rate of longitudinal cognitive decline in hispanic older adults. *The American Journal of Geriatric Psychiatry, 19*(5), 440–450. http://dx.doi.org.ezproxy.lib.uwm.edu/10.1097/JGP.0b013e3181e9b9a5

Fillenbaum, G. G., Kuchibhatla, M., Henderson, V. W., Clark, C. M., & Taussig, I. M. (2007). Comparison of performance on the CERAD neuropsychological battery of hispanic patients and cognitively normal controls at two sites. *Clinical Gerontologist: The Journal of Aging and Mental Health, 30*(3), 1–22. http://dx.doi.org.ezproxy.lib.uwm.edu/10.1300/J018v30n03_01

Fiske, A., Wetherell, J. L., & Gatz, M. (2009). Depression in older adults. *Annual Review of Clinical Psychology, 5*, 363–389. https://doi.org/10.1146/annurev.clinpsy.032408.153621

Gallagher-Thompson, D., Solano, N., Coon, D., & Areán, P. (2003). Recruitment and retention of Latino dementia family caregivers in intervention research: Issues to face, lessons to learn. *The Gerontologist, 43*(1), 45–51. http://dx.doi.org.ezproxy.lib.uwm.edu/10.1093/geront/43.1.45

Gallagher-Thompson, D., Tzuang, M., Hinton, L., Alvarez, P., Rengifo, J., Valverde, I., … Thompson, L. W. (2015). Effectiveness of a fotonovela for reducing depression and stress in Latino dementia family caregivers. *Alzheimer Disease and Associated Disorders, 29*(2), 146–153. https://ezproxy.lib.uwm.edu/login?url=https://search-proquest-com.ezproxy.lib.uwm.edu/docview/1689318933?accountid=15078

Gatewood-Colwell, G., Kaczmarek, M., & Ames, M. H. (1989). Reliability and validity of the Beck depression inventory for a White and Mexican-American grontic population. *Psychological Reports, 65*, 1163–1166. http://dx.doi.org.ezproxy.lib.uwm.edu/10.2466/pr0.1989.65.3f.1163

González, H. M., Haan, M. H., & Hinton, L. (2001). Acculturation and the prevalence of depression in older Mexican Americans: Baseline results of the Sacramento area Latino study on aging. *Journal of American Geriatric Society, 49*, 948–953. http://dx.doi.org.ezproxy.lib.uwm.edu/10.1046/j.1532-5415.2001.49186.x

González, P., Nuñez, A., Merz, E., Brintz, C., Weitzman, O., Navas, E. L., … Gallo, L. C. (2016). Measurement properties of the Center for Epidemiologic Studies Depression Scale (CES-D): Findings from HCHS/SOL. *Psychological Assessment*. Advance online publication. https://doi.org/10.1037/pas0000330

Hansen, M. C., & Cabassa, L. J. (2012). Pathways to depression care: Help-seeking experiences of low income Latinos with diabetes and depression. *Journal of Immigrant Minority Health, 14*, 1097–1106. http://dx.doi.org.ezproxy.lib.uwm.edu/10.1007/s10903-012-9590-x

Herrera, A. P., Lee, J. W., Nanyonjo, R. D., Laufman, L. E., & Torres-Vigil, I. (2009). Religious coping and caregiver well-being in Mexican-American families. *Aging and Mental Health, 1*, 84–91. http://dx.doi.org.ezproxy.lib.uwm.edu/10.1080/13607860802154507

Hinton, L., Haan, M., Geller, S., & Mungas, D. (2003). Neuropsychiatric symptoms in Latino elders with dementia or cognitive impairment without dementia and factors that modify their association with caregiver depression. *Gerontologist, 43*, 669–677. http://dx.doi.org.ezproxy.lib.uwm.edu/10.1093/geront/43.5.669

Jimenez, D. E., Cook, B., Bartels, S. J., & Alegría, M. (2013). Disparities in mental health service use of racial and ethnic minority elderly adults. *Journal of the American Geriatrics Society, 61*, 18–25. http://dx.doi.org.ezproxy.lib.uwm.edu/10.1111/jgs.12063

Kaiser Family Institute. (2013). Health coverage for the Hispanic population today and under the Affordable Care Act. Washington, DC. https://kaiserfamilyfoundation.files.wordpress.com/2013/04/84321.pdf

Katon, W., Lyles, C. R., Parker, M. M., Karter, A. J., Huang, E. S., & Whitmer, R. A. (2012). Association of depression with increased risk of dementia in patients with type 2 diabetes: The diabetes and aging study. *Archives of General Psychiatry, 69*(4), 410–417. http://dx.doi.org.ezproxy.lib.uwm.edu/10.1001/archgenpsychiatry.2011.154

Koerner, S. S., Shirai, Y., & Pedroza, R. (2013). Role of religious/spiritual beliefs and practices among latino family caregivers of mexican descent. *Journal of Latina/o Psychology, 1*(2), 95–111. http://dx.doi.org.ezproxy.lib.uwm.edu/10.1037/a0032438

Kok, R. M., & Reynolds, C. F. (2017). Management of depression in older adults: A review. *JAMA: Journal of the American Medical Association, 317*(20), 2114–2122. http://dx.doi.org.ezproxy.lib.uwm.edu/10.1001/jama.2017.5706

Kørner, A., Lauritzen, L., Abelskov, K., Gulmann, N., Brodersen, A. M., Wedervang, T., & Kjeldgaard, K. M. (2006). The geriatric depression scale and the Cornell scale for depression in dementia. A validity study. *Nordic Journal of Psychiatry, 60*(5), 360–364. http://dx.doi.org.ezproxy.lib.uwm.edu/10.1080/08039480600937066

Kroenke, K., & Spitzer, R. L. (2002). The PHQ-9: A new depression diagnostic and severity measure. *Psychiatric Annals, 32*(9), 509–515. http://dx.doi.org.ezproxy.lib.uwm.edu/10.3928/0048-5713-20020901-06

Kroenke, K., Spitzer, R. L., & Williams, J. B. W. (2001). The PHQ-9: Validity of a brief depression severity measure. *Journal of Internal Medicine, 16*, 606–613. http://dx.doi.org.ezproxy.lib.uwm.edu/10.1046/j.1525-1497.2001.016009606.x

Lewis-Fernández, L., Das, A. K., Alfonso, C., Weissman, M. M., & Olfson, M. (2005). Depression in U.S. Hispanics: Diagnostic and management considerations in family practice. *Journal of the American Board of Family Medicine, 18*, 282–296. https://doi.org/10.3122/jabfm.18.4.282

Liang, J., Xu, X., Quiñones, A. R., Bennet, J. M., & Ye, W. (2011). Multiple trajectories of depressive symptoms in middle and late life: Racial/ethnic variations. *Psychology and Aging, 26*, 761–777. http://dx.doi.org.ezproxy.lib.uwm.edu/10.1037/a0023945

Lopez, M. N., Quan, N. M., & Carvajal, P. M. (2010). A psychometric study of the geriatric depression scale. *European Journal of Psychological Assessment, 26*(1), 55–60. http://dx.doi.org.ezproxy.lib.uwm.edu/10.1027/1015-5759/a000008

Losada, A., Márquez-González, M. M., Romero-Moreno, R., Mausbach, B. T., López, J., Fernández-Fernández, V., & Nogales-González, C. (2015). Cognitive-behavioral therapy (CBT) versus Acceptance and Commitment Therapy (ACT) for dementia family caregivers with significant depressive symptoms: Results of a randomized clinical trial. *Journal of Clinical and Consulting psychology, 83*, 760–772. http://dx.doi.org.ezproxy.lib.uwm.edu/10.1037/ccp0000028

Lucas-Carrasco, R. (2012). Spanish version of the geriatric depression scale: Reliability and validity in persons with mild dementia. *International Psychogeriatrics, 24*, 1284–1290. http://dx.doi.org.ezproxy.lib.uwm.edu/10.1017/S1041610212000336

Marín, G. (1992). Issues in the measurement of acculturation among Hispanics. In K. F. Geisinger (Ed.), *Psychological testing of Hispanics* (pp. 235–251). Washington, DC: American Psychological Association.

Marín, G. (1993). Influence of acculturation on familialism and self-identification among Hispanics. In M. E. Bernal & G. P. Knight (Eds.), *Ethnic identity: formations and transmission among Hispanics and other minorities* (pp. 181–196). Albany, NY: State University of New York Press.

Morris, J. C., Heyman, A., Mohs, R. C., Hughes, J. P., van Belle, G., Fillenbaum, G., ... Clark, C. (1989). The consortium to establish a registry for Alzheimer's disease (CERAD). Part 1. Clinical and neuropsychological assessment of Alzheimer's disease. *Neurology, 39*, 1159–1165.

Mungas, D., Reed, B. R., Crane, P. K., Haan, M. N., & González, H. (2004). Spanish and English neuropsychological assessment scales (SENAS): Further development and psychometric characteristics. *Psychological Assessment, 16*(4), 347–359. http://dx.doi.org.ezproxy.lib.uwm.edu/10.1037/1040-3590.16.4.347

Mungas, D., Reed, B. R., Marshall, S. C., & González, H. M. (2000). Development of psychometrically matched English and Spanish language neuropsychological tests for older persons. *Neuropsychology, 14*(2), 209–223. http://dx.doi.org.ezproxy.lib.uwm.edu/10.1037/0894-4105.14.2.209

National Hispanic Council on Aging. (2012). *State of Hispanic older adults: An analysis and highlights from the field.* Washington, DC. http://www.nhcoa.org/wp-content/uploads/2012/10/State-of-Hispanic-Older-Adults-Brief-2012-.pdf

National Institute of Mental Health. (2016). *Mental health medications*. U.S. Department of Health and Human Services. http://www.nimh.nih.gov/health/topics/mental-health-medications/index.shtml

Noguera, A., Centeno, C., Carvajal, A., Tejedor, M. A. P., Urdiroz, J., & Martinez, M. (2009). Are you discouraged? Are you anxious, nervous, or uneasy? In Spanish some words could be better than others for depression and anxiety screening. *Journal of Palliative Medicine, 12*, 707–712. https://doi.org/10.1089/jpm.2009.0024

Novy, D. M., Stanley, M. A., Averill, P., & Daza, P. (2001). Psychometric comparability of English- and Spanish-language measures of anxiety and related affective symptoms. *Psychological Assessment, 13*(3), 347–355. http://dx.doi.org.ezproxy.lib.uwm.edu/10.1037/1040-3590.13.3.347

Ortiz, F., & Fitten, J. (2000). Barriers to healthcare access for cognitively impaired older Hispanics. *Alzheimer Disease and Associated Disorder, 14*, 141–150. http://dx.doi.org.ezproxy.lib.uwm.edu/10.1097/00002093-200007000-00005

Ortiz, F., Fitten, J., Cummings, J., Hwang, S., & Fonseca, M. (2006). Neuropsychiatric and behavioral symptoms in a community sample of Hispanics with Alzheimer's disease. *American Journal of Alzheimer's Disease*

& Other Dementias, 21, 263–273. http://dx.doi.org.ezproxy.lib.uwm.edu/10.1177/1533317506289350

Ortman, J. M., Velkoff, V. A., & Hogan, H. (2014). An aging nation: The older population in the United States. Current Population reports, P25–1140. Washington, DC: U.S. Census Bureau. http://www.census.gov/content/dam/Census/library/publications/2014/demo/p25-1140.pdf

Radloff, L. S. (1977). The CES-D scale: A self-report depression scale for research in the general population. *Applied Psychological Measurement, 1*, 385–401. http://dx.doi.org.proxy3.noblenet.org/10.1177/014662167700100306

Rapp, M. A., Schnaider-Beeri, M., Wysocki, M., Guerrero-Berroa, E., Grossman, H. T., Heinz, A., & Haroutunian, V. (2011). Cognitive decline in patients with dementia as a function of depression. *The American Journal of Geriatric Psychiatry, 19*(4), 357–363. http://dx.doi.org.ezproxy.lib.uwm.edu/10.1097/JGP.0b013e3181e898d0

Reuland, D. S., Cherrington, A., Watkins, G. S., Bradford, D. W., Blanco, R. A., & Gaynes, B. N. (2009). Diagnostic accuracy of spanish language depression-screening instruments. *Annals of Family Medicine, 7*(5), 455–462. http://dx.doi.org.ezproxy.lib.uwm.edu/10.1370/afm.981

Robison, J., Gruman, C., Gaztambide, S., & Blanks, K. (2002). Screening for depression in middle-aged and older Puerto Rican primary care patients. *Journal of Gerontology: Medical Sciences, 57*(M), 308–M314. https://doi.org/10.1093/gerona/57.5.M308

Santiago-Rivera, A. L., Benson-Flórez, G., Santos, M. M., & Lopez, M. (2015). Latinos and depression: Measurement issues and assessment. In K. F. Geisinger (Ed.), *Psychological testing of Hispanics* (2nd ed., pp. 255–271). Washington, DC: American Psychological Association.

Segal, D. L., Coolidge, F. L., Cahill, B. S., & O'Riley, A. A. (2008). Psychometric properties of the beck depression inventory-II (BDI-II) among community-dwelling older adults. *Behavior Modification, 32*(1), 3–20. http://dx.doi.org.ezproxy.lib.uwm.edu/10.1177/0145445507303833

Shirey, L., & Summer, L. (2003). *Older hispanic adults*. Washington, DC: Georgetown University Health Policy Institute. https://hpi.georgetown.edu/agingsociety/pubhtml/hispanics/hispanics.html

Siedlecki, K. L., Manly, J. J., Brickman, A. M., Schupf, N., Tang, M. X., & Stern, Y. (2010). Do neurological tests have the same meaning in Spanish speakers as they do un English speakers? *Neuropsychology, 24*, 402–411. https://doi.org/10.1037/a0017515

Simões, D. C., Lunet, N., Ginó, S., Chester, C., Freitas, V., Maruta, C., … Mendonça, A. D. (2016). Depression with melancholic features is associated with higher long-term risk for dementia. *Journal of Affective Disorders, 202*, 220–229. http://dx.doi.org.ezproxy.lib.uwm.edu/10.1016/j.jad.2016.05.026

Sloan, F. A., & Wang, J. (2005). Disparities among older adults in measures of cognitive function by race or ethnicity. *The Journals of Gerontology: Series B: Psychological Sciences and Social Sciences, 60*(5), P242–P250. http://dx.doi.org.ezproxy.lib.uwm.edu/10.1093/geronb/60.5.P242

Snowden, M. B., Atkins, D. C., Steinman, L. E., Bell, J. F., Bryant, L. L., Copeland, C., & Fitzpatrick, A. L. (2015). Longitudinal association of dementia and depression. *American Journal of Geriatric Psychiatry, 23*, 897–905. http://dx.doi.org.ezproxy.lib.uwm.edu/10.1016/j.jagp.2014.09.002

Sorrell, J. M. (2012). The patient protection and affordable care act: What does it mean for mental health services for older adults? *Journal of Psychosocial Nursing and Mental Health Services, 50*(11), 14–18. http://dx.doi.org.ezproxy.lib.uwm.edu/10.3928/02793695-20121003-04

Spitzer, R. L., Kroenke, K., & Williams, J. B. W. (1999). Validation and utility of a self-report versión of PRIME-MD: The PHQ Primary Care Study. *JAMA, 282*, 1737–1744. http://dx.doi.org.ezproxy.lib.uwm.edu/10.1001/jama.282.18.1737

Szapocznik, J., Rio, A., Perez-Vidal, A., Kurtines, W., Hervis, O., & Santisteban, D. (1986). Bicultural effectiveness training (BET): An experimental test of an intervention modality for families experiencing intergenerational/intercultural conflict. *Hispanic Journal of Behavioral Sciences, 8*(4), 303–330. http://dx.doi.org.ezproxy.lib.uwm.edu/10.1177/07399863860084001

Trujillo, M. A., Perrin, P. B., Panyavin, I., Peralta, S. V., Stolfi, M. E., Morelli, E., & Arango-Lasprilla, J. C. (2016). Mediation of family dynamics, personal strengths, and mental health in dementia caregivers. *Journal of Latina/o Psychology, 4*, 1–17. https://doi.org/10.1037/lat0000046

Turner, R., Hinton, L., Gallagher-Thompson, D., Tzuang, M., Tran, C., & Valle, R. (2015). Using an emic lens o understand how latino families cope with dementia behavioral problems. *American Journal of Alzheimer's Disease, 30*, 454–462. https://doi.org/10.1177/1533317514566115

U.S. Department of Human Health and Services (HHS). (2017). Older Adults: Health People 2020. Retrieved April 27, 2017 from: https://www.hhs.gov/

Valle, R., Garrett, M. D., & Velaquez, R. (2013). Developing dementia prevalence rates among Latinos: A local-attuned, data-based, service planning tool. *Population Ageing, 6*, 211–225. https://doi.org/10.1007/s12062-013-9084-1

Weisman, A., Feldman, G., Gruman, C., Rosenberg, R., Chamorro, R., & Belozersky, I. (2005). Improving mental health services for Latino and Asian immigrant elders. *Professional Psychology: Research and Practice, 36*(6), 642–648. http://dx.doi.org.ezproxy.lib.uwm.edu/10.1037/0735-7028.36.6.642

Weissberger, G. H., Salmon, D. P., Bondi, M. W., & Gollan, T. H. (2013). Which neuropsychological tests predict progression to Alzheimer's disease in Hispanics? *Neuropsychology, 27*(3), 343–355. http://dx.doi.org.ezproxy.lib.uwm.edu/10.1037/a0032399

Wiebe, J. S., & Penley, J. A. (2005). A psychometric comparison of the beck depression inventory-II in English and Spanish. *Psychological Assessment, 17*(4), 481–485. http://dx.doi.org.ezproxy.lib.uwm.edu/10.1037/1040-3590.17.4.481

Xiong, G. L., Filshtein, T., Beckett, L. A., & Hinton, L. (2015). Antipsychotic use in a diverse population with dementia: A retrospective review of the National Alzheimer's Coordinating Center Database. *The Journal of Neuropsychiatry and Clinical Neurosciences, 27*, 326–332. https://doi.org/10.1176/appi.neuropsych.15010020

Yesavage, J. A., Brink, T. L., Rose, T. L., Lum, O., Huang, V., Adey, M., & Leirer, V. O. (1983). Development and validation of a geriatric depression screening scale: A preliminary report. *Journal of Psychiatric Research, 17*(1), 37–49. http://dx.doi.org.ezproxy.lib.uwm.edu/10.1016/0022-3956(82)90033-4

Zahodne, L. B., Stern, Y., & Manly, J. J. (2014). Depression symptoms precede memory decline, but not vice versa, in non-demented adults. *Journal of American Geriatric Society, 62*, 130–134. https://doi.org/10.1111/jgs.12600

Zukerman, I. H., Ryder, P. T., Simoni-Wastila, L., Shaffer, T., Sato, M., Zhao, L., & Stuart, B. (2008). Racial and ethnic disparities in the treatment of dementia among medicare beneficiaries. *Journal of Gerontology: Social Sciences, 63B*, S328–S333.

7 Neurological, Psychiatric, and Affective Aspects of Dementia in Latinxs

Lee Ashendorf and Julie E. Horwitz

Abstract

This chapter reviews common neuropsychiatric symptoms, their expression among Latinx individuals with dementia, and their effect on caregiver burden in this population. As the understanding of dementia syndromes has progressed, increasing attention has been devoted to the neurological, psychiatric, and affective symptoms that commonly accompany neurodegeneration. Most large-scale studies have shown that at least one symptom or syndrome can be identified in almost all individuals diagnosed with dementia. These symptoms impose significant burden on caregivers and on the healthcare system.

Introduction

As medicine and healthcare continue to advance in the United States and worldwide, the proportion of the population classified as elderly (i.e., age 65 and older) is rapidly increasing, with a concurrent increase in individuals diagnosed with dementia. Furthermore, within the United States, Latinxs[1] continue to account for more of the nation's overall population growth than any other ethnic group (Krogstad, 2017), and the Latinx population is projected to comprise 18% of the older population by 2050 (www.aoa.gov/prof/Statistics/statistics.aspx). Although there remains a lack of adequate research on presentation, diagnosis, and treatment of dementia within the Latinx population—an endeavor which is complicated in part by the considerable heterogeneity of this population—the current chapter seeks to provide an overview of the nature of neuropsychiatric symptom presentation in individuals with dementia, with a specific focus on Latinx individuals. This discussion will particularly emphasize those symptoms which most negatively impact caregivers and will subsequently turn to how these symptoms can be best managed by treatment providers and caregivers.

L. Ashendorf (✉)
University of Massachusetts Medical School;
VA Central Western Massachusetts, Worcester, MA, USA
e-mail: Lee.Ashendorf@va.gov

J. E. Horwitz
Memorial Hospital, University of Colorado, Colorado Springs, CO, USA
e-mail: julie.horwitz@uchealth.org

[1] The term Latinx is used throughout the chapter to include the broad range of gender identities present among people of Latin American descent (see Santos, 2017).

H. Y. Adames, Y. N. Tazeau (eds.), *Caring for Latinxs with Dementia in a Globalized World*,
https://doi.org/10.1007/978-1-0716-0132-7_7

Neuropsychiatric Symptom Presentation in Dementia

Although changes in cognitive function—which are the hallmark of dementia syndromes—are typically the primary focus of concern among individuals with neurodegenerative conditions, the accompanying neuropsychiatric symptoms also significantly affect patient quality of life and are often the most distressing feature for families and caregivers. Most individuals with dementia or mild cognitive impairment (MCI) exhibit these behavioral and psychological symptoms. While many of these symptoms are not exclusive to dementia and occur in the general population, they are seen more frequently in individuals with dementia (Savva et al., 2009). Some cross-sectional prevalence estimates have approximated the occurrence of neuropsychiatric symptoms at 80% in dementia and 50% in MCI (Lyketsos et al., 2002; Rozzini et al., 2008). The Cache County Study found the 5-year point prevalence of at least one neuropsychiatric symptom (i.e., the proportion of individuals with dementia who exhibit at least one symptom during the 5 years of that study) to be a staggering 97% (Steinberg et al., 2008). The MAASBED Study found a similar 2-year point prevalence of 95% (Aalten, de Vugt, Jaspers, Jolles, & Verhey, 2005). These studies reflect the consistent finding in the literature that virtually all individuals with dementia will at some point experience or exhibit neuropsychiatric symptoms.

In the limited literature that focuses on neuropsychiatric symptoms among Latinxs, most studies appear to have found increased rates of neuropsychiatric symptoms overall in Latinxs as compared with non-Latinxs (Chen, Borson, & Scanlan, 2000; Hinton, Haan, Geller, & Mungas, 2003; Ortiz, Fitten, Cummings, Hwang, & Fonseca, 2006; Sink, Covinsky, Newcomer, & Yaffe, 2004). Sink et al. (2004) found the overall prevalence of dementia-related behaviors (i.e., overt behaviors such as agitation or psychotic symptoms, but not including depression or other internalized symptoms) to be higher among Latinxs. In their multiethnic sample, 46% of White participants exhibited at least four such behaviors, while 61% of Black and 57% of Latinx participants also did. One study indicated that among individuals with dementia, Latinxs are less likely than Whites to experience depression or apathy (Hargrave, Stoeklin, Haan, & Reed, 2000), but this finding appears to be in the minority.

In community samples without cognitive impairment, a wide variety of predictors of neuropsychiatric symptoms has been identified; however, many of these predictors do not seem to generalize to the dementia population. Even within a given dementia etiology, symptom presentation often differs between individuals. Although neurodegeneration itself can sometimes be the causative factor, this is not always the case, and the symptoms may or may not correspond to the neuroanatomical or neurochemical changes associated with the dementia syndrome. There are also likely contributions from baseline personality characteristics (von Gunten, Pocnet, & Rossier, 2009); for example, a systematic review of 18 studies suggested a positive relationship between premorbid neuroticism and overall challenging behavior, particularly aggression (Osborne, Simpson, & Stokes, 2010).

To account for the elevated incidence of neuropsychiatric symptoms among individuals with dementia, investigators have developed a broad range of specific explanatory models or more complex biopsychosocial models. For example, the Progressively Lowered Stress Threshold model (Smith, Gerdner, Hall, & Buckwalter, 2004) states that the progression of dementia reduces the individual's tolerance to stress and makes traditional environmental triggers more stressful and threatening. Within this model, reducing or ameliorating the triggering stimuli will help prevent or reduce the behavioral symptoms that otherwise result from them. More complex models, such as that summarized by Kales and colleagues (2015), posit that the neurodegenerative changes alter the way the individual interacts with his or her environment (including caregivers, social circles, and other environmental stressors) and also increase their vulnerability to stress in the context of these relationships. This heightened vulnerability to environmental stress, in turn, leads to the expression of neuropsychiatric symptoms.

As suggested above, neuropsychiatric symptoms impose substantial burden on patients and their families. For example, the presence of such symptoms is associated with an increased risk of hospitalization and increased financial concerns (Ayalon, Gum, Feliciano, & Arean, 2006; Lyketsos et al., 1997; Yaffe et al., 2002). Studies also seem to confirm that in both White and Latinx samples, individuals who experience more neuropsychiatric dementia symptoms also experience more significant functional disability (Hinton, Farias, & Wegelin, 2008), with the strongest associations, between symptom reports and functional disability, being with psychotic symptoms, depression, anxiety, and agitation. Importantly, neuropsychiatric symptoms also account for much of the decreased quality of life experienced not only by the patient but by the family and/or caregiver(s) as well. Caregivers, in general (including those caring for individuals with ailments other than dementia), experience a poorer sense of well-being, lower self-efficacy, and more extreme feelings of depression and stress when dealing with a care recipient with prominent neuropsychiatric symptoms. Multiple studies have observed that caregivers for individuals with dementia who had higher neuropsychiatric symptom burden experienced poorer mental health themselves (e.g., Brodaty, Woodward, Boundy, Ames, & Balshaw, 2014; Conde-Sala et al., 2014). These factors all render it important to identify and treat these symptoms as early as possible. Additionally, from the standpoint of the provider, identification of behavioral symptoms can potentially improve the accuracy of diagnostic classification and in turn help identify proper treatment and recommendations.

Given the allocentric nature of the Latinx family and community, the implications from such explanatory and biopsychosocial models highlight the necessity for culturally responsive care for Latinxs with dementia and their caregivers. Neurodegenerative processes that increase vulnerability to stress within family and/or caregiver relationships should be given special consideration for the Latinx population, as Latinxs are more likely to rely on family members for caregiving, as opposed to using outside palliative, hospital, or psychiatric care, and often depend on extended family and social networks to coordinate health-related decision-making for their family members. Given the connection between heightened vulnerability to relational stress as a result of dementia symptoms and the subsequent expression of additional neuropsychiatric symptoms and worsening mental health of caregivers, it is important for psychiatrists, the medical community, and the mental health profession to recognize the specific impact relational stress can have on Latinxs with dementia, their caregivers, and their family members. For example, interpersonal conflicts can trigger a range of symptoms, expressed through cultural idioms of distress, in Latinx families. *Respeto*, a cultural strength of Latinx family systems, involves the reliance on authority within the family (determined by a hierarchical family structure in which elders and parents are often the decision-makers). For those unfamiliar with *respeto*, professionals may seem perplexed by the involvement of family members in decisions regarding all aspects of care, including when to hospitalize or when to include medication in treatment. Thus, it is best for practitioners to utilize *respeto* within the healthcare profession when working with Latinxs with dementia and professionals. In addition, the consideration of other Latinx cultural values, including *familismo* and *personalismo*, should ideally be incorporated into treatment to help reduce relational stress and develop respectful and successful therapeutic alliances with Latinx families and patients.

Assessment of Neuropsychiatric Symptoms in Dementia

Neuropsychiatric symptoms are often evaluated using structured interviews or questionnaires, such as the Neuropsychiatric Inventory (NPI; Cummings et al., 1994). The NPI uses information from a caregiver or other collateral informants to assess symptoms within 12 domains: depression, euphoria, anxiety, apathy, agitation, delusions, hallucinations, disinhibition, irritability, motor behaviors, nighttime behaviors, and appetite/eating changes. These tend to be the classes of neuropsychiatric symptoms that most commonly appear in the dementia literature as well. In a study comparing NPI scores in Latinx and White individuals with dementia, Ortiz et al. (2006) found that the total NPI score was more than 50% higher in Latinxs. The study also noted that the most prevalent symptoms reported in Latinxs with dementia were apathy and anxiety disorders, whereas among Whites, depression was the second most common. However, other studies investigating neuropsychiatric symptoms in patients with dementia have indicated that depression was most common among Latinxs (Hinton et al., 2003).

In light of the broad range of neuropsychiatric symptoms that occur in dementia, some investigators have attempted to cluster the symptoms into factors or groups. Some classify them as "positive," reflecting the presence of atypical behaviors such as aggression, or "negative," such as apathy, reflecting the absence of typical behaviors or personality characteristics. As summarized by van der Linde, Dening, Matthews, and Brayne (2014), the most common groupings in the research literature tend to be affective symptoms, psychosis, hyperactivity, and euphoria. Only those groupings and domains most salient to caregivers will be discussed here.

Depression

Depression is arguably the most extensively researched neuropsychiatric symptom of dementia as it is seen with some frequency across a wide variety of dementia etiologies, including Alzheimer's disease (AD), vascular dementia, Lewy body dementia (LBD), frontotemporal dementia (FTD), and Parkinson's disease dementia (PDD). Additionally, Lyketsos et al. (2002) identified depression as the most common neuropsychiatric symptom among individuals with MCI, as it occurred in 20% of their MCI sample. Comparatively, depression was somewhat more prevalent among individuals with dementia (32%). Sink et al. (2004) found the same rate among White individuals, but they noted an increased prevalence (46%) among Latinxs. Over a 5-year period, 77% of participants with dementia in one study exhibited at least some signs of depression (Steinberg et al., 2008).

Depression is a well-established risk factor for dementia. In a Framingham Heart Study cohort, not only did participants who were depressed at a baseline assessment have a 50% increased risk of developing dementia during the ensuing 17 years but the relationship was dose dependent: the risk of eventual dementia increased with severity of depression (Saczynski et al., 2010). When adjusting for medical comorbidities, the risk of developing AD or vascular dementia may not be increased for individuals with past history of depressive symptoms, but it was doubled (AD) or more than tripled (vascular dementia) for those who continued to experience depressive symptoms in late life (Barnes et al., 2012). Causative theories have included the contributions of neuroanatomical or neurochemical changes related to chronic depression, as well as the adverse neurological effect of limited or reduced physical and social activity that is often seen in individuals who experience depression. Interestingly, depression has been shown to manifest differently in the context of a dementia syndrome. For example, individuals with depression and dementia tend to be less decisive and more ruminative; however, they also tend to experience less sleep disturbance and less frequent feelings of worthlessness than those with depression but without dementia (Lyness et al., 1997; Zubenko et al., 2003).

Apathy

Apathy is commonly defined as a lack of emotion, an absence of interest, or amotivation. Clinically, it is often misclassified or misdiagnosed as depression or as a symptom of depression (Landes, Sperry, & Strauss, 2005). It is fairly easy for laypersons to confuse it with anhedonia experienced as a symptom of depression, but apathy has been demonstrated to be an independent neuropsychiatric symptom (Levy et al., 1998). In one study, only about one-quarter of individuals who met criteria for either apathy or dementia could be diagnosed with both syndromes (Starkstein, Ingram, Garau, & Mizrahi, 2005), and approximately half of individuals with dementia and apathy do not have depression (Tagariello, Girardi, & Amore, 2009). In addition, manifestations of depression and apathy tend to differ over the course of a dementia syndrome, with one study suggesting that whereas affective symptom reporting tends to dissipate with more advanced dementia, amotivation tends to increase (Spalletta et al., 2004).

Even when controlling for depression, apathy has been shown to be independently associated with poor cognitive test performance (Levy et al., 1998; Onyike et al., 2007; Starkstein et al., 2005) and may serve as an early marker that can predict cognitive decline (Onyike et al., 2007; Vicini Chilovi et al., 2009). One study found the presence of apathy to confer an approximately 50% increase in mortality risk for geriatric inpatients (Hölttä et al., 2012). In addition, apathy is quite prevalent even in earlier stages of dementia, occurring in 15–17% of individuals with MCI (Lyketsos et al., 2002; Onyike et al., 2007). It has also been found to be a risk factor for conversion from MCI to dementia, whereas individuals with depression may demonstrate a reduced likelihood of conversion (Vicini Chilovi et al., 2009).

Regarding prevalence, apathy has been found to be present in about 30% of community samples with AD, 60% of outpatients with AD, and 30–50% of people with other dementia syndromes (van Reekum, Stuss, & Ostrander, 2005). It was easily the most common neuropsychiatric symptom in the Medical Research Council Cognitive Function and Ageing Study, occurring in 50% of that study's dementia sample (Savva et al., 2009). Of the various dementia syndromes, there is a particularly strong relationship to depression in individuals with AD, whereas apathy in FTD renders the individual less likely to be depressed; apathy in such individuals is instead concurrent with impulsive and compulsive behaviors (Chow et al., 2009). Apathy has also been found to be highly correlated with agitation symptoms (ibid). Apathy is often rated as the most severe of all neuropsychiatric symptoms (Steinberg et al., 2008) and has been found to significantly impact caregiver burden (e.g., Terum et al., 2017).

Anxiety

Anxiety symptoms occur in roughly half of individuals with dementia diagnoses (McCurry, Gibbons, Logsdon, & Teri, 2004; Steinberg et al., 2008) and have been found to be the third most common neuropsychiatric symptom type across all dementia etiologies (e.g., Terum et al., 2017); anxiety symptoms are highly comorbid with depression. Anxiety has been found to be more common in vascular dementia than in AD (Ballard et al., 2000). One study found that among elderly Latinxs with dementia, anxiety was experienced by approximately 34% (Hinton et al., 2003). Although there is an overall lack of research investigating the nature of specific neuropsychiatric symptoms in Latinx individuals with dementia, the culture-bound syndrome *ataque de nervios*, which is a term often used among Latinxs of Caribbean descent to describe an array of psychological symptoms often associated with anxiety, suggests that anxiety may be perceived and experienced differently in Latinxs with dementia and their caregivers. The extent to which differences may be experienced may also vary with level of acculturation. Regardless of how anxiety may be manifested or perceived, the presence of patient anxiety has been shown to significantly impact caregiver burden (Terum et al., 2017).

Psychosis

Up to 50% of individuals with AD develop psychotic symptoms at some point during the disease process (Jeste & Finkel, 2000). Delirium can also cause psychosis and other neuropsychiatric symptoms in older adults (Hodgson, Gitlin,

Winter, & Czekanski, 2011). In general, it is important to evaluate whether psychotic symptoms are a characteristic of the dementia syndrome or instead a sign of an incipient delirium resulting from medication side effects, nutritional deficiency, environmental factors, and/or an underlying treatable medical condition. Psychotic symptoms (delusions and hallucinations) tend to be equally prevalent across ethnic backgrounds (Hargrave et al., 2000), although one study found Latinxs to be at higher risk for hallucinations, which occurred in 53% of Latinxs vs. 42% of White participants (Sink et al., 2004).

Hallucinations

Hallucinations may occur in as many as 15% of individuals diagnosed with dementia (Savva et al., 2009). Visual hallucinations are a hallmark symptom of LBD and also occur in PDD. These hallucinations are complex and most typically characterized by well-formed visions of anonymous people; less frequently, they comprise animals or disembodied/floating faces or heads (Mosimann et al., 2006). Dorsal and ventral visual pathways of the brain have been implicated, which suggests parietal and occipital lobe involvement (ibid). Unlike those hallucinations typically experienced by individuals with primary psychiatric disorders, hallucinations in LBD and PDD are less frequently experienced as frightening. However, hallucinations were found to be among the three neuropsychiatric symptoms most strongly correlated with caregiver burden among caregivers of Argentinian individuals with probable AD (Allegri et al., 2006). Hallucinations have been found to occur in 42% of White individuals with dementia, 57% of Black individuals with dementia, and 53% of Latinxs with dementia (Sink et al., 2004).

Delusions

Delusions are reasonably common in dementia, occurring in one-third of individuals with AD (Mizrahi, Starkstein, Jorge, & Robinson, 2006). In AD, these delusions tend to occur in response to anosognosia, or a lack of awareness of one's own deficits. For example, one may forget having relocated an item; and, when it cannot be found in the place he or she expects, the presumption is that someone has stolen it. This belief can become deeply ingrained, particularly if the individual also has significant perseverative tendencies. Delusions of grandiosity can also occur in the setting of expansive mood (Mizrahi et al., 2006). Misidentification delusions, which involve an enduring belief that a person, place, or object has been altered, can occur in AD but are more commonly seen in LBD, occurring in about half of patients with the latter disease (Ballard et al., 1999). One such delusion, Capgras syndrome, is the belief that a familiar person has been replaced with a duplicate. It is more commonly experienced in primary psychiatric disorders but does occur in dementia syndromes as well, particularly in LBD but also AD and vascular dementia (ibid). Reduplicative paramnesias are characterized by a belief that a familiar location, usually one's house, has been duplicated and exists in two distinct locations. Other misidentification delusions include the "phantom boarder" delusion, which entails a belief that a stranger is living in one's home, and Fregoli syndrome, which involves a belief that a persecuting individual is able to assume the identities of different people whom one encounters. Some individuals with LBD also experience "television misidentifications," or the belief that characters on television are instead actually present in the individual's home; these delusions have been associated with particularly poor visuoperceptual functioning (Mori et al., 2000). Misidentification syndromes have been associated with disconnection of the brain's temporal and limbic areas from the frontal lobes, particularly when lateralized in the right hemisphere, potentially resulting in a disrupted sense of familiarity of the recalled person or place (Alexander, Stuss, & Benson, 1979). Delusions in dementia have also been associated with functional changes in the parietal lobe (Starkstein, Vazquez, & Petracca, 1994).

Given the breadth of research related to paranoia and the way it may manifest in dementia patients, it is important for practitioners to consider possible psychogenic causes vs. true neuropsychiatric symptoms in the Latinx population when examining whether symptoms are pathological. For example, healthy cultural suspicion, experienced by people of color and specifically undocumented immigrants, may be misinterpreted as paranoia in Latinxs with dementia. Given aggressive immigration laws, undocumented Latinxs with dementia—or caregivers and/or family members with undocumented status—may avoid the medical system, appear noncompliant, or be preoccupied with the safety of their family members. It is recommended that professionals working with Latinxs remain mindful of the ways that the medical community can inadvertently place undocumented Latinxs in danger and that practitioners be able to distinguish legitimate fears immigrants may have when encountering the medical system.

Disinhibition

Behavioral disinhibition is considered one of the hallmark signs of behavioral variant FTD (bvFTD) and is, in many cases, among the first noticeable signs of this neurodegenerative condition. Although often not displayed until the moderate stages in other types of dementia, disinhibition may also occur in individuals with AD, vascular dementia, and other neurodegenerative conditions. One study found that among elderly Latinxs with dementia living in the United States, disinhibition was experienced by approximately 32% (Hinton et al., 2003). An Argentina-based study of individuals with probable AD and their caregivers found a significant correlation between disinhibition and caregiver burden (Allegri et al., 2006); an association of disinhibition and increased caregiver burden has similarly been found in studies from other countries, including Japan (Kazui et al., 2016) and Korea (Shim, Kang, Kim, & Kim, 2016). In a Hong Kong-based study, researchers found that among individuals with AD, disruptive behaviors, including disinhibition as well as agitation/aggression, irritability, and aberrant motor behaviors, were the strongest predictors of depression and experienced burden in caregivers (Cheng, Kwok, & Lam, 2012). These authors noted that the rate of disruptive behaviors reported by Asian caregivers was higher than those reported by caregivers in Western-based studies and surmised that this may be partially attributable to the emphasis on harmony and consequently reduced tolerance of such behaviors in Asian societies; this notion, which has also been used to describe differing rates of various childhood psychiatric conditions across cultures, suggests that differences may also exist not only between Latinx and majority Western populations but also within different Latinx groups.

Aggression/Agitation/Irritability

Agitation encompasses a range of disruptive behaviors that include yelling, attention-seeking behavior, and resisting assistance or intervention. Aggression can be verbal or physical. In one study, behavioral agitation was observed in 30% of individuals with dementia (Lyketsos et al., 2002). Furthermore, another study found agitation and wandering to be the most prevalent symptoms seen in Latinxs with dementia, with agitation occurring in 48% (similar to the 45% rate seen in White participants) and wandering occurring in 63% (greater than the 58% found in White individuals) of participants (Sink et al., 2004). Irritability has been observed in roughly 30% of all individuals with dementia, at a rate approximately double that seen in the general population (Savva et al., 2009), and has been noted as a core characteristic of the behavioral variant of FTD (Edwards-Lee et al., 1997). "Combativeness" was found in 37% of Latinxs, 30% of Black participants, and 24% of White participants with dementia (Sink et al., 2004). Although not always related, agitation or irritability can reflect depression when exhibited by individuals with dementia (Lyketsos et al., 1999). Aggression has been suggested to occur in individuals who have intact dopamine-releasing nuclei, but experience damage to serotonin-releasing nuclei in the brain (Victoroff, Zarow, Mack, Hsu, & Chui, 1996).

Sundowning is an increase in agitation as the day passes. This phenomenon, which is experienced by as many as two-thirds of individuals with dementia, has often been associated with a disturbance in the circadian rhythm that occurs in dementia (Volicer, Harper, Manning, Goldstein, & Satlin, 2001). The severity of this phenomenon is moderated by a variety of environmental, social, and cognitive factors. As one may expect, agitation and aggression often place a strain on the relationship with caregivers and are frequently the proximal precipitants of placement in an inpatient or nursing care setting (Yaffe et al., 2002).

Sleep Disturbances

Sleep disturbances are very common in dementia, which is unsurprising given the likelihood that neurodegeneration would affect regions of the brain that regulate daily rhythms such as sleep. In the Cardiovascular Health Study, while sleep disturbance was not the most *prevalent* neurobehavioral symptom among people with cognitive impairment, it was rated to be the most clinically *significant* symptom for individuals with MCI and the second most significant for individuals with dementia (Lyketsos et al., 2002). It has also, though, been noted to be comparably prevalent in the general population (Savva et al., 2009). Prospectively, sleep disturbance and sleep-disordered breathing increase the risk of dementia. Interestingly, there is a significant association between genetic risk for AD and obstructive sleep apnea (OSA; Gottlieb et al., 2004), which further buttresses the apparent association between OSA and dementia. Rapid eye movement (REM) sleep behavior disorder, or excessive motor activity while dreaming, has a strong relationship with LBD and PDD, which develop in about half of individuals initially diagnosed with an idiopathic REM sleep behavior disorder; the sleep behavior can predate the onset of dementia by as many as 10–12 years (Postuma et al., 2009). In AD, anxiety and agitation can lead to frequent nighttime awakenings and wandering behavior (McCurry et al., 2004). These nighttime behaviors, in turn, interfere with caregivers' daily functioning and increase caregiver burden which can accelerate nursing home placement (Teri et al., 1999).

The Impact of Neuropsychiatric Symptoms on Caregiver Burden in the Latinx Population

Hinton and colleagues (2003) evaluated the nature of the relationship between neuropsychiatric symptoms in Latinxs with dementia and caregiver depression. Greater overall neuropsychiatric symptom expression in that study was associated with greater caregiver depression; this association was particularly strong when affective symptoms and apathy were displayed, when the patient was younger and at an earlier dementia stage (i.e., MCI), and when the caregiver was a spouse. Latinx caregivers have been found to frequently attribute symptoms to a cause other than the dementia process (Hinton, Chambers, & Velásquez, 2009). A variety of attributions were endorsed in that study, but the most common non-dementia causes of neuropsychiatric symptoms were thought to be medical conditions and interpersonal factors. Depression, for example, was rated as stemming from the dementia by only 11% of caregivers of individuals who had dementia with depression.

Management of Neuropsychiatric Symptoms in Dementia and the Role of the Caregiver

Although no cure has yet been developed for any of the known neurodegenerative conditions, intervention to address associated neuropsychiatric symptoms can significantly improve patient quality of life and reduce the burden for caregivers (e.g., Ayalon et al., 2006). Although primary care physicians or other prescribing providers may prescribe any of a number of medications to help minimize neuropsychiatric symptoms, such can also be managed via behavioral and/or environmental interventions, many of which may be implemented by a caregiver directly. These interventions may be grouped into three categories, including *unmet needs interventions*, which are based on the idea that neuropsychiatric symptoms in dementia serve to express an unmet need; *learning and behavioral interventions*, which presume that neuropsychiatric symptoms represent inadvertently reinforced behaviors and aim to ameliorate these maladaptive patterns of reinforcement; and/or *environmental vulnerability and reduced stress-threshold interventions*, which assume that neuropsychiatric symptoms emerge as the result of an individual's inability to cope with aspects of his or her environment (Ayalon et al., 2006). Although a systematic review of such interventions suggests that there is insufficient research to strongly support any single treatment, the available research appears encouraging regarding the potential efficacy of several of these types of interventions (Ayalon et al., 2006). For example, one study examined in this review suggested that simply providing hearing aids was effective in reducing problem behaviors in a group of individuals with mild to moderate AD (Palmer, Adams, Bourgeois, Durrant & Rossi, 1999), and another found that bright light therapy reduced agitation in a group of individuals with moderate to severe dementia (Lovell, Ancoli-Israel, & Gevirtz, 1995). Interventions geared toward changing maladaptive reinforcement patterns through use of behavioral functional analysis, contingency management, and caregiver training also showed promise for being efficacious in reducing agitation and aggression, as well as wandering (e.g., Bakke et al., 1994; McCallion, Toseland, & Freeman, 1999).

In addition to interventions targeting the patient, which may assist in improving patient quality of life as well as reducing caregiver burden by minimizing behaviors experienced as problematic, interventions which aim to reduce caregiver burden by directly addressing caregiver-related factors may also hold promise. For example, Trapp et al. (2015) sought to investigate the role of resilience, optimism, and sense of coherence (i.e., the degree to which one believes she/he can understand, manage, and derive meaning from a challenging life event) on mental and physical health-related quality of life (HRQOL) among 130 caregivers of dementia patients in Mexico and Argentina. This study found that these strengths collectively accounted for 58.4% of the variance in caregiver mental HRQOL. The authors concluded that interventions geared toward building these strengths in caregivers may assist in reducing the sense of burden among caregivers.

By integrating culturally responsive, strengths-based treatments for both Latinx dementia patients and their caregivers, interpersonal stress, caregiver burden, and complex neuropsychiatric symptoms can be better managed. For example, involving Latinx families as an active part of the treatment process (as opposed to solely the primary caregiver) incorporates the cultural values of *familismo* and *respeto* in treatment. Throughout treatment, it is important that the professional gather the family's perspective on the patient's current neuropsychiatric symptoms and incorporate any additional or traditional Latinx values and customs (i.e., *espiritismo*, *personalismo*, and traditional Latinx folk healing methods) into the patient's intervention, based on each Latinx family's unique heritage, belief systems, and worldview. As was demonstrated in previously mentioned studies, simple therapies that enhance communication and bring comfort to the patient should be considered within a specific Latinx family framework. In addition, caregiver and family education about the patient's current behavioral symptoms may help Latinx families work together to implement behavioral, medical, and culturally congruent practices at home to ease relational distress.

Conclusion

Although cognitive changes tend to be the primary focus in discussions of neurodegenerative conditions, the extant literature suggests that the accompanying neuropsychiatric symptoms may exert the most significant impact on patients and caregivers. The interventions focused on addressing these symptoms may, therefore, result in the biggest improvements in quality of life. Despite this, research on the nature of and interventions for neuropsychiatric symptoms among Latinx individuals with dementia is lacking. As noted above, the perceptions of such symptoms among caregivers and the extent to which such symptoms may impact caregiver quality of life are likely to vary based on a number of factors, such as the region from where an individual or family originally hails, immigration status, religious and/or spiritual beliefs, and the degree to which the individual and family have acculturated to the majority culture. Therefore, studies which include assessment of these and other related factors are likely to be of most utility in gathering an understanding of how dementia and associated symptoms are manifested, explained, and experienced by individuals of different cultures; in turn, this understanding is critical for informing development of effective treatments. Particularly given the increasing numbers of older Latinx individuals, including those with dementia, both within the United States and worldwide, this area of research is becoming increasingly important and relevant in ensuring proper assessment and treatment for the proliferating diverse patients entrusting health professionals with their care.

References

Aalten, P., De Vugt, M. E., Jaspers, N., Jolles, J., & Verhey, F. R. (2005). The course of neuropsychiatric symptoms in dementia. Part I: Findings from the two-year longitudinal Maasbed study. *International Journal of Geriatric Psychiatry, 20*, 523–530.

Alexander, M. P., Stuss, D. T., & Benson, D. F. (1979). Capgras syndrome: A reduplicative phenomenon. *Neurology, 29*, 334–339.

Allegri, R. F., Sarasola, D., Serrano, C. M., Taragano, F. E., Arizaga, R. L., Butman, J., & Loñ, L. (2006). Neuropsychiatric symptoms as a predictor of caregiver burden in Alzheimer's disease. *Neuropsychiatric Disease and Treatment, 2*, 105–110.

Ayalon, L., Gum, A. M., Feliciano, L., & Arean, P. A. (2006). Effectiveness of nonpharmacologic interventions for the management of neuropsychiatric symptoms in patients with dementia: A systematic review. *Archives of Internal Medicine, 166*, 2182–2188.

Bakke, B. L., Kvale, S., Burns, T., McCarten, J. R., Wilson, L., Maddox, M., & Cleary, J. (1994). Multicomponent intervention for agitated behavior in a person with Alzheimer's disease. *Journal of Applied Behavior Analysis, 27*, 175–176.

Ballard, C., Holmes, C., McKeith, I., Neill, D., O'Brien, J., Cairns, N., … Perry, R. (1999). Psychiatric morbidity in dementia with Lewy bodies: A prospective clinical and neuropathological comparative study with Alzheimer's disease. *American Journal of Psychiatry, 156*, 1039–1045.

Ballard, C., Neill, D., O'Brien, J., McKeith, I. G., Ince, P., & Perry, R. (2000). Anxiety, depression and psychosis in vascular dementia: Prevalence and associations. *Journal of Affective Disorders, 59*, 97–106.

Barnes, D. E., Yaffe, K., Byers, A. L., McCormick, M., Schaefer, C., & Whitmer, R. A. (2012). Midlife vs late-life depressive symptoms and risk of dementia: Differential effects for Alzheimer disease and vascular dementia. *Archives of General Psychiatry, 69*, 493–498.

Brodaty, H., Woodward, M., Boundy, K., Ames, D., & Balshaw, R. (2014). Prevalence and predictors of burden in caregivers of people with dementia. *The American Journal of Geriatric Psychiatry, 22*, 756–765.

Chen, J. C., Borson, S., & Scanlan, J. M. (2000). Stage-specific prevalence of behavioral symptoms in Alzheimer's disease in a multi-ethnic community sample. *The American Journal of Geriatric Psychiatry, 8*(2), 123–133.

Cheng, S. T., Kwok, T., & Lam, L. C. (2012). Neuropsychiatric symptom clusters of Alzheimer's disease in Hong Kong Chinese: Prevalence and confirmatory factor analysis of the neuropsychiatric inventory. *International Psychogeriatrics, 24*(9), 1465–1473.

Chow, T. W., Binns, M. A., Cummings, J. L., Lam, I., Black, S. E., Miller, B. L., … Van Reekum, R. (2009). Apathy symptom profile and behavioral associations in frontotemporal dementia vs dementia of Alzheimer type. *Archives of Neurology, 66*, 888–893.

Conde-Sala, J. L., Turro-Garriga, O., Calvo-Perxas, L., Vilalta-Franch, J., Lopez-Pousa, S., & Garre-Olmo, J. (2014). Three-year trajectories of caregiver burden in Alzheimer's disease. *Journal of Alzheimer's Disease, 42*, 623–633.

Cummings, J. L., Mega, M., Gray, K., Rosenberg-Thompson, S., Carusi, D. A., & Gornbein, J. (1994). The neuropsychiatric inventory comprehensive assessment of psychopathology in dementia. *Neurology, 44*, 2308–2308.

Edwards-Lee, T., Miller, B. L., Benson, D. F., Cummings, J. L., Russell, G. L., Boone, K., & Mena, I. (1997).

The temporal variant of frontotemporal dementia. *Brain: A Journal of Neurology, 120*, 1027–1040.

Gottlieb, D. J., DeStefano, A. L., Foley, D. J., Mignot, E., Redline, S., Givelber, R. J., & Young, T. (2004). APOE ε4 is associated with obstructive sleep apnea/hypopnea: The sleep heart health study. *Neurology, 63*, 664–668.

Hargrave, R., Stoeklin, M., Haan, M., & Reed, B. (2000). Clinical aspects of dementia in African-American, Hispanic, and white patients. *Journal of the National Medical Association, 92*, 15.

Hinton, L., Chambers, D., & Velásquez, A. (2009). Making sense of behavioral disturbances in persons with dementia: Latino family caregiver attributions of neuropsychiatric inventory domains. *Alzheimer Disease and Associated Disorders, 23*, 401–405.

Hinton, L., Farias, S. T., & Wegelin, J. (2008). Neuropsychiatric symptoms are associated with disability in cognitively impaired Latino elderly with and without dementia: Results from the Sacramento area Latino study on aging. *International Journal of Geriatric Psychiatry, 23*, 102–108.

Hinton, L., Haan, M., Geller, S., & Mungas, D. (2003). Neuropsychiatric symptoms in Latino elders with dementia or cognitive impairment without dementia and factors that modify their association with caregiver depression. *The Gerontologist, 43*, 669–677.

Hodgson, N., Gitlin, L. N., Winter, L., & Czekanski, K. (2011). Undiagnosed illness and neuropsychiatric behaviors in community-residing older adults with dementia. *Alzheimer Disease and Associated Disorders, 25*, 109.

Hölttä, E. H., Laakkonen, M.-L., Laurila, J. V., Strandberg, T. E., Tilvis, R. S., & Pitkälä, K. H. (2012). Apathy: Prevalence, associated factors, and prognostic value among frail, older inpatients. *Journal of the American Medical Directors Association, 13*, 541–545.

Jeste, D. V., & Finkel, S. I. (2000). Psychosis of Alzheimer's disease and related dementias: Diagnostic criteria for a distinct syndrome. *The American Journal of Geriatric Psychiatry, 8*, 29–34.

Kales, H. C., Gitlin, L. N., & Lyketsos, C. G. (2015). Assessment and management of behavioral and psychological symptoms of dementia. *BMJ, 350*, h369.

Kazui, H., Yoshiyama, K., Kanemoto, H., Suzuki, Y., Sato, S., Hashimoto, M., … Tanaka, T. (2016). Differences of behavioral and psychological symptoms of dementia in disease severity in four major dementia. *PLoS One, 11*, e0161092.

Krogstad, J. (2017). *U.S. Hispanic population growth has leveled off.* http://www.pewresearch.org/fact-tank/2017/08/03/u-s-hispanic-population-growth-has-leveled-off/

Landes, A. M., Sperry, S. D., & Strauss, M. E. (2005). Prevalence of apathy, dysphoria, and depression in relation to dementia severity in Alzheimer's disease. *The Journal of Neuropsychiatry and Clinical Neurosciences, 17*, 342–349.

Levy, M. L., Cummings, J. L., Fairbanks, L. A., Masterman, D., Miller, B. L., Craig, A. H., … Litvan, I. (1998). Apathy is not depression. *Journal of Neuropsychiatry and Clinical Neurosciences, 10*, 314–319.

Lovell, B. B., Ancoli-Israel, S., & Gevirtz, R. (1995). Effect of bright light treatment on agitated behavior in institutionalized elderly subjects. *Psychiatry Research, 57*, 7–12.

Lyketsos, C. G., Lopez, O., Jones, B., Fitzpatrick, A. L., Breitner, J., & DeKosky, S. (2002). Prevalence of neuropsychiatric symptoms in dementia and mild cognitive impairment: Results from the cardiovascular health study. *JAMA, 288*, 1475–1483.

Lyketsos, C. G., Steele, C., Baker, L., Galik, E., Kopunek, S., Steinberg, M., & Warren, A. (1997). Major and minor depression in Alzheimer's disease: Prevalence and impact. *Journal of Neuropsychiatry and Clinical Neurosciences, 9*(4), 556–561.

Lyketsos, C. G., Steele, C., Galik, E., Rosenblatt, A., Steinberg, M., Warren, A., & Sheppard, J. M. (1999). Physical aggression in dementia patients and its relationship to depression. *American Journal of Psychiatry, 156*, 66–71.

Lyness, J. M., Noel, T. K., Cox, C., King, D. A., Conwell, Y., & Caine, E. D. (1997). Screening for depression in elderly primary care patients: A comparison of the Center for Epidemiologic Studies—Depression Scale and the geriatric depression scale. *Archives of Internal Medicine, 157*(4), 449–454.

McCallion, P., Toseland, R. W., & Freeman, K. (1999). An evaluation of a family visit education program. *Journal of the American Geriatrics Society, 47*, 203–214.

McCurry, S. M., Gibbons, L. E., Logsdon, R. G., & Teri, L. (2004). Anxiety and nighttime behavioral disturbances: Awakenings in patients with Alzheimer's disease. *Journal of Gerontological Nursing, 30*, 12–20.

Mizrahi, R., Starkstein, S. E., Jorge, R., & Robinson, R. G. (2006). Phenomenology and clinical correlates of delusions in Alzheimer disease. *The American Journal of Geriatric Psychiatry, 14*, 573–581.

Mori, E., Shimomura, T., Fujimori, M., Hirono, N., Imamura, T., Hashimoto, M., ... & Hanihara, T. (2000). Visuoperceptual impairment in dementia with Lewy bodies. *Archives of Neurology, 57*, 489–493.

Mosimann, U. P., Rowan, E. N., Partington, C. E., Collerton, D., Littlewood, E., O'Brien, J. T., … McKeith, I. G. (2006). Characteristics of visual hallucinations in Parkinson disease dementia and dementia with lewy bodies. *The American Journal of Geriatric Psychiatry, 14*, 153–160.

Onyike, C. U., Sheppard, J.-M. E., Tschanz, J. T., Norton, M. C., Green, R. C., Steinberg, M., … Lyketsos, C. G. (2007). Epidemiology of apathy in older adults: The Cache County study. *American Journal of Geriatric Psychiatry, 15*, 365–375.

Ortiz, F., Fitten, L. J., Cummings, J. L., Hwang, S., & Fonseca, M. (2006). Neuropsychiatric and behavioral symptoms in a community sample of Hispanics with Alzheimer's disease. *American Journal of Alzheimer's Disease and Other Dementias, 21*, 263–273.

Osborne, H., Simpson, J., & Stokes, G. (2010). The relationship between pre-morbid personality and challenging behavior in people with dementia: A systematic review. *Aging & Mental Health, 14*, 503–515.

Palmer, C. V., Adams, S. W., Bourgeois, M., Durrant, J., & Rossi, M. (1999). Reduction in caregiver-identified problem behaviors in patients with Alzheimer disease post-hearing-aid fitting. *Journal of Speech, Language, and Hearing Research, 42*, 312–328.

Postuma, R. B., Gagnon, J. F., Vendette, M., Fantini, M. L., Massicotte-Marquez, J., & Montplaisir, J. (2009). Quantifying the risk of neurodegenerative disease in idiopathic REM sleep behavior disorder. *Neurology, 72*, 1296–1300.

Rozzini, L., Vicini Chilovi, B., Conti, M., Delrio, I., Borroni, B., Trabucchi, M., & Padovani, A. (2008). Neuropsychiatric symptoms in amnestic and nonamnestic mild cognitive impairment. *Dementia and Geriatric Cognitive Disorders, 25*, 32–36.

Saczynski, J. S., Beiser, A., Seshadri, S., Auerbach, S., Wolf, P. A., & Au, R. (2010). Depressive symptoms and risk of dementia: The Framingham heart study. *Neurology, 75*, 35–41.

Santos, C. E. (2017). The history, struggles, and potential of the term Latinx. *Latino/a Psychology Today, 4*(2), 7–14.

Savva, G. M., Zaccai, J., Matthews, F. E., Davidson, J. E., McKeith, I., & Brayne, C. (2009). Prevalence, correlates, and course of behavioral and psychological symptoms of dementia in the population. *The British Journal of Psychiatry, 194*, 212–219.

Shim, S. H., Kang, H. S., Kim, J. H., & Kim, D. K. (2016). Factors associated with caregiver burden in dementia: 1-year follow-up study. *Psychiatry Investigation, 13*, 43–49.

Sink, K. M., Covinsky, K. E., Newcomer, R., & Yaffe, K. (2004). Ethnic differences in the prevalence and pattern of dementia-related behaviors. *Journal of the American Geriatrics Society, 52*, 1277–1283.

Smith, M., Gerdner, L. A., Hall, G. R., & Buckwalter, K. C. (2004). History, development, and future of the progressively lowered stress threshold: A conceptual model for dementia care. *Journal of the American Geriatrics Society, 52*(10), 1755–1760.

Spalletta, G., Baldinetti, F., Buccione, I., Fadda, L., Perri, R., Scalmana, S., … Caltagirone, C. (2004). Cognition and behaviour are independent and heterogeneous dimensions in Alzheimer's disease. *Journal of Neurology, 251*, 688–695.

Starkstein, S. E., Ingram, L., Garau, M. L., & Mizrahi, R. (2005). On the overlap between apathy and depression in dementia. *Journal of Neurology, Neurosurgery & Psychiatry, 76*(8), 1070–1074.

Starkstein, S. E., Vazquez, S., & Petracca, G. (1994). SPECT study of delusions in Alzheimer disease. *Neurology, 44*, 2055–2059.

Steinberg, M., Shao, H., Zandi, P., Lyketsos, C. G., Welsh-Bohmer, K. A., Norton, M. C., … Tschanz, J. T. (2008). Point and 5-year period prevalence of neuropsychiatric symptoms in dementia: The Cache County study. *International Journal of Geriatric Psychiatry, 23*, 170–177.

Tagariello, P., Girardi, P., & Amore, M. (2009). Depression and apathy in dementia: Same syndrome or different constructs? A critical review. *Archives of Gerontology and Geriatrics, 49*, 246–249.

Teri, L., Ferretti, L. E., Gibbons, L. E., Logsdon, R. G., McCurry, S. M., Kukull, W. A., … Larson, E. B. (1999). Anxiety in Alzheimer's disease: Prevalence and comorbidity. *The Journals of Gerontology. Series A, Biological Sciences and Medical Sciences, 54*, M348–M352.

Terum, T. M., Andersen, J. R., Rongve, A., Aarsland, D., Svensboe, E. J., & Testad, I. (2017). The relationship of specific items on the neuropsychiatric inventory to caregiver burden in dementia: A systematic review. *International Journal of Geriatric Psychiatry, 32*, 703–717.

Trapp, S. K., Perrin, P. B., Aggarwal, R., Peralta, S. V., Stolfi, M. E., Morelli, E., … Arango-Lasprilla, J. C. (2015). Personal strengths and health related quality of life in dementia caregivers from Latin America. *Behavioural Neurology, 2015*, 507196.

van der Linde, R. M., Dening, T., Matthews, F. E., & Brayne, C. (2014). Grouping of behavioural and psychological symptoms of dementia. *International Journal of Geriatric Psychiatry, 29*, 562–568.

van Reekum, R., Stuss, D. T., & Ostrander, L. (2005). Apathy: Why care? *The Journal of Neuropsychiatry and Clinical Neurosciences, 17*, 7–19.

von Gunten, A., Pocnet, C., & Rossier, J. (2009). The impact of personality characteristics on the clinical expression in neurodegenerative disorders—A review. *Brain Research Bulletin, 80*, 179–191.

Vicini Chilovi, B., Conti, M., Zanetti, M., Mazzù, I., Rozzini, L., & Padovani, A. (2009). Differential impact of apathy and depression in the development of dementia in mild cognitive impairment patients. *Dementia and Geriatric Cognitive Disorders, 27*, 390–398.

Victoroff, J., Zarow, C., Mack, W. J., Hsu, E., & Chui, H. C. (1996). Physical aggression is associated with preservation of substantia nigra pars compacta in Alzheimer disease. *Archives of Neurology, 53*, 428–434.

Volicer, L., Harper, D. G., Manning, B. C., Goldstein, R., & Satlin, A. (2001). Sundowning and circadian rhythms in Alzheimer's disease. *American Journal of Psychiatry, 158*, 704–711.

Yaffe, K., Fox, P., Newcomer, R., Sands, L., Lindquist, K., Dane, K., & Covinsky, K. E. (2002). Patient and caregiver characteristics and nursing home placement in patients with dementia. *JAMA, 287*, 2090–2097.

Zubenko, G. S., Zubenko, W. N., McPherson, S., Spoor, E., Marin, D. B., Farlow, M. R., … Sunderland, T. (2003). A collaborative study of the emergence and clinical features of the major depressive syndrome of Alzheimer's disease. *American Journal of Psychiatry, 160*, 857–866.

Part III

Supportive Interventions and Services

8 Latino Families Living with Dementia: Behavioral Issues and Placement Considerations

Janette Rodriguez and Viviana Padilla-Martínez

Abstract

This chapter aims to provide a conceptual framework from which to approach common behavioral issues and placement concerns faced by Latino families who care for family members who have dementia, while also providing practical information regarding psychosocial interventions based on the available research findings and theory. First, the authors provide up-to-date research findings related to dementia in Latino families. Second, there is a discussion of existing models that address the various behavioral issues and placement problems these families face, as well as options available for dementia care services. Third, the authors discuss Latino values and strengths, as well as the intersection of various cultural and societal contributions to the challenges faced by these families, including culture, gender, race, power and privilege, and access to resources. Fourth, community psychology, liberation psychology, and mujerista psychology are introduced as important backdrops for understanding and serving these families. The chapter concludes with strategies for researchers, clinicians, and others to consider, as well as future directions.

Introduction

The purpose of this chapter is to increase knowledge, awareness, and skills as related to the common behavioral issues and placement concerns faced by Latino families when caring for family members who have dementia. The chapter presents a conceptual framework and practical information regarding psychosocial interventions, with attention to Latino lived experience, as well as relevant empirical studies and theory. In order to underscore the complexities being discussed, a case example and reflective check-ins are utilized. To accomplish the above-noted goals, the chapter provides both conceptual and practical information. First, an up-to-date introduction of the various dementia-related behavioral issues and placement roadblocks faced by Latino families will be discussed; this is explained with

This chapter was authored by Janette Rodriguez and Viviana Padilla-Martinez. The opinions expressed in this article are the authors' own and do not reflect the view of the Miami, Bay Pines, or Lee County VA Healthcare Centers/Systems, the Department of Veterans Affairs, or the United States government.

J. Rodriguez (✉)
Miami VA Healthcare System, Miami, FL, USA
e-mail: janette.rodriguez@va.gov

V. Padilla-Martínez
Bay Pines VA Healthcare System, Lee County VA Healthcare Center, Cape Coral, FL, USA
e-mail: viviana.padilla-martinez@va.gov

H. Y. Adames, Y. N. Tazeau (eds.), *Caring for Latinxs with Dementia in a Globalized World*,
https://doi.org/10.1007/978-1-0716-0132-7_8

non-stigmatizing and culturally relevant language to address the target population. Second, this chapter discusses the existing dementia care service options, along with a proposed model for conceptualizing service options. Third, Latino values and a Latino worldview are highlighted, along with the intersectionality of various cultural and societal contributions to the challenges faced by these families, including culture, gender, race, power and privilege, and access to resources. Fourth, the authors introduce community psychology, liberation psychology, and mujerista psychology as important backdrops for understanding and serving these families. Lastly, practical examples of strategies that can be utilized by researchers, clinicians, and others working with these families are discussed.

The material and topics discussed are best understood in the context of a case example that allows for and includes questions for *reader reflection*. Therefore, the chapter begins with the following case example and includes reflective "checkpoints" (i.e., steps along the way), during which readers can reflect on the examples and their broader applicability to caring for Latinos with dementia.

The Case of Pilar

Pilar is a 76-year-old Puerto Rican woman. She is slim, of medium height, and trigueña [a person with olive skin] (Adames & Chaves-Duenas, 2017a*) and wears her dark hair short. She was born in Puerto Rico and has six living brothers and sisters. She has five adult children and ten grandchildren; her closest child lives 3 hours away. She is divorced. She has many nieces and nephews. She owns her home and lives alone, but is next door to one of her sisters and brother-in-law; her sister often stays with her overnight. Another sister lives 10 minutes away. Her nieces and nephews live in various parts of the United States (U.S.) and Puerto Rico. Her primary language is Spanish, although she speaks some English as well. She has a history of hypertension which is well controlled with medication. She tends to have a sensitive stomach, and she makes aloe vera juice to remedy it. She also experiences pain from arthritis which she manages* via *the use of herbal ointments and creams. She has always loved to travel, to meet new people, and has enjoyed cooking, telling jokes to her family and close friends, giving advice, hosting family get-togethers, and gardening. She also has been very active in her church; she identifies herself as Catholic. She especially enjoys traveling to and from Puerto Rico and various U.S. mainland states where some of her children live. She completed eighth grade and has worked a wide range of blue-collar jobs throughout her life. She raised her children without a partner's financial or emotional support. She currently receives a small social security retirement check, as well as the financial support of her children. She receives her medical care through Medicare. She has always viewed herself as independent and hard-working. Others have viewed her as* amorosa, pero con caracter fuerte *[loving but has a strong personality] and somewhat of a "control freak."*

> ***Reader Reflection:*** *What are your initial thoughts and emotional reactions to Pilar and her family? What is your potential role in their lives – as a clinician, researcher, administrator, activist or advocate, lawmaker, teacher, or supporter?*

Six years ago, Pilar began to occasionally become disoriented when out and about on her own, although she had driven a car and taken public transportation for much of her adult life. Her family found this somewhat problematic, but despite one child being in the healthcare field, they did not perceive a need for a healthcare evaluation. Rather, they explained these experiences as her personality and "getting older." They suggested she not travel very far and that she keep her cell phone with her at all times. Several times, she got "lost" and called for "help" but could not explain where she was – the kindness of others helped her communicate to her family where they should pick her up. The situation worsened as her symptoms increased; Pilar's family was unaware that they might be one of many Latino families whose loved one was living with dementia.

Latinos in the United States and Dementia

The Latino population in the United States continues to thrive. From Latin American *novelas* adapted to create prime-time television series to Lin-Manuel Miranda's *Hamilton*, to our elected leaders, and to our workforce, Latinos are core to the fabric of the United States (U.S.). According to Colby and Ortman (2014), by the year 2060, those whose ethnicity is identified as Hispanic (using U.S. Census Bureau identifiers which typically capture and encompass the Latino population) will comprise 29% of the U.S. population. As such, attention to this large and traditionally underserved community is warranted. Moreover, there is a subgroup of Latinos that are also in need of urgent attention – Latino families living with dementia (LFLD). Of note, the authors propose the use of the term and associated acronym, Latino families living with dementia (LFLD), to discuss the needs of these families; this stems from a desire to use non-stigmatizing language that is based in core Latino values to be discussed later in the chapter.

As previously outlined, Pilar's family is a Puerto Rican (Latino) family living in the United States. Like many Latino families, they are an integral part of the United States. In Pilar's case, three of her children were full-time employees of various private businesses and nonprofit organizations in the United States. Pilar herself worked numerous blue-collar jobs in her primary state of residence prior to retiring. Several of her grandchildren were living in the United States. Pilar's father and daughter both served in the U.S. Armed Services (Army). Many of her siblings, cousins, nieces, nephews, grand-nieces/grand-nephews, and other family members comprise a part of the fabric noted above. Now, they are also a LFLD.

According to the medical model, dementia is not solely associated with older age. Further, it varies in etiology and presentation. However, there are many commonalities among the dementias, including cognitive and behavioral changes and functional decline. As related to Latinos, there is increasing evidence that this group has higher prevalence and incidence rates of dementia, as compared to other groups (Alzheimer's Association, 2010; Clark et al., 2005; Fitten et al., 2014). The Alzheimer's Disease Facts and Figures (Alzheimer's Association, 2010) special report on ethnicity and race indicates that Hispanics have a higher probability of developing dementia than both African-Americans and non-Latino Whites. As compared to non-Latino Whites, estimates indicate that older Hispanics are one-and-one-half times more likely to develop dementia. This difference remains in younger groups as well – that is, Hispanics aged 65–74 are at least three times more likely than non-Latino Whites to have cognitive impairment. Additionally, several studies have found that Latinos are likely to develop dementia significantly earlier than other ethnic/racial groups, and this is the case despite adjusting for other variables. For example, Clark et al. (2005) found that Latinos had symptom onset of dementia an average of 6.8 years earlier than Anglo patients. Fitten et al. (2014) found that despite subtype, genotype, gender, dementia severity, years of education, and history of chronic medical problems, Hispanics were younger at the time of diagnosis than non-Latino Whites. Given that these findings could not be accounted for by other variables, Fitten et al. (2014) recommend further studies. An additional study on dementia prevalence rates in a specific Latino subgroup found that Puerto Rican military veterans had a higher probability of being diagnosed with dementia as compared to mainland U.S. military veterans (Carrión-Baralt, Suárez-Pérez, Del Río, Moore, & Silverman, 2010). Further, they found that Puerto Rican military veterans also had higher prevalence rates of cardiovascular risk factors most strongly associated with dementia. Taken together, these incidence and prevalence findings further highlight the importance of providing attention to Latino families living with dementia.

Reader Reflection: *What are your thoughts on addressing the healthcare needs of Latinos in the United States?*

Furthering the need for attention to this group, Latinos also appear to have higher rates and/or likelihood of neuropsychiatric symptoms (e.g., behavioral and/or affective psychiatric symptoms associated with dementia; Ortiz, Fitten, Cummings, Hwang, & Fonseca, 2006; Sink, Covinsky, Newcomer, & Yaffe, 2004). For example, Sink et al. (2004) found that Latinos had a significantly higher likelihood than non-Latino Whites, of experiencing specific symptoms, such as hallucinations, unreasonable anger episodes, combativeness, and wandering. These higher rates of neuropsychiatric symptoms may further complicate dementia care; on the other hand, they may be an indicator that more targeted/tailored supports in terms of caregiving and intervention are required in order to address the needs of this population.

In a systematic review of the incidence and prevalence literature, Mehta and Yeo (2016) found that there are differences in dementia rates even among Hispanic subgroups in the United States. More specifically, dementia incidence rates for Hispanics from the Caribbean were significantly higher than those for Mexican-Americans. In this same study, the differences persisted when Caribbean Hispanics were compared to Japanese-Americans and non-Latino Whites. Therefore, they concluded that ethnic subgroups merit targeted research studies; given these findings, this appears to be particularly important for Latinos.

Another important finding in dementia care among Latinos is that related to use and/or initiation of pharmacological intervention. This is important as a better understanding of this issue may lead to consequent targeted interventions that can help address behavioral concerns and quality of life. While Latino families have a pattern of greater use and/or initiation of pharmacological interventions than other ethnic/racial minorities, such as Blacks (Carrión-Baralt et al., 2010; Xiong, Filshtein, Beckett, & Hinton, 2015), both Latino and Black families have fast discontinuation rates, as compared to White families (Thorpe et al., 2016). Premature discontinuation of psychopharmacological agents could negatively impact quality of life for both the person with dementia and caregivers, in terms of poor management of symptoms and consequent emotional distress, unsafe behaviors, and access to care facilities and external supports, as some may require some stabilization of symptoms for admission/acceptance in a program or for receipt of a service. On the other hand, it is also possible that an understanding of the reasons for higher discontinuation rates may lead to better psychosocial interventions (vs. psychopharmacological) that respect the preferences of the LFLD.

The findings thus far related to dementia in Latinos provide significant support for the notion that this ethnic group warrants further attention. The available research literature is limited and tends to examine Latino dementia issues as though Latinos are a homogeneous group. Although Latinos are a heterogeneous group based on key aspects such as country of origin, there are strong commonalities among the different Latino groups based on shared cultural values. This can help address this important area of need for Latino families living with dementia (LFLD): caregiving issues associated with caring for a person with dementia. An approach that considers the available information while also acknowledging an awareness of inherent differences, particularly those related to sociopolitical histories, is strongly encouraged (see Adames & Chavez-Dueñas, 2017b, 2017c). In this chapter, it is proposed that increasing knowledge about and exposure to this population, and to frameworks that are consistent with empowerment and promulgation of social change, can serve the purpose of advancing care for this population.

Dementia Common Behavioral Problems

Four years ago, Pilar started accusing others of not giving her what she paid for or providing the correct change. She also started asking questions repeatedly. Her family found this annoying but attributed it to her "strong personality" or that she was "just old." They criticized her for these behaviors, alternating with joking about them as a part of getting older.

Two years ago, her family realized she was losing weight and was probably not eating as much as before. They grew concerned about a possible medical problem leading to loss of appetite; her sister began to prepare, provide, and monitor all of her meals to ensure good health. She was also evaluated by a primary care physician who noted that she was pretty healthy for her age and maybe a bit grumpy and forgetful – "maybe a little senile." He prescribed medications to "help her" – that she'd feel better, be less grumpy, and eat more now. Her family considered that she was aging and had a strong personality – and this was probably normal, they told each other. Pilar started her medications. She was feeling better; her family agreed.

There are numerous problems associated with dementia, often referred to in the literature as "behavioral and psychological symptoms of dementia" (BPSD); and these can be difficult to manage, become financially stressful, and have a negative impact on the caregiver, as well as for the person with dementia (Buhr & White, 2006). These can be conceptualized as being the result of neurobiological factors, in conjunction with vulnerabilities including unmet basic needs (e.g., hunger or thirst), caregiving issues, or environmental triggers (Kales, Gitlin, & Lyketsos, 2015). While the onset and course of any predominant cognitive and behavioral problems will vary based on type of dementia (e.g., Alzheimer's dementia, dementia with Lewy bodies, frontotemporal dementia, vascular dementia), there is research to suggest that regarding behavioral problems, families find themselves needing to manage several common ones. People with dementia can experience difficulties with agitation, depression, apathy, repetitive questioning, psychosis, aggression, sleep problems, wandering, and other inappropriate behaviors (Finkel, Costa e Silva, Cohen, Miller, & Sartouis, 1996).

Pilar stopped taking her medications within the year. She quickly deteriorated, presenting with agitation, labile mood frequently alternating between tearfulness and anger, aggressive behaviors, wandering behaviors, and frequent accusations toward family members. She misplaced things often. These behaviors were in addition to getting lost, repeating questions, and apparent poor eating habits. She now told dirty jokes to strangers and not just at family get-togethers. The family went back to the physician who indicated she had some symptoms of dementia – he noted they were "mild."

Reader Reflection: *What other conditions, if any, could explain or account for these behaviors? How might family members understand distinctions such as "mild" dementia?*

Reader Reflection: *What are your beliefs and attitudes about aging? Are your beliefs and attitudes solely medically/scientifically informed, or do you hold beliefs and attitudes that are also informed by your own familial, cultural, spiritual, or societally based perspectives?*

Approaches to interventions can involve targeting the person with dementia, targeting caregivers, and targeting the environment. Brodaty and Arasaratnam (2012) note support for non-psychopharmacological intervention for persons with dementia living in the community that involves family caregivers and thus targets both the person with dementia and the caregivers. In some studies, the environment is also targeted. In their meta-analysis which involved the review of 23 trials, Brodaty and Arasaratnam (2012) found that non-pharmacological (or psychosocial) interventions that include the family caregiver may have comparable effects to pharmacological intervention. Further, non-pharmacological interventions may have a unique effect. The authors of the meta-analysis note that not only might the behavioral and psychological problems of the person with dementia be reduced but the caregiver's negative reactions may also be reduced. They found that a pattern emerged in terms of the interventions; that is, they noted that 9–12 sessions over 3–6 months were optimal when offered to the person with dementia and family caregiver, with these occurring in the home with follow-up via telephone. They also note that these were multiple-component interventions; that is, they utilized more than one strategy or technique for the intervention. These strategies or techniques included skills training, education, activity planning and environmental redesign, enhancing caregiver support, and self-care techniques for caregivers. Based on their findings, they recommend the use of non-pharmacological interventions that include tailored, multicomponent interventions, as well as periodic follow-up.

Similarly, in their review of a variety of dementia care issues for a community-dwelling population, Kales et al. (2015) also note the benefits of intervention at the caregiver level. They specifically note that tailored interventions that help the caregiver develop strategies for problem-solving are critical; that is, helping caregivers identify and address modifiable causes of symptoms is effective in reducing behavioral problems. Some examples they highlight are modifiable causes such as unmet needs or sensory deficits such as hunger, fear, and need for eyeglasses or hearing aids. As per Kales, Gitlin, and Lyketsos (2014), there are a number of strategies that include education and support, stress management skills, and problem-solving for behavioral problems. Others include improving communication, reducing complexity of the environment, and simplifying tasks for the person with dementia. There is strong support for multiple-component interventions, and they appear to provide significant benefit to caregivers (Belle et al., 2006: Brodaty & Arasaratnam, 2012). Given the unique cultural aspects, prevalence and incidence rates of dementia, and high discontinuation rates of pharmacological interventions among Latinos, tailored multicomponent non-pharmacological interventions with follow-up may be especially important for Latino families living with dementia, especially for those who are community dwelling.

Latino families tend to prefer home-based and/or community-based care and may be likely to perceive dementia-related behavioral difficulties as a normal part of aging; however, they also can experience high levels of distress and behavioral difficulties (Luchsinger et al., 2015); thus, preventive measures may be another, and perhaps best, approach for this population. Rather than allowing problems to increase in severity, it may be helpful to provide outreach, education, and support early on in dementia care. However, if, culturally, Latinos perceive the psychological and behavioral problems of dementia as "normal aging," these perceptions can limit planned actions of coping with and addressing the problems. If, from the onset, an understanding about the reason for the behaviors is acknowledged, it may make it easier to react in ways that may be more timely and helpful. Measures found to be appropriate for use with Latino populations, in order to assess and monitor behavioral changes, may be utilized to help begin the process. While an exhaustive review of assessment measures is beyond the scope of this chapter, the Behavioral Checklist is one measure that has been found to be appropriate for use with Latinos (Cox & Monk, 1996).

Existing Models of Dementia Care

Unfortunately, for Pilar and her family, they still did not know much about the difficulties they were having, about dementia, or about how to face these difficulties. Her sister was exhausted, and her children felt stressed and minimized the seriousness of the functional problems. But one day, one of Pilar's young grandchildren asked for a cup of water, and Pilar inadvertently served him soap instead. The family decided they would need to hire a señora [a woman who would help care for her] to assist Pilar.

> ***Reader Reflection:*** *What models of care are you familiar with that are culturally congruent for Latino families? To what degree and at what stage might Latinos look to governmental or other institutions to seek care for their family members with dementia?*

Three Basic Dementia Care Models: Internal, External, and Hybrid

In the United States, the term "dementia care model" is frequently used to describe service type. For example, adult day services (ADS) are considered one model of care, while long-term residential care (LTC) is considered another. Similarly, a distinction is made between informal and formal caregivers. The authors suggest that models of dementia care need not be limited to or based on service type alone, rather that the approach to care be defined in terms of cultural congruency. We propose that there be three basic and overarching models of care for Latino families who are caring for family members with dementia, including (a) internal, (b) external, and (c) hybrid. While several service type models related to dementia care exist, Latino families have historically been more oriented toward internal models of dementia care and/or, at minimum, have delayed institutionalization of the person with dementia (Mausbach et al., 2004). It is often, when common behavioral problems overwhelm their internal resources, that Latino families may be willing to explore other options, with a preference for a hybrid model over an external model.

Internal model The *internal model* of care involves Latino families and their family member who has dementia receiving their care via culturally congruent mechanisms. Based on this model, individuals with dementia receive their care while living at home with family (this includes non-blood kin), with the majority of assistance for activities of daily living and other needs provided by kin. This may also include traditional and culturally specific care, such as services from a religious or spiritual guide and folk healing interventions. This is often most consistent with Latino family preferences and values and can be most comfortable and familiar to the families. However, this can also lead to significant caregiver stress because of the additional care burden; and in the changing and challenging social systems, it may be difficult to provide care in this way in the long term due to the demands of modern life, including work, limited finances, and other diminishing resources.

External model For those receiving care via an *external model*, there is a significant shift, as Latino families receive care and support primarily from systems, services, and approaches that are external to their cultural values and/or traditions. In this situation, the person with dementia lives apart from other members of the Latino family. This may include an assisted living program or a long-term care or skilled nursing facility. Thus, the person with dementia receives a majority of their day-to-day assistance and support from non-kin. However, kin likely also provide some degree of support, especially in an assisted living facility. Members of the Latino family who are not identified as the person with dementia may also receive care and support via approaches that are external to them, such as participating in a support group or other clinical interventions. This approach is less consistent with traditional Latino values and may be perceived as directly in opposition to the norms and expectations of the Latino family. On the other hand, for Latino families who may have decreasing resources, this may be beneficial, especially if the adjunctive services of support are culturally congruent. However, if they are not culturally congruent, this approach may lead to additional stressors and difficulties.

Hybrid model The third model is a *hybrid model* of care wherein significant supports from both *internal* and *external* approaches to dementia care are in place. This may include the Latino family and the person with dementia living together, but having a visiting nurse or companion or housekeeper assist with caretaking duties or household tasks. Another possibility is that of the person with dementia participating in adult day services while living with other members of the Latino family. The hybrid model may also consist of respite visits, stays, and care when the Latino family caregivers are out of town or otherwise need assistance. This approach may be of most interest to Latino families who may be struggling to care for the person with dementia. Providers working with Latino families may wish to explore this option. However, oftentimes, the lack of financial resources can become a barrier to this type of care if they are not provided by public services.

With the señora attending to her needs, Pilar was now going to be in safe hands while her children and other family members worked and took care of errands and other life activities. However, each week or two, a new señora quit, indicating that Pilar was "impossible." Several of Pilar's family members, including her sister who provided the majority of the day-to-day support, were confident that their prayers would be heard; they fasted weekly and asked for prayers and healing and strength from their church congregation. However, it did not matter if she hailed from a home health agency or was a private referral – all the señoras would quit. All of Pilar's children lived quite a distance away given traffic and had full schedules with jobs and families, and her sister was exhausted and highly anxious on a daily basis. Pilar seemed angry frequently and resisted attempts to help her. Her children got into arguments with her, chastising her for not "complying" with their requests and recommendations. They thought she was being difficult, even though they acknowledged a diagnosis of "a little" or mild dementia. They weren't sure what to do about these behaviors, and they were incredibly frustrated. They didn't know what steps to take next.

Attributions and Placement

Latino families who care for a family member with dementia have been identified as having a wide range of attributions or explanations for dementia-related symptoms. These include dementia itself, medical conditions such as diabetes, interpersonal family conflicts or loss, mental illness (e.g., *nervios* [nerves] or *locura* [being "crazy"]), premorbid personality, stress, and the normal aging process (Ayalon, 2013; Connell, Scott Roberts, & McLaughlin, 2007; Gray, Jimenez, Cucciare, Tong, & Gallagher-Thompson, 2009; Hinton, Chambers, & Velasquez, 2009; Mahoney, Cloutterbuck, Neary, & Zhan, 2005). However, Latino families are also more likely than other groups to attribute symptoms to something other than dementia. This may have negative consequences and can lead to conflict about (a) whether the person with dementia has control over symptoms and/or (b) help-seeking; therefore, this has the potential to increase stress and burden (Hinton et al., 2009).

Latino families appear to have more informal care at home and more hours of informal care (Weiss, Gonzalez, Kabeto, & Langa, 2005) when compared to both non-Latino White and African-American groups. Additionally, Hispanic caregivers are less likely to place their relative in a care facility than non-Latino Whites (Schulz et al., 2004). However, there are factors that may serve as predictors of placement – including living alone, more functional and cognitive impairment, having at least one behavioral problem, higher burden levels, and older caregiver age.

Cox and Monk (1996) highlight dementia-related challenges among Hispanics, including distrust of outsiders, acceptance of stress as a normal and expected part of the familial role, and resistance to sharing familial problems with outsiders or admitting that care is too demanding. In fact, many Latino families may find that expressing or acknowledging associated distress or difficulties is unacceptable, as caring for elders with grace is culturally expected. And perhaps somewhat consistent with these cultural beliefs, long-term care staff appear to be in need of education and training, despite their status as formal caregivers (Ayalon, Arean, Bornfeld, &

> ***Reader Reflection:*** *How might Pilar's family's approach to care be conceptualized and described in terms of the proposed three models of dementia care?*

Beard, 2009). Fortunately, the Vital Outcomes Inspired by Caregiver Engagement (VOICE) Dementia Care Training Program, which is a multi-component intervention, has shown evidence of improvements in knowledge, attitudes, and self-efficacy among formal caregivers (Karlin, Young, & Dash, 2016).

Pilar's family met with their physician once again, hoping to receive different medications that would help. The nurse spoke with Pilar's family and noticed that Pilar's sister appeared tearful and less energized than usual; he also noticed one of Pilar's daughters appeared a bit irritable and short. He made it a point to talk with them a little more than the other family members. He concluded that while they were probably experiencing some caregiver stress, what he noticed was probably related to Latina women tending to be "dramatic." By the end of the visit, he was certain of this, as Pilar's family expressed confusion, disbelief, and perceived insult when the White nurse and White physician indicated that Pilar's mild dementia had worsened and it was time to "place" her in a "nursing home." The clinicians highlighted that this move would allow each of them to get some relief as well. The family refused to go back to the physician. However, there was a wait to see another physician, so they found a physician from the Dominican Republic whom they thought might be able to help them.

Culturally Sensitive Interventions for Behavioral Concerns

Researchers have identified several approaches to addressing behavioral problems, including ones that aim to be culturally sensitive, as well as some specifically targeting the Latino population. Kales et al. (2014) outline and summarize one such approach developed by a panel of national experts and comprised of various disciplines: the Describe, Investigate, Create, and Evaluate (DICE) approach. While this approach does include consideration of need for pharmacological intervention, it strongly encourages the use of tailored non-pharmacological interventions. The approach is aimed at addressing behavioral problems related to dementia with various strategies and in a tailored way. This tailoring can lend itself well to cultural sensitivity. Describing (*describe* step) and contextualizing the behavior is identified as the first step, while investigating (*investigate* step) the possible underlying causes of behavior is the second step. This includes patient factors, caregiver factors, environmental factors, and cultural factors. Using the *create* step, collaboration is with the care team and caregivers to develop and implement a plan to manage the most distressing symptoms. Lastly, *evaluate* involves assessing whether the intervention was effective, as well as any side effects. Of note, the DICE approach also assists in determining whether psychotropic medications are warranted. For example, if when *describing* or *investigating* it is determined that the behavior presents a safety issue, psychotropic intervention may be considered. While this approach is not specific to the Latino population, it does include caregivers in a collaborative way and involves assessing cultural context and factors that can be important in working to create a plan. In other words, the same plan is not applied to all families with a specific behavioral concern.

Reader Reflection: *How do you understand and explain the family's reaction? Consider your role in working with families living with dementia. How do you approach the topic of placement or need for increased level of care with families? What might you need to consider in working with Latino families?*

Another example of a culturally sensitive tool, which is specific to Latinos, is Gallagher-Thompson et al.'s (2015) *fotonovela*. Based on the identification of a clinical need to address stress and depression in Latino dementia caregivers who may also have low health literacy and less accurate knowledge of dementia, they developed a pictorial tool called a *fotonovela* to teach (a) coping skills for caregiver stress and (b) self-assessment of depression and (c) to encourage improved utilization of available resources. The *fotonovela* entitled *Together We Can! Facing Memory Loss as a Family* is a 16-page "picture book," available in Spanish and English, that has a dramatic storyline wherein Latino actors depict specific challenging scenes designed to illustrate key skills for managing difficult behaviors. The actors demonstrate the use of adaptive coping strategies, asking for help from other family members, and managing stress. The results of the study demonstrated that the *fotonovelas* designed for caregiver stress significantly reduced depressive symptoms as compared to the usual caregiver stress interventions, such as informational materials and brochures. A significant decrease in level of stress due to memory and behavioral problems was also reported. The caregivers using the *fotonovela* reported that the *fotonovela* was more helpful and that they referred to it more often than the usual interventions of informational materials. It appears as though culturally tailored *fotonovelas* can be an effective tool for Latino caregivers given their high unmet needs for assistance and various existing barriers in accessing resources.

Intersecting Cultural and Social Factors

Collectivist Versus Individualist Worldviews

A relationship exists between a collectivist orientation and Latino family cultural values toward dementia care. According to Triandis (1995), Hispanics are more collectivist in their social orientation as compared to a traditional, U.S. Anglo-majority culture which is represented by a more individualist orientation. Collectivism and individualism are complex constructs; however, at its core, collectivism prioritizes the needs of the in-group, which may include family, ethnic group, and religious group, among others (Oyserman, Coon, & Kemmelmeier, 2002). Individualism is described as the opposite since it focuses on the individual rather than the community needs; furthermore, individuals are independent from each other (Oyserman et al., 2002). The conversation below highlights these worldview differences of collectivism and individualism and underscores how important care discussions can be hampered because of the differences. For example, Pilar's sister may feel confused and not heard, and the White physician on call may feel frustrated that she is not receiving the information she needs to know.

Physician: *How are* ***you*** *doing?*
Pilar's Sister: ***We*** *are there…battling…in the struggle.*
Physician: *Okay, how are her dementia symptoms affecting* ***you****… Are* ***you*** *sleeping and eating okay?*
Pilar's Sister: *In our house, there is no sleeping because she wakes up at night, but you can be sure that* ***we have*** *good nutrition.*
Physician: *How is* ***she*** *doing – is* ***she*** *having any new problems? I need to hear about her difficulties now.*

The conversation above likely reflects a mismatch in worldviews on the collectivistic-individualistic continuum (Kernahan, Bettencourt, & Dorr, 2000). For many providers and researchers, especially those who hold primarily non-collectivist cultural views, their research questions, illness conceptualization, health assessment, and clinical intervention may often be based on a predominantly individualistic worldview. This may be incongruent with Latino family experiences and their approaches to facing dementia. Therefore, this chapter draws attention to and embraces a conceptualization that is overtly based on the idea that Latinos have cultural roots in collectivist cultures. This means that a problem or difficulty or illness is not solely an individual one with potential impact on one another; rather, dementia is a Latino family concern. When working with or attempting to address the needs of Latino families, a shift in conceptualization is strongly encouraged beginning with considered use of word and language choices. Using Pilar's family as an example, providers will want to respond accordingly when Pilar's sister uses "we" language, as this is an indicator regarding her worldview. In the example below, this physician notices how Pilar's sister uses language and adapts her questions accordingly, promoting further discussion.

Physician: *How are* **you** *doing?*

Pilar's Sister: ***We*** *are there…battling…in the struggle.*

Physician: *Okay, I can tell you are in this together. Tell me about any worries in terms of symptoms. Any problems with sleep or appetite?*

Pilar's Sister: *In our house, there is no sleeping because she wakes up at night, but you can be sure that we are having good nutrition.*

Nurse: *Alright, neither of you is getting much sleep, but nutrition is not a problem – you both eat well. What other behaviors or difficulties are you all experiencing?*

Pilar's Sister: *Lately, we've been not well. We've been arguing too much. She has been quite argumentative with me – she accuses me of stealing her cosmetics and mail. That's why we are here…to see if you can prescribe some medication to help us with this.*

While a shift away from an individualistic conceptualization to a collectivistic one is deemed important, this chapter also highlights several values and guiding frameworks as being critical when attempting to address the needs of Latino families.

Values

Despite the heterogeneity among Latinos, there are shared cultural values that unite this group and that impact perspectives and current status of dementia care among Latino families. In exploring how Latino families confront dementia, it is imperative to contextualize the challenges as well as the strengths within the Latino culture and consider social factors that impact this process. As previously highlighted, Latino families mostly hold a collectivist worldview where facing dementia is not viewed as an individual process, but rather a process that includes family members and non-kin relationships. In the spirit of this broader concept of collectivism, despite the many differences among Latinos, there are several cultural values commonly shared by Latino communities that are critical to consider in health research, as well as in health prevention and intervention. These values include *familismo*, *compadrazco*, *marianismo*, *machismo*, *respeto*, *personalismo*, *simpatía*, and *fatalismo*.

A key value is *familismo* – the idea that the family is of utmost importance (Turner et al., 2015) and a strong sense of commitment to family functioning as the main system of support (Montoro-Rodríguez, Small, & McCallum, 2006). This centrality on family fosters connections and decisions that are made around family values. Historically, Latino families have had multiple generations living in the same household, often with at least one woman available for caregiving and home-oriented duties at all times. However, as social resources shrink and globalization increases (i.e., family members may be more dispersed geographically), more Latino families will be less represented in household environments that support full-time caregiving for a person with dementia. These now non-traditional Latino configurations may be in conflict with the value of *familismo*, which has been found to exist even in third-generation Latinos, despite a decline due to generational changes (Landale, Oropesa, & Bradatan, 2006).

When discussing *familismo*, it is important to note that within Latino communities, the extended family is considered the nuclear family (i.e., there is no distinction between extended and nuclear). Additionally, the family includes individuals who are not blood relatives. Part of this extended family are *compadres* and *comadres* – individuals who become part of the family despite no blood relationship. The value of *compadrazco* (Montoro-Rodríguez et al., 2006) refers to extending family relation responsibilities (including care for family members) to non-kin individuals. This mainly focuses on godparents, neighbors, or individuals with whom the family shares a relationship.

In addition to the above-noted values, Latino families have a strong sense of *respeto*, *personalismo*, and *simpatía* (Castellanos & Gloria, 2016), which influence family dynamics. *Respeto* refers to respecting authority and seniority, including deference to adults, especially to those from their families. This means that within Latino families there is a hierarchy. *Personalismo* is the value of treating others with dignity and that each person has their value. *Simpatía* refers to the focus on relationships and kindness. These values impact how Latino families perceive the task-oriented, "business-like," and formal culture of providers in the healthcare system. Latino families tend to value "formal friendliness," handshakes, and manifest displays of confidence. Without *simpatía*, engagement and satisfaction with care could decrease.

All the values briefly outlined above impact how Latino families manage the diagnosis and care of a relative with dementia. Equally important, these also influence how family members relate to healthcare professionals, the help they seek, how they talk to and with providers, and the relationship with home and day care service workers and long-term care staff, among others. Thus, for Latino families, this can become a complex challenge to navigate when non-Latino providers are working with Latino families living in United States and/or when research and intervention approaches do not consider the critical issues of culture.

In considering Latino culture, the concept of *fatalismo* (Hovey & Morales, 2006) is one that impacts how Latino families might interpret symptoms and understand and accept a diagnosis and how care is delivered in general. *Fatalismo* is the idea that personal events or life situations are explained by a higher power and/or have a purpose and that may be immutable. This has the potential to impact the acceptance process and could increase denial and avoidance. Research findings reflect a range of common attributions related to symptoms of dementia, including normal signs of aging, family tragedies, emotional distress, migration, and other narratives (Turner et al., 2015). According to a National Hispanic Council on Aging (NHCOA; 2013) study, Latino families often hold misconceptions regarding the etiology of Alzheimer's disease; these include attributions of dementia due to long-standing personality traits, stress, and "thinking too much."

In Latino families living with dementia, the value of *familismo* is often present right alongside *marianismo*. Together, they serve as a guide for families when making decisions regarding the care and responsibility for Latinos with dementia. *Marianismo* is a value wherein society assigns roles and attitudes for women who are reflective of attributes commonly considered associated with the Virgin Mary – saintly and sacrificial. In fact, it is often a woman in the family – a daughter, sister, or daughter-in-law – who provides the majority of the caretaking for a family member with dementia. Given that in many Latino communities women are the primary caregivers (Luchsinger et al., 2015), the responsibility line/path often is as follows, in descending order: wife > daughter >sister > female family > *comadres* or non-kin women. When this order or responsibility is not available or functional, men may be considered for participation in primary caretaking. It should be noted that male gender also plays an important role within Latino families, especially due to *machismo* and *caballerismo*. *Machismo* is a strong pride in masculinity, and this is embedded in the overall culture. In Latino culture, *machismo* has both negative and positive connotations, as being "macho" or hypermasculinity might mean having an overbearing attitude toward others, especially females, but this may also refer to protective, decisive, and gentleman-like behaviors. *Caballerismo* focuses on an emotional connection and social responsibility (Arciniega, Anderson, Tovar-Blank, & Tracey, 2008). Although *machismo* and *caballerismo* have similar cultural roots, these differ in the sense that *machismo* is perceived as negative since it is centered on dominance and *caballerismo* is perceived as positive as it focuses on nurturing (Pardo, Weisfeld, Hill, & Slatcher, 2012). Overall, there is a tendency wherein males have power over women and characteristics associated with the social construct of being a male, such as being strong and directive, are preferred.

Language and Acculturation

Latino families encounter a wide range and unique set of challenges when facing dementia. Acculturation, language, and generational differences interact with family dynamics; and the family cultural dynamics affect how family members interact with the healthcare system. One of the barriers faced by this community may be language proficiency as there may be different levels of language proficiency among members of a family, with wide variations of the following: some speaking English and a native language, others speaking English only, and yet others having limited English language skills. Regarding language, there is also the possibility that dementia will first impact the language skills in the person's acquired/second language. This has the potential for major repercussions, especially when the Latino with dementia lives in an English language-based residential program or is treated by English-speaking providers.

Regarding level of acculturation, because some members of the family may be more acculturated than others, this has the potential to impact values and belief systems, as well as a family's lifestyle. It is important to account for generational differences, especially when working with older Latinos who have younger caregivers. Older generations of Latinos may have a stronger sense of *familismo* than younger generations; therefore, in younger generations, *familismo* may extend to nuclear family more so than to the "extended" family (Landale, Oropesa, & Bradatan, 2006). Younger generations, who are often more acculturated, might identify more with different cultural values than older Latinos. For example, older Latinos may take care of their family members at home for longer periods of time than younger generations. In addition to commonly held values by Latinos, there are many other contextual factors that challenge Latino families when facing dementia, including power and privilege, stigma, discrimination, and access to resources.

> ***Reader Reflection:*** *How might U.S.-based care facilities begin to incorporate the culturally sensitive and linguistically congruent aspects that made Pilar's Puerto Rico-based residential treatment care a success for her and her family?*

Social Inequalities: Discrimination, Bias, Power, Privilege, Access, and Stigma

Pilar's family found only one Latino physician in the area who could see them. There was a long wait; they suspected that Medicare was the culprit. Eventually, Pilar and her family met with the physician. He gave them short forms to complete, offered in both English and Spanish, which provided him with some information regarding the difficulties they were experiencing. He went on to describe for them, in both Spanish and English, what he knew and understood about dementia and how this was different from typical changes related to aging. Further, he asked the nurse to provide them with literature, the information for a support group in the area, and to learn a bit more about what they had done thus far and any concerns they had about care and safety. The nurse was also instructed to provide them with some tips for helping Pilar stay safe. Pilar's family did some research, spoke to co-workers, and called the office back, asking the doctor to confirm the diagnosis. They requested "testing." They explained that they learned of testing that some patient families find beneficial in completing. The physician noted two problems. First, Medicare would probably not pay for the testing since he was already certain of the condition; and testing is expensive. Second, he noted that his Spanish-speaking patients usually cannot get tested because the doctors he knows have told him they don't have the ability to test Spanish-speaking patients. Also, Pilar's limited English would not be enough.

> ***Reader Reflection:*** *In Pilar's family, which Latino values are evident? For example, what does familismo look like in their family? Respeto? Other values?*

Latinos often experience significant barriers to healthcare, including limited financial resources, lack of healthcare coverage, low socio-economic status, language barriers, transportation limitations, and many other social inequities and limited access, including lack of culturally sensitive and culturally appropriate services and often limited funding and attention in the scholarly and scientific realms. In fact, Fennell, Feng, Clark, and Mor (2010) note that elderly Hispanics are more likely than their non-Hispanic White peers to reside in nursing homes that are characterized by severe deficiencies in performance, understaffing, and poor care. Moreover, given the economic challenges facing many Latino families, it becomes difficult to avoid long-term care even for those that wish to do so.

As many Latinos are immigrants, fear related to immigration status of any family member can also present as a barrier to dementia care. This is a particular concern during U.S. historical periods of anti-immigration sentiment that negatively impacts health policy and practice. In such instances, access to care and help-seeking behaviors can witness declines due to fear associated with immigration status. In addition, immigration status can affect family members' residential status (of needing to live in different states and countries) which can impact how families make decisions and divide responsibilities.

Latino families may not receive much of the information that is available about dementia, either because it is not available in the Spanish language, or is not written in a way that can be understood by Latinos of varying educational backgrounds. This lack of information can extend beyond explanations of the dementia condition, to information about healthcare services, payment systems, and health insurance. Older Latinos tend to be less active in healthcare systems and encounter significant barriers to preventive health access. For example, 25% of older Latinos with a chronic condition are uninsured, while less than 30% have their medications covered (Center on an Aging Society, 2003). These disparities impact Latino families in general, but especially older Latinos. Overall, older Latinos come from lower socioeconomic backgrounds, lack health coverage, and experience limited access to resources (i.e., social assistance, especially for older Latinos living alone).

In addition to the systemic factors noted above, diagnostic screenings and evaluations (e.g. neuropsychological testing), along with interventions, are likely to be culturally biased given that there are few specifically developed for use with Latinos and that are appropriately adapted, culturally and linguistically. Further, there is a shortage of bilingual, Spanish-English speaking staff, as well as bi-cultural professionals to diagnose and treat older Latinos, especially in the field of neuropsychology (Geisinger, 2015). Together, these factors present serious limitations to culturally sensitive and responsive assessment and intervention with this vulnerable and underserved population.

Latino families who provide care for family members with dementia tend to seek and/or receive a diagnosis and treatment later than non-Latinos. Stigma around having dementia or having a family member with dementia can delay the process of diagnosis and treatment, especially when making decisions. The stigma toward dementia might include viewing the family member as *loco* [crazy] or "just having aging issues." Stigma can impact the process of acceptance and denial of dementia and might impact when family members begin receiving formal dementia care. Also, stress related to managing the situation may increase without the proper help. Latino family cultural mistrust of

U.S.-based care systems can also be a factor, as many Latino families may have historically faced significant discrimination and racism. Some Latino families might trust and participate in other health treatments, including ethno-medical ones such as visiting *curanderos* [traditional cultural healer] or using "healing hands" in church. Combined, these factors may prevent Latino families from seeking allopathic dementia care.

Latinos have strongly held stigma as related to dementia (Chin, Negash, & Hamilton, 2011), as they fear that having dementia means being *loco*, a highly undesirable cultural characterization. While at first glance, this may appear to be based on internal and cultural ways of thinking, there may be other explanations for this perception. For example, using a liberation psychology perspective may help explore and consider how oppression might impact Latino families. For example, perhaps it is the experience and history of bias, racism, discrimination, and historical oppression and stereotypes that create an expectancy of Latinos as "crazy," "dangerous," and "bad people," thus leading to self-stigma. Latinos may believe that use of systems or institutions or facilities – external or hybrid models of care – will lead to biased, discriminatory, and racist interactions. Further, Latinos are heterogeneous in many ways, including race, and therefore it is likely that Latino families of African or Indigenous backgrounds would be particularly impacted by racism and bias (for detailed discussion, see Chavez-Dueñas, Adames, & Organsita, 2014). In fact, there is some evidence that Latinos receive poorer institutional care in certain geographical areas (Escarce & Kapur, 2006); and therefore, this may not be an unfounded fear which prevents the use of critical external resources and, understandably, promotes cultural mistrust.

Pilar's family highly preferred an internal model of care, but eventually utilized an external model after experimenting with a hybrid approach. After several visits and discussions with their new physician, Pilar's family reflected on the financial strain of hiring caregivers, the worry about her safety, and the symptoms that continued to worsen. With the help of a Mexican social worker who shared her own family experience and outcome of dementia treatment, Pilar's family decided to try a residential placement. Because Pilar's children and grandchildren all had full-time jobs or school and her sister was feeling unable to provide adequate or sufficient care, Pilar entered a residential program. After starting at the residential program, they found Pilar complaining frequently about the "American food" and observed her to have limited socialization and increased irritability. Her peers and caregivers were primarily English speaking. They also didn't feel comfortable with many of the staff, as they felt that some of them were not very respectful, as evidenced by them not spending much time greeting them or chatting with them when they visited. They also felt that the building and environment felt very hospital-like and unpleasant. When they raised their concerns, they learned that there was nothing more that the care facility could/would do for Pilar and, they noted, she was one of the "difficult ones." The care facility suggested that Pilar's family might be "too attached" and that it would be "healthy" if they spent less time there, for both Pilar and family.

Pilar's family eventually opted to move to a long-term care facility nearer to Pilar's family and friends in Puerto Rico. The facility is "home-like," serves her familiar foods, and speaks her primary language. She even gets to participate in church services; a priest visits each week. She asks to go "home" (i.e., *to the United States) every day, but enjoys socializing with others in Spanish and tending to the garden. The behavioral problems decreased, and she seems to enjoy visits from her family. She doesn't remember all of her children anymore. Pilar's family enjoys the small talk and the "café con leche" (coffee drink) they are served when they arrive to visit; they have benefited immensely from talking with other families and the social worker who facilitates a support group who also teaches skills for interacting most effectively with the person with dementia. The center incorporates music, prayer, and other media to explore the thoughts, feelings, and needs of Pilar's family. Pilar's family and friends visit frequently and without complaints from the staff. The placement decision was a difficult one for Pilar's family to make, yet they are pleased that Pilar is safe, and they are satisfied with their path on their journey of coping and managing the challenges associated with dementia and the care of their family member.*

Liberation, Community, and Mujerista Psychology Frameworks

Liberation, community, and mujerista psychology are all frameworks considered invaluable in furthering research and service for Latino families. All three of these provide for research, prevention, intervention, and, most importantly, social change considerations. They are also all based on the tenet that there has been a history of oppression for many Latino families as associated with Eurocentric-based Western psychology.

Liberation Psychology

Liberation psychology can serve as a model for working with marginalized or oppressed groups as it resonates with values of social justice (Burton & Kagan, 2005). It involves both the external agents and the groups that are oppressed. In liberation psychology, which has roots in Latin America, the process of *concientizacion* becomes central. *Concientizacion* refers to developing a critical consciousness (Freire, 1972) or awareness of sociopolitical realities. In this way, through dialogue and consequent awareness of oppression, new possibilities for action emerge. Of note, this is not intended to occur for solely the group being served, but also the person or entities providing services.

Realismo-crítico and *de-ideologization* are core concepts. *Realismo-crítico* requires that theory not be considered the driving force for action; instead, the problems themselves can lead to theory, and the process of *de-ideologization* can be utilized to address issues such as *fatalismo* and individualism (Burton & Kagan, 2005). *De-ideologization*, as discussed by Martín-Baró (1996), means exploring existing ideology in order to disengage from predefined ideas and theories that that do not serve the oppressed/liberated (as cited in Burton & Kagan, 2005). Liberation psychology embraces interdependence, interdisciplinary work, dialogue, empowerment and participation, attention to power dynamics, participatory spaces/interactions, critical analysis of the purpose of the work including examination of who benefits from the status quo, and listening to voices that have previously been silenced (Burton & Kagan, 2005). It is implemented via various methods including those of community psychology, such as participatory action research. However, it is not limited to those strategies; in fact, the framework has been used by clinicians, researchers, activists, scholars, and more. While liberation psychology does not prescribe specific clinical interventions, clinicians may find that incorporating these concepts into dialogue with those they serve will help with empowerment. In the case example of Pilar and her family, inquiring about and helping the family explore their experience and their understanding of the first residential placement, with attention to exploration and understanding of issues of possible racism, discrimination, access, stigma, and other sociopolitical issues, may be of benefit to Pilar and her family as it can lead to changes in self-advocacy and can culminate in a decision to change care centers. This type of intervention can be done in family work or group work as well.

Community Psychology

Community psychology embraces approaches that serve to empower, critically consider, and then intervene in a situation or problem. Regarding Latinos, this means considering both similarities and differences that might bind individuals together or those that make them unique (Sánchez, Rivera, Liao, & Mroczkowski, 2017). The sociopolitical histories of many Latinos include colonization and slavery experiences, similar or distinct languages of origin (including indigenous languages), and shared values which are core aspects of many Latino backgrounds. Latinos also have similarities and differences for issues of race, culture, religion, and economic background/status, to name but a few. For example, there are group differences in economic backgrounds in that Latino poverty rates are higher than the U.S. national rate, 27% vs. 15%, respectively (Sánchez et al., 2017). Given this stark reality, the concept of intersecting identities becomes important when discussing Latinos, particularly given the many aspects of oppression that may impact them including classism, racism, sexism, heterosexism, and anti-immigrant rhetoric. In order to address some of the challenges associated with developing Latino-based interventions for dementia that are "one size fits all," use of an ecological model that intervenes at multiple levels (Sánchez et al., 2017) can be embraced. This is especially relevant to program development endeavors.

Resnicow, Soler, Braithwaite, Ahluwalia, and Butler (2000) noted that a Latino-based community psychology needs to have both cultural sensitivity and tailoring. They outline several strategies. For example, a "surface structural adaptation" would be represented by using a group's native language, while a "deep structure adaption" would incorporate values, history, and environmental components. This is consistent with the description of Pilar's family experience of ultimately choosing and settling on long-term placement in Puerto Rico for Pilar. The family not only desired surface adaptations (e.g., Spanish over English being spoken in the care facility) but also more deep adaptations consistent with their values. Community psychology values community collaboration and citizen participation, yet individual empowerment is also salient, in the form of experientially teaching techniques (vs. didactic presentation) for leadership skills/opportunities, social support, and general skills development. Resnicow et al. (2000) noted that because one of the goals of community psychology is to promote social change for marginalized populations, practitioners of community psychology should consider models and intervention approaches that have not typically been used, including liberation psychology. For clinicians and other interventionists, the use of the local community, family, and other related resources becomes particularly relevant. For Pilar's family and other similar families, helping them develop skills for self-advocacy, providing increased knowledge and access to relevant information, and encouraging creative and novel community-based approaches can mean the difference between success and failure in coping with dementia.

Mujerista Psychology

In their book *Womanist and Mujerista Psychologies*, Bryant-Davis and Comas-Diaz (2016) describe mujerista ideologies and approaches. They highlight something rarely noted explicitly in the literature – the immense value of lived experience in scholarship and practice. Regarding *mujerismo*, they describe an interdisciplinary perspective and inclusion as critical components. They further note that *mujerismo* is both an alternative feminism and an expansion of feminism. They emphasize that this includes a focus on the centrality of community, mutual caring, and global solidarity. Their approach inherently requires use of participatory, healing, and liberation methods, including participatory action research, photovoice, activism, narrative therapy, bibliotherapy, group counseling and/or informal support building, community resources, and cultural healing practices.

Consistent with some components of the above frameworks, Turner et al. (2015) used qualitative anthropological methods to analyze meaning-centered themes on how Latino families who care for family members with dementia perceive and manage behavior problems. Using focus group data, they identified indigenous approaches to management of dementia behaviors and challenges. Five indigenous approaches were extrapolated including (1) acceptance, (2) love, (3) patience, (4) adaptability, and (5) establishing

Table 8.1 Guiding questions: reflective practice with Latino families living with dementia

Is there a culturally congruent group intervention you might develop that incorporates the values of Latino families living with dementia (LFLD) while also exploring stigma, racism, and other factors that impact decision-making and intervention around behavioral issues and placement, as well as caregiving concerns?
Might you attend to language and worldviews when discussing options or providing information?
Do you stop and check for stigma, racism, and discrimination in your work with LFLD?
Have you implemented a healthcare provider program to train others to provide culturally congruent or tailored services?
Have you developed or implemented policies to provide guidance for and encouragement of use of culturally congruent or tailored services in your institution?
Has your research considered incorporating participatory action research elements or photovoice as methods?
Does your team know about and routinely share resources related to LFLD?
Have you adapted written material to serve this population?

routines of care. Challenges included issues with providers, problems with family members, limited knowledge of resources, emotional distress, and financial strain. They noted that ethnographic studies may be able to provide crucial information not available in larger-scale studies. They also noted the importance of both the *emic* (meaning-centered approach done from within the group) and *etic* (from outside or from the researcher's point of view) approaches in their study and how this may be utilized to inform culturally tailored interventions. They also suggested that these interventions might address the identified challenges by targeting both the individual and the system.

The above-noted frameworks and orientations provide guidance for exploring and understanding Latino families, their needs in caring for a family member with dementia, and how to best serve families with both cultural "humility," i.e., an interpersonal stance wherein an individual is oriented and open toward others (Hook, Davis, Owen, Worthington, & Utsey, 2013), and cultural sensitivity and congruency.

In summary, when exploring how to approach behavioral and placement concerns for Latinos with dementia and their families, we endorse the utilization of multiple frameworks as critical. We consider the following to be the cornerstone to this work: ensuring *awareness of the values* of Latino families; facilitating *exploration of ideology and systems* that negatively impact the families; engaging in *interventions at multiple levels* (for the person with dementia, the family, and the system); considering *novel approaches that are inclusive, supportive, and affirming*; and appreciating individuals' *lived experiences*. Table 8.1 provides additional reflective questions related to practical strategies for approaching work with Latino families and their family member who experiences dementia.

Reader Reflection: *Consider how familiar you are with the three frameworks and how confident you would be incorporating liberation psychology, community psychology, and/or mujerista psychology in your work with Latino families whose family member experiences dementia. How would you measure the success or outcomes of using any/all of these ideologies in dementia care?*

Strengths, Implications, and Recommendations

Latinos are a marginalized and oppressed group that faces significant sociopolitical and socioeconomic inequities, ones that impact their daily lives and certainly affect their journey of coping with dementia. Yet, Latinos' strengths have also allowed them to thrive. Cultural activities such as living in community, being with family, and having spiritual support provide Latino families with empowerment to move forward and provide them with the strength to take care of their families. Castellanos and Gloria (2016) identified that strengths have to be seen with a cultural lens when working with Latino families. They highlighted the interconnection between mind, body, and spirit as a source of strength for Latino families. They concluded that in order to achieve *bienestar* [well-being], the values of *fortaleza* [inner strength] and *voluntad* [will] must be considered.

Latinos' shared values are a core aspect of the culture, and there needs to be an understanding of these in order to effectively address the issues related to Latino families who manage issues of dementia. The use of available research is also important. The utilization of frameworks that are consistent with the experiences and needs of this population, such as liberation psychology, community psychology, and mujerista psychology, is critical to those interested in this work. While there is no one specific intervention and strategy to be applied, the recommended frameworks, the reviewed literature, and a model for understanding preferences can assist with advancing care for Latino families who face the challenges of common dementia-related behavioral and placement issues.

For researchers, scholars, and academics, participatory action research, photovoice, and other similar methods can be a component of the most meaningful work. This can transform interventionists, as well as Latino families, thus leading to the needed change. For Latino families, this may be the key to understanding some of the research findings related to higher rates of behavioral symptoms.

For the clinician or other providers of services, incorporating the multiple layers of an ecological perspective can facilitate the expression and empowerment of the voice of Latino families. The development and implementation of psychosocial interventions that are multi-component and tailored are empowering.

When activists, administrators, ethics officers, public servants, teachers, voters, and policy-makers use systemic interventions – which are based on Latino values and contributions, including Latino lived experience and de-construction of Eurocentric approaches to care – everyone benefits from increasing their knowledge, awareness, and active participation toward social change and justice for Latino families facing dementia care concerns. For all of those wanting to learn how to provide more effective help for Latino families, the work needs to be more interdisciplinary and, most importantly, more interdependent.

References

Adames, H. Y., & Chavez-Dueñas, N. Y. (2017a). *Cultural foundations and interventions in Latino/a mental health: History, theory and within group differences*. New York, NY: Routledge.

Adames, H. Y., & Chavez-Dueñas, N. Y. (2017b). The diverse historical roots of today's Latinos/as: Learning from our past to move into the future. In H. Y. Adames & N. Y. Chavez-Dueñas (Eds.), *Cultural foundations and interventions in Latino/a mental health: History, theory, and within group differences* (pp. 3–31). New York, NY: Routledge.

Adames, H. Y., & Chavez-Dueñas, N. Y. (2017c). The history of Latinos in the United States: Journeys of hope, struggle, and resilience. In H. Y. Adames & N. Y. Chavez-Dueñas (Eds.), *Cultural foundations and interventions in Latino/a mental health: History, theory, and within group differences* (pp. 53–77). New York, NY: Routledge.

Alzheimer's Association. (2010). 2010 Alzheimer's disease facts and figures. *Alzheimer's & Dementia, 6*(2), 158–194.

Arciniega, G. M., Anderson, T., Tovar-Blank, Z. G., & Tracey, T. J. (2008). Toward a fuller conception of machismo: Development of a traditional machismo and caballerismo scale. *Journal of Counseling Psychology, 55*(1), 19–33.

Ayalon, L. (2013). Re-examining ethnic differences in concerns, knowledge, and beliefs about Alzheimer's disease: Results from a national sample. *International Journal of Geriatric Psychiatry, 28*(12), 1288–1295.

Ayalon, L., Arean, P., Bornfeld, H., & Beard, R. (2009). Long term care staff beliefs about evidence based practices for the management of dementia and agitation. *International Journal of Geriatric Psychiatry, 24*(2), 118–124.

Belle, S. H., Burgio, L., Burns, R., Coon, D., Czaja, S. J., Gallagher-Thompson, D., … Resources for Enhancing Alzheimer's Caregiver Health (REACH) II Investigators. (2006). Enhancing the quality of life of dementia caregivers from different ethnic or racial groups: A randomized, controlled trial. *Annals of Internal Medicine*, 145, 727–738.

Brodaty, H., & Arasaratnam, C. (2012). Meta-analysis of nonpharmacological interventions for neuropsychiatric symptoms of dementia. *American Journal of Psychiatry, 169*(9), 946–953.

Bryant-Davis, T. E., & Comas-Díaz, L. E. (2016). *Womanist and mujerista psychologies: Voices of fire, acts of courage*. Washington, DC: American Psychological Association.

Buhr, G. T., & White, H. K. (2006). Difficult behaviors in long-term care patients with dementia. *Journal of the American Medical Directors Association, 7*(3), 180–192.

Burton, M., & Kagan, C. (2005). Liberation social psychology: Learning from Latin America. *Journal of Community & Applied Social Psychology, 15*, 63–78.

Carrión-Baralt, J. R., Suárez-Pérez, E., Del Río, R., Moore, K., & Silverman, J. M. (2010). Prevalence of dementia in Puerto Rican veterans is higher than in mainland US veterans. *Journal of the American Geriatrics Society, 58*(4), 798–799.

Castellanos, J., & Gloria, A. (2016). *SOMOS Latina/os – ganas, comunidad, y el espíritu: La fuerza que llevamos por dentro* (Latina/os – drive, community, and spirituality: The strength within). In C. A. Downey & E. C. Chang (Eds.), *Positive psychology in racial and ethnic minority groups: Theory, research, assessment, and practice*. Washington, DC: American Psychological Association.

Center on an Aging Society. (2003). *Older Hispanic Americans: Less care for chronic conditions*. Georgetown University. Retrieved from www.aging-society.org

Chavez-Dueñas, N. Y., Adames, H. Y., & Organista, K. C. (2014). Skin-color prejudice and within-group racial discrimination: Historical and current impact on Latino/a populations. *Hispanic Journal of Behavioral Sciences, 36*(1), 3–26. https://doi.org/10.1177/0739986313511306

Chin, A. L., Negash, S., & Hamilton, R. (2011). Diversity and disparity in dementia: The impact of ethnoracial differences in Alzheimer disease. *Alzheimer Disease and Associated Disorders, 25*(3), 187–195.

Clark, C. M., DeCarli, C., Mungas, D., Chui, H. I., Higdon, R., Nuñez, J., … Ferris, S. (2005). Earlier onset of Alzheimer disease symptoms in Latino individuals compared with Anglo individuals. *Archives of Neurology, 62*(5), 774–778.

Colby, S. L., & Ortman, J. M. (2014). *Projections of the Size and Composition of the U.S. Population: 2014 to 2060*, Current Population Reports, P25-1143, Washington, DC: U.S. Census Bureau. Retrieved from https://www.census.gov/content/dam/Census/library/publications/2015/demo/p25-1143.pdf

Connell, C. M., Scott Roberts, J., & McLaughlin, S. J. (2007). Public opinion about Alzheimer disease among blacks, Hispanics, and whites: Results from a national survey. *Alzheimer Disease and Associated Disorders, 21*(3), 232–240.

Cox, C., & Monk, A. (1996). Strain among caregivers: Comparing the experiences of African American and Hispanic caregivers of Alzheimer's relatives. *International Journal of Aging & Human Development, 43*(2), 93–105.

Escarce, J., & Kapur, K. (2006). Hispanic families in the United Stated: Family structure and process in an era of family change. In M. Tienda & F. Mitchell (Eds.), *Hispanics and the future of America*. Washington, DC: National Academies Press.

Fennell, M. L., Feng, Z., Clark, M. A., & Mor, V. (2010). Elderly Hispanics more likely to reside in poor-quality nursing homes. *Health Affairs (Project Hope), 29*(1), 65–73.

Finkel, S. I., Costa e Silva, J., Cohen, G., Miller, S., & Sartouis, N. (1996). Behavioral and psychological signs and symptoms of dementia: A consensus statement on current knowledge and implications for research and treatment. *International Psychogeriatrics, 8*(3), 497–500.

Fitten, L. J., Ortiz, F., Fairbanks, L., Bartzokis, G., Lu, P., Klein, E., … Ringman, J. (2014). Younger age of dementia diagnosis in a Hispanic population in Southern California. *International Journal of Geriatric Psychiatry, 29*(6), 586–593.

Freire, P. (1972). *Pedagogy of the oppressed* (M. B. Ramos, Trans.) New York: Herder and Herder (Original work published 1968).

Gallagher-Thompson, D., Tzuang, M., Hinton, L., Alvarez, P., Rengifo, J., Valverde, I., … Thompson, L. W. (2015). Effectiveness of a fotonovela for reducing depression and stress in Latino dementia family caregivers. *Alzheimer Disease and Associated Disorders, 29*(2), 146–153.

Geisinger, K. (2015). *Psychological testing of Hispanics: Clinical and intellectual assessment* (2nd ed.). Washington, DC: American Psychological Association.

Gray, H. L., Jimenez, D. E., Cucciare, M. A., Tong, H., & Gallagher-Thompson, D. (2009). Ethnic differences in beliefs regarding Alzheimer disease among dementia family caregivers. *The American Journal of Geriatric Psychiatry, 17*(11), 925–933.

Hinton, L., Chambers, D., & Velasquez, A. (2009). Making sense of behavioral disturbances in persons with dementia: Latino family caregiver attributions of neuropsychiatric inventory domains. *Alzheimer Disease and Associated Disorders, 23*(4), 401–405.

Hook, J. N., Davis, D. E., Owen, J., Worthington, E. L., & Utsey, S. O. (2013). Cultural humility: Measuring openness to culturally diverse clients. *Journal of Counseling Psychology, 60*(3), 353–366.

Hovey, J., & Morales, L. (2006). Religious/spiritual beliefs: *Fatalismo*. In Y. Jackson (Ed.), *Encyclopedia of multicultural psychology* (pp. 410–411). Thousand Oaks, CA: Sage Publications.

Kales, H. C., Gitlin, L. N., & Lyketsos, C. G. (2014). The time is now to address behavioral symptoms of dementia. *Generations, 38*(3), 86–95.

Kales, H. C., Gitlin, L. N., & Lyketsos, C. G. (2015). Assessment and management of behavioral and psychological symptoms of dementia. *British Medical Journal, 350*, h369.

Karlin, B. E., Young, D., & Dash, K. (2016). Empowering the dementia care workforce to manage behavioral symptoms of dementia: Development and training outcomes from the VOICE dementia care program. *Gerontology & Geriatrics Education, 38*(4), 375–391.

Kernahan, C., Bettencourt, B. A., & Dorr, N. (2000). Benefits of allocentrism for the subjective well-being of African-Americans. *Journal of Black Psychology, 26*(2), 181–193.

Landale, N., Oropesa, S., & Bradatan, C. (2006). Hispanic families in the United States: Family structure and process in an era of family change. In M. Tienda & F. Mitchell (Eds.), *Hispanics and the future of America*. Washington, DC: National Academies Press.

Luchsinger, J. A., Tipiani, D., Torres-Patino, G., Silver, S., Eimicke, J. P., Ramirez, M., … Mittelman, M. (2015). Characteristics and mental health of Hispanic dementia caregivers in New York City. *American Journal of Alzheimer's Disease and Other Dementias, 30*(6), 584–590.

Mahoney, D. F., Cloutterbuck, J., Neary, S. R., & Zhan, L. (2005). African American, Chinese, and Latino family caregivers' impression of the onset and diagnosis of dementia: Cross cultural similarities and differences. *The Gerontologist, 45*(6), 783–792.

Mausbach, B. T., Coon, D. W., Depp, C., Rabinowitz, Y. G., Wilson-Arias, E., Kraemer, H. C., … Gallagher-Thompson, D. (2004). Ethnicity and time to institutionalization of dementia patients: A comparison of Latina and Caucasian female family caregivers. *Journal of the American Geriatrics Society, 52*(7), 1077–1084.

Martín-Baró, I. (1996). Toward a liberation psychology. In A. Aron & S. Corne (Eds.), *Writings for a liberation psychology*. New York, NY: Harvard University Press.

Mehta, K. M., & Yeo, G. W. (2016). Systematic review of dementia prevalence and incidence in US race/ethnic populations. *Alzheimer's & Dementia, 16*, 1552–5260.

Montoro-Rodríguez, J., Small, J. A., & McCallum, T. J. (2006). Hispanic/Latino American families with focus on Puerto Ricans. In G. Yeo & D. Gallagher-Thompson (Eds.), *Ethnicity and the dementias* (2nd ed., pp. 287–309). New York, NY, US: Routledge/Taylor & Francis Group. xxii, 390 pp.

National Hispanic Council on Aging. (2013). *Executive summary: Attitudes, level of stigma, and level of knowledge about Alzheimer's disease among Hispanic elderly adults and caregivers, and Alzheimer's-related challenges for caregivers*. Retrieved from http://www.nhcoa.org/wp-content/uploads/2013/05/NHCOA-Alzheimers-Executive-Summary.pdf

Ortiz, F., Fitten, L. J., Cummings, J. L., Hwang, S., & Fonseca, M. (2006). Neuropsychiatric and behavioral symptoms in a community sample of Hispanics with Alzheimer's disease. *American Journal of Alzheimer's Disease and Other Dementias, 21*(4), 263–273.

Oyserman, D., Coon, H. M., & Kemmelmeier, M. (2002). Rethinking individualism and collectivism: Evaluation of theoretical assumptions and meta-analyses. *Psychological Bulletin, 128*(1), 3–72.

Pardo, Y., Weisfeld, C., Hill, E., & Slatcher, R. B. (2012). Machismo and marital satisfaction in Mexican American couples. *Journal of Cross-Cultural Psychology, 44*, 1–17. https://doi.org/10.1177/0022022112443854

Resnicow, K., Soler, R., Braithwaite, R. L., Ahluwalia, J. S., & Butler, J. (2000). Cultural sensitivity in substance use prevention. *Journal of Community Psychology, 28*, 271–290.

Sánchez, B., Rivera, C., Liao, C. L., & Mroczkowski, A. L. (2017). Community psychology interventions and U.S. Latinos and Latinas. In M. Bond, I. Serrano-García, C. Keys, & M. Shinn (Eds.), *APA handbook of community psychology: Methods for community research and action for diverse groups and issues* (Vol. 2, pp. 491–506). Washington, DC: American Psychological Association.

Schulz, R., Belle, S. H., Czaja, S. J., McGinnis, K. A., Stevens, A., & Zhang, S. (2004). Long-term care placement of dementia patients and caregiver health and well-being. *JAMA, 292*(8), 961–967.

Sink, K. M., Covinsky, K. E., Newcomer, R., & Yaffe, K. (2004). Ethnic differences in the prevalence and pattern of dementia-related behaviors. *Journal of the American Geriatrics Society, 52*(8), 1277–1283.

Thorpe, C. T., Fowler, N. R., Harrigan, K., Zhao, X., Kang, Y., Hanlon, J. T., … Thorpe, J. M. (2016). Racial and ethnic differences in initiation and discontinuation of antidementia drugs by medicare beneficiaries. *Journal of the American Geriatrics Society, 64*(9), 1806–1814.

Triandis, H. C. (1995). *Individualism and collectivism*. Boulder, CO: Westview.

Turner, R. M., Hinton, L., Gallagher-Thompson, D., Tzuang, M., Tran, C., & Valle, R. (2015). Using an emic lens to undersatand how Latino families cope with dementia behavioral problems. *American Journal of Alzheimer's Disease and Other Dementias, 30*(5), 454–462.

Weiss, C. O., Gonzalez, H. M., Kabeto, M. U., & Langa, K. M. (2005). Differences in amount of informal care received by non-Hispanic whites and Latinos in a nationally representative sample of older Americans. *Journal of the American Geriatrics Society, 53*(1), 146–151.

Xiong, G. L., Filshtein, T., Beckett, L. A., & Hinton, L. (2015). Antipsychotic use in a diverse population with dementia: A retrospective review of the national Alzheimer's coordinating center database. *The Journal of Neuropsychiatry and Clinical Neurosciences, 27*(4), 326–332.

Educational, Diagnostic, and Supportive Psychosocial Interventions for Latinos with Dementia

9

Caroline Rosenthal Gelman, Mari Umpierre, and Karla Steinberg

Abstract

This chapter describes Latinos'* unique experiences of Alzheimer's disease and related dementias (ADRD) that pose particular challenges for the provision of relevant and accessible psychosocial interventions. It presents specific examples of educational, diagnostic, and supportive psychosocial interventions for Latinos with ADRD and their caregivers. It concludes with a discussion of future directions in the development of psychosocial interventions that target both persons with ADRD and their caregivers.

*As used in the 2010 Census, Hispanic or Latino refers to a person of Cuban, Mexican, Puerto Rican, Central or South American, or other Spanish culture or origin regardless of race. The terms Hispanic and Latino are often used interchangeably, and there is currently no agreement as to which is more acceptable, although different individuals and groups have a preference. Because the people we work with more often use the term Latino, we employ it here, unless, when referring to specific literature, the cited authors have chosen a different term. Furthermore, Latinos are an extremely heterogeneous group, varying by race, ethnicity, religion, country of origin, socioeconomic status, and other factors. However, because of their common aggregation in the literature, we will also use the term Latino without further subdivision.

C. Rosenthal Gelman (✉)
Silberman School of Social Work, Hunter College, City University of New York, New York, NY, USA
e-mail: cgelman@hunter.cuny.edu

M. Umpierre · K. Steinberg
The Icahn School of Medicine at Mount Sinai, New York, NY, USA
e-mail: mari.umpierre@mountainsinai.org; karlsbadd329@gmail.com

Introduction

Psychosocial interventions targeted to Latinos with dementia and their families must address several aspects unique to Latinos and their experience of Alzheimer's disease and related dementias (ADRD). These critical factors include large and rapidly growing numbers of Latino older adults, high rates of dementias among this population, comparably lower levels of knowledge about ADRD and related diagnostic and treatment services than other groups, and significant language, cultural, geographic, financial, and immigrant status barriers to accessing the resources that do exist. In this chapter, we review aspects of Latinos' experiences of ADRD that pose particular challenges for the provision of relevant and accessible psychosocial interventions. We present a framework for assessing and understanding psychosocial interventions as culturally relevant to an increasingly large and heterogeneous group. Finally, we describe the range of psychosocial interventions for Latinos with ADRD and their caregivers along the con-

H. Y. Adames, Y. N. Tazeau (eds.), *Caring for Latinxs with Dementia in a Globalized World*,
https://doi.org/10.1007/978-1-0716-0132-7_9

tinuum of interventions from health promotion through interventive and supportive initiatives targeting both persons with ADRD and their caregivers, providing examples of each.

A Large and Growing Population in Need of Psychosocial Supports

Latinos constitute the fastest-growing subpopulation among the aged in the United States. This group is predicted to grow from 3.6 million in 2014 to 21.5 million by 2060 (FIFA, 2016). Accompanying this tremendous growth will be a dramatic increase in the number of older Latinos suffering from Alzheimer's disease and related dementias—from under 200,000 to as many as 1.3 million by 2050 (Alzheimer's Association, 2010).

Latino elders may be at greater risk of ADRD than the population as a whole—and at as much as twice the risk as European Americans—because of the prevalence of several known or suspected risk factors for Alzheimer's disease (AD) (Livney et al., 2011). Such factors include low levels of education, which must be accounted for in the development of interventions. In 2014, 54% of Latino older adults were high school graduates, compared to 84% of all older adults. Latino older adults also experience a greater prevalence of vascular diseases (Daviglus et al., 2012; Haan et al., 2003) and different susceptibility genes in various Hispanic subgroups (Tang et al., 1998) that may increase their risk of ADRD. Thus, interventions need to be broad in scope and reach to accommodate these increasing numbers.

Specific ADRD Presentation Requiring Tailored Interventions

The specific way that AD presents in Latinos also necessitates distinct interventions. Dementia-related behaviors such as combativeness, wandering, and hallucinations, often the most troubling and difficult to address, have been found to be more likely in African Americans and Latinos than European Americans (Sink, Covinsky, Newcomer, & Yaffe, 2004). Furthermore, Latinos may develop Alzheimer's symptoms on average nearly 7 years earlier than their non-Latino counterparts (Clark et al., 2004), making it more likely that they and their caregivers will be younger, with multiple familial and work roles to fulfill. Thus, interventions need to target caregiver support, in-home services, and community-level information and resources.

Significant Psychosocial Needs in Addition to ADRD

Latino older adults tend to have significant social service needs, greater than those of other aging groups. For example, the poverty rate for Latinos 65+ in 2013 was 20.4%, double that of all older Americans (ACL, 2015). A comparison between a national sample of non-Hispanic and Hispanic elders found that the latter fared worse on almost every measure of health and well-being, including self-reported health, satisfaction, and functional status (Andrews, Lyons, & Rowland, 1992). Several studies comparing the functional status of elderly Latinos and European Americans have found that Latinos are significantly more disabled (Jette, Crawford, & Tennstedt, 1996; Tennstedt, Chang, & Delgado, 1998). Interviews with Latino AD families (Hinton & Levkoff, 1999; Rosenthal Gelman, 2010b) underscore families' experiences of poverty, trauma, and discrimination. This suggests that, in addition to tailoring supportive interventions with case management components that address ADRD caregiving families' concrete needs, macro-level approaches to addressing broader issues such as poverty, discrimination, lack of access to housing and education, and immigrant rights for Latinos are also needed.

Lack of Information About ADRD and ADRD Services

The belief among many Latinos that even significant memory loss and confusion are part of normal aging, which delays recognition and treatment of actual medical conditions, has been frequently reported in the literature (Hinton, Franz, Yeo, & Levkoff, 2005; Mahoney, Cloutterbuck, Neary, & Zhan, 2005). A study of knowledge and beliefs regarding AD among a large sample of European American, African American, and Hispanic adults found that in general the public does not clearly distinguish between normal memory loss and AD; but this finding is particularly true of African Americans and Hispanics (Connell, Roberts, & McLaughlin, 2007). In another study comparing knowledge of AD among European, Latino, Asian, and African American elders, low levels of knowledge regarding AD were prevalent across all groups, but Latinos and Asians demonstrated the greatest gaps (Ayalon & Areán, 2004). While 78% of older European Americans answered at least half of the questions correctly, only 53% of the African Americans, 20% of the Asians, and 22% of the Latino older adults did so. Findings regarding whether perceptions and behaviors surrounding AD are due to cultural beliefs or to the poverty and low levels of formal education often accompanying minority status are equivocal and require further study. What is clear, however, is the need for outreach, education, and information regarding normal aging, signs of ADRD, and available diagnostic and intervention services.

Structural Barriers to Services

The importance of structural barriers such as linguistic, transportation, and financial obstacles has also been noted in the literature. Language proficiency and economic status significantly delay early diagnosis and treatment for older Hispanics with memory or cognitive problems (Ortiz & Fitten, 2000). Conversely, efforts to provide culturally attuned information and services regarding AD to Latinos result in reduction of barriers to care and increased knowledge and service utilization, as reported in various descriptions of targeted educational interventions and programs (see, e.g.,, Aranda, Villa, Trejo, Ramirez, & Ranney, 2003; Valle, Yamada, & Matiella, 2006).

Latinos' Caregiving Burden

Caring for a person with AD is often experienced as burdensome and distressing for families. Caregivers can experience serious mental and physical health effects from the chronic stress of dementia caregiving. This includes increased rates of depression and anxiety, role stress, family conflict, poorer self-rated health, alterations in immune functioning, and even increased mortality (Pinquart & Sörensen, 2003).

Latino caregivers of older adults face special challenges for a variety of reasons. They are caring for vulnerable older adults experiencing the effects of multiple jeopardy—the cumulative disadvantage over a lifetime described above (Stoller & Gibson, 1994). Latino caregivers are also more likely to be younger and also caring for children under 18, in addition to being poorer, less educated, underemployed, and in worse mental and physical health than White counterparts (NAC/AARP, 2015). Finally, Latino, female, and immigrant caregivers are more likely to have inflexible work situations that complicate caregiving (Lahaie, Earle, & Heymann, 2013).

Positive Aspects of Caregiving

There are positive aspects to caregiving, including a feeling of connection to the care recipient and pride in providing good care (Harmell, Chattillion, Roepke, & Mausbach, 2011), and some studies have shown that Latinos appraise caregiving as less burdensome than White caregivers do (Coon et al., 2004). Nevertheless, studies conducted specifically with Latino dementia caregivers support a picture of significant caregiving burden among this population (Dilworth-Anderson, Williams, & Gibson, 2002; Gallagher-Thompson et al., 2000; Pinquart & Sörensen, 2005). For example, despite the value of family in Latino culture, some Latino caregivers report insufficient support from extended family (John & McMillan, 1998; Rosenthal Gelman, 2014) and high levels of self-reported depression (Luchsinger et al., 2015; Polich & Gallagher-Thompson, 1997). Most studies also report a higher prevalence of depression among Latinos when compared to other groups of caregivers (Adams, Aranda, Kemp, & Takagi, 2002; Pinquart & Sörensen, 2005). Thus, because of the special challenges faced by Latino family caregivers, they may have a particularly strong need for psychosocial interventions designed to support them in their caregiving role.

Understanding the Value of *Familismo*

It is an oft-repeated tenet in the literature that Latinos' value of *familismo*, "a strong identification and attachment of individuals with their families (nuclear and extended), and strong feelings of loyalty, reciprocity and solidarity among members of the same family" (Sabogal et al., 1987), leads to greater actual involvement of extended family in the care of sick members and reduced perception of caregiving burden, since providing care is expected, valued, and culturally syntonic (see, e.g.,, John, Resendiz, & De Vargas, 1997). Indeed, as noted, some studies have reported lower appraisals of stress and greater perceived benefits of caregiving by Latinos (Coon et al., 2004, Depp et al., 2005).

Research has explicitly explored the role of *familismo* in caregiving. There are some reports that Latino elders actually have infrequent contact with family (Ruiz & Ransford, 2012). Other research has uncovered a relationship between familism and caregiver outcomes that is inconsistent across ethnic groups or, in contrast to the assumption that it would lead to greater acceptance and better caregiver outcomes, positively correlated with distress (Knight & Sayegh, 2010). A likely explanation is variation in the experience of *familismo* (Rosenthal Gelman, 2014; Smith-Morris, Morales-Campos, Alvarez, & Turner, 2013). For some caregivers, familism facilitates caregiving in the traditional, expected manner by providing meaning and support for their role. Other caregivers disavow its current relevance. Yet others feel a contrast between familism, which they may value in a general, abstract way, and more personal, immediate negative feelings they are experiencing from caregiving. Thus, it is important to recognize that *familismo*, a powerful construct still central to many Latinos, is exerting influence on caregiving in ways previously not fully considered.

This more nuanced understanding of *familismo* is at odds with the simple, comforting version of all Latino families fully desirous and capable of caring for older adults without complaint or struggle. The unsubstantiated view that Latinos do not want or require formal services or support for themselves in providing care may be inhibiting service providers from developing and extending appropriate resources to this population. Thus, healthcare and social service providers must consider the diversity among the needs and experiences of Latinos with AD and their caregivers and the importance of tailoring health information, policies, and services accordingly. While indeed some Latino families are able to fully care for their older members, others cannot or are doing so at great cost to themselves and, ultimately, the care recipient. Development of services incorporating a more complex understanding of the role of *familismo* might entail supporting a caregiver reluctant to admit a fragile relative to a long-term care facility to make this medically necessary decision, while for another Latino caregiver cultural sensitivity might entail acknowledgement of their reluctance to become a caregiver, despite community expectations. We turn now to a description of a range of resources and services, both for Latinos with ADRD and their family caregivers.

Culturally Competent Psychosocial Interventions for Older Latinos and Their Families: From Outreach to Caregiver Support

In this section, we will discuss psychosocial interventions for Latinos with ADRD and their family caregivers. We will use as a framework the arc of interventions from (1) prevention and information (2) through accessible and timely comprehensive evaluation and diagnosis, (3) to disease management services for the caregiver-care receiver dyad, and 4) to caregiver-focused, culturally attuned interventions to manage symptoms and the experience of caring for someone with AD. We will present and discuss specific examples of each type of intervention (see Fig. 9.1).

Outreach, Engagement, and Health Communication Tools

General health literacy influences an individual's level of engagement in healthcare (Boucher, Guadalupe, Lara, & Alejandro, 2014). Gaps in dementia health literacy as well as misconceptions regarding cognitive health have been identified as barriers to dementia diagnosis and care, as well as participation in research for older Latinos (Ayalon & Areán, 2004; Rosenthal Gelman, 2010b). The need for outreach, education, and information regarding normal aging, signs of ADRD, dementia diagnosis assessments, care services, and disease management specifically designed for Latinos has been well established in the literature (Connell et al., 2007; Mahoney et al., 2005). In this section, we will describe examples of successful health communication strategies for improving outreach, engagement, and service utilization among older Latinos and their families.

In the field of health communication, entertainment-education (EE) (Singhal & Rogers, 2012) has been widely used with Latinos (Cabassa, Molina, & Baron, 2010; Hernandez & Organista, 2013; Sood, 2002; Sood & Storey, 2013). EE is defined as the process of purposely designing and implementing a media message to entertain, educate, and activate audiences to make changes in their own lives. Videos and printed materials that use entertainment-education strategies have been successful in helping to educate and engage Latinos in various types of health issues such as depression and

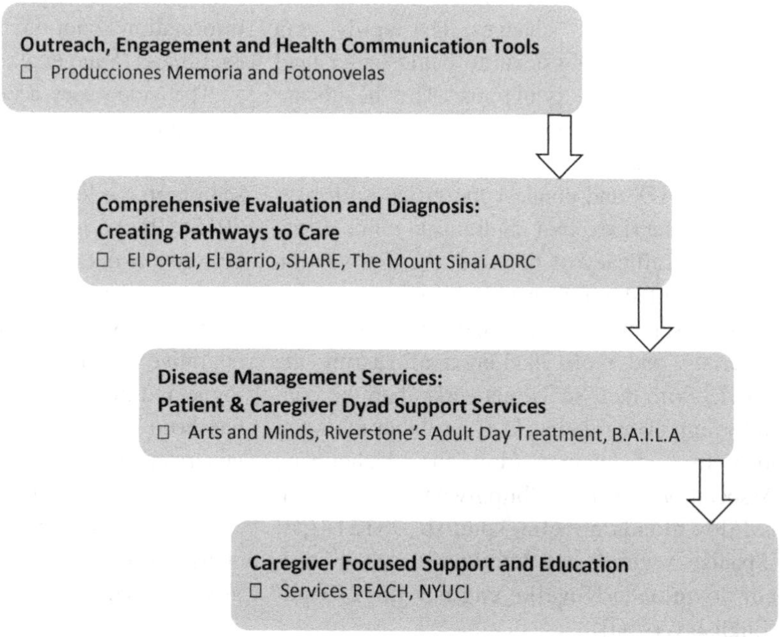

Outreach, Engagement and Health Communication Tools
- Producciones Memoria and Fotonovelas

Comprehensive Evaluation and Diagnosis: Creating Pathways to Care
- El Portal, El Barrio, SHARE, The Mount Sinai ADRC

Disease Management Services: Patient & Caregiver Dyad Support Services
- Arts and Minds, Riverstone's Adult Day Treatment, B.A.I.L.A

Caregiver Focused Support and Education
- Services REACH, NYUCI

Fig. 9.1 Examples of psychosocial interventions for Latinos and their caregivers

breast cancer (Borrayo, 2004; Gallagher-Thompson et al., 2015; Unger, Cabassa, Molina, Contreras, & Baron, 2012). Recent research reports that an EE approach is particularly useful with Spanish-speaking Latinos in the United States because this group watches more television than other populations studied and is more likely to act on information garnered through this medium (Borrayo, 2004; Wilkin & Ball-Rokeach, 2006; Wilkin et al., 2007).

There are several examples of EE for Latinos applied directly to content on AD and related dementias. Valle and his colleagues (2006) describe the development of two *fotonovelas* or "photo novels" – a popular form of reading material in Latin American countries that tells a story and provides photos to advance the narrative. Participants demonstrated increased knowledge of AD and satisfaction with this format, which is cost-efficient but requires basic literacy and visual acuity.

Gallagher-Thompson et al. (2015) developed a *fotonovela* intervention – *Together We Can! Facing Memory Loss as a Family* – to address the psychosocial needs of Latino caregivers. The *fotonovela* promotes the reader's identification with the characters and their life circumstances. In the story, Latino actors portray a family in which the grandmother presents with memory loss and personality changes. The family, as a group, consults a physician to better understand the grandmother's symptoms. The healthcare professionals, also portrayed by Latino actors, engage the grandmother in care, educate the family about AD, and connect the group with psychosocial services. In a randomized clinical trial to test the efficacy of this tool, the intervention group reported a greater reduction in depressive symptoms, more frequent use of health education materials, and more likelihood of sharing the booklet with their social networks than the control group. The *fotonovela* is available to download free of charge from the Alzheimer's Association's website (http://www.alz.org/espanol/downloads/novella_spanish_081213.pdf [Spanish version] and http://www.alz.org/espanol/downloads/Novella_english_081213.pdf [English version]).

Grigsby, Unger, Molina, and Baron (2016) evaluated the effectiveness of a dementia *novela* presented in an audiovisual format in which the still pictures from the novela are presented on a screen and narrated by voice actors. This format capitalizes on the popularity of *fotonovelas* but can be used with people with low literacy skills or visual impairments. This audiovisual *novela* led to improved dementia attitudes, beliefs, and knowledge. It did not, however, lead to increased belief in the benefits of screening (Grigsby et al., 2016).

To identify best practices and preferred health communication tools for Latino older adults, the Alzheimer's Disease Research Center at Icahn School of Medicine at Mount Sinai (MSADRC) developed a community academic partnership with Latino seniors, researchers, and other community stakeholders (Sano, Sewell, Umpierre, Neugroschl, & Cedillo, 2016). This work led to the identification of videos as preferred health communication tools for Latino elderly in the East Harlem (EH) community, Spanish as the preferred language to receive health information, and peers as trusted advisors for information pertaining to cognitive health and dementia screenings. The group produced a health education video using a short telenovela format with community seniors as cast members, to disseminate information about memory evaluations and their usefulness (Sano et al., 2016).

The video uses a culturally congruent informal approach to address a serious topic and is tailored to resonate with the experience of Latino seniors in the local community (Umpierre et al., 2015). It portrays an informal conversation between two Latinas in a senior center. Using easy-to-understand nonmedical terms in Caribbean Spanish, accurate information about cognitive health evaluations is part of the dialogue and indirectly provided to the audience. It has been presented in community settings, has generated positive reactions from the audience, and is available online at https://www.youtube.com/watch?v=1irhcTNsvsg to be used free of charge in cognitive health education and community outreach.

All four of these examples of EE strategies address existing barriers such as language and literacy levels to create materials that provide information about cognitive health and ADRD in relevant and accessible ways. Increasing knowledge of ADRD and relevant diagnostic and interventive services is critical to prevention and early intervention. Just as important are accessible resources for diagnosis and treatment for older Latinos and their families.

Comprehensive Evaluation and Diagnosis: Creating Pathways to Care

The MSADRC successfully engages older Latinos in dementia research and care using a "go to them" model (Li et al., 2016) and research participation as a pathway into dementia evaluations and care. In collaboration with the Mount Sinai Hospital's Health Education Department, the MSADRC provides free memory health talks and memory screenings to older adults at senior centers, senior housing projects, and other community settings. Bilingual and bicultural staff members conduct presentations, screenings, and informal consultations in English and Spanish to engage seniors. Research memory evaluations are provided in English and/or Spanish at the MSADRC campus or at community senior centers. Research evaluation participants receive a diagnostic impression and a care plan with prevention and/or treatment recommendations. The care plan is shared with the participant's primary care doctor and research partner or caregiver. As part of any study in which they enroll, participants return to the center yearly for cognitive health monitoring. To ensure follow-up with yearly visits, phone calls, letters, and informal reminders are provided. Those who choose not to participate in the study obtain information and referrals to geriatric primary or psychiatric care as needed for dementia evaluations.

Nationally, minority participation in clinical research is less than 18% (George, Duran, & Norris, 2014). However at the MSADRC, 25% of research participants are minorities, and many are Latinos (Neugroschl, 2016). The greater success in engaging Latinos at the MSADRC is an example of how maintaining a strong ongoing presence in the local community, using a "go to them" model (Li et al., 2016), can have a positive impact. The model integrates culturally relevant outreach practices such as preparing all information and recruitment materials in English and Spanish, substituting the word dementia with the term cognitive health to minimize social stigma, (Fitten, Ortiz, & Pontón, 2001) and integrating an informal, personable, respectful style of communication (*personalismo* and *respeto*, two common Latino cultural values and behavior codes) into all presentations, screenings, informal consultations, and recruitment interventions to promote the establishment of trusting relationships between seniors and research staff.

For example, during community lectures, each senior is invited to meet privately with a bilingual research assistant (RA) for a memory screening. The same RA invites each senior to participate in the MSADRC satellite study, coordinates visits to the center, or visits to the senior's home to complete recruitment procedures. This personable style of interaction promotes the establishment of a trusting relationship between Latino seniors and bilingual, bicultural RAs. Relationships with staff have been reported as a key factor that motivates participants to enter and to stay in the AD research program (Neugroschl, Sano, Luo, & Sewell, 2014).

AD trials are moving toward earlier intervention in non-demented participants (Neugroschl & Sano, 2010; Neugroschl et al., 2014). Research participation may be a viable pathway into timely dementia diagnosis and care for this vulnerable group. This is a strong possibility given the broadening scope of Alzheimer's research efforts, the increasing need to engage nonclinical populations in prevention studies, and the receptivity of older Latinos to participate in research when recruited in the community through culturally sensitive methods.

Another useful strategy to engage Latinos in diagnostic and treatment services is through establishing structured and formal collaborations between dementia-specific care providers. For example, El Portal Latino is a dementia-specific interorganizational community-based collaborative led by the Los Angeles County Alzheimer's Association. This group of agencies engages Latinos and other minorities in disease management services targeting caregivers and older adults. The group provides case management, counseling, education, training, referrals, transportation, adult day care, in-home respite, legal assistance, and diagnostic consultations. Services are delivered in a coordinated and flexible manner to prevent fragmentation and to ensure that patients and caregivers enter and stay in care (Aranda et al., 2003).

El Portal Latino helps identify persons with dementia (PWD) even when they are engaged in primary medical care. Language barriers and lack of knowledge often result in miscommunication with healthcare providers, and this may delay a formal dementia diagnosis. Without a clear diagnosis of dementia, psychosocial care for patients and support for their caregivers is also delayed. El Portal Latino was designed with these challenges in mind and addresses the specific needs of vulnerable Latinos in need of coordinated, flexible, and integrated medical and psychosocial care by providing a single portal of access to a range of bilingual, culturally relevant services.

El Barrio SHARE: Supporting Natural Helpers with ADRD Resources and Education

The idea of collaborating with existing AD services and creating an accessible single point of entry for Latinos is taken a step further in a program in development in East Harlem (EH), New York, also known as El Barrio (the neighborhood). This program, El Barrio SHARE (Supporting Natural Helpers with ADRD Resources and Education), will train residents naturally embedded in the community – such as storeowners and hairstylists – to engage and provide culturally tailored information and resources about AD to Latino elders and their families (Rosenthal Gelman, Giunta, & Glushefski, 2017). Led by Silberman Aging, a Hartford Center of Excellence in Geriatric Social Work within the Silberman School of Social Work at Hunter College, a cross-section of stakeholders, including individuals who work or live in EH (both younger and older adults), local merchants, volunteers, social service providers, urban planners, and academics, have collaborated in developing the program. El Barrio SHARE will recruit and train natural helpers on healthy aging, dementia, engagement techniques, and providing information on a single entry point of service staffed by a bilingual and bicultural social worker. The expectation is that such training will increase knowledge among natural helpers, reduce stigma and misconceptions around AD, and lead to earlier identification of dementia among Latino elders.

Disease Management Services

Adult day treatment programs, in general, are designed to provide care and socialization opportunities to PWD in combination with respite for caregivers. The Riverstone program, located in a traditionally Latino community in New York City, is an example of a day program designed to meet the needs of older Latinos with mild to moderate dementia and their caregivers. Spanish-speaking PWD participate in activities conducted in Spanish and in English and in culturally relevant ways designed to encourage socialization, promote independence, and increase well-being. For example, music therapy uses classic material from Latin America and the Caribbean.

Recent literature indicates that physical activity such as dance helps to promote and maintain brain health and may improve cognition even in persons with AD. Among Latinos, physical activity such as practicing sports or going to the gym is associated with youth. However, dancing is a culturally acceptable and important social activity for older adults and across generations (Melillo et al., 2001). Thus, there are multiple examples of programs across the country utilizing dance to engage Latinos in a form of exercise that is physically and cognitively challenging as well as culturally acceptable. It also promotes joyful interactions by PWD and their family or paid caregivers (Marquez et al., 2014).

Other cultural and artistic approaches have also been tailored for work with Latinos with AD and their families. Arts & Minds consists of analysis of art, immediately followed by "hands-on art making" by participants, who include both the caregiver and the care recipient with AD. Acknowledging the demographic makeup of East Harlem, Arts & Minds has adapted its content and implementation of programs by offering the programs in Spanish, ideal for those who may have learned English as a second language, only to lose it due to their AD diagnosis, as well as those members in the community who are Spanish speaking only. Arts & Minds has partnered with El Museo del Barrio, which houses art reflective of the diverse NYC Latino community.

Arts & Minds also hires teaching artists who are native Spanish speakers representative of the various Latino cultures in the classroom, offering coffee and playing music that is reflective of the students in the classroom as they are engaged in the art-making process. From Colombian *cumbias* to Dominican *bachata* and to the Puerto Rican *boleros* of Los Panchos, the participants are surrounded by the sounds, flavors, and visual topography of their own culture(s). Additionally, Arts & Minds also substitutes the highly charged term *demencia*, which means "crazy" to many native Spanish speakers, with *deficiencias de memoria* or *problemas de la memoria*, which translate to "memory deficiencies" and "problems with memory," respectively.

Caregiver-Focused Support and Education Services

A final area of services designed for Latinos with AD focuses on support for their caregivers. Interventions for caregivers generally are effective in producing clinically meaningful improvements in psychological well-being. Multifaceted, individualized interventions appear to have the strongest impact (Pinquart & Sörensen, 2006; Sörensen, Pinquart, & Duberstein, 2002). The Resources for Enhancing Alzheimer's Caregiver Health (REACH) project, funded beginning in 1995 by NIH, tested several interventions with Latinos at various sites (Gallagher-Thompson, Areán, Rivera, & Thompson, 2001; Gallagher-Thompson et al., 2003; Gitlin et al., 2003). In the second wave of this project, REACH II, specific aspects of interventions found to be efficacious were combined into a multicomponent psychosocial behavioral intervention tested at five sites among Hispanic/Latino, White European, and African American caregivers. The intervention consisted of nine in-home and three telephone sessions over 6 months and five telephone support group sessions. Hispanic/Latino caregivers experienced modest but clinically meaningful improvement compared to those in a minimal support control condition in levels of depression, burden, self-care, social support, and patient problem behaviors (Belle et al., 2006; Llanque & Enriquez, 2012).

The New York University Caregiver Intervention (NYUCI) has demonstrated improved caregiver and care recipient outcomes, including postponed placement of patients in nursing homes (Mittelman, Ferris, Shulman, Steinberg, & Levin, 1996; Mittelman, Haley, Clay & Roth, 2006) and increased emotional and psychological well-being for caregivers (Mittelman et al., 1995, Mittelman, Roth, Coon, & Haley, 2004). It has three components. The first is two individual and four family counseling sessions in the space of 4 months focusing on (1) education, (2) enhancement of social support for the primary caregiver, (3) promotion of communication, (4) problem solving, (5) patient behavior management strategies, and (6) concrete planning. The second component, beginning after counseling completion and continuing indefinitely, consists of weekly attendance at an existing support group. The third component, ad hoc counseling, consists of ongoing consultation, case management, and referrals at the behest of the caregiver or any participating family member.

In focusing on the entire family, as well as in offering ongoing support and case management, the NYUCI appears well suited to addressing the significant psychosocial needs of many Latino caregivers. It has been pilot tested with Latinos (Rosenthal Gelman, 2010a, 2010b) and is currently being tested in a randomized controlled trial with this population (Luchsinger et al., 2012, 2016). Finding effective interventions addressing Latino caregivers' unique needs will not only reduce caregiver burden but delay institutionalization of care recipients and thus have a positive impact on public health.

Conclusion

Latinos are a large and rapidly growing population of older adults with high rates of dementia, relatively low levels of knowledge about ADRD and related diagnostic and treatment services, and significant language, cultural, geographic, financial, and immigrant status barriers to accessing the resources that do exist. Psychosocial interventions for Latinos with AD and their caregivers must be tailored to address these unique needs and obstacles. We have described a range of existing and developing interventions that demonstrate this can be accomplished in innovative and culturally sensitive ways.

Building on this base, future interventions need to be broad in scope and reach to accommodate the increasing numbers and diversity of the Latino population. Furthermore, given the psychosocial stressors impinging on this group, in addition to including case management components that address ADRD caregiving families' concrete needs, macro-level approaches addressing poverty; discrimination; lack of access to housing, education, and healthcare; and immigrant rights for Latinos are also necessary. Addressing such disparities is likely to lead to increased brain health in this population. Another currently underutilized approach includes encouraging Latinos and other so-called minority older adults and their caregivers to utilize research as a pathway to interventions and support. This will have the added benefit of increasing health and service providers' knowledge and understanding of these populations' experiences and needs. Comprehensive approaches at micro-, mezzo- and macro-levels that build and expand on the programs and services discussed in this chapter, from outreach to caregiver support, can serve as a foundation for the development of much-needed relevant, accessible resources for this underserved and important population.

References

Adams, B., Aranda, M., Kemp, B., & Takagi, K. (2002). Ethnic and gender differences in distress among Anglo American, African American, Japanese American, and Mexican American spousal caregivers of persons with dementia. *Journal of Clinical Geropsychology, 8*, 279–301.

Administration for Community Living (ACL). (2015). *A statistical profile of Hispanic older Americans aged 65+*. Washington, DC: Administration on Aging.

Alzheimer's Association. (2010). 2010 Alzheimer's disease facts and figures. *Alzheimer's & Dementia, 6*(2), 158–194.

Andrews, J. W., Lyons, B., & Rowland, D. (1992). Life satisfaction and peace of mind: A comparative analysis of elderly Hispanic and other elderly Americans. *Clinical Gerontologist, 11*(3/4), 21–42.

Aranda, M. P., Villa, V. M., Trejo, L., Ramirez, R., & Ranney, M. (2003). El Portal Latino Alzheimer's project: Model program for Latino caregivers of Alzheimer's disease-affected people. *Social Work, 48*(2), 259–271.

Ayalon, L., & Areán, P. A. (2004). Knowledge of Alzheimer's disease in four ethnic groups of older adults. *International Journal of Geriatric Psychiatry, 19*(1), 51–57.

Belle, S. H., Burgio, L., Burns, R., Coon, D., Czaja, S., Gallagher-Thompson, D., … Martindale-Adams, J. (2006). Enhancing the quality of life of dementia caregivers from different ethnic or racial groups: A randomized, controlled trial. *Annals of Internal Medicine, 145*(10), 727–738.

Borrayo, E. A. (2004). Where's Maria? A video to increase awareness about breast cancer and mammography screening among low-literacy Latinas. *Preventive Medicine, 39*, 99–110.

Boucher, N. A., Guadalupe, E., Lara, L., & Alejandro, M. (2014). Health care and end-of-life decisions: Community engagement with adults in East Harlem. *Journal of Community Health, 39*(6), 1032–1039.

Cabassa, L. J., Molina, G. B., & Baron, M. (2010). Depression fotonovela: Development of a depression literacy tool for Latinos with limited English proficiency. *Health Promotion Practice, 20*, 1–8.

Clark, C. M., DeCarli, C., Mungas, D., Chui, H., Higdon, R., Nuñez, J., … Ferris, S. (2004). Latino patients with Alzheimer's disease have an earlier age of symptom onset compared to Anglos. *Neurobiology of Aging, 25*(Supplement 2), S106.

Connell, C. M., Roberts, J. S., & McLaughlin, S. J. (2007). Public opinion about Alzheimer disease among blacks, Hispanics, and whites: Results from a national survey. *Alzheimer Disease and Associated Disorders, 21*(3), 232–240.

Coon, D. W., Rubert, M., Solano, N., Mausbach, B., Kraemer, H., Arguelles, T., … Gallagher-Thompson, D. (2004). Well-being, appraisal, and coping in Latina and Caucasian female dementia caregivers: Findings from the REACH study. *Aging and Mental Health, 8*(4), 330–345.

Daviglus, M. L., Talavera, G. A., Avilés-Santa, M. L., Allison, M., Cai, J., Criqui, M. H., … LaVange, L. (2012). Prevalence of major cardiovascular risk factors and cardiovascular diseases among Hispanic/Latino individuals of diverse backgrounds in the United States. *Journal of the American Medical Association (JAMA), 308*(17), 1775–1784.

Depp, C., Sorocco, K., Kasl-Godley, J., Thompson, L., Rabinowitz, Y., & Gallagher-Thompson, D. (2005). Caregivers self-efficacy, ethnicity, and kinship differences in dementia caregivers. *American Journal of Geriatric Psychiatry, 13*(9), 787–794.

Dilworth-Anderson, P., Williams, I. C., & Gibson, B. E. (2002). Issues of race, ethnicity and culture in caregiving research: A 20-year review (1980–2000). *The Gerontologist, 42*(2), 237–272.

Federal Interagency Forum on Aging-related Statistics (FIFA). (2016). *Older Americans 2016: Key indicators of well-being*. Hyattsville, MD: Author.

Fitten, L. J., Ortiz, F., & Pontón, M. (2001). Frequency of Alzheimer's disease and other dementias in a community outreach sample of Hispanics. *Journal of the American Geriatrics Society, 49*, 1301–1308. https://doi.org/10.1046/j.1532-5415.2001.49257.x

Gallagher-Thompson, D., Areán, P., Coon, D., Menendez, A., Takagi, K., Haley, W. E., … Szapocznik, J. (2000). Development and implementation of intervention strategies for culturally diverse caregiving populations. In R. Schulz (Ed.), *Handbook on dementia caregiving: Evidence-based interventions for family caregivers* (pp. 151–185). New York, NY: Springer.

Gallagher-Thompson, D., Areán, P., Rivera, P., & Thompson, L. W. (2001). Reducing distress in Hispanic family caregivers using a psychoeducational intervention. *Clinical Gerontologist, 23*, 17–32.

Gallagher-Thompson, D., Coon, D. W., Solano, N., Ambler, C., Rabinowitz, Y., & Thompson, L. W. (2003). Change in indices of distress among Latino and Anglo female caregivers of elderly relatives with dementia: Site-specific results from the REACH national collaborative study. *The Gerontologist, 43*(4), 580–591.

Gallagher-Thompson, D., Tzuang, M., Hinton, L., Alvarez, P., Rengifo, J., Valverde, I., … Thompson, L. W. (2015). Effectiveness of a Fotonovela for reducing depression and stress in Latino dementia family caregivers. *Alzheimer Disease and Associated Disorders, 29*(2), 146–153. https://doi.org/10.1097/WAD.0000000000000077

George, S., Duran, N., & Norris, K. (2014). A systematic review of barriers and facilitators to minority research participation among African Americans, Latinos, Asian Americans, and Pacific Islanders. *American Journal of Public Health, 104*, e16–e31.

Gitlin, L. N., Belle, S. H., Burgio, L. D., Czaja, S. J., Mahoney, D., Gallagher-Thomspon, D., … Ory, M. G. (2003). Effects of multicomponent interventions on caregiver burden and depression: The REACH multisite initiative at 6-month follow-up. *Psychology and Aging, 18*(3), 361–374.

Grigsby, T. J., Unger, J. B., Molina, G. B., & Baron, M. (2016). Evaluation of an audio-visual novella to improve beliefs, attitudes and knowledge toward dementia: A mixed-methods approach. *Clinical Gerontologist, 40*(2), 130–138.

Haan, M., Mungas, D. M., Gonzalez, H. M., Ortíz, T. A., Acharya, A., & Jagust, W. J. (2003). Prevalence of dementia in older Latinos: The influence of type 2 diabetes mellitus, stroke and genetic factors. *Journal of the American Geriatric Society, 51*, 169–177.

Harmell, A. L., Chattillion, E. A., Roepke, S. K., & Mausbach, B. T. (2011). A review of the psychobiology of dementia caregiving: A focus on resilience factors. *Current Psychiatry Reports, 13*(3), 219–224.

Hernandez, M. Y., & Organista, K. C. (2013). Qualitative exploration of an effective depression literacy fotonovela with at risk Latina immigrants. *American Journal of Community Psychology, 52*, 224. https://doi.org/10.1007/s10464-013-9587-1

Hinton, L., Franz, C. E., Yeo, G., & Levkoff, S. E. (2005). Conceptions of dementia in a multiethnic sample of family caregivers. *Journal of the American Geriatrics Society, 53*, 1405–1410.

Hinton, W. L., & Levkoff, S. (1999). Constructing Alzheimer's: Narratives of lost identities, confusion and loneliness in old age. *Culture, Medicine and Psychiatry, 23*, 453–475.

Jette, A. M., Crawford, S. L., & Tennstedt, S. L. (1996). Toward understanding ethnic differences in late-life disability. *Research on Aging, 18*, 292–309.

John, R., & McMillan, B. (1998). Exploring caregiver burden among Mexican-Americans: Cultural prescriptions, family dilemmas. *Journal of Aging and Ethnicity, 1*, 93–111.

John, R., Resendiz, R., & De Vargas, L. W. (1997). Beyond familism?: Familism as explicit motive for eldercare among Mexican American caregivers. *Journal of Cross-Cultural Gerontology, 12*, 145–162.

Knight, B. G., & Sayegh, P. (2010). Cultural values and caregiving: The updated sociocultural stress and coping model. *Journal of Gerontology: Psychological Sciences, 65B*(1), 5–13.

Lahaie, C., Earle, A., & Heymann, J. (2013). An uneven burden: Social disparities in adult caregiving responsibilities, working conditions, and caregiver outcomes. *Research on Aging, 35*(3), 243–274.

Li, C., Neugroschl, J., Umpierre, M., Martin, J., Huang, Q., Zeng, X., ... Sano, M. (2016). Recruiting U.S. Chinese elders into clinical research for dementia. *Alzheimer Disease and Associated Disorders, 30*(4), 345–347. https://doi.org/10.1097/wad.0000000000000162

Livney, M. G., Clark, C. M., Karlawish, J. H., Cartmell, S., Negrón, M., Nuñez, J., ... Arnold, S. E. (2011). Ethnoracial differences in the clinical characteristics of Alzheimer's disease at initial presentation at an urban Alzheimer's disease center. *The American Journal of Geriatric Psychiatry, 19*(5), 430–439.

Llanque, S. M., & Enriquez, M. (2012). Interventions for Hispanic caregivers of patients with dementia: A review of the literature. *American Journal of Alzheimer's Disease and Other Dementias, 27*(1), 23–32.

Luchsinger, J., Mittelman, M., Mejia, M., Silver, S., Lucero, R. J., Ramirez, M., ... Teresi, J. A. (2012). The northern Manhattan caregiver intervention project: A randomised trial testing the effectiveness of a dementia caregiver intervention in Hispanics in New York City. *BMJ Open, 2*(5), e001941.

Luchsinger, J. A., Burgio, L., Mittelman, M., Dunner, I., Levine, J. A., Kong, J., ... Teresi, J. A. (2016). Northern Manhattan Hispanic caregiver intervention effectiveness study: Protocol of a pragmatic randomised trial comparing the effectiveness of two established interventions for informal caregivers of persons with dementia. *BMJ Open, 6*(11), e014082.

Luchsinger, J. A., Tipiani, D., Torres-Patiño, G., Silver, S., Eimicke, J. P., Ramirez, M., ... Mittelman, M. (2015). Characteristics and mental health of hispanic dementia caregivers in New York City. *American Journal of Alzheimer's Disease and Other Dementias, 30*(6), 584–590.

Mahoney, D. F., Cloutterbuck, J., Neary, S., & Zhan, L. (2005). African American, Chinese, and Latino family caregivers' impressions of the onset and diagnosis of dementia: Cross-cultural similarities and differences. *The Gerontologist, 45*(6), 783–792.

Marquez, D. X., Wilbur, J., Hughes, S., Berbaum, M. L., Wilson, R., Buchner, D. M., & McAuley, E. (2014). B.A.I.L.A. – A Latin dance randomized controlled trial for older Spanish-speaking Latinos: Rationale, design, and methods. *Contemporary Clinical Trials, 38*(2), 397–408.

Melillo, K. D., Williamson, E., Houde, S. C., Futrell, M., Read, C. Y., & Campasano, M. J. (2001). Perceptions of older Latino adults regarding physical fitness, physical activity, and exercise. *Gerontological Nursing, 27*(9), 38–46.

Mittelman, M. S., Ferris, S. H., Shulman, E., Steinberg, G., Ambinder, A., Mackell, J., & Cohen, J. (1995). A comprehensive support program: Effect on depression in spouse-caregivers of AD patients. *The Gerontologist, 35*, 792–802.

Mittelman, M. S., Ferris, S. H., Shulman, E., Steinberg, G., & Levin, B. (1996). A family intervention to delay nursing home placement of patients with Alzheimer disease: A randomized controlled trial. *Journal of the American Medical Association, 276*, 1725–1731.

Mittelman, M. S., Haley, W. E., Clay, O. J., & Roth, D. L. (2006). Improving caregiver well-being delays nursing home placement of patients with Alzheimer's disease. *Neurology, 67*, 1593–1599.

Mittelman, M. S., Roth, D. L., Coon, D. W., & Haley, W. E. (2004). Sustained benefit of supportive intervention for depressive symptoms in Alzheimer's caregivers. *American Journal of Psychiatry, 161*, 850–856.

National Alliance for Caregiving/American Association of Retired Persons (NAC/AARP). (2015). *Family caregiving in the U.S. 2015*. Washington, DC: The NAC and AARP Public Policy Institute.

Neugroschl, J. (2016). Attitudes and perceptions of research in aging and dementia in an urban minority population. *Journal of Alzheimer's Disease, 53*(1), 69–72.

Neugroschl, J., & Sano, M. (2010). Current treatment and recent research in Alzheimer disease. *Mount Sinai Journal of Medicine, 77*(1), 3–16.

Neugroschl, J., Sano, M., Luo, X., & Sewell, M. (2014). Why they stay: Understanding research participant retention in studies of aging, cognitive impairment and dementia. *Journal of Gerontology & Geriatric Research, 3*(4), 1000170.

Ortiz, F., & Fitten, L. J. (2000). Barriers to healthcare access for cognitively impaired older Hispanics. *Alzheimer Disease and Associated Disorders, 14*(3), 141–150.

Pinquart, M., & Sörensen, S. (2003). Differences between caregivers and noncaregivers in psychological health and physical health: A meta-analysis. *Psychology and Aging, 18*, 250–267.

Pinquart, M., & Sörensen, S. (2005). Ethnic differences in stressors, resources, and psychological outcomes of family caregivers: A meta-analysis. *The Gerontologist, 345*, 90–106.

Pinquart, M., & Sörensen, S. (2006). Helping caregivers of persons with dementia: Which interventions work and how large are their effects? *International Psychogeriatrics, 18*(4), 577–595.

Polich, T., & Gallagher-Thompson, D. (1997). Preliminary study investigating psychological distress among female Hispanic caregivers. *Journal of Clinical Geropsychology, 3*, 1–15.

Rosenthal Gelman, C. (2010a). Learning from recruitment challenges: Barriers to diagnosis, treatment, and research participation for Latinos with symptoms of Alzheimer's disease. *Journal of Gerontological Social Work, 53*(1), 94–114.

Rosenthal Gelman, C. (2010b). "La lucha": The experience of Latino family caregivers of patients with Alzheimer's disease. *Clinical Gerontologist, 33*(3), 181–193.

Rosenthal Gelman, C. (2014). Familismo and its impact on the family caregiving of Latinos with Alzheimer's disease: A complex narrative. *Research on Aging, 36*(1), 39–70.

Rosenthal Gelman, C., Giunta, N., & Glushefski, R. (2017). *Community based participatory research with older adults in East Harlem.* (symposium organizer). Contribution *East Harlem's Latinos and Alzheimer's disease: Community-engaged identification of needs and development of solutions.* Society for Social Work and Research, New Orleans, LA, January 13, 2017.

Ruiz, M. E., & Ransford, H. E. (2012). Latino elders reframing familismo: Implications for health and caregiving support. *Journal of Cultural Diversity, 19*(2), 50–57.

Sabogal, F., Marin, G., Otero-Sabogal, R., Marin, B. V. O., & Perez-Stable, E. J. (1987). Hispanic familism and acculturation: What changes and what doesn't? *Hispanic Journal of Behavioral Sciences, 9*, 397–412.

Sano, M., Sewell, M., Umpierre, M., Neugroschl, J., & Cedillo, G. (2016). Barriers to cognitive evaluations among elderly Lations: Using community-based participatory research to create health education videos. *Alzheimer's & Dementia: The Journal of the Alzheimer's Association, 12*(7), P791–P792.

Singhal, A., & Rogers, E. (2012). *Entertainment-education: A communication strategy for social change*. New York, NY: Routledge.

Sink, K. M., Covinsky, K. E., Newcomer, R., & Yaffe, K. (2004). Ethnic differences in the prevalence and pattern of dementia-related behaviors. *Journal of the American Geriatrics Society, 52*, 1277–1283.

Smith-Morris, C., Morales-Campos, D., Alvarez, E. A. C., & Turner, M. (2013). An anthropology of familismo: On narratives and description of Mexican/immigrants. *Hispanic Journal of Behavioral Sciences, 35*(1), 35–60.

Sood, S. (2002). Audience involvement and entertainment-education. *Communication Theory, 12*, 153–172.

Sood, S., & Storey, D. (2013). Increasing equity, affirming the power of narrative and expanding dialogue: The evolution of entertainment education over two decades. *Critical Arts: South-North Cultural and Media Studies, 27*, 9–35. https://doi.org/10.1080/02560046.2013.767015

Sörensen, S., Pinquart, M., & Duberstein, P. (2002). How effective are interventions with caregivers? An updated meta-analysis. *The Gerontologist, 42*, 356–372.

Stoller, E., & Gibson, R. (1994). *Worlds of difference*. Thousand Oaks, CA: Pine Forge Press.

Tang, M. X., Stern, Y., Marder, K., Bell, K., Gurland, B., Lantigua, R., … Mayeux, R. (1998). The APOE-epsilon4 allele and the risk of Alzheimer's disease among African Americans, whites, and Hispanics. *Journal of the American Medical Association, 279*(10), 751–755.

Tennstedt, S. L., Chang, B., & Delgado, M. (1998). Patterns of long-term care: A comparison of Puerto Rican, African-American, and non-Latino white elders. *Journal of Gerontological Social Work, 30*(1/2), 179–199.

Umpierre, M., Meyers, L. V., Ortiz, A., Paulino, A., Rodriguez, A. R., Miranda, A., … McKay, M. M. (2015). Understanding Latino parents' child mental health literacy: Todos a bordo/all aboard. *Research on Social Work Practice, 25*(5), 607–618. https://doi.org/10.1177/1049731514547907. http://dx.doi.org.tcsedsystem.idm.oclc.org/

Unger, J. B., Cabassa, L. J., Moina, G. B., Contreras, S., & Baron, M. (2012). Evaluation of a fotonovela to increase depression knowledge and reduce stigma among Hispanic adults. *Journal of Immigrant and Minority Health, 14*(2), 398–406.

Valle, R., Yamada, A. M., & Matiella, A. C. (2006). Fotonovelas: A health literacy tool for educating Latino older adults about dementia. *Clinical Gerontologist, 30*(1), 71–88.

Wilkin, H. A., & Ball-Rokeach, S. J. (2006). Reaching at risk groups: The importance of health storytelling in Los Angeles Latino media. *Journalism, 7*(3), 299–320.

Wilkin, H. A., Valente, T. W., Murphy, S., Cody, M. J., Huang, G., & Beck, V. (2007). Does entertainment-education work with Latinos in the United States? Identification and the effects of a telenovela breast cancer storyline. *Journal of Health Communication, 12*, 455–469.

Cognitive Rehabilitation for Maintenance of Function in Latinos with Dementia

10

Shawneen R. Pazienza and Erin E. Andrews

Abstract

Numerous health, cultural, and socioeconomic factors are associated with dementia in older adult Latinos, a rapidly growing demographic in the United States (U.S.). Dementia involves significant cognitive deficits that interfere with daily functioning. Cognitive rehabilitation is an important intervention that can prevent, delay, mediate, or repair aspects of cognitive deterioration through changes in brain pathways, or neuroplasticity, especially in those with mild cognitive impairment (MCI) and mild to moderate diagnoses of dementia. However, there is scant research as to how Latinos with dementia can best benefit from cognitive rehabilitation. Latinos are underrepresented in research participant samples of dementia patients, and the results of cognitive rehabilitation trials involving other types of neurocognitive issues (i.e., traumatic brain injury, cerebrovascular accident) are not always generalizable to dementia. Despite this, it is well established that culture can affect patients' experience and understanding of their impairment, expectations, engagement, and progress in rehabilitation. By applying Latino cultural strengths, Latino patients with dementia can benefit from an individually and culturally tailored cognitive rehabilitation treatment plan. Without engaging in stereotyping and with full awareness of the rich within-group diversity among Latinos, there are several cultural concepts that are central to Latino communities that should be considered within the context of cognitive rehabilitation for treatment of dementia, including family/*familismo*, respect/*respeto*, personal relationship/*personalismo*, trust/*confianza*, spirit/*espiritu*, and presence/*presentismo*, among others. Integration of these important cultural implications can lead to optimal treatment outcomes for Latinos with dementia.

S. R. Pazienza (✉)
Central Texas Veterans Health Care System, Austin, TX, USA
e-mail: shawneen.pazienza@va.gov

E. E. Andrews
Central Texas Veterans Health Care System, University of Texas at Austin Dell Medical School, Texas A&M Health Science Center College of Medicine, Austin, TX, USA
e-mail: erin.andrews2@va.gov

Introduction

According to the fifth edition of the *Diagnostic and Statistical Manual of Mental Disorders (DSM-5)*, an individual is considered to meet diagnostic criteria for major neurocognitive disorder (i.e., dementia) when there is evidence of significant cognitive decline from a previous

H. Y. Adames, Y. N. Tazeau (eds.), *Caring for Latinxs with Dementia in a Globalized World*,
https://doi.org/10.1007/978-1-0716-0132-7_10

level of performance in at least one cognitive domain and deficits interfere with the individual's independence in daily activities (American Psychiatric Association, 2013). Patients with dementia, regardless of the dementia's etiology, experience significant deficits in cognitive functioning that interfere with their independence in daily activities (ADLs) such as functional mobility (i.e., transferring ability), self-feeding, bathing or showering, personal hygiene and grooming, dressing, and toileting, among other basic tasks (Troyer, 2017). Also, adversely impacted by the effects of dementia, instrumental activities of daily living (IADLs) require advanced skill levels and include home management, meal preparation and cleanup, shopping for groceries and necessities, managing money, taking prescribed medications, and using the telephone or other forms of communication (Troyer, 2017).

The goal of this chapter is to describe the application of cognitive rehabilitation as an intervention for maintenance of function in Latino populations aimed to inform a broad range of medical and rehabilitation professionals. The objectives of this chapter are to first explain the various stages and strategies of cognitive rehabilitation applied to dementia, then to explore how cognitive rehabilitation interventions can be appropriately utilized culturally with Latino populations, and finally to address the role of caregivers in cognitive rehabilitation for maintenance of function for Latinos with dementia. While we hope that some of these approaches will be useful to different types of providers and family members, cognitive rehabilitation should be provided by trained and qualified professionals for whom cognitive rehabilitation is in the scope of practice (i.e., rehabilitation psychologists, neuropsychologists, speech language pathologists, or occupational therapists).

The rates of diagnoses of cognitive and neurological disorders in Latinos have increased across the world (Arango-Lasprilla, 2012). Prevalence varies among Latino subgroups, but in all groups there is significant association with type 2 diabetes mellitus and cerebral vascular accident (CVA). Latinos tend to develop symptoms of dementia earlier than Whites, and with increases in lifespan, rates of dementia among older Latino adults are expected to grow (Alvarez, Rengifo, Emrani, & Gallagher-Thompson, 2014). It is important to note the increased rates of dementia and associated etiologies among Latinos to understand the significance of identifying and developing culturally relevant interventions to maintain functioning in this growing and diverse population.

This chapter is intended to address cognitive rehabilitation for a multiplicity of Latinos, but it is important to note there are differences among Latino subgroups. Latinos are not monolithic in their experience of the disease process. For example, while Latinos of Caribbean descent in the United States are at significantly higher risk of dementia than White Americans, Latinos with Mexican heritage have similar rates to their White counterparts (Mayeda, Glymour, Quesenberry, & Whitmer, 2016). Numerous factors such as acculturation, immigration, socioeconomic status, obesity, access to healthcare, and educational level are associated with dementia in Latinos (Ottenbacher et al., 2009). Genetic studies also suggest that allele differences (i.e., forms of a gene arising from mutation) among different racial and ethnic groups may contribute to differing etiologies of dementia (Mayeda et al., 2016). Further investigation of these topics will be crucial, as some estimates predict a sixfold increase in the older adult Latino American population in the United States by the year 2050, representing 20% of the U.S. population (Alvarez et al., 2014; Vincent & Velkoff, 2010).

Assessment and Identification of Dementia

Determination of cognitive decline can be based on either concern of the individual, a knowledgeable informant or a clinician, or demonstration of substantial cognitive impairment through standardized neuropsychological assessment or other quantified clinical evaluations. Comprehensive neuropsychological assessment is preferred, as its benefits go beyond diagnostic clarification. For instance, a comprehensive neuropsychological assessment can (1) assist in identifying the areas of impairment that may interfere with daily functioning and (2) identify areas of relative strength or preserved functioning that can support performance in daily activities and which can be engaged to assist in the use of compensatory strategies. Cognitive and psychological assessment can assist in treatment planning by identifying the level of assistance needed and the ways in which independence in functioning can be maximized (Sohlberg & Mateer, 2017). A comprehensive assessment may also include exploration of emotional or psychological factors that may impact the patient's rehabilitation.

The Neuroscience of Cognitive Rehabilitation: How It Works

Evidence exists that cognitive rehabilitation activities work by creating changes in brain pathways (Smith et al., 2009; van Paasschen et al., 2013). The brain is often referred to as "plastic," indicating that it can change in structure and function as the result of learning and experience, and plasticity is thought to be the outcome of activity in the spaces between nerve cells called synapses (Smith et al., 2009; Sohlberg & Mateer, 2017). Some participants in cognitive rehabilitation demonstrate new activation on functional magnetic resonance imaging (fMRI) brain scans during encoding (learning), the first step of memory when the brain performs an initial analysis of content, as well as during memory recall tasks wherein patients are asked to recollect content previously presented to them (Belleville et al., 2011; Sohlberg & Mateer, 2017). The distribution, i.e., pattern, of fMRI activation during specific cognitive rehabilitation activities has shown to be consistent with the areas of the brain where the skill being measured is based (Belleville et al., 2011; Clare et al., 2010). Participants who were taught language strategies demonstrated activation of the brain's left temporal lobe, where language is processed, while spatial and object memory strategies activated the right prefrontal and parietal areas of the brain and skill learning activated the cerebellum and basal ganglia (Belleville et al., 2011).

One implication of these findings is that cognitive rehabilitation can change brain activity by altering synaptic activity in specific locations in the brains of those at high risk of dementia due to mild cognitive impairment. Importantly, positive correlations have been observed between the changes in brain activation and performance and behavior on daily tasks, suggesting that changes in brain structure may have consequences for real-life functioning (Clare et al., 2010). Further, limited research suggests that cultural differences in behavior may have neural correlates; in other words, preliminary evidence indicates that differences in brain functioning on fMRI are associated with cultural group membership (Adams et al., 2010; Chiao & Blizinsky, 2009). These findings would support the importance of not only individualized but also culturally congruent cognitive rehabilitation interventions.

Cognitive Rehabilitation for Dementia

Evidence suggests that although dementia is characterized by cognitive deterioration over time, cognitive rehabilitation treatments can improve cognitive functioning, particularly in those with mild cognitive impairment (MCI) and mild dementia (Hopper et al., 2013). These interventions are receiving increasing attention as a means of delaying cognitive decline, restoring function, and compensating for deficits (Walton, Mowszowski, Lewis, & Naismith, 2014).

Patients with mild cognitive impairment (MCI) can benefit from cognitive rehabilitation treatment. A diagnosis of MCI is warranted when an individual experiences cognitive impairment that does not interfere with one's ability to independently perform daily functional tasks. People with MCI, particularly those with deficits in memory, are considered at increased risk of developing Alzheimer's disease (AD). For this reason, MCI is often considered a prodromal phase of AD (Gauthier et al., 2006). Thus, providing cognitive rehabilitation to those with MCI may reduce the risk of developing dementia, delay progression into dementia, or help mediate the effects of declining cognition (Greenaway, Duncan, & Smith, 2012; Willis et al., 2006). Research suggests that patients with MCI who participate in cognitive rehabilitation experience improvement on objective and subjective measures of memory (Jean, Bergeron, Thivierge, & Simard, 2010) as well as attention (Gagnon & Belleville, 2012).

The review studies of cognitive rehabilitation interventions indicate that largest positive effects are observed in those individuals with mild to moderate dementia, with smaller effects reported in people with severe dementia; one hypothesis is that those with more significant dementia may have benefited less from treatment because of more severe deficits and impairment in multiple cognitive domains (de Werd, Boelen, Rikkert, & Kessels, 2013).

Application of Cognitive Rehabilitation to Latinos with Dementia

Less is known about how cultural differences in families may affect the cognitive rehabilitation process in general, specifically for Latinos with dementia. Health disparities exist in the provision of cognitive rehabilitation among economically, racially, culturally, and linguistically diverse patients; indeed, access to disability rehabilitation services is limited for Spanish-speaking people in the United States and other countries (Arango-Lasprilla, 2012; Arango & Kreutzer, 2010). Cultural and linguistic differences may influence the process of cognitive rehabilitation, yet very few studies have examined outcomes of cognitive rehabilitation among Latinos and/or Spanish speakers (Niemeier & Arango, 2007; Tzuang, Owusu, Spira, Albert, & Rebok, 2017).

Most interventions described in the literature, including computerized programs such as cognitive vitality training (Gooding et al., 2016), are in English, developed with primarily White participant data, and are therefore questionable regarding their ability to be appropriately adapted for diverse cultural groups (Tzuang et al., 2017). Given the underrepresentation of Latinos in cognitive rehabilitation research, it is difficult to ascertain if empirically supported interventions are adequately comparable. Tzuang et al. (2017) point out that simple comparison studies between racial and ethnic groups will not suffice; using Whites as the norm and ignoring within-minority group differences and mediating factors, such as gender, will do little to advance evidence-based interventions.

Another related problem in examining the utility of cognitive rehabilitation among Latinos with dementia is that most studies that do include Latinos and/or Spanish-speaking participants are focused on other etiologies of cognitive dysfunction, such as acquired brain injury, cerebral vascular accidents, or serious mental illness such as schizophrenia (Arango-Lasprilla, 2012; Cicerone et al., 2011). These differences are important given that these conditions affect cognitive functioning but typically do not result in continued cognitive deterioration. Dementia involves continued decline, making it difficult to apply findings from other neurocognitive disorders, although some results may extrapolate to MCI and dementia. Clearly, further research on Latinos with dementia is warranted to explore intergroup differences and the cultural relevance of existing interventions and to establish empirical support for interventions developed specifically for Latino populations.

While there is a paucity of research about specific treatment models and outcomes of cognitive rehabilitation for dementia among Latino patients, it is widely accepted that culture will largely influence a patient's presentation of symptoms, understanding of their illness, expectations for recovery, engagement in rehabilitation services, and overall outcome of treatment (Lequerica & Krch, 2014). Cognitive rehabilitation is considered to be most effective when the treatment plan is individualized based on the patient's needs (Haskins et al., 2012; Lam et al., 2010); therefore, careful consideration of a patient's cultural characteristics and values should lead to the best treatment outcomes for Latino patients. It is also important to keep in mind the existence of within-group heterogeneity (Gelman, 2010), understanding that information about cultural values should not be presumed to reflect the experience of all Latinos.

There is evidence that cognitive rehabilitation works among Spanish-speaking individuals, as was seen in Guardia-Olmosa, Esparcia, Morales, and Ferred (2012) meta-analysis of studies that included Spanish-speaking cognitive rehabilitation participants. Although methodologically hampered by the small amount of studies that met inclusion criteria and the differing approaches to measurement and technique among individual studies, the results yielded statistically significant improvements in health-related quality of life with moderate effect sizes, associated with improvements in cognitive functioning. These findings were consistent regardless of etiology and included studies with patients with dementia. However, in studies where longitudinal data were available, the effect diminished over time, which may indicate that initial gains are temporary (Guardia-Olmosa et al., 2012).

Goals of Cognitive Rehabilitation

The primary goal of cognitive rehabilitation is to reduce the negative impact of cognitive decline on a patient's daily functioning. The process of cognitive rehabilitation is goals driven, which is established in a collaborative manner between the patient, the clinician, and at times a caregiver. Typically, goals are focused on increasing a patient's safety, increasing independence, and improving quality of life (Cicerone et al. 2011; Sohlberg & Mateer 2017). When setting goals, providers should heed the importance of the Latino value *confianza* [trust], which emphasizes the importance of building assurance and intimacy within interpersonal relationships. It will be of benefit to Latino patients for rehabilitation treatment providers to take extra time in building rapport and establishing trust in the patient-provider relationship, and it has been suggested that establishing rapport with Latino patients should be considered one of the crucial first steps in developing a treatment plan (Añez, Silva, Paris, & Bedregal, 2008).

Progress toward goals is typically achieved in a stepwise manner, with gains in short-term goals or objectives building into progress toward a patient's long-term goals. For example, a patient with mild to moderate dementia may have a goal of increased independence in managing his or her ADLs. Achievement of a long-term goal of increased independence may first rely on success in intermediary short-term goals, such as successful independent use of a calendar or a pillbox for medication. A balance is needed between fostering hope (e.g., the possibility of increasing independence in management of a schedule and medications) and continuing to help the patient evaluate his or her progress toward the goal (e.g., success in independent use of a calendar and pillbox). Monitoring progress will help determine whether the long-term goal is achievable or modification is indicated. It is critical to create a connection between the short-term intermediary goals and the patient's long-term goals. Otherwise, the value of the therapeutic activities may be lost.

Patients who participate in cognitive rehabilitation may benefit from reminders of how therapeutic tasks are relevant to the progression toward their identified long-term goals. A cultural variable to consider in the context of cognitive rehabilitation is the concept of time and how time is perceived and used within Latino communities. The concept of time is often thought of as a culturally dependent phenomenon, as different cultures have varying perspectives on time and importance of speed (Ardila, 2005; Brickman, Cabo, & Manly, 2006). For example, speed and quality may be considered contradictory constructs within some cultures; these communities believe quality arises from a slow and careful process (Ardila, 2005). Additionally, the value of speed and efficiency is a largely Westernized value. Time, within Latino cultures, is often focused on the present moment rather than the past or future – a concept referred to as *presentismo* (Talamantes & Sanchez-Reilly, 2010). In contrast to non-Hispanic, White patients, Latinos are less likely to consider the ways in which the present moment might be incorporated into the future, so providers may need to be culturally aware and sensitive in helping patients link short- and long-term goals. In the context of cognitive rehabilitation, Latino patients may approach rehabilitation tasks and goal-setting with more of a focus on the present moment (Arredondo, Gallardo-Cooper, Delgado-Romero, & Zapata, 2015). This temporal difference has implications for the treatment dynamic. In this regard, Latino patients and their families may have goals and expectations regarding treatment outcomes that are more present focused than future focused. One potential example is a Latino patient or family who expects immediate results from cognitive rehabilitation. Providers who are aware of this cultural dynamic can more easily collaborate with patients and their families by setting and focusing on several smaller, short-term goals that can give the patient a more immediate sense of accomplishment.

Engagement in Cognitive Rehabilitation

Cultural factors should also be considered when engaging Latinos with dementia in cognitive rehabilitation. *Espiritu* [spirit], which includes the concepts of religion and spirituality, is an important aspect of the Latino culture; and it is likely that a patient's religious and spiritual belief systems will impact their attitudes and beliefs regarding disability and their motivation for treatment (Lequerica & Krch, 2014). Religion and spirituality often provide a framework for conceptualizing why things happen, and many patients turn to their religious and spiritual frameworks as a means of coping with difficult situations (Herrera, Lee, Nanyonjo, Laufman, & Torres-Vigil, 2009). Generally speaking, patients may engage in either positive or negative religious coping, labeled as such to reflect either the positive or negative outcomes associated with these methods of coping (Ano & Vasconcelles, 2005). For example, positive religious coping may include seeking religious direction, benevolent religious appraisal (e.g., redefining the stressor as an opportunity for spiritual growth), and collaborative spiritual support. In contrast, negative religious coping may involve demonic or punishing religious reappraisal (e.g., redefining the stressor as the work of the devil or punishment from God), pleading for direct intercession, or spiritual discontentment (Pargament, Koenig, & Perez, 2000). A study by Dunn and O'Brien (2009) focused on the use of religious coping within a Latino population found that the participants reported moderate to high levels of positive religious coping, including a tendency toward benevolent religious appraisal and collaborative religious coping (i.e., the participants reported looking for strength and support in God and felt they were collaborating with God to address their situation). Latino patients have also been found to have particularly strong religious social support networks (Dunn & O'Brien, 2009) which would likely be an asset in the context of cognitive rehabilitation. Negative religious coping has also been observed in Latino populations. The same study by Dunn and O'Brien found that, although participants generally reported low to moderate levels of negative religious coping, there was indication of a high level of pleading for direct intercession by God which has been associated with negative outcomes (Pargament et al., 2000). Additionally, Latino patients have been observed to exhibit religious fatalism (e.g., belief that the stressor is God's will and therefore nothing can be done about it), which has been linked to decreased motivation to participate in treatment (Dilworth-Anderson & Gibson, 2002; Sayegh & Knight, 2012). Spiritual and religious beliefs can also impact the conceptualization of the dementing illness or injury, with some Latino patients or families potentially interpreting the symptoms of dementia as the presence or possession of an evil spirit, which could negatively impact the patient's access to appropriate diagnosis and treatment (Sayegh & Knight, 2012).

Stages of Cognitive Rehabilitation

Cognitive rehabilitation generally takes place over three primary stages: *acquisition*, *application*, and *adaptation* (Sohlberg & Mateer, 2017). In the *acquisition* stage, the primary task is orientation to treatment and the initial introduction of rehabilitation strategies. At this stage, treatment providers should be aware of cultural considerations in working with Latinos. The term *personalismo* reflects the value of personal relationship among Latinos, the expectation of interpersonal closeness, and the emphasis of the value of people over objects (Añez et al., 2008; Furman et al., 2009). In the context of cognitive rehabilitation, it would be beneficial for rehabilitation specialists to be aware of this value and the ways in which *personalismo* may affect treatment. For example, Latino patients may be more likely than non-Hispanic patients to inquire about the details of the rehabilitation treatment provider's personal life. In this context, providing a few professionally appropriate disclosures may be beneficial for the patient-provider relationship and assist in developing and maintaining rapport and fostering alliance in the treatment (Furman et al., 2009).

The value of *respeto* emphasizes the importance of mutual respect for one another, including respect specifically for elders and for professionals (Alvarez et al., 2014; Añez et al., 2008). Given that respect is highly valued in Latino culture, rehabilitation treatment providers should be careful to ensure that they are demonstrating respect for their Latino patients. Use of more formal language, for example, may be one way to verbally demonstrate respect. It is advisable to ask a Latino patient how he or she prefers to be addressed, i.e., by their family/last name or by first name. Clinicians should also be aware that Latino patients might avoid openly disagreeing with professionals out of concern of being disrespectful (Añez et al., 2008). It is important to note that the value of *personalismo* may also involve a desire to maintain harmonious relationships with others, which may lead a patient to avoid expressing disagreement or taking an assertive stance with others (Antshel, 2002). In the context of cognitive rehabilitation, this may mean that Latino patients may not assert themselves in regard to their treatment needs or goals or may not be explicit in voicing their concerns. To mitigate the possibility of a Latino patient withholding dissent, providers should establish a dynamic whereby the patient is routinely invited and encouraged to share feedback, especially for treatment areas of concern.

During the *application* stage, patients begin to utilize strategies to engage in simple tasks, often within the context of a therapy session or other in-clinic settings. The goal of repetitive practice of strategies is to increase the effectiveness of their use and decrease the patient's reliance on external assistance. Internalization of a learned strategy occurs when the skill becomes automated, i.e., independent skill use without cueing.

The goal of the *adaptation* phase is to promote transfer of skills to more functional and everyday tasks outside the context of the clinic and in a variety of personally relevant situations or settings. This may include tasks that are less structured, more novel, and more complex than those practiced in the clinic setting. Generalization occurs when skills that are taught and practiced in the clinical setting can be applied to other similar tasks or settings. Generalization to real-life settings is most likely to occur when the rehabilitation services resemble the patient's real-life experience as closely as possible (Geusgens, Winkens, van Heugten, Jolles, & van den Heuvel, 2007). As a result, Latino patients should practice cognitive rehabilitation skills in settings and contexts that are congruent with their cultural experiences. For example, the linguistic environment of the patient should drive the decision for a bilingual patient to use mnemonic (i.e., "memory") strategies in English or Spanish. Based on the characterization and level of impairment and the patient's unique needs and goals, it is important to note that not all patients may reach each level of treatment.

Cognitive Rehabilitation Strategies

External Versus Internal

Cognitive rehabilitation strategies are either *external* or *internal* in nature. *External* strategies are those that are externally available to the patient (Cicerone et al., 2011). This may include devices, such as calendars, auditory or visual cueing systems (e.g., alarms, smartphone applications), and other task-specific aids (e.g., pillboxes). The provision of cues by others, such as a caregiver or therapist, is also considered an external strategy. A commonly used external strategy is the use of a memory book. Depending on the level of impairment and needs of the patient, a memory book may include orientation information such as current date, location, pictures and/or names of relevant individuals (e.g., nursing staff), and personal messages (e.g., notes from family during their last visit). Memory books are used on a daily basis and include sections detailing the patient's daily schedule, a memory log, and a to-do list. The patient is expected to use the memory log to write notes regarding the activities of their day. The to-do list can include tasks that the patient is expected to complete in the future and may be transferred to the patient's schedule as needed. Depending on the level of impairment, a daily schedule may be completed by the patient or a caregiver.

Other *external* reminder systems can assist in cueing patients to complete specific tasks. For example, a patient with the goal of increasing independence in managing his or her medications may place a note on his or her coffee maker as a reminder to take medications in the morning. The research suggests that formal memory support systems, including calendar/appointment systems, to-do lists, and regular journaling of the day's events, along with structured training, practice, and homework assignments, can lead to significant improvements in reported functional ability and sense of self-efficacy (Greenaway et al., 2012).

All of these strategies are heavily reliant on the ability to read and write. Providers should be aware of lower mean rates of formal education and higher rates of illiteracy in elderly Latinos, particularly among immigrants (Farias, Mungas, Hinton, & Haan, 2011). As a result, it is important not to over-rely on literacy-based approaches. McDougall et al. (2010) found that minority participants in their sample ($n = 265$), which was represented by 17% Hispanic, made significantly more improvements compared to Whites on a measure of visual memory following a health promotion and memory training intervention. Use of pictures or other representative visual stimuli may serve as a substitution for written cues for those individuals without the ability to read and write. One possible adaptation is use of a communication system developed for people with autism, the picture exchange communication system (PECS). PECS is used to communicate through exchanging pictures of a desired item with caregivers in order to receive the item (Ziomek & Rehfeldt, 2008). It is also possible to use the pictures to create short sentences (e.g., "I want, "I see," and "I have"). The pictures can be kept with or near the patient (i.e., in a pocket or wheelchair pouch). Instead of memory books with text, representative pictures can be substituted. Alternatively, Latinos immersed in oral tradition may be more responsive to the repetition of meaningful oral material, rather than the use of external aids. Family and community members can be excellent sources of culturally congruent traditions, phrases, and symbols that are salient for the patient in his or her natural environment (Judd & DeBoard, 2007, 2009).

In working with Latino patients, the rehabilitation provider should be mindful to make appropriate adjustments in treatment planning that are more inclusive of the patient's family members. Whereas non-Hispanic, White patients tend to receive their dementia care in assisted living or nursing facilities, Latino patients are more likely to receive their care at home and are more likely to have their family members be directly involved in their rehabilitation care (Alvarez et al., 2014; Taylor & Jones, 2014). For this reason, cognitive rehabilitation strategies may be more likely to work if they are developed around the family's needs. For example, the use of a personalized memory book or calendar system may not appropriately reflect Latino patients' collectivistic value system, which places greater emphasis on group interests, belongingness, and conforming to in-group expectations (Furman et al. 2009). For example, the development of a shared family calendar rather than a personal calendar may more appropriately reflect the patient's values and could lead to improved treatment outcomes.

For more severely impaired patients, reality orientation may be a useful strategy. Reality orientation involves presenting information intended to reorient the patient with dementia throughout the day (for a comprehensive overview, see Woods, 2014). Examples of reality orientation strategies may include a prominently displayed whiteboard with information regarding the date, day of week, and location. This can be used alone or in conjunction with another cognitive rehabilitation program (e.g., prospective memory process training; Sohlberg & Mateer, 2017). There is indication that reality orientation can have a positive effect on a patient's cognitive functioning and behavior; although the results tend to be short-lived, they should still be included as a part of a long-term treatment program (Woods, 2014).

Internal cognitive strategies are those that come from within the patient, either through self-cueing or other action sequences that assist the patient in approaching the task at hand. Organizational strategies, such as the use of acronyms, can assist in simplifying the storage and retrieval of information. For example, the familiar acronym "ROYGBIV" helps facilitate the learning and recall of the colors of the rainbow in the accurate order (*r*ed, *o*range, *y*ellow, *g*reen, *b*lue, *i*ndigo, *v*iolet). In this acronym, each letter serves as a cue for the correct color and reduces the burden on the memory system. Semantic clustering is another internal strategy that helps facilitate the organization of information into meaningful clusters (Sohlberg & Mateer, 2017). A shopping list, for example, can be sorted into specific categories (e.g., produce, meat, dairy, boxed goods). In this situation, the categories serve as a cue for recalling the sorted grocery items and reduce the cognitive burden of having to recall the shopping list from memory. Association techniques can also assist in improving memory by linking two or more items or tasks together to facilitate better recall (Sohlberg & Mateer, 2017). For example, name-face association training can increase memory for names by pairing verbal information with visual stimuli. Other examples of association include linking tasks, such as pairing the taking of medication with brushing teeth (when medication is to be taken twice per day) or with a meal (when medication is to be taken three times per day). Paired associations and other nonverbal techniques are especially useful when working with Latinos who have lower levels of formal education and/or inability to read and write.

Selection of internal or external strategies depends on the severity of impairment and the specific needs and goals of the patient. In general, patients with more significant cognitive impairment (i.e., moderate to severe) will benefit more from use of external rather than internal strategies. Additionally, the earlier phases of cognitive rehabilitation will initially rely on more external strategies, particularly as the provider begins the work of introducing the tasks and providing the initial instructions for strategy use to the Latino patient. As a patient makes progress toward the treatment goals and makes gains in internalization and generalization of cognitive rehabilitation tasks, the types of strategies will become increasingly more internal in nature. More complex cognitive tasks will likely involve a combination of internal and external strategies.

Compensatory Versus Restorative

The strategies learned in cognitive rehabilitation may be *compensatory* or *restorative* in nature (Sohlberg & Mateer, 2017). *Compensatory* strategies aim to teach patients ways to compensate for their deficits and often involve a "work-around." An example of a compensatory strategy for memory loss may be having patients refer to a daily planner rather than expecting them to recall a schedule from memory. Again, providers may consider substituting pictures in the daily planner for those with literacy limitations. Compensatory strategies do not aim to improve functioning in a particular cognitive domain, such as memory. Rather, the goal of a compensatory strategy is to improve daily functioning or performance in light of cognitive decline. *Restorative* strategies, on the other hand, aim to improve functioning in an identified cognitive domain rather than mere compensation. Depending on the patient's needs and level of impairment, restorative strategies may be more beneficial than compensatory strategies, particularly among patients with significant memory impairment who may forget to use compensatory strategies on their own or require additional external aids or prompting (Huckans et al., 2013). However, for caregivers in Latino families, it may be comfortable and appropriate to offer cues regularly to the patient. Because of the emphasis on collectivism among many Latino communities, patients may strive less for the Westernized concept of independence and more toward interdependence.

Implicit and procedural memory strategies (i.e., the learning "how") are often more persevered in dementia and may assist in making use of compensatory tools more habitual and lead to better outcomes (Sohlberg & Mateer, 2017). Examples of restorative strategies include errorless learning, vanishing cues, and spaced retrieval (de Werd et al., 2013; El Haj, Kessels, & Allain, 2016). One of the most commonly used restorative strategies is errorless learning (Dunn & Clare, 2007). In errorless learning, errors are prevented by providing the accurate response before an error can be made. An example of an errorless learning trial would be to make a statement ("My name is Javier") followed by an immediate question pertaining to that statement ("What is my name?"). In errorless learning, guessing is discouraged, and the patient is always provided with the accurate information before an error occurs. When an error does occur, it is immediately corrected. Due to impairment in explicit memory, patients with dementia may not recognize that they have made an error and are at risk of consolidating the error into their implicit memory (Kessels & Hensken, 2009). By preventing errors in the first place, errorless learning keeps inaccurate information from being stored into memory.

Other restorative strategies include spaced retrieval and vanishing cues (El Haj et al., 2016). During a spaced retrieval task, the patient is asked to recall newly learned information after gradually lengthening delays (de Werd et al., 2013; Hopper et al., 2013). When the patient makes no mistakes between the learning phases, the patient is provided with the correct response after hesitation or after indication that he or she does not know how to respond. For example, after being told the provider's name, the patient may be asked to recall that name at increasingly lengthy intervals (e.g., immediately after, 15 seconds later, 30 seconds later, 1 minute later, etc.). The length of the intervals may be, initially, determined by the patient's needs or severity of impairment and, later, adjusted based on the patient's performance. The goal of spaced retrieval is to increase the duration of time that the information is retained. Vanishing cues can also be used to increase the complexity of the task (Hopper et al., 2013). Vanishing cues tasks involve the systematic reduction of cues that assist in memory retrieval (e.g., slowly removing the letters of the provider's name which serve as cues when the patient is unable to freely recall the name). Spaced retrieval and vanishing cues can include errorless learning (e.g., patients are provided with the correct response before an error can be made) or may involve effortful processing by encouraging the use of trial and error (Haslam, Moss, & Hodder, 2010).

There is evidence to support the benefits of errorless learning in patients with dementia across a variety of daily tasks and skills. In a review paper, de Werd et al. (2013) examined 26 studies wherein participants with dementia were taught practical tasks such as the use of devices (e.g., mobile phone, coffee maker, microwave), face-name associations, and orientation skills (e.g., use of calendar, following directions/routes) using either errorless learning or effortful processing. Despite variance in treatment frequency and duration, 17 studies demonstrated significant gains through errorless learning compared to effortful learning or nontreatment conditions, and the majority (17 out of 20) of studies that conducted follow-up evaluations found maintenance of the learning effects of errorless learning.

Different combinations of errorless learning tasks may be more or less helpful depending on the task or skill to be trained. Mastering of procedural tasks, such as use of a cellular telephone, calendar, or alarm reminder system, can be achieved through errorless learning by use of implicit, automatic strategies (e.g., motor memory or procedural memory wherein a specific motor task is consolidated into memory through repetition) as well as conscious, explicit memory of how the task should be performed (Kessels & Hensken, 2009). Based on their review study on errorless learning, de Werd et al. (2013) suggested that learning of procedural tasks may also benefit from modeling, provision of verbal instruction, and use of a stepwise approach, particularly in the acquisition phase of treatment. In the following stages, the use of vanishing cues may be helpful in decreasing the patient's degree of reliance on external assistance. For non-procedural learning tasks, such as face-name association, verbal instruction and spaced retrieval may result in the most benefit. Different rehabilitation tasks may also be more or less helpful depending on the severity of dementia and complexity of the task. Dunn and Clare (2007) have even suggested that effortful processing may be more helpful than errorless learning in assisting patients with mild to moderate dementia to learn complex information and achieve complex tasks. It was suggested that effortful processing, in contrast to errorless learning, may lead to better retention of information via a deeper level of information processing.

Factors That Impact Cognitive Rehabilitation

There can be a great amount of diversity among patients with dementia in regard to the severity of impairment, the domains of cognitive functioning that have been affected, and the specific needs of the person. For this reason, cognitive rehabilitation works best when an individualized approach is utilized (Lam et al., 2010). An individualized approach may include flexibility in the length of the sessions, the frequency of the treatment, and the duration of the rehabilitation program (e.g., number of sessions) (Cicerone et al., 2011; de Werd et al., 2013). The selection of strategies (e.g., internal vs. external, errorless or effortful) should be determined by the patient's particular needs and goals, with review of performance informing the direction of treatment. The availability of caregiver support may also determine the trajectory of treatment and which cognitive rehabilitation strategies may be most useful. Although individuals with cognitive impairment may benefit from participation in group-based cognitive rehabilitation services, there is indication that individuals who participate in individual treatment experience greater improvement, likely due to the program being more tailored to fit their particular needs and increased individual attention (Lam et al., 2010).

The Role of Caregivers

Caregivers and family members are an important aspect of the cognitive rehabilitation process. Latinos with dementia may experience significant benefit when caregivers are engaged in treatment (Alvarez et al., 2014). *Familismo* refers to the emphasis on family within the Latino culture, which generally expands beyond immediate family and includes extended family (Añez et al., 2008; Campos, Ullman, Aguilera, & Dunkel Schetter, 2014). In many Latino families, *la tercera edad*, or the final stage of life, is a time for family caregiving and comfort and also representative of collectivism, as responsibilities are passed down to the next generation (Flores, Hinton, Barker, Franz, & Velasquez, 2009). As a result, providers can expect high levels of family involvement in the dementia caregiving of the Latino family's elder.

Latino patients and their family members may subscribe to more traditional gender roles that could influence the patient's needs and goals in the context of cognitive rehabilitation (Lequerica & Krch, 2014). In cases of dementia caregiving, significant responsibilities often fall to women in the family, following the cultural ideal of *marianismo*, wherein women are considered spiritually stronger than men and are therefore better positioned to care for and maintain the dignity of the elder care recipient (Flores et al., 2009; Miville, Mendez, & Louie, 2017). It is important to note that, although adherence to traditional gender roles is a Latino cultural ideology, the extent to which individual families and family members behaviorally mirror this ideal varies considerably (Miville et al., 2017). In addition, caregiving responsibilities can offer a sense of purpose, and many Latina caregivers skillfully navigate their bicultural identities and balance role conflicts (Flores et al., 2009). Should Latino patients elect to focus on rehabilitation goals congruent with traditional gender roles (Sander, Clark, & Pappadis, 2010), these values should be reflected in treatment planning and goal-setting.

The cultural respect that Latino children (including adult children) have for their elders should also be considered, as this may impact the perspective that family members have regarding the patient with dementia. For example, Latino family members may tend to minimize the decline that they see in their loved one's functioning, as their underlying respect for the patient may lead them to rationalize or attempt to normalize the changes in cognition or behavior (Alvarez et al., 2014). As an example, a Latino dementia caregiver might state "Anybody can forget about a bill" or "We all forget names sometimes," exemplifying the way in which a family member may attempt to normalize or rationalize symptoms of dementia out of respect for the elder Latino patient.

In a research study by Neely, Vikström, and Josephsson (2009), individuals with dementia who engaged in collaborative rehabilitation tasks with their caregiver experienced more positive benefits, including an increase in memory recall and episodic memory, than patients who were trained alone. What is known as "collaborative training" led to increased participation in a functional task that resembled daily life (e.g., recalling items) and reduced dependence on the caregiver. Additionally, caregivers who participated in collaborative training were observed to provide assistance of cueing only as needed, as well as encouragement (described as "supportive acts"). It was also suggested that hands-on caregiver participation in the cognitive rehabilitation process helped to facilitate the transfer of learning into other tasks for the patients with dementia. Finally, the research indicates that caregivers who participate in the rehabilitation process may experience improvement in their own moods and that although their self-report of burden does not decrease, it remains stable (Neely et al., 2009).

There is also indication that family caregivers who participate in support groups may experience significant benefit in regard to mood and are able to care for the patient with dementia for longer periods of time. Reynoso-Vallejo (2009) described a four-stage process of developing a support group for Latino caregivers for patients with dementia. First, he conducted focus groups to solicit participant input into the format, content, and structure of the group. Next, he developed and implemented the support group intervention. Third, culturally applicable materials developed from the group were distributed to the families involved. Finally, a network of family caregivers was established as an ongoing source of support in the community (Reynoso-Vallejo, 2009). This approach, of partnering with the community while demonstrating an authentic regard for cultural considerations, can serve as a model for caregiver support group development at local levels.

Lehan et al. (2012) found that Mexican families of traumatic brain injury survivors who had greater levels of adaptability and cohesion reported better family communication and increased family satisfaction. This is congruent with the broader research base in cognitive rehabilitation in which appropriate and effective caregiver involvement and family support are robust predictors of positive outcome (Prigatano, 2013; Sohlberg & Mateer, 2017). While the applicability to broader groups of Latinos with a range of cognitive impairments resulting from dementia is unknown, Vickrey et al. (2007), for example, found that Latino caregivers reported a sense of meaning for their roles and enjoyment in spending time with their care recipients. By integrating families and communities into the cognitive rehabilitation process, providers are enabling "natural helpers" (Judd & DeBoard, 2007, p. 358) who can care for the patient at home, which in the case of Latino cultures is often more comfortable and culturally congruent. Given the physical, emotional, social, and behavioral impact of dementia on caregivers and families, the impact illustrated by family support in other groups, and the relative strength of collectivism and multigenerational dwelling among many Latino cultures, it is judicious to acknowledge the effects of Latino family systems on cognitive rehabilitation outcomes.

Differences from Other Treatment Modalities for Cognitive Decline

Use of drug therapy in combination with cognitive rehabilitation can have a positive effect on a patient's performance in daily tasks. Acetylcholinesterase inhibitors (AChE-Is) such as donepezil and rivastigmine have demonstrated positive effects on the cognitive and behavioral functioning of patients with dementia (Palmer et al., 2015). There is indication that combining AChE-I treatment with cognitive rehabilitation leads to more positive outcomes than the use of AChE-I alone (Bottino et al., 2005). There is also indication of positive impact of combined medication and rehabilitation treatment on mood (Rozzini et al., 2007).

Another modality of treatment for cognitive decline is cognitive stimulation, which involves immersing a patient into a cognitively stimulating environment with the intention of improving cognitive functioning. This may include thoughtful discussions or debates, engagement in problem-solving activities, or participating in psychoeducation (Vidovich & Almeida, 2011). Research suggests that patients with dementia receive significant benefit from being placed in a cognitively stimulating environment with regular interpersonal interactions and that cognitive stimulation can improve subjective quality of life (Spector, Gardner, & Orrell, 2011). Cognitive stimulation can happen in a wide range of contexts, such as at home or within the community, and need not occur within a formal treatment setting, thus making it a good fit for Latinos with dementia who are more likely to experience their care within their homes or communities (Alvarez et al., 2014).

Cognitive training is a form of intervention that involves a set of standardized, repeatedly performed tasks that are intended to enhance cognitive functioning (Vidovich & Almeida, 2011). Cognitive training differs from cognitive rehabilitation in a number of important ways (Bahar-Fuchs, Clare, & Woods, 2013). In cognitive training, treatment occurs in the context of structured tasks and environments, whereas cognitive rehabilitation focuses on real-world settings such as home or community life, such as consistently taking medication or attending religious worship services. The target of intervention in cognitive training is cognitive impairment, while the target of intervention in cognitive rehabilitation is participation restriction, or involvement in preferred activities and valued social roles (Bahar-Fuchs et al., 2013; WHO, 2001). In cognitive training, the focus of intervention is on isolated cognitive abilities and processes; in cognitive rehabilitation, the focus is on groups of cognitive abilities that impact performance in daily tasks, such as memory interventions to enable a patient to consistently and correctly take prescribed medications. The goals of cognitive training and cognitive rehabilitation also differ. In cognitive training, the goal is to improve performance within a specific cognitive domain or task. In contrast, the goal of cognitive rehabilitation is improved performance in functioning in regard to collaboratively selected goals. Although the availability of cognitive training is plentiful, particularly through web-based computer programs (e.g., CogniPlus, Lumosity, MyBrainTrainer), there is evidence that cognitive training has limited positive effects for patients with dementia (Bahar-Fuchs et al., 2013; Simpson, Camfield, Pipingas, Macpherson, & Stough, 2012). Although patients who participate in cognitive training programs experience measurable improvement in performance on the training tasks, these benefits fail to generalize to real-life settings (Owen, et al., 2010).

Increasingly, physical activity, and in particular aerobic exercise, is proving not only to serve a preventative and protective role against cognitive decline and dementia but also as a form of treatment to improve cognition and to reduce the rate of cognitive decline in those with MCI and dementia (Ahlskog, Geda, Graff-Radford, & Petersen, 2011). An early meta-analysis (Heyn, Abreu, & Ottenbacher, 2004) of 12 randomized controlled trials (RCTs) demonstrated improved cognitive performance following adoption of an exercise routine, with changes documented as soon as 2 weeks after starting. Later RCTs indicated that the positive effects of exercise could be maintained in patients with Alzheimer's disease at 18 months and that physical activity was as effective as an AChE-I (Ahlskog et al., 2011). In a prospective, longitudinal study of older Mexican-Americans, Ottenbacher et al. (2014) found a statistically significant positive association between physical activity and cognitive function after adjusting for age, sex, marital status, education, and comorbid health conditions; the greatest gains in cognitive decline were in memory and less pronounced in other aspects of cognition. Suspected mechanisms include improvements in neuroplasticity as a result of exercise and the protective effect of physical activity for vascular risk factors associated with dementia (Ahlskog et al., 2011). Finally, emerging research in complementary and alternative treatment approaches has provided some support for the use of music and dancing to improve cognitive, social, and emotional functioning in patients with dementia (Horowitz, 2013; Rylatt, 2012). These modalities hold promise to integrate culturally salient customs into treatment for dementia for Latinos. Music therapy has shown to improve communication behavior, expression of positive emotions, and well-being in patients with dementia (Schall, Haberstroh, & Pantel, 2015). Kim et al. (2011) found cognitive improvements following dance therapy in elderly patients with risk factors for dementia. Music and dance provide an opportunity for the involvement of family caregivers and patient enjoyment of traditional Latino music and dance.

Conclusion

Cognitive rehabilitation can be an effective intervention for maintenance of function among Latinos with dementia when implemented in an individualized and culturally sensitive manner by medical and rehabilitation professionals. Cultural considerations should be integrated into each of the stages and strategies of cognitive rehabilitation by involving patients and their families to the fullest extent and tailoring the treatment to the values of the patient, his or her family, and their community. Although awareness of Latino cultural variables will certainly help cognitive rehabilitation treatment providers better understand their Latino patients, it is the application of Latino cultural principles into the practice of cognitive rehabilitation that will ensure that Latino patients are receiving the best possible care. A truly culturally integrated, cognitive rehabilitation framework will include careful consideration of how important Latino values, such as *familismo*, *respeto*, *personalismo*, *confianza*, *espiritu*, and *presentismo*, will impact goal-setting, treatment planning, and treatment implementation. Fortunately, the prevailing emphasis on an individualized approach in cognitive rehabilitation lends itself well in efforts toward cultural competency in working with Latino patients.

Although we anticipate that the integration of Latino cultural values into cognitive rehabilitation services will have a positive impact on Latino patients with dementia, and we can make suppositions regarding the ways in which rehabilitation services can be individualized to meet the needs of a Latino population, there remains a gap in the available research regarding cognitive rehabilitation treatment outcomes among Latinos with dementia. Future research efforts should consider these clinical realities coupled with the burgeoning population of elder Latinos in the United States. Given significant intergroup differences, including immigrant status, dominant language, biculturalism, and intersections of multiple diverse identities, empirical focus must pivot toward exploring cultural differences among Latinos and with emphasis on other factors that may affect treatment engagement, treatment, and outcome.

References

Adams, R. B., Rule, N. O., Franklin, R. G., Wang, E., Stevenson, M. T., Yoshikawa, S., … Ambady, N. (2010). Cross-cultural reading the mind in the eyes: An fMRI investigation. *Journal of Cognitive Neuroscience, 22*(1), 97–108. https://doi.org/10.1162/jocn.2009.21187

Ahlskog, J. E., Geda, Y. E., Graff-Radford, N. R., & Petersen, R. C. (2011). Physical exercise as a preventive or disease-modifying treatment of dementia and brain aging. *Mayo Clinic Proceedings, 86*(9), 876–884. https://doi.org/10.4065/mcp.2011.0252

Alvarez, P., Rengifo, J., Emrani, T., & Gallagher-Thompson, D. (2014). Latino older adults and mental health: A review and commentary. *Clinical Gerontologist, 37*(1), 33–48. https://doi.org/10.1080/07317115.2013.847514

American Psychiatric Association. (2013). *Diagnostic and statistical manual of mental disorders* (5th ed.). Arlington, VA: American Psychiatric Publishing. https://doi.org/10.1176/appi.books.9780890425596

Añez, L. M., Silva, M. A., Paris, M., Jr., & Bedregal, L. E. (2008). Engaging latinos through the integration of cultural values and motivational interviewing principles. *Professional Psychology: Research and Practice, 39*(2), 153–159. https://doi.org/10.1037/0735-7028.39.2.153

Ano, G. G., & Vasconcelles, E. B. (2005). Religious coping and psychological adjustment to stress: A meta-analysis. *Journal of Clinical Psychology, 61*(4), 461–480. https://doi.org/10.1002/jclp.20049

Antshel, K. M. (2002). Integrating culture as a means of improving treatment adherence in the Latino population. *Psychology, Health & Medicine, 7*(4), 435–449. https://doi.org/10.1080/13548500210000015258

Arango, J. C., & Kreutzer, J. (2010). Racial and ethnic disparities in functional, psychosocial, and neurobehavioral outcomes after brain injury. *Journal of Head Trauma Rehabilitation, 25*, 128–136. https://doi.org/10.1097/htr.0b013e3181d36ca3

Arango-Lasprilla, J. C. (2012). Spanish speakers with neurological and psychiatric disabilities: Relevant factors related to rehabilitation. *NeuroRehabilitation, 30*, 9–11. https://doi.org/10.3233/NRE-2012-0722

Ardila, A. (2005). Cultural values underlying psychometric cognitive testing. *Neuropsychology Review, 15*(4), 185–195. https://doi.org/10.1007/s11065-005-9180-y

Arredondo, P., Gallardo-Cooper, M., Delgado-Romero, E. A., & Zapata, A. L. (2015). Latino worldviews and cultural values. *Culturally Responsive Counseling with Latinas/os*, 15–30. https://doi.org/10.1002/9781119221609.ch2

Bahar-Fuchs, A., Clare, L., & Woods, B. (2013). Cognitive training and cognitive rehabilitation for mild to moderate Alzheimer's disease and vascular dementia. *Cochrane Database of Systematic Reviews, 6*. https://doi.org/10.1002/14651858.CD003260.pub2

Belleville, S., Clément, F., Mellah, S., Gilbert, B., Fontaine, F., & Gauthier, S. (2011). Training- related brain plasticity in subjects at risk of developing Alzheimer's disease. *Brain, 134*(6), 1623–1634. https://doi.org/10.1093/brain/awr037

Bottino, C. M., Carvalho, I. A., Alvarez, A. M. M., Avila, R., Zukauskas, P. R., Bustamante, S. E., … Camargo, C. H. (2005). Cognitive rehabilitation combined with drug treatment in Alzheimer's disease patients: A pilot study. *Clinical Rehabilitation, 19*(8), 861–869. https://doi.org/10.1191/0269215505cr911oa

Brickman, A. M., Cabo, R., & Manly, J. J. (2006). Ethical issues in cross-cultural neuropsychology. *Applied Neuropsychology, 13*(2), 91–100. https://doi.org/10.1207/s15324826an1302_4

Campos, B., Ullman, J. B., Aguilera, A., & Dunkel Schetter, C. (2014). Familism and psychological health: The intervening role of closeness and social support. *Cultural Diversity and Ethnic Minority Psychology, 20*(2), 191–201. https://doi.org/10.1037/a0034094

Chiao, J. Y., & Blizinsky, K. D. (2009). Culture-gene coevolution of individualism-collectivism and the serotonin transporter gene. *Proceedings of the Royal Society B: Biological Sciences, 277*(1681), 529–537. https://doi.org/10.1098/rspb.2009.1650

Cicerone, K. D., Langenbahn, D. M., Braden, C., Malec, J. F., Kalmar, K., Fraas, M., … Azulay, J. (2011). Evidence-based cognitive rehabilitation: Updated review of the literature from 2003 through 2008. *Archives of Physical Medicine and Rehabilitation, 92*(4), 519–530. https://doi.org/10.1016/j.apmr.2010.11.015

Clare, L., Linden, D. E., Woods, R. T., Whitaker, R., Evans, S. J., Parkinson, C. H., … Rugg, M. D. (2010).

Goal-oriented cognitive rehabilitation for people with early-stage Alzheimer disease: A single-blind randomized controlled trial of clinical efficacy. *The American Journal of Geriatric Psychiatry, 18*(10), 928–939. https://doi.org/10.1097/jgp.0b013e3181d5792a

de Werd, M. M., Boelen, D., Rikkert, M. G. O., & Kessels, R. P. (2013). Errorless learning of everyday tasks in people with dementia. *Clinical Interventions in Aging, 8*, 1177. https://doi.org/10.2147/cia.s46809

Dilworth-Anderson, P., & Gibson, B. E. (2002). The cultural influence of values, norms, meanings, and perceptions in understanding dementia in ethnic minorities. *Alzheimer Disease & Associated Disorders, 16*, S56–S63. https://doi.org/10.1097/00002093-200200002-00005

Dunn, J., & Clare, L. (2007). Learning face–name associations in early-stage dementia: Comparing the effects of errorless learning and effortful processing. *Neuropsychological Rehabilitation, 17*(6), 735–754. https://doi.org/10.1080/09602010701218317

Dunn, M. G., & O'Brien, K. M. (2009). Psychological health and meaning in life: Stress, social support, and religious coping in Latina/Latino immigrants. *Hispanic Journal of Behavioral Sciences, 31*(2), 204–227. https://doi.org/10.1177/0739986309334799

El Haj, M., Kessels, R. P., & Allain, P. (2016). Source memory rehabilitation: A review toward recommendations for setting up a strategy training aimed at the "what, where, and when" of episodic retrieval. *Applied Neuropsychology: Adult, 23*(1), 53–60. https://doi.org/10.1080/23279095.2014.992071

Farias, S. T., Mungas, D., Hinton, L., & Haan, M. (2011). Demographic, neuropsychological, and functional predictors of rate of longitudinal cognitive decline in Hispanic older adults. *The American Journal of Geriatric Psychiatry, 19*(5), 440–450. https://doi.org/10.1097/jgp.0b013e3181e9b9a5

Flores, Y. G., Hinton, L., Barker, J. C., Franz, C. E., & Velasquez, A. (2009). Beyond familism: A case study of the ethics of care of a Latina caregiver of an elderly parent with dementia. *Health Care for Women International, 30*(12), 1055–1072. https://doi.org/10.1080/07399330903141252

Furman, R., Negi, N. J., Iwamoto, D. K., Rowan, D., Shukraft, A., & Gragg, J. (2009). Social work practice with Latinos: Key issues for social workers. *Social Work, 54*(2), 167–174. https://doi.org/10.1093/sw/54.2.167

Gagnon, L. G., & Belleville, S. (2012). Training of attentional control in mild cognitive impairment with executive deficits: Results from a double-blind randomised controlled study. *Neuropsychological Rehabilitation, 22*(6), 809–835. https://doi.org/10.1080/09602011.2012.691044

Gauthier, S., Reisberg, B., Zaudig, M., Petersen, R. C., Ritchie, K., Broich, K., … Cummings, J. L. (2006). Mild cognitive impairment. *The Lancet, 367*(9518), 1262–1270. https://doi.org/10.1016/s0140-6736(06)68542-5

Gelman, C. R. (2010). Learning from recruitment challenges: Barriers to diagnosis, treatment, and research participation for Latinos with symptoms of Alzheimer's disease. *Journal of Gerontological Social Work, 53*(1), 94–113. https://doi.org/10.1080/01634370903361847

Geusgens, C. A., Winkens, I., van Heugten, C. M., Jolles, J., & van den Heuvel, W. J. (2007). Occurrence and measurement of transfer in cognitive rehabilitation: A critical review. *Journal of Rehabilitation Medicine, 39*(6), 425–439. https://doi.org/10.2340/16501977-0092

Gooding, A. L., Choi, J., Fiszdon, J. M., Wilkins, K., Kirwin, P. D., van Dyck, C. H., … Rivera Mindt, M. (2016). Comparing three methods of computerised cognitive training for older adults with subclinical cognitive decline. *Neuropsychological Rehabilitation, 26*(5–6), 810–821. https://doi.org/10.1080/09602011.2015.1118389

Greenaway, M. C., Duncan, N. L., & Smith, G. E. (2012). The memory support system for mild cognitive impairment: Randomized trial of a cognitive rehabilitation intervention. *International Journal of Geriatric Psychiatry, 28*(4), 402–409. https://doi.org/10.1002/gps.3838

Guardia-Olmosa, J., Esparcia, A. J., Morales, A. U., & Ferred, E. G. (2012). Neuropsychological rehabilitation and quality of life in patients with cognitive impairments: A meta-analysis study in Spanish-speaking populations. *NeuroRehabilitation, 30*, 35–42. https://doi.org/10.1016/s2171-2069(15)70002-5

Haskins, E. C., Cicerone, K., Dams-O'Conner, K., Eberle, R., Langenbahn, D., & Shapiro-Rosenbaum, A. (2012). *Cognitive rehabilitation manual: Translating evidence-based recommendations into practice*. Reston, VA: American Congress of Rehabilitation Medicine.

Haslam, C., Moss, Z., & Hodder, K. (2010). Are two methods better than one? Evaluating the effectiveness of combining errorless learning with vanishing cues. *Journal of Clinical and Experimental Neuropsychology, 32*(9), 973–985. https://doi.org/10.1080/13803391003662686

Herrera, A. P., Lee, J. W., Nanyonjo, R. D., Laufman, L. E., & Torres-Vigil, I. (2009). Religious coping and caregiver well-being in Mexican-American families. *Aging and Mental Health, 13*(1), 84–91. https://doi.org/10.1080/13607860802154507

Heyn, P., Abreu, B. C., & Ottenbacher, K. J. (2004). The effects of exercise training on elderly persons with cognitive impairment and dementia: A meta-analysis. *Archives of Physical Medicine and Rehabilitation, 85*, 1694–1704. https://doi.org/10.1016/j.apmr.2004.03.019

Hopper, T., Bourgeois, M., Pimentel, J., Qualls, C. D., Hickey, E., Frymark, T., & Schooling, T. (2013). An evidence-based systematic review on cognitive interventions for individuals with dementia. *American Journal of Speech-Language Pathology, 22*(1), 126–145. https://doi.org/10.1044/1058-0360(2012/11-0137)

Horowitz, S. (2013). The healing power of music and dance. *Alternative and Complementary Therapies, 19*(5), 265–269. https://doi.org/10.1089/act.2013.19502

Huckans, M., Hutson, L., Twamley, E., Jak, A., Kaye, J., & Storzbach, D. (2013). Efficacy of cognitive rehabilitation therapies for mild cognitive impairment (MCI) in older adults: Working toward a theoretical model and evidence-based interventions. *Neuropsychology Review, 23*(1), 63–80. https://doi.org/10.1007/s11065-013-9230-9

Jean, L., Bergeron, M. È., Thivierge, S., & Simard, M. (2010). Cognitive intervention programs for individuals with mild cognitive impairment: Systematic review of the literature. *The American Journal of Geriatric Psychiatry, 18*(4), 281–296. https://doi.org/10.1097/jgp.0b013e3181c37ce9

Judd, T., & DeBoard, R. (2007). Natural recovery: An ecological approach to neuropsychological recuperation. In B. P. Uzell, M. Ponton, & A. Ardila (Eds.), *International Handbook of Cross-cultural Neuropsychology* (p. 341). https://doi.org/10.4324/9780203936290.ch0019

Judd, T., & DeBoard, R. (2009). Community-based neuropsychological rehabilitation in the cosmopolitan setting. *Neuropsychological Rehabilitation, 19*(6), 841–866. https://doi.org/10.1080/09602010903024943

Kessels, R. P., & Hensken, L. M. (2009). Effects of errorless skill learning in people with mild-to-moderate or severe dementia: A randomized controlled pilot study. *NeuroRehabilitation, 25*(4), 307–312. https://doi.org/10.3233/NRE-2009-0529

Kim, S. H., Kim, M., Ahn, Y. B., Lim, H. K., Kang, S. G., Cho, J. H., … Song, S. W. (2011). Effect of dance exercise on cognitive function in elderly patients with metabolic syndrome: A pilot study. *Journal of Sports Science & Medicine, 10*(4), 671–678.

Lam, L. C., Lui, V. W., Luk, D. N., Chau, R., So, C., Poon, V., … Fung, A. (2010). Effectiveness of an individualized functional training program on affective disturbances and functional skills in mild and moderate dementia—A randomized control trial. *International Journal of Geriatric Psychiatry, 25*(2), 133–141. https://doi.org/10.1002/gps.2309

Lehan, T. J., Stevens, L. F., Arango-Lasprilla, J. C., Díaz Sosa, D. M., & Espinosa Jove, I. G. (2012). Balancing act: The influence of adaptability and cohesion on satisfaction and communication in families facing TBI in Mexico. *NeuroRehabilitation, 30*(1), 75–86. https://doi.org/10.3233/NRE-2012-0729

Lequerica, A., & Krch, D. (2014). Issues of cultural diversity in acquired brain injury (ABI) rehabilitation. *NeuroRehabilitation, 34*(4), 645–653.

Mayeda, E. R., Glymour, M. M., Quesenberry, C. P., & Whitmer, R. A. (2016). Inequalities in dementia incidence between six racial and ethnic groups over 14 years. *Alzheimer's & Dementia 12*(3), 216–224.

McDougall, G. J., Becker, H., Pituch, K., Acee, T. W., Vaughan, P. W., & Delville, C. L. (2010). The SeniorWISE study: Improving everyday memory in older adults. *Archives of Psychiatric Nursing, 24*(5), 291–306. https://doi.org/10.1016/j.apnu.2009.11.001

Miville, M. L., Mendez, N., & Louie, M. (2017). Latina/o gender roles: A content analysis of empirical research from 1982 to 2013. *Journal of Latina/O Psychology, 5*(3), 173–194. https://doi.org/10.1037/lat0000072

Neely, A. S., Vikström, S., & Josephsson, S. (2009). Collaborative memory intervention in dementia: Caregiver participation matters. *Neuropsychological Rehabilitation, 19*(5), 696–715. https://doi.org/10.1080/09602010902719105

Niemeier, J., & Arango, J. C. (2007). Toward improved provider practice competency in rehabilitation services for ethnically diverse survivor's of traumatic brain injury. *Journal of Head Trauma Rehabilitation, 22*, 75–84. https://doi.org/10.1097/01.htr.0000265095.06565.7b

Ottenbacher, A. J., Al Snih, S., Bindawas, S. M., Markides, K. S., Graham, J. E., Samper-Ternent, R., … Ottenbacher, K. J. (2014). Role of physical activity in reducing cognitive decline in older Mexican-American adults. *Journal of the American Geriatric Society, 62*, 1786–1791. https://doi.org/10.1111/jgs.12978

Ottenbacher, K. J., Graham, J. A., Al Snih, S., Raji, M., Samper-Ternent, R., Ostir, G. V., & Markides, K. S. (2009). Mexican Americans and frailty: Findings from the Hispanic established populations epidemiologic studies of the elderly. *American Journal of Public Health, 99*(4), 673–679. https://doi.org/10.2105/ajph.2008.143958

Owen, A. M., Hampshire, A., Grahn, J. A., Stenton, R., Dajani, S., Burns, A. S., … Ballard, C. G. (2010). Putting brain training to the test. *Nature, 465*(7299), 775–778. https://doi.org/10.1038/nature09042

Palmer, J. B., Albrecht, J. S., Park, Y., Dutcher, S., Rattinger, G. B., Simoni-Wastila, L., … Zuckerman, I. H. (2015). Use of drugs with anticholinergic properties among nursing home residents with dementia: A national analysis of Medicare beneficiaries from 2007 to 2008. *Drugs & Aging, 32*(1), 79–86. https://doi.org/10.1007/s40266-014-0227-8

Pargament, K. I., Koenig, H. G., & Perez, L. M. (2000). The many methods of religious coping: Development and initial validation of the RCOPE. *Journal of Clinical Psychology, 56*(4), 519–543. https://doi.org/10.1002/(SICI)1097-4679(200004)56:4<519::AID-JCLP6>3.0.CO;2-1

Prigatano, G. P. (2013). Challenges and opportunities facing holistic approaches to neuropsychological rehabilitation. *NeuroRehabilitation, 32*(4), 751–759. https://doi.org/10.3233/NRE-130899

Reynoso-Vallejo, H. (2009). Support group for Latino caregivers of dementia elders: Cultural humility and cultural competence. *Ageing International, 34*(1–2), 67–78. https://doi.org/10.1007/s12126-009-9031-x

Rozzini, L., Costardi, D., Chilovi, B. V., Franzoni, S., Trabucchi, M., & Padovani, A. (2007). Efficacy of cognitive rehabilitation in patients with mild cognitive impairment treated with cholinesterase inhibitors. *International Journal of Geriatric Psychiatry, 22*(4), 356–360. https://doi.org/10.1002/gps.1681

Rylatt, P. (2012). The benefits of creative therapy for people with dementia. *Nursing Standard, 26*(33), 42–47. https://doi.org/10.7748/ns2012.04.26.33.42.c9050

Sander, A. M., Clark, A., & Pappadis, M. R. (2010). What is community integration anyway?: Defining meaning following traumatic brain injury. *Journal of Head Trauma Rehabilitation, 25*(2), 121–127. https://doi.org/10.1097/htr.0b013e3181cd1635

Sayegh, P., & Knight, B. G. (2012). Cross-cultural differences in dementia: The sociocultural health belief model. *International Psychogeriatrics, 25*(04), 517–530. https://doi.org/10.1017/s104161021200213x

Schall, A., Haberstroh, J., & Pantel, J. (2015). Time series analysis of individual music therapy in dementia: Effects on communication behavior and emotional well-being. *GeroPsych: The Journal of Gerontopsychology and Geriatric Psychiatry, 28*(3), 113–122. https://doi.org/10.1024/1662-9647/a000123

Simpson, T., Camfield, D., Pipingas, A., Macpherson, H., & Stough, C. (2012). Improved processing speed: Online computer-based cognitive training in older adults. *Educational Gerontology, 38*(7), 445–458. https://doi.org/10.1080/03601277.2011.559858

Smith, G. E., Housen, P., Yaffe, K., Ruff, R., Kennison, R. F., Mahncke, H. W., & Zelinski, E. M. (2009). A cognitive training program based on principles of brain plasticity: Results from the improvement in memory with plasticity-based adaptive cognitive training (IMPACT) study. *Journal of the American Geriatrics Society, 57*(4), 594–603. https://doi.org/10.1111/j.1532-5415.2008.02167.x

Sohlberg, M. M., & Mateer, C. A. (2017). *Cognitive rehabilitation: An Integrative Neuropsychological Approach* (2nd ed.). [kindle DX version]. Available from Amazon.com

Spector, A., Gardner, C., & Orrell, M. (2011). The impact of cognitive stimulation therapy groups on people with dementia: Views from participants, their carers and group facilitators. *Aging & Mental Health, 15*(8), 945–949. https://doi.org/10.1080/13607863.2011.586622

Talamantes, M. & Sanchez-Reilly, S. (2010). Health and health care of Hispanic/Latino American older adults. In Periyakoil V.S. (ed). *eCampus-Geriatrics*, Stanford CA. Retrieved from: https://geriatrics.stanford.edu/ethnomed/latino.html

Taylor, E., & Jones, F. (2014). Lost in translation: Exploring therapists' experiences of providing stroke rehabilitation across a language barrier. *Disability and Rehabilitation, 36*(25), 2127–2135. https://doi.org/10.3109/09638288.2014.892636

Troyer, A. K. (2017). Activities of daily living (ADL). *Encyclopedia of Clinical Neuropsychology*, 1–3. https://doi.org/10.1007/978-3-319-56782-2_1077-2

Tzuang, M., Owusu, J. T., Spira, A. P., Albert, M. S., & Rebok, G. W. (2017). Cognitive training for ethnic minority older adults in the united States: A review. *The Gerontologist.* Advance online publication. https://doi.org/10.1093/geront/gnw260

van Paasschen, J., Clare, L., Yuen, K. S., Woods, R. T., Evans, S. J., Parkinson, C. H., ... Linden, D. E. (2013). Cognitive rehabilitation changes memory-related brain activity in people with Alzheimer disease. *Neurorehabilitation and Neural Repair, 27*(5), 448–459. https://doi.org/10.1177/1545968312471902

Vickrey, B. G., Strickland, T. L., Fitten, L. J., Adams, G. R., Ortiz, F., & Hays, R. D. (2007). Ethnic variations in dementia caregiving experiences: Insights from focus groups. *Journal of Human Behavior in the Social Environment, 15*(2–3), 233–249. https://doi.org/10.1300/j137v15n02_14

Vidovich, M., & Almeida, O. P. (2011). Cognition-focused interventions for older adults: The state of play. *Australasian Psychiatry, 19*(4), 313–316. https://doi.org/10.3109/10398562.2011.579973

Vincent, G. K., & Velkoff, V. A. (2010) The next four decades: The older population in the United States: 2010–2050. Current population reports 2010. US Retrieved from https://www.census.gov/prod/2010pubs/p25-1138.pdf

Walton, C. C., Mowszowski, L., Lewis, S. J., & Naismith, S. L. (2014). Stuck in the mud: Time for change in the implementation of cognitive training research in ageing? *Frontiers in Aging Neuroscience, 6.* https://doi.org/10.3389/fnagi.2014.00043

Willis, S. L., Tennstedt, S. L., Marsiske, M., Ball, K., Elias, J., Koepke, K. M., ... Wright, E. (2006). Long-term effects of cognitive training on everyday functional outcomes in older adults. *Journal of the American Medical Association, 296*(23), 2805–2814. https://doi.org/10.1001/jama.296.23.2805

Woods, B. (2014). What can be learned from studies on reality orientation? In G. M. Jones & B. M. Miesen (Eds.), *Care-Giving in Dementia: Volume 1: Research and Applications* (pp. 121–126). New York, NY: Routledge.

World Health Organization. (2001). *International classification of functioning, disability, and health.* Geneva, Switzerland: Author.

Ziomek, M. M., & Rehfeldt, R. A. (2008). Investigating the acquisition, generalization, and emergence of untrained verbal operants for mands acquired using the picture exchange communication system in adults with severe developmental disabilities. *The Analysis of Verbal Behavior, 24*(1), 15–30. https://doi.org/10.1007/bf03393054

11 Behavioral Symptoms of Dementia in Latinos: Pharmacological, Non-pharmacological, and Ethnocultural Healthcare Interventions

Diana Lynn Woods

Abstract

Although it can be argued that the general knowledge base of how to effectively treat the behavioral symptoms of dementia (BSDs) is less than optimal, it is even more challenging to understand how to best treat BSDs in ethnic/racial minority groups, as in the case of Latinos. In this chapter, the inclusion of indigenous and other folk-based frameworks is discussed, as well as the potential for a greater accounting for cultural differences in the treatment of BSDs. This chapter provides an overview of the predominant interventions utilized in healthcare to treat patients with dementia, including Western-based methods of pharmacological and non-pharmacological approaches. In the case of Latinos, the ethnocultural and traditional healing systems are promoted as those that can be integrated in the Western approaches. Central to appropriate use of pharmacological interventions are factors such as those that are epigenetic, gene-environment interactions, and genetic variations in different populations. When ethnic/racial status is considered for a Latino patient, it is the realm of ethnopharmacology that answers the call for more specific and tailored consideration of medications (e.g., psychotropics, for an aging Latino population). Non-pharmacological interventions are also reviewed and discussed for their range of effectiveness and their limitations being based on differing research protocols that cannot always provide predictive or definitive improved treatment outcomes. A call is made for more Latinos as research participants, as well as more interdisciplinary research studies for BSDs.

D. L. Woods (✉)
Azusa Pacific University, Azusa, CA, USA;
University of California Los Angeles,, Los Angeles, CA, USA
e-mail: dwoods@apu.edu

Introduction

An estimated 44 million people across the world have Alzheimer's disease (AD), the most prevalent form of dementia (Prince et al., 2016). In the United States (U.S.), 5.2 million people were diagnosed with dementia (Alzheimer's Association, 2014), with those over 85 years of age accounting for more than 80% of those diagnosed. As ethnic and racial diversity increases in the United States, a larger proportion of persons over the age of 85 with dementia will be minority elders. While the proportion of Whites aged 85 and older is projected to decrease from 90% to 81% by 2050, the proportion of Hispanics[1] aged

[1] Latino and Hispanic are used interchangeably throughout this chapter.

H. Y. Adames, Y. N. Tazeau (eds.), *Caring for Latinxs with Dementia in a Globalized World*,
https://doi.org/10.1007/978-1-0716-0132-7_11

85 and older is projected to increase from 5% to 15% (U.S. Department of Health and Human Services, 2014). Prevalence rates of dementia among Hispanics have been less commonly measured; however, one study estimated dementia prevalence of 8% of Caribbean Latinos in the 65–74 age group, 28% in the 75–84 age group, and 63% in those aged 85 and older (Gurland et al., 1999). Another study indicated that Mexican-American men had significantly higher prevalence of cognitive impairment than did non-Hispanic White men (Cano, Samper-Ternent, Al Snih, Markides, & Ottenbacher, 2012).

The goal of this chapter is to broadly discuss behavioral symptoms of dementia (BSDs) within the aging Latino population. To achieve this goal, the chapter addresses the following: (1) the assessment of BSDs, (2) existing pharmacological interventions for BSD, (3) the role of gene-environment interactions and epigenetic factors of BSD in the aging Latino population, and (4) non-pharmacological interventions for BSD in the older Latino population, including the role of ethnocultural approaches particular to Latinos.

Behavioral symptoms of dementia (BSDs), also referred to as behavioral and psychological symptoms of dementia (BPSDs), is the term used to describe the wide variety of behavioral symptoms experienced by persons with dementia (PWD) during the course of their illness (Trivedi et al., 2012). Occurring in up to 90% of individuals with dementia (Kales et al., 2011; Lyketsos et al., 2011), these behaviors include agitation and restlessness, screaming, wandering, repetitive questioning, resistance to care, mood disturbances, and sleep disturbances. For the purpose of this chapter, behavioral symptoms of dementia will be referred to as BSDs. Regardless of whether persons with dementia (PWD) receive care at home, in assisted living facilities (ALFs), or in nursing homes (NHs), the management of BSDs is one of the most difficult and costly of care problems, consuming vast amounts of time and effort (Ellison, Harper, Berlow, & Zeranski, 2008; Muangpaisan, Intalapaporn, & Assantachai, 2008). These behaviors are associated with numerous poor health outcomes for the caregivers (Janevic & Connell, 2001) and for the person with dementia (Maust et al., 2015), including functional decline, depression (Martin et al., 2010), increased risk of cardiovascular events, and increased mortality (Maust et al., 2015). For example, the data suggest that the highest prevalence of restlessness and agitation occurs in mild to moderate dementia, as opposed to severe dementia (Borroni et al., 2006), and that individuals are most likely to live in the community, either at home or in assisted living facilities. These behaviors can result in fear and avoidance in caregivers and others in close proximity, embarrassment for affected older adults, and nursing home placement (Hart et al., 2003; Kales, Chen, Blow, Welsh, & Mellow, 2005). Moreover, BSDs contribute to staff turnover in residential care settings (Cohen-Mansfield, 1997; Sloane, Davidson, Knight, Tangen, & Mitchell, 1999) and accidents (e.g., falls, tripping) involving both elders and care providers (Beattie, Song, & LaGore, 2005). Moreover, the BSDs of restlessness and agitation are strongly associated with reduced caregiver quality of life (Shin et al., 2014; Smith, Buckwater, Kang, Ellingrod, & Schultz, 2008).

While medications, especially psychotropic medications, are still used extensively to ameliorate BSDs, they are an unsatisfactory solution; and when used, medications often create more problems than they solve (Ballard et al., 2009; Maust et al., 2015; Polsky, Doshi, Bauer, & Glick, 2006). Given the limited efficacy of psychotropic medications in addressing BSDs, several organizations and specialty groups such as the *International Association of Gerontology and Geriatrics*, *Centers for Medicare and Medicaid Services* (CMS), *National Alzheimer's Project Association* (NAPA), *American Geriatrics Society*, and *American Association for Geriatric Psychiatry* all currently advocate for the use of non-pharmacological interventions as the first-line treatment for the management of behavioral symptoms (Alzheimer's Association Expert Advisory Workgroup on NAPA, 2012; Tolson et al., 2011). These professional organizations seek to minimize the use of psychotropic medications, especially in nursing homes. They also advocate for greater research and education on non-pharmacological strategies for managing BSDs.

Cultural Factors of Behavioral Symptoms of Dementia

There is a paucity of information about ethnic differences in the manifestation of behavioral symptoms of dementia mainly because ethnic minority elders are underrepresented in clinical trials and other research studies (Cooper, Tandy, Balamuralli, & Livingston, 2010). In fact, many Hispanics are not diagnosed until the later stages of the disease and experience a delay of up to 7 years before any evaluation of symptoms (Griffith and Lopez, 2009). Yet, Sink, Covinsky, Newcomer, and Yaffe (2004) found that Hispanics were at increased risk of exhibiting hallucinations, expressing unreasonable anger, being combative, and wandering as compared to Whites.

Cultural and ethnic considerations can influence the interpretation of behaviors. Traditional cultural values underlying the Latino culture and how elders are treated may in part (Adames, Chavez-Dueñas, Fuentes, Salas, & Perez-Chavez, 2014) explain some of the differences in Latinos' manifestations of BSDs. Culturally, younger people are expected to show respect, *respeto*, and behave in a deferential manner toward their Latino elders which may delay any recognition and acknowledgement of cognitive or behavioral symptoms. For example, Griffith and Lopez (2009) found that 65% of Hispanic caregivers dismissed symptoms of dementia or Alzheimer's dementia (AD) in part because of adherence to a "folk model" supporting the belief that the behaviors may be due to psychological stress or the process of normal aging. The folk model may include the notion that dementia is a form of mental illness or a culturally specific syndrome and thus unavoidable or a result of fate (Avant, 2013; Griffith & Lopez, 2009; Kramer, 2002).

Another but most frequently documented cultural difference for Hispanic caregivers, compared to White caregivers, is the high value placed on the extended family, and a strong sense of duty to family, known as *familismo*. Moreover, Napoles, Chadiha, Eversley, and Moreno-John (2010) found that Hispanic caregivers have more positive views of caregiving and a greater sense of spirituality. In contrast, several studies found evidence of worse mental health among Hispanic caregivers compared to Whites (Napoles et al., 2010), with depression being more common in Hispanic caregivers (Janevic & Connell, 2001). In addition, Mexican-American caregivers of persons with dementia report greater distress, poorer self-rated health, increased somatic complaints, and increased sensitivity to behavioral disturbances (Harwood, Barker, Ownby, & Duara, 2000). These seemingly contradictory findings can, in part, be explained by the heterogeneity of the Hispanic/Latino culture (e.g., Cuban, Puerto Rican, Mexican, Latin American). Differences in cultural values and norms including the cultural embeddedness of caregiving, cultural determinants of caregiving, ethnic group identity, and level of acculturation influence the caregiving experience. While caregiving is generally assigned based on female gender, these experiences are further influenced by educational level and social support. Educational level, social support, and level of acculturation tend to produce more positive caregiving experiences with better mental health outcomes reported (Adames et al., 2014).

Several cultural-bound syndromes that may influence behavioral interpretation have been identified by Mezzich and colleagues (1996) and subsequently incorporated in the *Diagnostic and Statistical Manual* (American Psychological Association, 2010). To illustrate, for Puerto Ricans, *ataque de nervios* is a culture-dependent syndrome and is an expression of malaise in a subpopulation (Guarnaccia, Rubio-Stipec, & Canino, 1989). Other examples of culture-bound syndromes present in Latin America countries include *atontado* found in Tuxtlas and Veracruz, Mexico; *brujería* in Latin America; *colerina* in Distas, Yucatan; *el bla* in Miskitos as well as in both Honduras and Nicaragua; and *syndrome de la Nevada*, as found in Peru. Many of these culture-bound syndromes are frequently associated with the loss of spiritual strength or loss of the soul (Guarnaccia, et al., 1989). While these syndromes are generally limited to specific cultural areas and are localized in highly diverse U.S. areas such as Los Angeles, several ethnic enclaves retain their original cultural practices and thus may not seek assistance early in the dementia trajectory (Latin American Guide for Psychiatric Diagnosis [GLDP from the Spanish name] Latin American Psychiatric Association [APAL], 2004).

As ethnic groups relocate to the United States, many retain their main beliefs and practices related to healing. Moreover, traditional and folk healing systems are intertwined with religious beliefs. Broadly these include (a) a sense that the cause of an illness is external to the individual and is spiritual in nature [God's will, *el mal ojo* (the evil eye)] (Talamantes, Trejo, Jimenez and Gallagher-Thompson, 2006); (b) that the group participates in the healing of an individual, often treated within the family and community; and (c) that morality functions as a cause and path for recovery, integral to healing and the healing process and which depends primarily on nonverbal and symbolic interaction (Koss-Chioino, 1992). These systems of healing do not subscribe to a mind-body distinction and are frequently found in Western concepts of healing. These differences in healing practices have been described as alternate paradigms, *etic* (i.e., scientific) vs. *emic* (i.e., insider, relativistic).

Three main healing systems exist within the Hispanic/Latino community in the United States: *espiritismo* (spiritism) associated with Puerto Ricans, *Santería* associated with Cubans, and *curanderismo* associated with Mexican and Latin American communities. These systems derive from a synthesis of beliefs and practices related to neocolonial history: *espiritismo* (spiritism) is the synthesis of French European and Afro-Caribbean traditions, *Santería* from Catholicism and West African traditions, and *curanderismo* from Catholicism and Mexican Indigenous traditions. Whereas Western thought considers the body, mind, and spirit as holistic healing, underlying all of these Hispanic/Latino traditions is the notion that illness is a result of being "out of balance," i.e., within the individual vis-à-vis within the cosmos; thus, healing occurs when harmony and balance are restored. *Espiritismo* (spiritism) encompasses all of three realms of body, mind, and spirit by calling on the Holy Spirit. *Santería* incorporates the concepts of gods or saints, while *curanderismo* through the *curanderas* (healers) endeavors to restore balance using the spiritual realm and advocating moral practices. Practices vary within each system (Koss-Chioino, 1992).

Indigenous healers do not distinguish between physical, emotional, and social origins of illness. All illness is viewed as an issue of balance. For example, some distress may be viewed as "bewitchment," i.e., being bewitched by spirits and therefore out of balance with the spiritual realm. In *espiritismo* (spiritism), the healer does not intervene, but by use of image and metaphor and communing with their spirit guides serves as an instrument to bring about the change necessary for healing. In *Santería*, the santeros or priests prescribe herbal remedies and amulets to restore health and guard against future injury. In *curanderismo*, the curanderas (healers) consider their ability a gift from God and frequently apprentice with an elderly relative.

Gray, Jimenez, Cucciare, Tong, and Gallagher (2009) completed a qualitative study to examine the ethnic differences in knowledge, attitudes, and beliefs about dementia in women caring for a family member with dementia. In general, Hispanic caregivers thought that dementia was a normal part of aging, was a result of a difficult life, or was a result of outside forces such as *el mal ojo* (the evil eye). Others believe that the person with dementia is bewitched (Woods, 1999, unpublished interview data). This view of being bewitched or being affected by the evil eye has significant consequences for both the person with dementia and their families as it carries significant stigma.

Hispanic/Latino caregivers seek help from practitioners of folk and traditional healing, as well as from Western biomedicine, and some of the differences in help seeking are related to the education and income of the caregivers (Gray et al.). Folk and traditional healers tend to view dementia from a holistic perspective, i.e., an imbalance between the individual and their cosmos. The *curanderas* (healers) tend to incorporate Western biomedicine and traditional healing when providing treatment. In *espiritismo* (Spiritism), the healer focuses on restoring balance with the spiritual realm by communing with spirit guides and the use of image and metaphor. An intriguing example of the blending of treatment systems can be found in the study of Soto-Espinosa and Koss-Chioino (2017) wherein doctors were found to encourage their patients to use folk and traditional healing, in conjunction with Western biomedicine. The researchers interviewed 74 Puerto Rican doctors using semi-structured interviews and found that several of the physicians incorporated integrative healing in their practices in a search for new ways to heal and/or to find a pathway to higher morality. These approaches to care are embedded in deeply held belief systems and bring a holistic perspective to healing, not unlike the folk and traditional systems.

Pharmacological Interventions

Assessment of any behavioral symptoms and behavioral change is critical prior to treating BSD symptoms. However, potentially regardless of healing system, while those with untreated BSDs show a faster progression of their dementia, it is not clear that treating these symptoms earlier helps slow the trajectory of dementia (Rabins et al., 2013) as several factors can influence behavioral symptoms. These factors include underlying acute medical conditions such as bladder or respiratory infections or the side effects of medications, cultural and ethnic factors, psychiatric illness, and environmental factors. Because persons with dementia are disproportionately affected by pain and undiagnosed illness that can contribute to BSDs, these must be ruled out. Galik (2016) provides three steps to assess BSDs which include the descriptions of (1) the behavior and onset, (2) frequency and any aggravating or relieving factors through assessment of any underlying medical conditions, and (3) use of specific rating scales to determine cognitive, functional, and physical capabilities. Kales, Gitlin, and Lyketsos (2015) describe a similar assessment of BSDs referred to as DICE (describe, investigate, create, and evaluate). Regardless of the method used, assessment is essential. Once the assessment is completed, non-pharmacological interventions may be considered alone or in combination with appropriate pharmacological interventions.

Pharmacological interventions continue to be widely used and used inappropriately to alleviate BSD, despite limited efficacy (Kales et al., 2011; Lyketsos et al., 2011). Generally, the class of medications used for treating BSD is psychotropic medications, including antipsychotics, antidepressants, and sedative hypnotics (Ballard et al., 2009). Psychotropic medications are frequently overused for those with dementia (Kales et al., 2011; Lyketsos et al., 2011). In practice, 84% of nursing home residents and 28% of community-dwelling persons with dementia are prescribed psychotropic medications (Maust, Langa, Blow, & Kales, 2017). In addition, 30% of persons with dementia take antipsychotic medications that are associated with a higher risk of mortality and cerebrovascular events (Maust et al., 2015).

There tends to be an assumption of universality in prescribing practices and in an individual's experience of the medication. This view, largely influenced by the dominant Caucasian population in whom the drugs were tested (Chen, Chen, & Lin 2008), does not consider any cultural or ethnic factors that may influence an individual's response to the medication despite the empirical evidence that describes significant differences in response to psychotropic drugs in different ethnic groups (Cacabelos, Cacabelos, & Torrellas, 2014; McDowell, Coleman, & Ferner, 2006).

Ethnopharmacology, a relatively new field, considers ethnocultural differences and genetic variations that can affect an individual's response(s) to a specific drug or class of drugs. Genetic variations influence both the pharmacokinetics (i.e., absorption, metabolism, distribution, elimination) and pharmacodynamics (i.e., mechanism of action and effects at the target site) of medications. The aspect of pharmacokinetics most studied is metabolism, i.e., enzymes and transport receptor protein binding, while the aspect of pharmacodynamics most studied is the intensity of therapeutic and adverse effects (Lin, Chen, Yu, Lin, & Smith, 2003). For Latinos, ethnically determined genetic variations in enzymes needed for drug metabolism have a profound effect on the timing of medication metabolism and the optimal therapeutic effect.

The Use of Psychotropics in Dementia Treatment and the Role of Genetic Variation

Most psychotropic drugs, such as antidepressants, antipsychotics, and sedative hypnotics, are commonly prescribed for persons with dementia for depression and for behavioral disturbances, such as agitation and sleep problems, but with suboptimal results (Ballard et al., 2009; Kales et al., 2015). For Latino older adults, of whom up to 63% over the age of 85 experience dementia (U.S. Department of Health and Human Services, 2014), the results are even less optimal as they are made complicated by aging changes that affect drug pharmacokinetics and pharmacodynamics (Mentes, Cadogan, Woods, & Phillips, 2015). While outcomes may be complicated by psychosocial adversity and comorbidities, genetic differences can play a major role. Poorer outcomes are not inevitable if these ethnocultural differences are considered when prescribing medications and assessing a medication's optimal effect (Murphy & McMahon, 2013). Genetic testing is advisable prior to beginning new drugs, such as psychotropic medications, to determine metabolism rate that will guide prescribing practices.

Psychotropic medications are metabolized in the liver in an oxidation phase (phase 1) and a conjugation phase (phase 2). Most studies have focused on the oxidative phase and the role of cytochrome P450 (CYP) enzymes and enzyme subgroups (Lin et al., 2003). Several studies have indicated that genetic differences in the CYP enzymes have profound implications for altered drug responses in ethnic older adults, which are further complicated for those with dementia (Lin & Smith, 2000; Silva, 2013; Solus, et al., 2004). These issues apply specifically to those with dementia for whom psychotropic medications are prescribed.

The principal enzymes with polymorphic variants involved in phase 1 reactions are the cytochrome P450 monooxygenases (CYP3A4/CYP3A5/CYP3A7, CYP2E1, CYP2D6, CYP2C19, CYP2C9, CYP2C8, CYP2B6, CYP2A6, CYP1B1, CYP1A1/CYP1A2). Keltner and Folks (2001) noted that the genetic ability to produce these enzymes varies by ethnicity. For example, genetic changes in specific CYP alleles affect the rate of drug metabolism, which in turn affects the plasma level of the drug and the bioavailability of the drug. Among these enzymes, CYP2D6, CYP2C9, CYP2C19, and CYP3A4/CYP3A5 are the most relevant CYP polymorphisms when considering ethnic differences in the use of psychotropic drugs (Cacabelos et al., 2014). Approximately 18% of neuroleptics are metabolized by CYP1A2 enzymes, 40% of CYP2D6, and 23% of CYP3A4; 24% of antidepressants are metabolized by CYP1A2 enzymes, 5% of CYP2B6, 38% of CYP2C19, 85% of CYP2D6, and 38% of CYP3A4; 7% of benzodiazepines are major substrates of CYP2C19 enzymes, 20% of CYP2D6, and 95% of CYP3A4 (Cacabelos et al., 2014). Given that, in practice, more than 84% of persons with dementia are prescribed psychotropic medications, the issue of CYP genomics is an important consideration as the same drug can be metabolized differently dependent on the individual's genetic profile.

Sixty percent of currently prescribed psychotropic medications are metabolized by CYP2D2 so that any changes in this enzyme result in differential metabolism. There are 141 CYP2D6 allelic variants which result in extensive (EM), intermediate, poor (PM), and ultrarapid metabolizer (UM) types, characterized by normal, intermediate decreased, and very rapid metabolism of a drug. For example, those with two functional copies of CYP2D6 metabolize psychotropic medications more rapidly (rapid metabolizers [RMs]), while those with fewer functional copies metabolize these drugs more slowly (slow or poor metabolizers [PMs]), resulting in higher blood serum levels of the drugs.

Interethnic differences exist in the prevalence of PM and UM types. In fact, Isaza, Henao, Lopez, and Cacabelos (2000) found specific differences in the CYP2D6 gene in the population of Colombia. Cacabelos (2012) has also noted differences in Europe such that a north-south gradient exists in the frequency of the PM type (e.g., 6–12% in the southern European countries including Spain and only 2–3% in the northern European countries). Considering that increased age is associated with a decline in the ability of the liver to detoxify drugs, older adults in general are at higher risk of being poor metabolizers, increasing the vulnerability to inappropriate dosing.

Persons with Alzheimer's disease (AD) have an increased risk of altered drug responses to psychotropic medications. In persons with AD, 56.38% are EM, while 27.66%, 7.45%, and 8.51% are IM, PM, and UM, respectively (Cacabelos, 2012). One of the genes associated with AD is the apolipoprotein E (APOE) ε4 allele, an established genetic risk factor for mild cognitive impairment (MCI) and both familial and late-onset AD (Haan, Shemanski, Jagust, Manolio, & Kuller, 1999; Payami et al., 1997; Yaffe, Cauley, Sands, & Browner, 1997). Others, including Cacabelos et al. (2010) and Takeda et al. (2010), have found that those who carry the APOE4 allele are the worst responders to conventional drug treatment. In fact, when CYP2D6 and the APOE ε4/ε4 genotype were combined, the result was that in the presence of APOE4, there was a resulting conversion from someone who metabolized the drug very well (EM) to someone who metabolized the drug poorly (PM) (Cacabelos et al., 2010). This has significant clinical implications for prescribing psychotropic medications, especially for Latinos with AD who are 1.5 times more likely than Caucasians to have AD (U.S. Department of Health and Human Services, 2014).

The Beers Criteria for Potentially Inappropriate Medication Use in Older Adults (American Geriatrics Society [AGS], 2012), created in 1991 as a prescribing guide for clinicians, recommend that psychotropic medications, such as antipsychotics, be used with caution and not be prescribed for behavior disturbances associated with dementia unless non-pharmacological measures have been tried and have failed. The Beers Criteria recommend avoiding use of all benzodiazepines, e.g., lorazepam (Ativan), in all older adults (American Geriatric Society, 2012). These concerning issues are compounded in ethnic minority older adults given genetic differences and lack of research supporting therapeutic effects.

Plasma protein binding must also be considered with psychotropic medications. These medications are largely lipophilic, meaning that they are soluble in fat and depend on plasma proteins for their transport. The plasma proteins generally regarded as most important to solubility are the glycoproteins and albumin, which are genetically determined and vary across ethnic groups. Normal aging changes, such as an alteration in the muscle/fat ratio of an increase in fatty tissue, mean that more of the psychotropic drugs are in solution. This, coupled with changes in albumin, can result in unintended drug responses frequently assessed as adverse medication events.

Epigenetic Factors Associated with Medications for BSD

An individual's response to medications is multifaceted, involving an interaction among genes, environment, and culture. Environmental factors such as diet, pollution, herbal supplements, and tobacco can play a clinically significant role in the pharmacokinetics (i.e., absorption, metabolism, distribution, elimination) of psychotropic medications because they modify gene expression and thus deserve consideration. Ethnocultural factors must be addressed as they can affect drug response.

The P450 enzymes are particularly sensitive to changes in diet and other environmental factors, such as tobacco. For example, several micronutrients exert effects on the expression of metabolizing enzymes such as CYP1A2, an enzyme linked to the metabolism of antidepressants. This enzyme is induced by charbroiled beef and cruciferous vegetables, while corn and certain citrus fruits such as grapefruit can inhibit CYP3A4 (Fuhr, Klittich, & Staib, 1993; Oesterheld & Kallepalli, 1997). Ethnic groups vary widely in their consumption of potential inducers and inhibitors which can, in turn, affect the metabolism of numerous psychotropic medications.

Environmental factors such as smoking have a profound effect on the metabolism of psychotropic medications since it induces many of the P450 cytochromes. The polycyclic aromatic hydrocarbons (PAHs) present in tobacco smoke induce several hepatic (i.e., liver-based) cytochrome P450 (CYP) isozymes. For example, CYP1A2 induction has been a main reason for the decrease in plasma concentrations of several antidepressants (trazodone), antipsychotics (clozapine), and anxiolytics (lorazepam, alprazolam) (Zevin & Benowitz, 1999). It has become clear that cigarette smoking affects the pharmacokinetics of numerous psychotropic drugs. Within the Latino/Hispanic subpopulations, the rate of smoking varies widely. For example, up to 50% of Cuban men and 35% of Cuban women report smoking more than 20 cigarettes per day, compared to Mexican and Puerto Rican men and women (Kaplan, Bangdiwala, & Barnhart, 2014). Thus, it is critical that healthcare providers carefully consider smoking when prescribing psychotropic medications.

In summary, current studies indicate that the consideration of ethnocultural differences and genetic variations that affect a person's response to a specific drug, or class of drugs, is critically important when prescribing medications. Genetic variations influence both the pharmacokinetics and pharmacodynamics of medications. This is even more important when psychotropic medications are used since most of these medications are metabolized in the liver using cytochrome P450 and because ethnic differences can drastically alter the metabolism from potentially an ultra-metabolizer or extensive metabolizer to a poor metabolizer. In addition, it is known that increased age is associated with a decline in the ability of the liver to detoxify drugs further complicating the issue. Thus older adults, in general, are at higher risk of being poor metabolizers and experiencing increased vulnerability to inappropriate dosing.

In addition to ethnic variation in CYP450, epigenetic factors must be considered. Diet that is culturally determined, tobacco smoking, and other environmental issues can affect the function of CYP 450. To add to these complications, persons with dementia who carry the APOE4 allele are at increased risk of an altered drug response. To date, the interaction of these factors in ethnic older adults with dementia has not been studied. Given the dearth of research studies that include ethnic older adults and specifically Latinos, psychotropic drugs should be prescribed with caution. The emphasis should be placed on non-pharmacological interventions as the first-line of treatment for BSDs.

Non-pharmacological Interventions for BSD in Latinos with Dementia

Non-pharmacological interventions are recommended as the first line of treatment for behavioral disturbances (BSDs) as part of a comprehensive approach to dementia care to encourage person-centered care and to optimize quality of life. There have been several advances in research in non-pharmacological interventions, although there is currently no consensus about how to categorize non-pharmacological interventions or specific, evidence-based recommendations for their use. Most of the studies of non-pharmacological interventions have focused on the persons with dementia or on caregivers, rather than the environment (Kales et al., 2015). A study by Desai, Schwartz, and Grossberg (2012) assessed the quality and weight of the evidence for non-pharmacological findings;

however, none of the studies rose to *Grade A* recommendations for consistent, high-quality randomized controlled trials (RCTs). However, several rose to *Grade B* recommendations for consistent, low-quality RCTs, using the Oxford Centre for Evidence-Based Medicine guidelines (see www.cebm.net). Nonetheless, Brodaty and Burns (2012) point out that "a lack of quality research" does not necessarily connote a lack of efficacy. Reliance on RCTs excludes possibilities that arise from the naturalistic setting. In fact, several studies of non-pharmacological interventions report anecdotal and qualitative evidence but which is difficult to capture quantitatively, particularly with high-quality research in this field being labor intensive and expensive, especially with the required large sample sizes necessary to conduct rigorous clinical trials.

Several types of non-pharmacological interventions show efficacy for the treatment of BSDs, although the literature review for this chapter found only one article that discussed the importance of cultural and ethnic preferences and was based on using individualized music (Gerdner, 2015). It appears that there is next to no literature on ethnic responses to non-pharmacological interventions. Galik (2016) organized the discussion of non-pharmacological interventions into five categories: sensory stimulation, cognitive stimulation and training, emotion-oriented interventions, physical activity and exercise, and behavioral education and training. Kales et al. (2015) organized their discussion of the topic into three categories: interventions for the person with dementia, interventions for the caregiver, and interventions for the environment. The next section will discuss non-pharmacological interventions directed toward the person with the dementia as well as non-pharmacological interventions directed toward family caregivers. Other issues related to the environment, such as ambient noise, lighting levels, and consistent routines, are not addressed in this chapter.

Non-pharmacological Interventions Directed Toward Persons with Dementia

O'Neil et al. (2011) completed a systematic review of evidence-based, non-pharmacological interventions for behavioral symptoms of dementia for the Veterans Administration (VA). All studies reviewed included adults with mild, moderate, or severe dementia. Behavioral symptoms included apathy, agitation, disruptive vocalizations, aggression, disturbed sleep, wandering, impulsivity, disinhibition, depression, inappropriate sexual behavior, chronic/intermittent hallucinations, and delusions. Several different non-pharmacological interventions were described in the review including cognitive/emotion-oriented interventions (e.g., reminiscence therapy, simulated presence therapy, and validation therapy), sensory stimulation interventions (e.g., acupuncture, aromatherapy, light therapy, massage/touch therapy, music therapy, Snoezelen multisensory stimulation, and transcutaneous electrical nerve stimulation [TENS]), behavior management techniques, other psychosocial interventions (e.g., animal-assisted therapy and exercise), and various interventions targeting a specific behavioral symptom (e.g., wandering, agitation, and inappropriate sexual behavior). No harm was reported in using behavior management techniques for persons with dementia nor when these techniques were implemented by caregivers. O'Neil et al. (2011) concluded that overall, clear and consistent evidence supporting the use of the various non-pharmacological interventions for BSDs is lacking. Variability in outcomes, assessment tools, and interventions made it difficult to compare or to establish specific, effective components of interventions that are broadly classified as behavior management techniques. Additionally, findings across studies are inconsistent even when similar interventions and outcomes are being assessed.

O'Neil et al. (2011) also concluded that overall, non-pharmacological treatments for the behavioral symptoms associated with dementia warrant further investigation in the form of large-scale, randomized clinical trials (RCTs) of adequate duration. Several small studies suggest that some of the interventions, stimulation or sensory-oriented approaches (i.e., light therapy, aromatherapy, and massage/touch therapy), show greater promise than emotion-oriented approaches and may have potential benefits for individuals with dementia. Since the 2011 report, several VAs across the United States have implemented sensory-oriented approaches such as aromatherapy, light therapy, massage/touch therapy, music therapy, and Snoezelen multisensory stimulation (Beiter, 2014).

Persons with dementia have multiple behavioral symptoms associated with their dementia (e.g., sleep awakening, aggression, wandering, depression) that may respond to different interventions. Multicomponent studies may be one way to address the multiple needs of patients with dementia. Thus, a multicomponent strategy to reduce the impact of behavioral symptoms should be included in further research, although the issue then becomes which component or combination is the most effective. Future studies should include RCTs with clearly defined interventions, with a focus on duration and frequency of the intervention, as well as adverse effects. Validated tools should be used to measure outcomes; adequate blinding of participants, raters, and providers is also necessary to assure valid, replicable results. O'Neil et al. (2011) noted that the findings should be interpreted with caution given the overall limitations in the bodies of evidence. Overall, there are few, good-quality RCTs with much of the evidence coming from small studies often with great variability in how behavioral symptoms are defined, the duration of intervention, measuring instruments used, and the outcome measures utilized to determine the effectiveness of the various interventions.

Examples of Non-pharmacological Interventions for BSD in Dementia

The following are examples of studies of non-pharmacological interventions for BSD in dementia. There was minimal ethnic diversity in all the following studies.

Gitlin (2008) found that a "tailored activities program" (TAP) (i.e., selecting activities specifically tailored to the person with dementia's abilities, interests, and roles) reduced BSDs after six 90-minute home visits and two 15-minute follow-up telephone calls over a 4-month period, in addition to offering caregiver training.

Music, which is another type of intervention, has been used extensively as a non-pharmacological intervention to decrease BSD with significant results (Gerdner, 2000; Holmes, Knights, Dean, Hodkinson, & Hopkins, 2006; Sung, Lee, Li, & Watson, 2012; Svansdottir & Snaedal, 2006). In early work, Gerdner (2000) found that individualized music (i.e., music tailored to the individual's like and listening habits) compared to classical music decreased BSDs. More recently, Gerdner (2015) emphasized the importance of ethnicity in the assessment of individualized music, highlighting a case study of an 83-year-old, Hispanic male whose agitation decreased significantly when "Guitarras Mexicanas" was played for a 6-week period, compared to classical music. During the "Guitarras Mexicanas" music intervention, his BSDs decreased to 4.3 behaviors compared to 23.3–25.3 per 10-minute increment at baseline. This is in contrast to the classical music intervention which resulted in 13.5 behaviors compared to 23.3–25.3 per 10-minute increment at baseline. Holmes et al. (2006) and Sung et al. (2012) found a decrease in apathy ($p < 0.0001$), anxiety ($F = 8.98$; $p = 0.004$, Cohen's $d = 0.90$), and anxiety and agitation ($p = 0.02$), respectively, when the person with dementia actively participated in music-based interventions.

Aromatherapy, specifically lavender and *Melissa officinalis* oil (e.g., lemon balm), has been used in residential settings to decrease BSDs. A randomized crossover study by Lin et al. (2003) found that lavender oil diffused near the person's pillow alleviated agitation ($p < 0.001$). Burns et al. (2011), in an RCT, found no difference in BSDs when lemon balm was compared to donepezil and placebo, although all three groups showed an improvement on the Neuropsychiatric Inventory (NPI) (Cummings,

1997) and the Pittsburgh Agitation Scale (PAS) (Rosen et al., 1994). Yang et al. (2015) compared the use of lavender oil alone with aroma-acupressure (e.g., lavender oil on acupressure sites) in persons with dementia and a control group (N = 186) over a 4-week period in a Taiwanese population. According to a reported power analysis, a total sample N = 108 was required to achieve an effect size 0.25. The aromatherapy and aroma-acupressure groups showed a significant reduction in agitation ($p = 0.01$ and $p < 0.01$, respectively), compared to the control group. However, in this study, no effect size was reported for outcomes.

Pleasurable activities and physical activity have shown positive results in both residential and community-dwelling persons with dementia. Early studies by Teri and Logsdon (1991) and Teri, Huda, Gibbons, Young, and van Leynseele (2005) show positive effects of pleasurable activities and exercise on BSDs. Subsequent studies (Forbes et al., 2008; Thuné-Boyle, Iliffe, Cerga-Pshoja, Lowery, & Warner, 2012) support these findings, showing a decrease in depression and an improvement on mood; however, the population was predominately Caucasian.

In one RCT, touch therapy showed a significant decrease in restlessness. Using a three-group experimental design, Woods, Craven, and Whitney (2005) examined the effect of touch on restlessness in nursing home residents with dementia. There was a significant difference in restlessness in the intervention group ($p = 0.036$) compared to the control group.

De Oliveira et al. (2015) reviewed several non-pharmacological interventions directed toward the persons with dementia. These included occupational activities, music, aromatherapy, physical exercise, light therapy, touch therapy, and cognitive rehabilitation. Most of the studies reviewed were studies conducted in residential settings rather than the community. Results of this review indicated that tailoring the intervention to the participant's interests and skills resulted in the largest effect size for decreasing disruptive behaviors such as shadowing ($p = 0.003$, Cohen's $d = 3.10$) and repetitive questioning ($p = 0.023$, Cohen's $d = 1.22$) while increasing positive behaviors such as greater engagement ($p = 0.029$, Cohen's $d = 0.61$) and pleasure ($p = 0.045$, Cohen's $d = 0.69$). Overall, agitation responded to music, aromatherapy, bright light, and touch, with studies indicating promising results.

Desai et al. (2012) reviewed several non-pharmacological interventions for BSDs. One notable finding from this review is that large effect sizes were also obtained when activity interventions were tailored to the individual's interest and functional ability. In part, these conclusions were drawn based on the results from several nursing research studies (Buettner, Fitzsimmons, Atav, & Sink, 2011; Kolanowski, Litaker, Buettner, Moeller, & Costa, 2011; Kovach, Calamia, Walsh, & Ginsburg, 2004). Smith, Kolanowski, Buettner, & Buckwalter, 2009). For example, in a residential setting with 78 participants, Kovach et al. (2004) used an intervention titled Balancing Arousal Controls Excesses (BACE) that is focused on individual arousal states to decrease agitation. The mean change for the treatment group was a decrease of 8.43 points in the treatment group. Buettner et al. (2011) studied the effect of mentally stimulating activities on 77 persons with dementia residing in the community exhibiting apathy. The results indicated significantly lower levels of apathy ($p < 0.001$) and depression ($p < 0.001$) post-intervention. Kolanowski et al. (2011) examined the effect of individualized activities and personality-style intervention on agitation, mood, passivity, and engagement in 128 nursing home residents with dementia. Engagement, agitation, passivity, and alertness all improved during the intervention; however, this improvement was not sustained after the intervention was completed. These authors suggest that an interdisciplinary team approach meets with the best success.

Non-pharmacological Interventions for BSD Directed Toward the Family Caregiver

The majority of individuals with dementia live in the community; and, of these, approximately 75% have care provided by a family member or friend (Alzheimer's Association, 2014). It is critical to engage the caregiver in care given that several studies show that problem-solving with the family caregiver results in decreased BSDs for the person with dementia, improved caregiver confidence, decreased sense of burden, and improved quality of life. Studies such as TAP (Gitlin, 2008) demonstrate that engaged caregivers felt that they were "on duty" less ($p = 0.001$, $d = 0.74$).

The Advancing Caregiver Training (ACT) study (Gitlin, Winter, Dennis, Hodgson, & Hauck, 2010) used health professionals working with caregivers over 11 visits to help them identify triggers for BSDs and train them on how to modify these or modify their own behavior. The study found significant decrease in caregiver upset, improved communication with the person with dementia, and enhanced caregiver confidence. Similarly, the study entitled the Care of Persons with Dementia in Their Environment (COPE) found decreased functional dependence in the person with dementia and improved caregiver well-being after 12 contacts by health professionals over a 4-month period (Gitlin, 2008).

Brodaty and Arasaratnam (2012) completed a meta-analysis of non-pharmacological interventions for BSDs by reviewing family caregiver interventions in the community that included interventions directed toward caregiver skill training, caregiver education and enhanced support, activity planning and environmental redesign, self-care techniques, and collaboration with health professionals. This meta-analysis reviewed 23 studies. Several studies ranked at level II, denoting that the evidence was obtained from at least one well-designed RCT (e.g. large, multisite RCT), and level III, denoting that evidence was obtained from well-designed controlled trials without randomization (i.e., quasi-experimental). Of the 17 studies reviewed, the pooled effect size fell within the medium range ($d = 0.34$) indicating a positive overall effect on BSDs for interventions directed toward the caregiver, while 13 studies showed a small effect size ($d = 0.15$) indicating that these interventions were effective, albeit weak, for improving caregiver's responses to BSDs. Overall, the results were largely positive. The duration of the interventions ranged from 6 weeks to 24 months. The follow-up ranged from 3 months to 24 months. There were no reported adverse effects. These authors concluded that the influence that caregivers have on the occurrence and severity of BSDs is underestimated. The meta-analysis indicated that caregiver interventions can significantly reduce BSDs, in addition to reducing the caregiver's negative response to these BSDs.

Few non-pharmacological interventions directed toward family caregivers have included different ethnic or racial groups. One RCT conducted by Belle et al. (2006) did include three ethno-racial groups (i.e., Latino, African American, White or Caucasian) in an RCT to test the effects of a multicomponent intervention on quality of life and caregiver depression. In this two-group experimental study, 624 caregivers (212 Hispanic or Latino, 219 White or Caucasian, 211 African American) were randomly assigned to either the experimental or control group. Hispanic or Latino participants were more highly educated in the intervention than the control group participants. The intervention was made up of several components that included providing information, didactic instruction, role-playing, problem-solving, skills training, stress management techniques, and telephone support groups that took place through 12 in-home and telephone sessions over a 6-month period. The design allowed for tailoring to meet the specific needs of the caregiver and to be culturally sensitive and flexible. The social support component focused on decision-making, managing tasks, and decreasing social isolation. There was a statistically significant improvement in quality of life, measured by indicators of depression, caregiver burden, social support, self-care, and care recipient behaviors for Hispanic and Latino caregivers ($p < 0.001$). Moreover, depression was lower in the intervention group (12.6% vs. 22.7%); $p = 0.001$). Hispanic/Latinos reported less depression and fewer problem behaviors for the care recipient. The authors surmise that the significant difference for the Hispanic/Latino group can be explained by the increased social support and training.

Although the studies mentioned above have focused on family caregivers in the community, multicomponent studies such as Teri et al. (2005) using Staff Training in Assisted Living Residences (STAR) were effective in training residential care staff. Residents in the assisted living facilities showed reduced levels of affective and behavioral distress compared to residents in the control group. This model has been used effectively in the United Kingdom and piloted in Brazil (da Silva Serelli et al., 2016).

Conclusion

Clearly, one size does not fit all when it comes to interventions for behavioral symptoms of dementia (BSDs), whether for individuals with dementia or their caregivers and for patients and caregivers of different ethnic and racial backgrounds. It is critical for healthcare providers to enquire and become knowledgeable about the folk and traditional healing that many Hispanic/Latino groups utilize and for which they place their faith. Only then can there be fruitful communication between traditional healers (e.g., *curanderas*) and healthcare providers using Western biomedicine that can benefit the patient and the patient's family.

There are several hundred research studies that have examined the effects of a variety of non-pharmacological interventions for BSDs and for caregiver well-being, all showing degrees of positive results. These studies show promise. However, as O'Neil et al. (2011), De Oliveira et al. (2015), and Brodaty and Arasaratnam (2012) have noted, several of the studies present methodological challenges. These issues need to be addressed to move the science further. Future research should focus on larger randomized trials, using within-individual longitudinal designs, aiming for sustainability of outcomes over time and the use of blinding in the measurement of outcomes. Moreover, it is critical to include different ethnic groups in these studies so that there can begin to be a determination as to whether there is a variation in the manifestation of BSD by ethnic group and whether there are varied responses to specific non-pharmacological interventions, including specific, tailored interventions.

As for pharmacological interventions for BSDs, psychotropic drugs must be used with extreme caution in older adults from ethnic minority groups. Further research is needed that includes Latino older adults and specifically examines different responses to psychotropic medications. Education for healthcare professionals is essential to fully understand the risks of prescribing psychotropic medications to Latino older adults with dementia, especially considering the dearth of evidence of efficacy and perhaps its detrimental effects as seen in primarily White populations examined in earlier studies. Practitioner education about the various non-pharmacological interventions that can be used for BSDs is also important so that the current recommendations can be put into practice. Overall, the emphasis needs to be on including older Latinos with dementia in research studies and designing studies that include an interdisciplinary perspective.

References

Adames, H. Y., Chavez-Dueñas, N. Y., Fuentes, M. A., Salas, S. P., & Perez-Chavez, J. (2014). Integration of Latino/a cultural values into palliative health care: A culture centered model. *Palliative & Supportive Care, 12*(2), 149–157. https://doi.org/10.1017/S147895151300028X

Alzheimer's Association. (2014). *2013 Alzheimer's disease facts and figures*. Retrieved January 21, 2014, from http://www.alz.org/alzheimers_disease_facts_and_figures.asp

Alzheimer's Association Expert Advisory Workgroup on NAPA. (2012). Workgroup on NAPA's scientific agenda for a national initiative on Alzheimer's disease. *Alzheimer's Dementia, 8*(4), 357–371.

American Geriatrics Society 2012 Beers Criteria Update Expert Panel. (2012). American geriatrics society updated beers criteria for potentially inappropriate medication use in older adults. *Journal of the American Geriatrics Society, 60*, 616–631. https://doi.org/10.1111/j.1532-5415.2012.03923. Available online at http://www.americangeriatrics.org/files/documents/beers/2012BeersCriteria_JAGS.pdf

American Psychological Association. (2010). *Diagnostic and statistical manual-IV (DSM-IV)* (5th ed.). Washington, DC: APA.

Asociacion Psiquiatrica de America Latina (Latin American Psychiatric Association) [APAL]. (2004). *Guia Latinoamericana de Diagnostico Psiquiatrica. (Latin American Guide for Psychiatric Diagnosis) [GLADP]*. Mexico: Universidad de Guadalajara.

Avant, J. S. (2013). *Untangling cultural differences in behavioral, physiological, and psychological symptoms of dementia and Alzheimer's disease* (Unpublished research paper). University of Nevada, Las Vegas.

Ballard, C. G., Gauthier, S., Cummings, J. L., Brodaty, H., Grossberg, G. T., Robert, P., & Lyksetos, C. G. (2009). Management of agitation and aggression associated with Alzheimer disease. *Nature Reviews Neurology, 5*(5), 245–255. https://doi.org/10.1038/nrneurol.2009.39

Beattie, E., Song, J., & LaGore, S. (2005). A comparison of wandering behavior in nursing homes and assisted living facilities. *Research and Theory for Nursing Practice, 19*(2), 181–196. https://doi.org/10.1891/rtnp.19.2.181.66797

Beiter, D. (2014). *VHA voices, better care through better relationships*. US Department of Veterans Affairs. Moderator, Work Shop

Belle, S. H., Burgio, L., Burns, R., Coon, D., Czaja, S. J., Gallagher-Thompson, D., … Martindale-Adams, J. (2006). Enhancing the quality of life of dementia caregivers from different ethnic or racial groups: A randomized, controlled trial. *Annals of Internal Medicine, 145*(10), 727–738.

Borroni, B., Grassi, M., Agosti, C., Costanzi, C., Archetti, S., Franzoni, S., … Padovani, A. (2006). Genetic correlates of behavioral endophenotypes in Alzheimer disease: Role of COMT, 5-HTTLPR and APOE polymorphisms. *Neurobiology of Aging, 27*(11), 1595–1603. https://doi.org/10.1016/j.neurobiolaging.2005.09.029

Brodaty, H., & Arasaratnam, C. (2012). Meta-analysis of nonpharmacological interventions for neuropsychiatric symptoms of dementia. *The American Journal of Psychiatry, 169*(9), 946–953. https://doi.org/10.1176/appi.ajp.2012.11101529

Brodaty, H., & Burns, K. (2012). Nonpharmacological management of apathy in dementia: A systematic review. *The American Journal of Geriatric Psychiatry, 20*(7), 549–564. https://doi.org/10.1097/JGP.0b013e31822be242

Buettner, L. L., Fitzsimmons, S., Atav, S., & Sink, K. (2011). Cognitive stimulation for apathy in probable early-stage Alzheimer's. *Journal of Aging Research, 2011*(1–6), 1. https://doi.org/10.4061/2011/480890

Burns, A., Perry, E., Holmes, C., Francis, P., Morris, J., Howes, M. J. R., … Ballard, C. (2011). A double-blind placebo-controlled randomized trial of Melissa officinalis oil and donepezil for the treatment of agitation in Alzheimer's disease. *Dementia and Geriatric Cognitive Disorders, 31*(2), 158–164.

Cacabelos, R. (2012). Pharmacogenomics of the central nervous system (CNS) drugs. *Drug Development Research, 73*, 461–476. https://doi.org/10.1002/ddr.21039

Cacabelos, R., Cacabelos, P., & Torrellas, P. (2014). Personalized medicine of Alzheimer's disease. In S. Padmanabahn (Ed.), *Handbook of pharmacogenomics and stratified medicine* (pp. 563–611). London, UK: Academic Press of Elsevier.

Cacabelos, R., Fernandez-Novoa, L., Martinez-Bouza, R., McKay, A., Lombardi, J., Corzo, L., … Alvarez, A. (2010). Future trends in the pharmacogenomics of brain disorders and dementia: Influence of *APOE* and *CYP2D6* variants. *Pharmaceuticals, 3*(10), 3040–3100. https://doi.org/10.3390/ph3103040

Cano, C., Samper-Ternent, R., Al Snih, S., Markides, K., & Ottenbacher, K. J. (2012). Frailty and cognitive impairment as predictors of mortality in older Mexican Americans. *The Journal of Nutrition, Health & Aging, 16*(2), 142–147.

Chen, C., Chen, C. Y., & Lin, K. M. (2008). Ethnopsychopharmacology. *International Review Psychiatry, 20*(5), 452–459. https://doi.org/10.1080/09540260802515997

Cohen-Mansfield, J. (1997). Turnover among nursing home staff. A review. *Nursing Management, 28*(5), 59–64.

Cooper, C., Tandy, A. R., Balamurali, T., & Livingston, G. (2010). A systematic review and meta-analysis of ethnic differences in use of dementia treatment, care, and research. *American Journal of Geriatric Psychiatry, 18*(3), 193–203. https://doi.org/10.1097/JPG.0b013e31819caf

Cummings, J. L. (1997). The neuropsychiatric inventory: Assessing psychopathology in dementia patients. *Neurology, 48*(5 Suppl 6), S10–S16.

da Silva Serelli, L., Reis, R. C., Laks, J., de Padua, A. C., Bottino, C. M., & Caramelli, P. (2016). Effects of the staff training for assisted living residences protocol for caregivers of older adults with dementia: A pilot study in the Brazilian population. *International Geriatrics & Gerontology*, online March, *27*, 2017. https://doi.org/10.1111/ggi.12742

de Oliveira, A. M., Radanovic, M., de Mello, P. C. H., Buchain, P. C., Vizzotto, A. D. B., Celestino, D. L., … Forlenza, O. V. (2015). Nonpharmacological interventions to reduce behavioral and psychological symptoms of dementia: A systematic review. *Biomedical Research International, 2015*, 1–9. https://doi.org/10.1155/2015/218980

Desai, A. K., Schwartz, L., & Grossberg, G. T. (2012). Behavioral disturbance in dementia. *Current Psychiatry Reports, 14*(4), 298–309.

Ellison, J. M., Harper, D. G., Berlow, Y., & Zeranski, L. (2008). Beyond the "C" in MCI: Noncognitive symptoms in amnestic and non-amnestic mild cognitive impairment. *CNS Spectrums, 13*(1), 66–72. https://doi.org/10.1017/S1092852900016175

Forbes, D., Forbes, S., Morgan, D. G., Markle-Reid, M., Wood, J., & Culum, I. (2008). Physical activity

programs for persons with dementia. *Cochrane Database of Systematic Reviews, 3*, CD006489.
Fuhr, U., Klittich, K., & Staib, H. (1993). Inhibitory effect of grapefruit juice and its bitter principal naringenin on CYP1A2 dependent metabolism of caffeine in man. *British Journal of Clinical Pharmacology, 35*, 431–436. https://doi.org/10.1111/j.1365-2125.1993.tb04162.x
Galik, E. (2016). Treatment of dementia: Non-pharmacological approaches. In M. Boltz & J. E. Galvin (Eds.), *Dementia care: An evidence approach* (pp. 97–112). Cham, Switzerland: Springer.
Gerdner, L. A. (2000). Effects of individualized versus classical "relaxation" music on the frequency of agitation in elderly persons with Alzheimer's disease and related disorders. *International Psychogeriatrics, 12*(1), 49–65.
Gerdner, L. A. (2015). Ethnicity is an inherent criterion for assessment of individualized music for persons with Alzheimer's disease. *Clinical Gerontologist, 38*(2), 179–186.
Gitlin, L. N. (2008). Tailored activities to manage neuropsychiatric behaviors in persons with dementia and reduce caregiver burden: A randomized pilot study. *American Journal of Geriatric Psychiatry, 16*, 229–239.
Gitlin, L. N., Winter, L., Dennis, M. P., Hodgson, N., & Hauck, W. W. (2010). Targeting and managing behavioral symptoms in individuals with dementia: A randomized trial of a nonpharmacological intervention. *Journal of the American Geriatric Society, 58*, 1465–1474.
Gray, H. L., Jimenez, D. E., Cucciare, M. A., Tong, H., & Gallagher-Thompson, D. (2009). Ethnic differences in beliefs regarding Alzheimer's disease among dementia family caregivers. *American Journal of Geriatric Psychiatry, 17*, 925–933.
Griffith, P., & Lopez, O. (2009). Disparities in the diagnosis and treatment of Alzheimer disease in African American and Hispanic patients: A call to action. *Generations, 33*(1), 37–46.
Guarnaccia, P. J., Rubio-Stipec, M., & Canino, G. (1989). Ataques de nervios in the Puerto Rican diagnostic interview schedule: *The impact of cultural categories on psychiatric epidemiology. Culture, Medicine and Psychiatry, 13*(3), 275–295.
Gurland, B. J., Wilder, D. E., Lantigua, R., Stern, Y., Chen, J., Killeffer, E. H., & Mayeux, R. (1999). Rates of dementia in three ethnoracial groups. *International Journal of Geriatric Psychiatry, 14*(6), 481–493.
Haan, M. N., Shemanski, L., Jagust, W. J., Manolio, T. A., & Kuller, L. (1999). The role of APOE{epsilon}4 in modulating effects of other risk factors for cognitive decline in elderly persons. *Journal of the American Medical Association, 282*(1), 40–46.
Hart, D. J., Craig, D., Compton, S. A., Critchlow, S., Kerrigan, B. M., McIlroy, S. P., & Passmore, A. P. (2003). A retrospective study of the behavioral and psychological symptoms of mid and late phase Alzheimer's disease. *International Journal of Geriatric Psychiatry, 18*(11), 1037–1042. https://doi.org/10.1002/gps.1013
Harwood, D. G., Barker, W. W., Ownby, R. L., & Duara, R. (2000). Relationship of behavioral and psychological symptoms to cognitive impairment and functional status in Alzheimer's disease. *International Journal of Geriatric Psychiatry, 15*(5), 393–400.
Holmes, C., Knights, A., Dean, C., Hodkinson, S., & Hopkins, V. (2006). Keep music live: Music and the alleviation of apathy in dementia subjects. *International Psychogeriatrics, 18*(4), 623–630.
Isaza, C. A., Henao, J., Lopez, A. M., & Cacabelos, R. (2000). Isolation, sequencing and genotyping of the drug metabolizer CYP2D2 gene in the Columbian population. *Methods and Findings in Experimental Clinical Pharmacology, 22*(9), 695–705.
Janevic, M. R., & Connell, C. M. (2001). Racial, ethnic, and cultural differences in the dementia caregiving experience: Recent findings. *Gerontologist, 41*(3), 334–347.
Kales, H. C., Chen, P., Blow, F. C., Welsh, D. E., & Mellow, A. M. (2005). Rates of clinical depression diagnosis, functional impairment, and nursing home placement in coexisting dementia and depression. *The American Journal of Geriatric Psychiatry, 13*(6), 441–449. https://doi.org/10.1097/00019442-200506000-00002
Kales, H. C., Gitlin, L. N., & Lyketsos, C. G. (2015). Assessment and management of behavioral and psychological symptoms. *BMJ, 350*, 1–16. https://doi.org/10.1136/bmj.h369
Kales, H. C., Kim, H. M., Zivin, K., Valenstein, M., Seyfried, L. S., Chiang, C., … Blow, F. C. (2011). Risk of mortality among individual antipsychotics in patients with dementia. *American Journal of Psychiatry, 169*, 71–79. https://doi.org/10.1176/appi.ajp.2011.11030347. Epub 2011 Oct 31.
Kaplan, R. C., Bangdiwala, S. I., & Barnhart, J. M. (2014). Smoking among U.S. Hispanic/Latino adults: The Hispanic community health study/study of Latinos. *American Journal of Preventive Medicine, 46*(5), 496–506.
Keltner, N. L., & Folks, D. G. (2001). *Psychotropic drugs*. St Louis, MO: St Louis Mosby Press.
Kolanowski, A., Litaker, M., Buettner, L., Moeller, J., & Costa, P. T., Jr. (2011). A randomized clinical trial of theory-based activities for the behavioral symptoms of dementia in nursing home residents. *Journal of the American Geriatrics Society, 59*(6), 1032–1041.
Koss-Chioino, J. D. (1992). *Women as healers, women as patients: Mental health care and traditional healing in Puerto Rico*. San Francisco, CA: Westview Press.
Kovach, B. T., Calamia, K. T., Walsh, J. S., & Ginsburg, W. W. (2004). Treatment of multicentric reticulohistiocytosis with etanercept. *Archives of Dermatology, 140*(8), 919–921.
Kramer, E. (2002). Cultural factors influencing the mental health of Asian Americans. *Western Journal of Medicine, 176*, 227–231.

Lin, K. M., Chen, C. H., Yu, S. H., Lin, M. T., & Smith, M. (2003). *Psychopharmacology: Ethnic and cultural perspectives*. Hoboken, NJ: Wiley.

Lin, K. M., & Smith, M. W. (2000). Psychopharmacotherapy in the context of culture and ethnicity. In P. Ruiz (Ed.), *Ethnicity and psychopharmacology* (pp. 1–36). Washington, DC: American Psychiatric Press.

Lyketsos, C., Carillo, M., Ryan, J. M., Khachaturian, A. S., Trzepacz, P., Amatniek, J., ... Miller, D. S. (2011). *Alzheimer's Dementia, 7*(5), 532–539. https://doi.org/10.1016/j.jalz.2011.05.2410

Martin, J. L., Fiorentino, L., Jouldjian, S., Josephson, K. R., & Alessi, C. A. (2010). Sleep quality in residents of assisted living facilities: Effect on quality of life, functional status, and depression. *Journal of the American Geriatrics Society, 58*(5), 829–836. https://doi.org/10.1111/j.1532-5415.2010.02815.x

Maust, D. T., Kim, H. M., Seyfried, L. S., Chiang, C., Kavanagh, J., Schneider, L. S., & Kales, H. C. (2015). Antipsychotics, other psychotropics, and the risk of death in patients with dementia: Number needed to harm. *JAMA Psychiatry, 72*(5), 438–445. https://doi.org/10.1001/jama.2014.3018

Maust, D. T., Langa, K. M., Blow, F. C., & Kales, H. C. (2017). Psychotropic use and associated neuropsychiatric symptoms among patients with dementia in the USA. *International Journal of Geriatric Psychiatry, 32*(2), 164–174. https://doi.org/10.1002/gps.4452

McDowell, S. E., Coleman, J. J., & Ferner, R. E. (2006). Systematic review and meta-analysis of ethnic differences in risks of adverse reactions to drugs used in cardiovascular medicine. *BMJ, 332*(7551), 1177–1181.

Mentes, J., Cadogan, M., Woods, L., & Phillips, L. (2015). Evaluation of the nurses caring for older adults young scholars program. *The Gerontologist, 55*(Suppl_1), S165–S173.

Mezzich, J. E., Kleinman, A., Fabrega, H. F., & Parron, D. L. (Eds.). (1996). *Culture and psychiatric diagnosis: A DSM-IV perspective*. Washington, DC: American Psychiatric Press.

Muangpaisan, W., Intalapaporn, S., & Assantachai, P. (2008). Neuropsychiatric symptoms in the community-based patients with mild cognitive impairment and the influence of demographic factors. *International Journal of Geriatric Psychiatry, 23*(7), 699–703. https://doi.org/10.1002/gps.1963

Murphy, E., & McMahon, F. J. (2013). Pharmacogenetics of antidepressants, mood stabilizers, and antipsychotics in diverse human populations. *Discovery Medicine, 16*(87), 113.

Napoles, A. M., Chadiha, L., Eversley, R., & Moreno-John, G. (2010). Developing culturally sensitive dementia caregiver interventions: Are we there yet? *American Journal of Alzheimer's Disease & Other Dementias, 25*(5), 389–406. https://doi.org/10.1177/1533317510370957

O'Neil, M. E., Freeman, M., Christensen, V., Telerent, R., Addleman, A., & Kansagara, D. (2011). *A systematic evidence review of non-pharmacological interventions for behavioral symptoms of dementia*. Washington, DC: Department of Veterans Affairs.

Oesterheld, J., & Kallepalli, B. R. (1997). Grapefruit juice and clomipramine: Shifting metabolitic ratios. *Journal of Clinical Psychopharmacology, 17*(1), 62–63. https://doi.org/10.1097/00004714-199702000-00019

Payami, H., Grimslid, H., Oken, B., Camicioli, R., Sexton, G., Dame, A., ... Kaye, J. (1997). A prospective study of cognitive health in the elderly (Oregon brain aging study): Effects of family history and apolipoprotein E genotype. *American Journal of Human Genetics, 60*(4), 948–956.

Polsky, D., Doshi, J. A., Bauer, M. S., & Glick, H. A. (2006). Clinical trial-based cost-effectiveness analyses of antipsychotic use. *American Journal of Psychiatry, 163*(12), 2047–2056. https://doi.org/10.1176/appi.ajp.163.12.2047

Prince, M., Ali, G. C., Guerchet, M., Prina, A. M., Albanese, E., & Wu, Y. T. (2016). Recent global trends in the prevalence and incidence of dementia, and survival with dementia. *Alzheimer's Research & Therapy, 8*(1), 23.

Rabins, P., Schwartz, S., Black, B. S., Cocoran, C., Fauth, E., Mielke, M., ... Tschanz, J. (2013). Predictors of progression to severe Alzheimer's disease in an incidence sample. *Alzheimer's Dementia, 9*(2), 204–207. https://doi.org/10.1016/j.jalz.2012.01.003

Rosen, J., Burgio, L., Kollar, M., Cain, M., Allison, M., Fogleman, M., ... Zubenko, G. S. (1994). The Pittsburg agitation scale: A user friendly instrument for rating agitation in dementia patients. *American Journal of Geriatric Psychiatry, 2*(1), 52–59.

Shin, H. Y., Han, H. J., Shin, D. J., Park, H. M., Lee, Y. B., & Park, K. H. (2014). Sleep problems associated with behavioral and psychological symptoms as well as cognitive functions in Alzheimer's disease. *Journal of Clinical Neurology, 10*(3), 203–209. https://doi.org/10.3988/jcn.2014.10.3.203

Silva, H. (2013). Ethnopsychopharmacology and pharmacogenomics. *Advances in Psychosomatic Medicine, 33*, 88–96. https://doi.org/10.1159/000348741

Sink, K. M., Covinsky, K. E., Newcomer, R., & Yaffe, K. (2004). Ethnic differences in the prevalence and pattern of dementia-related behaviors. *Journal of the American Geriatrics Society, 52*(8), 1277–1283.

Sloane, P. D., Davidson, S., Knight, N., Tangen, C., & Mitchell, C. M. (1999). Severe disruptive vocalizers. *Journal of the American Geriatrics Society, 47*(4), 439–445. https://doi.org/10.1111/j.1532-5415.1999.tb07236.x

Smith, M., Buckwalter, K. C., Kang, H., Ellingrod, V., & Schultz, S. K. (2008). Dementia care in assisted living: Needs and challenges. *Issues in Mental Health Nursing, 29*(8), 817–838. https://doi.org/10.1080/01612840802182839

Smith, M., Kolanowski, A., Buettner, L. L., & Buckwalter, K. C. (2009). Beyond bingo: Meaningful activities for persons with dementia in nursing homes. *Annals of Long-Term Care, 17*(7), 22–30.

Solus, J. F., Arietta, B. J., Harris, J. R., Sexton, D. P., Steward, J. Q., McMunn, C., ... Dawson, E. P. (2004). Genetic variation in eleven phase I drug metabolism genes in an ethnically diverse population. *Pharmacogenomics, 5*(7), 895–931.

Soto-Espinosa, J., & Koss-Chioino, J. (2017). Doctors who integrate spirituality and CAM in the clinic: The Puerto Rican case. *Journal of Religious Health, 56*, 149–157.

Sung, H. C., Lee, W. L., Li, T. L., & Watson, R. (2012). A group music intervention using percussion instruments with familiar music to reduce anxiety and agitation of institutionalized older adults with dementia. *International Journal of Geriatric Psychiatry, 27*(6), 621–627.

Svansdottir, H. B., & Snaedal, J. (2006). Music therapy in moderate and severe dementia of Alzheimer's type: A case–control study. *International Psychogeriatrics, 18*(4), 613–621.

Takeda, M., Martinez, R., Kudo, T., Tanaka, T., Okochi, M., Tagami, S., ... Cacabelos, R. (2010). Apolipoprotein E and central nervous system disorders: Reviews of clinical findings. *Psychiatry Clinical Neuroscience, 64*(6), 592–607. https://doi.org/10.1111/j.1440-1819.2010.02148.x

Talamantes, M., Trejo, L., Jimenez, D., & Gallafher-Thompson, D. (2006). Working with Mexican American families. In G. Yeo & D. Gallagher-Thompson (Eds.), *Ethnicity and dementias* (pp. 327–340). New York, NY: Taylor & Francis.

Teri, L., Huda, P., Gibbons, L., Young, H., & van Leynseele, J. (2005). STAR: A dementia-specific training program for staff in assisted living residences. *The Gerontologist, 45*(5), 686–693. https://doi.org/10.1093/geront/45.5.686

Teri, L., & Logsdon, R. G. (1991). Identifying pleasant activities for Alzheimer's disease patients: The pleasant events schedule-ad. *The Gerontologist, 31*(1), 124–127. https://doi.org/10.1093/geront/31.1.124

Thuné-Boyle, I. C. V., Iliffe, S., Cerga-Pashoja, A., Lowery, D., & Warner, J. (2012). The effect of exercise on behavioral and psychological symptoms of dementia: Towards a research agenda. *International Psychogeriatrics, 24*(7), 1046–1057.

Tolson, D., Rolland, Y., Andrieu, S., Aquino, J.-P., Beard, J., Benetos, A., ... Morley, J. E. (2011). The international association of gerontology and geriatrics: A global agenda for clinical *research and quality of care in nursing homes. Journal of American Medical Director's Association, 12*(3), 184–189.

Trivedi, D., Goodman, C., Gage, H., Baron, N., Scheibl, F., Iliffe, S., ... Drennan, V. (2012). The effectiveness of inter-professional working for older people living in the community: A systematic review. *Health & Social Care in the Community, 21*(2), 113–128.

U.S. Department of Health and Human Services. (2014). *Racial and ethnic disparities in Alzheimer's disease: A literature review*. Washington, DC.

Woods, D. L. (1999). *Perceptions of dementia in Botswana*. Unpublished interview.

Woods, D. L., Craven, R. F., & Whitney, J. (2005). The effect of therapeutic touch on behavioral symptoms of persons with dementia. *Alternative Therapies in Health & Medicine, 11*(1), 66–74.

Yaffe, K., Cauley, J., Sands, L., & Browner, W. (1997). Apolipoprotein E phenotype and cognitive decline in a prospective study of elderly community women. *Archives of Neurology, 54*(9), 1110–1114. https://doi.org/10.1001/archneur.1997.00550210044011

Yang, M. H., Lin, L. C., Wu, S. C., Chiu, J. H., Wang, P. N., & Lin, J. G. (2015). Comparison of the efficacy of aroma-acupressure and aromatherapy for the treatment of dementia-associated agitation. *BMC Complementary and Alternative Medicine, 15*(1), 93.

Zevin, S., & Benowitz, N. L. (1999). Drug interactions with tobacco smoking. *Clinical Pharmacokinetics, 36*(6), 425–439. https://doi.org/10.2165/00003088-199936060-00004

Intersectionality as a Practice of Dementia Care for Sexual and Gender Minoritized Latinxs

12

Hector Y. Adames, Nayeli Y. Chavez-Dueñas, Silvia P. Salas, and Claire R. Manley

Abstract

Latinxs continue to be disproportionately impacted by dementia. Despite higher prevalence rates of dementia within the Latinx community, there is a dearth of literature that explicitly addresses the ways in which multiple forms of oppression (e.g., racism, ethnocentrism, sexism, cis-sexism, heterosexism, nativism) uniquely impact older sexual and gender minoritized (SGM) Latinxs with dementia. Using an intersectional framework, this chapter discusses the challenges and needs of older SGM Latinxs living with dementia. Namely, a discussion on the necessity of conceptualizing *weak*, *strong*, and *transformative* intersectionality in the lives of SGM Latinxs with dementia, as well as determining the institutions that, both alone and combined, serve to hinder this population's well-being and access to equitable, humane treatment in healthcare is addressed. This chapter concludes with recommendations for healthcare providers and policy makers working with members of the SGM Latinx aging population with a focus on multisystemic levels of support, advocacy, and individual care. This chapter summarizes how the three forms of intersectionality (i.e., weak, strong, and transformative) can advance effective healthcare solutions for this population that requires greater visibility in all sectors of health policy, research, and practice.

H. Y. Adames (✉) · N. Y. Chavez-Dueñas
C. R. Manley
The Chicago School of Professional Psychology, Chicago, IL, USA
e-mail: HAdames@thechicagoschool.edu; NChavez@thechicagoschool.edu; clara.rose.manley@gmail.com

S. P. Salas
University of Wisconsin-Milwaukee, Milwaukee, WI, USA
e-mail: spsalas@uwm.edu

Introduction

I am as I am and you are as you are. Let's build a world where I can be, without having to cease being me, where you can be, without having to cease being you, and where neither you nor I will force one another to be like either me or you

(Marcos, 2001, p. 169).

As we entered the new millennium, the United States of America (USA) became one of the most multiracial, multiethnic, and diverse societies in the world. Latinxs,[1] one of the groups contributing to the diverse fabric of the USA,

[1] To include and center the broad range of gender identities present among individuals of Latin American descent, the term Latinx is used throughout this chapter.

H. Y. Adames, Y. N. Tazeau (eds.), *Caring for Latinxs with Dementia in a Globalized World*,
https://doi.org/10.1007/978-1-0716-0132-7_12

are currently the largest (57 million) ethnic minoritized[2] group, comprising 18% of the total USA population. The growth in the Latinx population is expected to continue in the upcoming decades with projections indicating that by 2030, 20% of the total USA population will be of Latinx descent (U.S. Census Bureau, 2004). Similarly, according to the Administration on Aging (2015), Latinxs also represent the fastest-growing aging population in the USA, with 3.6 million individuals at, or above, the age of 65. The same report describes that between 2008 and 2030, the Latinx aging population will increase by 224% compared to a 65% increase for non-Latinx Whites. Furthermore, according to the Centers for Disease Control and Prevention (CDC, 2015), the life expectancy of Latinxs is projected to increase from 80 to 87 by 2050. Considering these demographic shifts, Latinxs will undoubtedly face changes related to age-associated conditions such as the loss of cognitive functioning – including dementia – which disproportionately affect Latinxs. Figures reported by the Alzheimer's Association (2016b) posit that Latinxs are 1.5 times more likely to develop dementia of the Alzheimer's type (7.5% between ages 65 and 74; 27.9% between ages 75 and 84; 62.9% over age 85) than non-Latinx Whites (2.9% between 65 and 74; 10.9% between ages 75 and 84; 30.2% over age 85).

Despite the prevalence of dementia within the Latinx community, there is a dearth of literature that explicitly addresses the impact of within-group differences on caregiving for Latinxs with dementia. To illustrate, the literature on dementia care with Latinxs focuses predominantly on ethnicity and culture (e.g., Angel & Angel, 2015; González-Sanders & Fortinsky, 2012; Vega, Markides, Angel, & Torres-Gil, 2015) while neglecting the role of race, gender, and sexual orientation. In many ways, the literature on dementia care with Latinx is saturated with unidimensional paradigms that purport to offer an understanding of the multidimensional experiences of this heterogeneous group. Hence, the geriatric literature on Latinxs relies heavily on monolithic frameworks when studying and describing this diverse community. Several reasons may help explain why most of the existing literature on Latinxs fails to address the heterogeneity that exists within this population, including researchers and clinicians' tendencies to (a) simplify models and frameworks for parsimonious reasons, (b) rely on historically reductionistic methodologies, and (c) control for membership in categories that are not of interest to the discussion at hand (Betancourt & López, 1993; Cole, 2009). In response to some of the above-noted critiques, a growing number of scholars have argued for the need to view and understand the experiences of historically marginalized individuals through an intersectional framework (Bowleg, 2013; Grzanka, 2014; Warner & Shields, 2013) which focuses on the ways in which systems of oppression (e.g., racism, ethnocentrism, sexism) impact individuals (Crenshaw, 1989). Using an intersectional framework, this chapter contextualizes dementia care for sexual and gender minoritized (SGM) Latinxs. To help frame this analysis, this chapter begins with a brief overview of the theory of *intersectionality* as a prelude to a discussion of how elderly SGM Latinxs are impacted by more than one domain of oppression (e.g., racism, ethnocentrism, heterosexism, cis-sexism). This chapter concludes with recommendations for healthcare providers working with members of the SGM Latinx aging population with a focus on multisystemic levels of support, advocacy, and individual care, addressing the three forms of intersectionality, i.e., weak, strong, and transformative, to advance effective healthcare solutions for this population that requires greater visibility in all sectors of health policy, research, and practice.

[2]Instead of "minority," we use "minoritized" throughout the chapter to signify the subordination of People of Color and Queer and Trans People of Color in the USA (Harper, 2012). "Minority" is identity based, whereas "minoritized" centers how historically oppressed groups are impacted by systems of oppression (Adames, Chavez-Dueñas, Sharma, & La Roche, 2018).

A Brief Introduction to Intersectionality Theory

Intersectionality theory, first introduced by USA Black feminists and Women of Color social justice activists and scholars (Collins, 2000/2009; Combahee River Collective, 1977/1995; Crenshaw, 1989, 1991), focuses on the ways in which systems of oppression (e.g., racism, ethnocentrism, nativism, sexism) impact individuals who hold membership in multiple socially constructed groups (Collins, 2000/2009; Crenshaw, 1989, 1991). Three forms of intersectionality, including *weak*, *strong*, and *transformative*, are identified in the literature (Dill & Kohlman, 2011; Shin et al., 2017). *Weak* intersectionality refers to the focus on multiple identities or intersecting identities (Marecek, 2016) in theory, research, and practice (Dill & Kohlman, 2011; Grzanka, 2014) and thus highlights its role in intrapersonal dynamics and interpersonal experiences. While many professionals in psychology and the health sciences utilize a weak or identity-focused conceptualization of intersectionality, this focus departs from the original principles outlined by the Black feminists and Women of Color social justice activists and scholars who coined the term and introduced the framework (see Combahee River Collective, 1977/1995; Crenshaw, 1989). In a recent special issue on intersectionality published by the American Psychological Association's (APA) *Journal of Counseling Psychology*, Moradi and Grzanka (2017) call for a "moratorium on using multiple or intersecting 'identities' language as a euphemism for intersectionality" (p. 506). Instead, scholars and clinicians are encouraged to implement a *strong* conceptualization of intersectionality, which underscores how interlocking systems of inequities uniquely impact individuals who hold membership in different social groups (Dill & Kohlman, 2011; Grzanka, 2014; Moradi & Grzanka, 2017). Finally, *transformative* intersectionality explicitly calls for scholars, psychologists, and other healthcare providers to work toward dismantling oppressive systems by engaging in social justice action (Shin et al., 2017). Scholars argue that "for scholarship to be considered truly intersectional, it must include a critique of the structural inequalities, which construct and reify the complex relationships between privileged and marginalized social identities" (Shin et al., 2017, p. 259).

While intersectionality has primarily been used in critical legal studies, sociology, and Black Women's studies, a growing number of scholars are beginning to integrate this framework into psychological theory, research, and practice (e.g., Adames et al., 2018; Cole, 2009; Lewis & Neville, 2015). For instance, using an intersectional framework, Lewis, Williams, Peppers, and Gadson (2017) empirically tested a model to understand the influence of interdependent gendered racial (i.e., intersection of sexism and racism) microaggression on the mental and physical health of Black women. They found that gendered racial microaggressions significantly predicted both self-reported poor mental and physical health outcomes. This study provides evidence of how racism and sexism as an interdependent form of oppression uniquely impact the well-being of Black women. Another intersectional study by Jerald, Cole, Ward, and Avery (2017) tested the ways in which simultaneously racialized and gendered stereotypes impacted Black women. The results revealed that when participants were aware that others held negative stereotypes about their social group membership (i.e., women and Black) they reported (a) negative mental health outcomes, (b) lower self-care behaviors, and (c) coping by engaging in substance use. Finally, new treatment literature in psychology is emerging that describes how a *strong* intersectional stance can serve as a way to help clients give themselves permission to stop blaming within for experiencing symptoms that result from holding membership in multiple oppressed groups (see Adames, Chavez-Dueñas, & Organista, 2016; Adames et al., 2018; Chavez-Dueñas & Adames, in press; Chavez-Dueñas, Adames, Perez-Chavez, & Salas, 2019). In the following section, intersectionality theory is applied to aging SGM Latinxs with dementia.

Aging Among Sexual and Gender Minoritized Communities: Context and Challenges

Historical Context

In order to contextualize and understand the lives and needs of the aging SGM population, a sociohistorical perspective is necessary. Currently, the SGM older adult population is comprised of three different generations (i.e., Baby Boomer, Silent, and the Greatest or GI Generation). Each cohort came of age during historical periods wherein prejudice, stigma, and discrimination against the SGM community were institutionalized through several USA federal laws and policies (Choi & Meyer, 2016; Fredriksen-Goldsen & Muraco, 2010). Much of the literature on SGM older adults predominately addresses the concerns of the GI and Silent generation, groups that lacked recognition and rights throughout much of their lifetime (Mackenzie, 2009). Thus, a lifelong experience of systemic oppression gave rise to health disparities and adverse health outcomes for members of this community (Fredriksen-Goldsen & Kim, 2017; U.S. Department of Health and Human Services, 2012). For instance, prior to the Civil Rights Movement and the Stonewall Inn Riots of 1960s, these generations lived in a society that pathologized and criminalized same-gender attraction and stigmatized same-gender relationships as well as identification with the SGM community, which led many individuals to conceal their identities (Fredriksen-Goldsen & Muraco, 2010; Morrow, 2001). As adults, members of these generational cohorts of SGM often experienced harassment at the hands of law enforcement and other people in positions of authority, including healthcare professionals who labeled their behavior, as well as their sexual and gender identities, as a sociopathic personality disturbance – a practice that ended in 1973 (Institute of Medicine, 2011; Silverstein, 2009).

Despite the changes that took place in the 1970s, stigma and prejudice continued to impact the SGM community. During the height of the human immunodeficiency virus (HIV) and the acquired immunodeficiency syndrome (AIDS) pandemic, many in the SGM community experienced psychological and physical threats from society which caused them to live in fear. Moreover, the SGM community disproportionately experienced the loss of many lives to the HIV/AIDS pandemic (Fredriksen-Goldsen, Jen, Bryan, & Goldsen, 2016). During the *Gay Liberation Movement*, younger SGM Baby Boomers fought against heterosexism by openly expressing their identities. SGM Baby Boomers began to acquire a personal history of activism through participation in social movements, differentiating them from the Greatest/GI Generation (Fredriksen-Goldsen & Muraco, 2010; McGovern, 2014). Despite continued societal stigma and discrimination, the *Gay Liberation Movement* has seen some progress in its access to civil and human rights (Goldsen et al., 2017). For instance, attitudes toward individuals who identify as SGM have changed in recent years, with an increase in the percentage of people reporting being more accepting of gay and lesbian people compared to previous decades. To illustrate, in 2006, approximately 51% of adults reported that lesbian and gay individuals should be accepted in society compared to 63% in 2013 (Pew Research Center, 2013). Similarly, during the same period, the number of adults who identified as members of the SGM community also increased. Together, the shift in attitudes and the increase in people who identify as a SGM have led to a number of systemic changes. One of the most significant changes includes the landmark U.S. Supreme Court decision *Obergefell v. Hodges* (2015), after James Obergefell filed a lawsuit for not being allowed to place his name on the death certificate of his husband who died from amyotrophic lateral sclerosis (ALS), a progressive neurodegenerative disease (History, 2017). The Supreme Court decision ruled that all state bans on same-sex/same-gender marriage were unconstitutional, thus legalizing same-sex/same-gender marriage and providing SGM individuals "equal dignity in the eyes of the law" (p. 28).

While much progress has been made, transgender and gender nonconforming (TGNC) individuals continue to bear the burden of transnegativity. Many are diagnosed as psychologically disordered when this aspect of their identity causes clinical distress and impairment in functioning (American Psychiatric Association, 2017). Consequently, the TGNC community continues to endure some of the most grueling systemic injustices, invisibility, and health disparities that persist into old age (Auldridge, Tamar-Mattis, Kennedy, Ames, & Tobin, 2012). Given the historical challenges faced by the aging SGM community, distrust in health institutions and healthcare providers continues to prevail (Auldridge et al., 2012; McGovern, 2014). Thus, it is important for healthcare providers to acknowledge the historical forces and forms of oppression that influenced the life circumstances of the aging SGM community and consider how their unique experiences may increase their vulnerability to victimization and discrimination in old age.

Challenges Related to Aging

Health and well-being, economic security, and social connections are among the cornerstones for healthy aging, yet a lifetime of oppression, discrimination, and isolation has created substantial barriers to these foundations, including a lack of (a) social and legal recognition, (b) reliance on chosen family, and (c) a competent and inclusive healthcare for SGM older adults (Movement, Advancement Project [MAP] & SAGE, 2017). To illustrate, a study conducted by Fredriksen-Goldsen et al. (2011) found that older SGM adults reported an average of four incidents of victimization and discrimination over the course of their lives due to their sexual orientation, gender identity, and/or gender expression. Within this study, transgender adults aged 50 and older reported an average of six or so incidents over the course of their lives. Among older adults who are members of both an ethnic minoritized and SGM group, even higher lifetime levels of discrimination related to sexual orientation and/or gender identity and expression are reported (MAP & SAGE, 2017). Lifetime experiences of discrimination, coupled with internalized heterosexism, are associated with poor mental health and physical health and greater disability among older SGM individuals (Dilley, Simmons, Boysun, Pizacani, & Stark, 2010; Fredriksen-Goldsen et al., 2013, 2014). Moreover, relative to heterosexual individuals, people of sexual minoritized status also have poorer general health (Conron, Mimiaga, & Landers, 2010; Wallace, Cochran, Durazo, & Ford, 2011). Among gay and bisexual men, as well as transgender women, elevated rates of HIV are commonly observed (Centers for Disease Control and Prevention [CDC], 2013; Schulden et al., 2008). In addition, lesbian and bisexual women are more likely to be overweight or obese (Boehmer, Bowen, & Bauer, 2007; Case et al., 2004; Dilley et al., 2010). Given the multiple forms of oppression that lesbian and bisexual women face (e.g., sexism, heterosexism, and racism for women of color), such statistics on obesity in these communities indicate the marked physical and psychological implications of enduring intersecting oppression over one's lifetime. Finally, some literature indicates that transgender individuals are at an increased risk for various health conditions due to negative effects of long-term hormone treatment (Feldman & Bockting, 2003; Williams & Freeman, 2007).

Challenges Related to Accurate Representation

While the SGM elder population represents a diverse segment of the aging USA population, no census data exists for this population. The available data on the SGM community predominately comes from research conducted by nongovernmental organizations and professional associations which estimate that 2.6–4.4 million SGM adults over the age of 60 reside in the USA (Choi & Meyer, 2016; McGovern, 2014). These numbers constitute a considerable segment of the estimated 3.5% or 11 million SGM adults in the USA (Gates, 2017). By 2060, the number of SGM older adults will more than double to an estimated 5.2–8.8 million (Fredriksen-Goldsen, Kim, Goldsen, Shiu, &

Emlet, 2016). However, the available data on the size of the SGM population is murky given the divergent ways in which sexuality and gender are defined and measured, as well as the differences in survey methodologies (Gates, 2011). To illustrate, estimates of people who report any lifetime same-gender sexual behavior (8.2%) and any same-gender sexual attraction (11%) are substantially higher than estimates of those who identify as lesbian, gay, or bisexual.

Despite the growing SGM elderly population, this group remains an under-researched segment of the aging population (Fredriksen-Goldsen & Kim, 2017). For instance, the Institute of Medicine (2011) identified sexual orientation and gender identity as major gaps in health disparity research. Unfortunately, the vast majority of available studies in the area of SGM older adults has disproportionately focused on gay men and lesbians, with few studies focused on bisexual or transgender elders or people over the age of 85 (Addis, Davies, Greene, MacBride-Stewart, & Shepherd, 2009; Choi & Meyer, 2016; Donahue & McDonald, 2005; Hughes, Harold, & Boyer, 2011; Institute of Medicine, 2011). Furthermore, while one in five SGM older adults are People of Color (SGM-POC), a proportion expected to double by 2050 (MAP & SAGE, 2017), studies on SGM-POC older adults have rarely been the focus of research (Kim & Fredriksen-Goldsen, 2017). Hence, most research on the extent to which sexual orientation and gender identity affect the aging experiences of individuals is centralized through a White lesbian and gay narrative, with very little known about SGM-POC elders (Institute of Medicine, 2011; Kim, Jen, & Fredriksen-Goldsen, 2017). Given the expected increase in the SGM elder population, the field of gerontology and aging research must work to reflect the unique aging experiences of members from this community (McGovern, 2014) while also mirroring the diversity of the USA. It is not surprising that there is a dearth of literature specifically focusing on the aging SGM Latinx population who face significant challenges in old age related to multiple forms of oppression (e.g., racism, ethnocentrism, nativism, heterosexism).

Challenges Unique to the Transgender Community

Similar to sexual minoritized individuals, transgender older adults are often subjected to discrimination, hostility, and even violence in healthcare settings meant to support successful and healthy aging (Auldridge et al., 2012). However, despite the myriad of challenges faced by members of the transgender community, the literature on aging transgender, independent of sexual orientation, is sparse (Witten, 2015). The few studies available have reported that 70% of transgender adults age 65 and older delay medical transition to avoid discrimination in employment, 13% reported abusing drugs to cope with mistreatment, and 16% reported attempting suicide at least once in their lifetime (Grant et al., 2011). Hence, due to the high level of discrimination and oppression they face, transgender older adults are often forced to hide and silence aspects of their gender identity (Auldridge et al., 2012; MAP & SAGE, 2017; Witten, 2015). One contributing factor known to increase transgender older adults' likelihood of hiding their gender identity relates to the increased risk of violence and homicide acted against this community. Over the past 5 years, rates of violence against members of the transgender community have been deemed an "epidemic" by the Human Rights Campaign (HRC, 2017). According to HRC, transgender People of Color (TPOC) are disproportionately impacted by violence, which can occur as a result of a multiplicity of vulnerable statuses encountering discriminatory attitudes of the dominant culture. According to HRC, 2017 was one of the deadliest years for TPOC, surpassing the previous record of fatal violence in 2016. In addition, the homicide rate for transgender individuals is only expected to increase (HRC, 2017). Effects of violence against transgender individuals include higher risk for sexual and physical violence, suicidal ideation, and suicide attempts for both trans women and trans men, as well as a higher risk of substance use to cope with exposure to violence (Testa et al., 2012). Intersecting vulnerable identities (e.g., transgender women of color) increase the risk of homicide and contribute to lower rates of reporting violence to the police, as police responses to reports have contributed to the homicide rate of this vulnerable population (HRC, 2017; Testa et al., 2012). As described by the HRC, "intersections of racism, sexism, homophobia and transphobia conspire to deprive [Transgender Women of Color] of employment, housing, healthcare and other necessities," making this population especially vulnerable to violence from a variety of perpetrators (HRC, 2017, para. 2). There is a dearth of research on rates of violence against older transgender People of Color, although report briefs from the *Violence Against Women and Girls* (VAWG, 2016) resource guide, which constitutes a conglomerate of partnerships between organizations such as the *World Bank*, the *Global Women's Institute*, the *Inter-American Development Bank*, and the *International Center for Research on Women*, suggest that older trans women across the world report higher rates of loneliness, violence, poorer mental health, and multiple suicide attempts (VAWG, 2016) and that, generally, transgender individuals are 25 times more likely to attempt suicide than their cisgender counterparts (VAWG, 2015). Given these distressing statistics, it is imperative that healthcare providers take transgender violence, including the impact of lifelong histories of violence, into account when working with older transgender individuals in any healthcare setting.

The Aging Latinx Population in the USA

In the next few decades, the Latinx population will continue growing with projections estimating that by 2040, 25% of the USA population will be of Latinx descent (U.S. Census, 2011). While the Latinx community has the youngest median age of all ethnic groups (27 years vs. 37 years), the proportion of older adults who identify as Latinx is also increasing (Angel, 2009). For instance, as of 2015, 3.6 million individuals over the age of 65 identified as Latinx; however, the *Administration on Aging Report* (2015) posits that between 2008 and 2030, the Latinx aging population will increase by 224%. Moreover, by 2060, approximately 21.5 million Latinxs will be older than 65 years of age, totaling 22% of the elderly USA population (Administration on Community Living, 2017). In addition to being the fastest-growing aging population in the USA, Latinxs also have a longer life expectancy than their African American and non-Latinx White American counterparts (CDC, 2015). In fact, Latinx life expectancy is projected to increase from 80 to 87 by 2050 (CDC, 2015). Given the strong and continuous growth in the Latinx elder population, it is important to understand some of the unique challenges this group faces in the USA.

Financial Challenges

The Latinx elderly community is a diverse group of individuals who face significant challenges related to maintaining economic security, social support, and access to quality healthcare (National Hispanic Council on Aging [NHCOA], 2016; Santiago-Rivera, Adames, Chavez-Dueñas, & Benson-Florez, 2016). These challenges stem from difficulties during their early years of life related to disparities in education, income, and English language fluency which impact access to resources in old age. For instance, Latinxs tend to lag behind other ethnic groups in terms of educational attainment (Krogstad, 2016). As a result, Latinxs tend to be overrepresented in occupations that are low wage and high risk. When Latinxs work in higher status occupations, they tend to earn lower salaries compared to their White counterparts (Bureau of Labor Statistics, 2011). Finally, many Latinx older adults are first-generation immigrants who tend to have lower levels of English language fluency than their USA-born Latinx counterparts (Taylor, Lopez, Martinez, & Velasco, 2012). Lack of English fluency further impacts Latinxs' ability to obtain well-paying jobs that offer benefits (e.g., health insurance, pension plans). Consequently, in their later years, Latinx older adults are less likely to have personal savings, property investments, retirement accounts, and other interest-accumulating savings accounts (Kochhar, 2004; Kochhar, Fry, & Taylor, 2011). These financial realities were compounded by the economic downturn of 2005–2009 which led to a net worth fall of 66% for Latinx in the USA (NHCOA, 2016). The loss in wealth, resulting from the economic downturn, further limited the ability for members of the Latinx community to prepare for old age. Lack of sufficient wealth makes economic security in postretirement years increasingly difficult for members of the Latinx community. It is not surprising that by 2013, nearly 20% of Latinx older adults lived in poverty, which is double the rate of the general older population (Administration on Community Living, 2017; DeNavas-Walt, Proctor, & Smith, 2011). Economic insecurity contributes to Latinx elder's dependency on social security to meet even their most basic necessities (e.g., food, shelter). To illustrate, a significant proportion of married (40%) and unmarried (60%) Latinx elders rely on social security income for 90% of their income (Administration on Community Living, 2017).

Health and Mortality Rates

Among the Latinx community, it is estimated that the death rates are lower for both males and females of all age groups compared to the general population (CDC, 2015). Moreover, the gap between Latinx and non-Latinx Whites tends to increase with age such that Latinx death rates are 26–30% below those of non-Latinx Whites between the ages of 80–84 and 85–89 (Martin & Soldo, 1997). While there appears to be a lower mortality rate among all Latinx in the USA, scholars explain that these figures are likely due to the significantly low mortality rates among Latinx immigrants (Wallace & Villa, 2003). This theory is corroborated by research findings indicating that mortality rates of USA-born Latinxs are not significantly different than those from non-Latinx Whites (Wallace & Villa, 2003). However, the lower mortality rates of Latinxs do not generalize to all diseases or conditions. For instance, Latinxs of all genders have a higher mortality rate for diabetes, with older adults being twice as likely to die from this chronic condition compared to the general population (Wallace & Villa, 2003). Moreover, while Latinx older adults may have lower mortality rates and longer life expectancies, the incidence and prevalence of medical conditions that impair functioning and quality of life is higher among this population (Wallace & Villa, 2003). In fact, it is estimated that approximately 85% of Latinx older adults have at least one chronic health condition (NHCOA, 2016). Latinxs also have a higher prevalence of arthritis, cognitive impairment, cardiovascular disease (CVD), and hypertension (NHCOA, 2016). Additionally, research indicates that Latinxs are disproportionately impacted by dementia, with the incidence of Alzheimer's disease and other related neurodegenerative diseases being significantly higher among Latinxs compared to non-Latinx Whites (Manly & Mayeux, 2004). For instance, according to figures reported by the Alzheimer's Association (2016b), Latinxs are 1.5 times more likely to develop dementia of the Alzheimer's type.

Older Sexual and Gender Minoritized Latinxs: The Cost of Experiencing Multiple Forms of Oppression

Facing multiple forms of oppression, along with living in social isolation, can have a devastating impact on the lives and well-being of SGM Latinx older adults (Services & Advocacy for Gay, Lesbian, Bisexual, & Transgender Elders [SAGE], National Hispanic Council on Aging [NHCOA], & Diverse Elders Coalition [DEC], 2014). For instance, older SGM are often devalued, denigrated, stigmatized, or penalized for their non-heteronormative behavior, identity, and romantic relationships, all of which constitute heterosexism (Kitzinger, 2000). These experiences with heterosexism create barriers to healthcare, increase psychological distress, and make them vulnerable to abuse (Szymanski & Moffitt, 2012). A report from the *Institute of Multigenerational Health* indicates that fewer than one in four SGM elderly adults in the USA disclose their sexual orientation to a healthcare provider (Fredriksen-Goldsen et al., 2011) believing that they will be treated with less dignity and respect (MetLife Mature Market Institute and the Lesbian & Gay Aging Issues Network of the American Society of Aging, 2010). Consequently, SGM older adults experience a unique set of difficulties that makes them more vulnerable to a wide range of psychological stressors when compared to their counterparts (American Psychological Association, n.d.). To illustrate, SGM older adults are more likely to experience elder abuse and neglect, live alone without companionship, not have children to help with caregiving, have lower self-esteem, and delay seeking treatment for health-related problems, which in turn results in greater levels of psychological distress (Espinoza, 2011; Fredriksen-Goldsen et al., 2011; Szymanski & Carr, 2008). While the relationship between heterosexism and psychological distress is well-established in the empirical literature (Szymanski & Moffitt, 2012), most of the existing studies exploring the impact of oppression on SGM older adults have failed to examine the many additional ways in which members from this group are "othered," i.e., treated poorly because of who they are as persons.

Older SGM Latinxs are disproportionately affected by mental health issues (Díaz, Ayala, Bein, Henne, & Marin, 2001). Compared to the general SGM population, SGM Latinxs are more likely to report symptoms of depression, high levels of stress (Fredriksen-Goldsen et al., 2011), and substance abuse (Krehely, 2009). Beyond the many challenges experienced by the aging SGM Latinx community, several unique barriers exist for this community in accessing healthcare provision, including linguistic barriers, lack of cultural competence in health providers, and lack of health insurance (Livingston, Minushkin, & Cohn, 2008). For instance, in a needs assessment of Latinx SGM, older adults indicated that "accessing community services and supports, and benefiting fully from Social Security, Medicare, and Medicaid, is more difficult for them than for other populations" (SAGE, NHCOA, DEC, 2014, p. 15). Older Latinx adults who are members of the SGM community are particularly susceptible to experiencing a combination of interlocking systems of oppression including heterosexism, racism, nativism, and ethnocentrism. As a result, older SGM Latinxs report verbal harassment, employment discrimination, and physical violence (Díaz et al., 2001). They also report higher numbers of victimization and neglect (Fredriksen-Goldsen, 2011). In fact, 70% reported feeling that their sexual orientation hurt their family. Over a lifetime, discrimination is harmful to mental health (Fredriksen-Goldsen et al., 2011; Fredriksen-Goldsen, Kim, & Barkan, 2012). For instance, in one study, 82% reported being victimized at least once during their lifetime due to being a member of the SGM community (Fredriksen-Goldsen et al., 2011, 2012). Discrimination can also happen due to ageism. The compounding effects of already vulnerable SGM Latinxs combined with ageism can create an even more hostile and silencing environment.

In addition to discrimination based on gender identity, expression, and sexual orientation, SGM older Latinxs may also be othered due to their ethnicity, race, and immigration status. While no previous studies exploring the impact of multiple forms of oppression on the physical and mental health of SGM older Latinxs have been identified, there is a growing body of literature suggesting that heterosexism, racism, ethnocentrism, and nativism are detrimental to Latinxs (Adames et al., 2018; Chavez-Dueñas et al., 2019). The next section provides a brief overview on how each of these systemic forces of oppression impacts the lives of aging SGM Latinx (see Fig. 12.1).

Racism

Latinxs are a racially heterogeneous group of people who can belong to any and all races (e.g.,

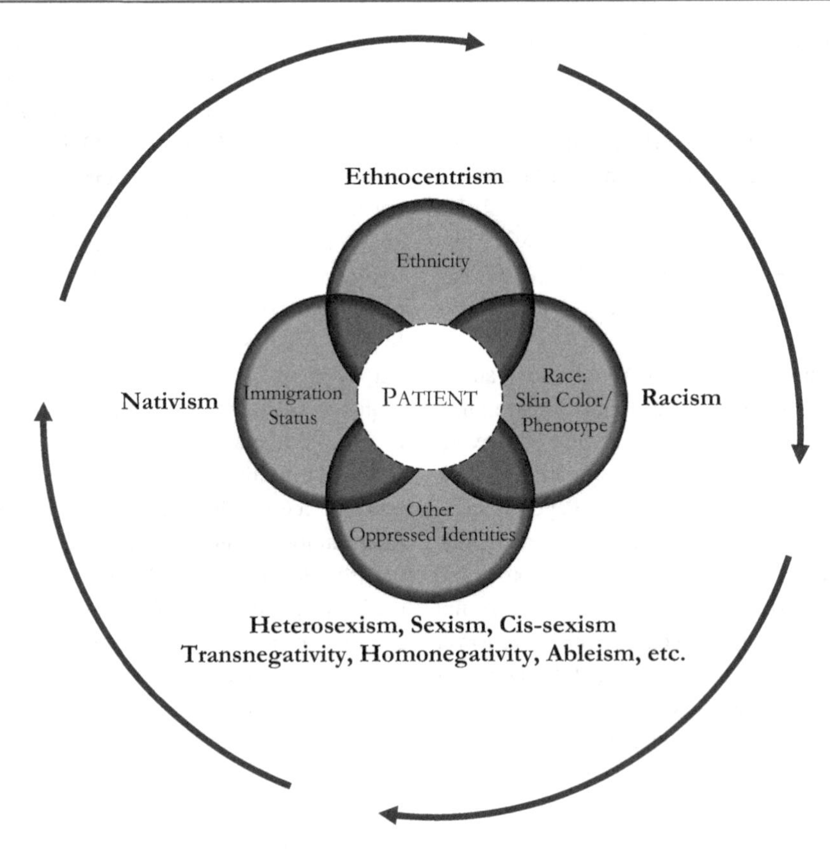

Fig. 12.1 A visual model of both weak and strong intersectionality. The model considers the ways in which overlapping systems of oppression impact the well-being, health, and resources available to patients. The model also aims to assist healthcare providers think about their own assumptions of how they conceptualize intersectionality in practice which will impact the language and type of questions they employ in their clinical and research practice. In the model, weak intersectionality (focus on multiple identities) is depicted by the constructs in the inner overlapping circles; strong intersectionality (interlocking systems of oppression) is depicted by the outer systems outside the circles. The circling arrows symbolize the importance of how different forms of oppression and discrimination are likely to shift depending on context. (From Adames et al. (2018). Copyright 2018 by the American Psychological Association. Adapted with permission)

Black, Indigenous, White). Hence, not all Latinxs are equally impacted by racism. For instance, Latinxs who are darker skinned and have non-European phenotypes (e.g., African, Indigenous) are especially vulnerable to experiencing racism (Araújo & Borrell, 2006; Telzer & Vazquez Garcia, 2009). In fact, research indicates that among the Latinx community, being darker and less European looking can have a negative impact on mental health (Capielo Rosario, Adames, Chavez-Dueñas, & Renteria, 2019; Montalvo, 2005; Montalvo & Codina, 2001; Perreira & Telles, 2014; Ramos, Jaccard, & Guilamo-Ramos, 2003), educational attainment, occupational status (Espino & Franz, 2002), and income (Arce, Murguia, & Frisbie, 1987; Montalvo & Codina, 2001). Thus, when considering social determinants of health, it is evident that racism impacts physical and mental health through its influence on access to economic, educational, and other social resources, all of which are known to have negative repercussions in old age for SGM Latinx.

Ethnocentrism

Latinxs in the USA are also othered due to their cultural differences which include traditional cultural values, customs, and language. Hence,

Latinxs are impacted by systemic oppression related to ethnocentrism, an ideology that perceives the beliefs, values, and practices of one's own ethnic group as superior to all others (Bizumic, Duckitt, Popadic, Dru, & Krauss, 2009). For instance, Latinx culture is characterized by a strong emphasis on the family unit (familism), collectivist ideologies (focus on "we" vs. "I"), relationship oriented (*personalismo*), Spanish language, strong connection to spirituality, and other unique cultural characteristics of this group. Thus, Latinxs are also othered because of their behaviors and beliefs. In fact, there is evidence to support the perpetual ways in which ethnocentrism impacts Latinxs. A study by Mukherjee, Molina, and Adams (2013) described how a sample of White American individuals of European descent from the Midwest support harsher treatment toward Latinx immigrants, regardless of their citizenship or immigration status, when compared to immigrants from Europe or Canada. While SGM Latinxs face unique experiences related to heterosexism, racism, and ethnocentrism, the unique impact resulting from simultaneously experiencing these forms of oppression is not fully captured when these systems are examined separately. Moreover, as the SGM Latinx community ages, they also have to contend with the negative forces related to ageism. Hence, it is critical that healthcare providers working with older SGM Latinxs understand the complexities and challenges related to treating this community.

Nativism

While members of the Latinx community are often characterized as foreigners, the majority, or 65% (36 million), are born in the USA. The remainder, 35% (19 million), are immigrants who were born in Latin America. Latinx immigrants are individuals who traveled to the USA in search of better economic opportunities, to escape violence, or to reunite with their families. Historically, access to economic opportunities and resources for immigrants have depended on the sociopolitical climate as well as on the financial stability of the USA economy. Therefore, in times of economic growth and prosperity, Latinx immigrants have been welcomed and used for their labor, but during times of economic downturn, this same community faced increased nativist ideologies marked by laws and policies that criminalize their presence in the USA (Chavez-Dueñas & Adames, 2018; Cornelius, 2009; Flores et al., 2008). Consequently, SGM Latinx immigrants, particularly those who are undocumented, are vulnerable to abuse, exploitation, and criminalization (Kronick & Hargis, 1998; National Council of La Raza, 1990; Suarez-Orozco & Suarez-Orozco, 2001; Velasquez, 1993). Moreover, undocumented Latinx immigrants often find it more difficult to secure well-paying jobs that offer benefits (Migration Policy Institute, 2013). As a result, the undocumented population is faced with high rates of poverty (33%), low levels of formal education (25% obtaining a high school diploma or General Equivalency Diploma (GED); Migration Policy Institute, 2013), and low rates of health insurance (40%; Migration Policy Institute, 2013). The lack of access to health insurance, coupled with economic hardship, serves as additional barriers to healthcare for members of this community.

Dementia Among Sexual and Gender Minoritized Population

One of the most pernicious cognitive diseases related to old age is dementia: an umbrella term used to define a group of degenerative, progressive, and incurable brain diseases that cause loss of memory and cognitive functioning, greatly impacting daily life. While there are many different types of dementia – e.g., vascular dementia (VaD), frontotemporal lobar degeneration (FTLD), and Lewy body dementias (LBD), including dementia with Lewy bodies (DLB) and Parkinson's disease dementia (PDD) – dementia of the Alzheimer's type (AD) is the most common, accounting for 60% to 80% of all dementia cases (Alzheimer's Association, 2016a). The statistics of people with AD are overwhelming. Currently, more than five million individuals are living with the disease, and every 66 seconds, someone develops AD (Alzheimer's Association, 2016a). These alarming figures make AD the sixth leading cause of death in the USA, more than prostate and breast cancers combined (Alzheimer's Association, 2016a; CDC, 2015). Over time, dementia results in marked behavioral changes that limit the patient's ability to problem-solve and perform simple, everyday tasks and activities (Wu, Vega, Resendez, & Jin, 2016).

In addition to age, research has also identified several modifiable risk factors for developing dementia including depression, cardiovascular disease (CVD), smoking, obesity, lower levels of formal education, and limited social or cognitive engagement (Alzheimer's Association, 2016a; Caraci, Copani, Nicoletti, & Drago, 2010). Some of these risk factors are known to be more prevalent among SGM older adults, increasing their risk of developing cognitive impairment, AD, or other types of dementia (Fredriksen-Goldsen, Jen, et al., 2016). Given the lack of research attention placed on the unique needs of the aging SGM population (Alzheimer's Association, 2017), the CDC placed a *Call to Action* in their *State of Aging & Health in America* (CDC, 2013) for changes in policies, research, and services for this growing segment of the population. Following the CDC's lead, the *Aging With Pride: National Health, Aging, Sexuality, and Gender Study* (Fredriksen-Goldsen et al., 2017) became the first longitudinal national study to investigate SGM aging, health, and well-being, including Alzheimer's disease and other types of dementias among a demographically diverse sample of 2450 SGM adults aged 50–100 years (Fredriksen-Goldsen, Jen, et al., 2016).

Dementia Risk Factors

The aging SGM community faces risk factors that place them at higher risk for developing cognitive impairment, Alzheimer's disease, or other dementias. The *Aging With Pride Study* has provided researchers with insight into how dementia-related disorders specifically impact the SGM population, including the following: 10% of participants reported severe or extreme cognitive difficulties, 38% reported moderate cognitive impairment, and 77% reported mild cognitive impairment in at least one area (Fredriksen-Goldsen, Jen, et al., 2016; Fredriksen-Goldsen, Kim, et al., 2016). Interestingly, cognitive difficulties were elevated among SGM older adults who were African American, Latinx, HIV positive, and male and those who identified their sexual orientation and/or gender as "other" (Fredriksen-Goldsen, Jen, et al., 2016; Fredriksen-Goldsen, Kim, et al., 2016). Moreover, transgender individuals are just as likely as cisgender people to be diagnosed with a form of dementia; however, scholars argue that transgender individuals may be at increased risk due to the various psychosocial stressors they encounter (Witten, 2014). Additionally, more women than men are diagnosed with Alzheimer's or other dementias (e.g., 3.3 million or two-thirds of those diagnosed; Alzheimer's Association, 2017). Given that more women are impacted by dementia, lesbian and bisexual women are likely to be disproportionately affected when compared to gay and bisexual men (Westwood, 2015). Of note, many of these studies continue to use binary categories when measuring or accounting for gender, leaving us to question the validity and generalizability of results.

Challenges and Unique Needs of Sexual and Gender Minorities with Dementia

SGM people living with dementia share many of the experiences and concerns of their non-SGM counterparts living with dementia; however, they also face unique challenges associated with being a member of a SGM group (Westwood & Price, 2016). One of the most challenging aspects for SGM older adults is having to conceal their identity as a way to protect themselves from hate (e.g., homonegativity,[3] transnegativity[4]). Moreover, fear of discrimination at the hands of healthcare providers may further lead to identity concealment, which puts SGM elders at a higher risk for social isolation and loneliness (Kim & Fredriksen-Goldsen, 2017). The fear of being discriminated against by healthcare providers is grounded in the unfortunate historical and current reality of hate and prejudice toward members of this community among healthcare providers (Kim & Fredriksen-Goldsen, 2017).

An Intersectional Approach to Caring for Sexual and Gender Minoritized Latinxs with Dementia

The literature reviewed in this chapter thus far outlines the multiple ways in which members of the SGM Latinx community are stigmatized, devalued, and dehumanized and discusses unique implications of such oppression on both physical and mental health. Unfortunately, these challenges are exacerbated for members of this community who are living with dementia, a group that is rapidly growing. For instance, according to the Alzheimer's Association, there are currently over 350,000 SGM people in the USA with dementia (Alzheimer's Association, 2012). Given that dementia eventually impairs an individual's ability to carry out simple activities of daily living (e.g., driving, self-care, household chores), those diagnosed with dementia often rely on the care provided by people in their social network such as family and friends. To place this into context, approximately 83% of caregiving help provided to USA adults come from family members, friends, or other unpaid caregivers (Alzheimer's Association, 2016a). Sadly, SGM individuals with dementia are less likely to have family members as caregivers (Fredriksen-Goldsen et al., 2011). A study of older SGM adults by Fredriksen-Goldsen et al. (2013) reported that nearly 60% of the participants lived alone and approximately 63% were neither partnered nor married. Moreover, about 15% of participants surveyed had offspring, and of those, 60% reported that their children were not available to help them when needed. Disparities in social support are also reflected by a study conducted by MetLife Mature Market Institute and The Lesbian & Gay Aging Issues Network of the American Society of Aging (2010) which reported that members of the SGM community are twice as likely to age single and live alone and are four times less likely to have children. SGM Latinx older adults also report feeling excluded and isolated from their various communities (e.g., families, other Latinxs) and that they are often unable to count on them for support. Despite the central cultural value of *familismo*, SGM Latinx can become estranged from their relatives who may exhibit homonegativity and transnegativity. Moreover, religion and spirituality, which can often serve as important forms of social support for Latinx, can also become additional source of alienation given the history of homonegativity in organized religion and religious institutions.

Lastly, the needs of older SGM Latinx with dementia are not only unique but compounded by the complex multiple and interlocking forms of oppression they experience. Hence, members of this community will necessitate multisystemic levels of support, advocacy, and care. In the next section, recommendations for healthcare providers working with members of the SGM Latinx aging population with dementia are provided. These recommendations are designed to address the impact of the intersecting forms of oppression on this community.

[3] *Homonegativity* is a psychological construct proposed by Hudson and Ricketts (1980) to describe the negative attitudes toward queer people which better captures the discrimination and hate toward sexual- and gender-diverse individuals.

[4] Similar to homonegativity, *transnegativity* places the focus on the hate experienced by members of the transgender and gender nonconforming community.

Identity-Based Intersectionality (Weak): Interpersonal Interactions That Honor the Patient's Authentic Self

I am as I am, and you are as you are...

Weak intersectionality provides an understanding of the complex intersecting identities from an individualistic perspective, taking into account how these unique identities form the constellational makeup of an individual, who may experience both privileged and marginalized identities. In this section, we provide recommendations tailored to identity-based concerns that impact Latinx SGM individuals and the interaction between such identity memberships and encounters with healthcare professionals seeking help for the assessment of dementia.

Healthcare's Role in Working with Intersecting Identities

Healthcare institutions are encouraged to approach training of healthcare providers through an intersectionality framework, ensuring that students, volunteers, and professionals are trained to consider how their own biases may impact healthcare delivery to SGM Latinx patients. It is also recommended that healthcare providers develop an understanding of Latinx cultural values (e.g., *familismo*, *personalismo*, *respeto*, *dignidad*) and learn how to integrate these values into the services they provide. Other areas of consideration specific to the Latinx community include acculturation, immigration status, race, ethnic identity, religion/spirituality, age, HIV status, disability, gender, gender identity, and sexual orientation.

- *Familismo* is a key traditional cultural value to honor in the delivery of aging services for SGM Latinx with dementia. *Familismo* involves broad networks of support that extend beyond the nuclear family to include aunts, uncles, grandparents, and other close family friends (Adames, Chavez-Dueñas, Fuentes, Salas, & Perez-Chavez, 2014). Recognize that for SGM Latinxs, family can be nuclear or extended, and also include non-blood relatives. Patients should be allowed to identify and define what family is for them.
- *Respeto* is a Latinx cultural value that becomes apparent in relationships with authority figures (e.g., elders, professionals, spiritual leaders, teachers, etc.) and involves a system of expected respect determined by hierarchical structure (Adames et al., 2014). Because of this, *respeto* will be central to relationships of healthcare providers working with Latinx patient with dementia. Recognize that regardless of how a Latinx patient identifies related to SGM status(es), internalized cultural values related to authority, age, and gender roles likely still impact their life, decisions, and relationships to some degree. Healthcare providers should be able to balance respect for the SGM Latinx patient, their possible life partner(s), and their family members.
- *Personalismo* is a value that places considerable emphasis on personal interactions with people (Adames et al., 2014). Latinxs tend to be relationship oriented and prefer informal and supportive interactions to formal and professional interactions (Carteret, 2011). Thus, warm and supportive interactions with aging healthcare providers are imperative. Healthcare providers may need to consider that families may lean toward more harmonious relationships with healthcare providers that are free of conflict. Hence, families may need time to congregate after meeting with a healthcare provider to further discuss disagreements and inconsistencies (Adames et al., 2014) with how they view dementia care. In these situations, the healthcare provider should follow up with the family to address questions or concerns regarding a patient's care plan.
- Among Latinx elders, the *dignidad* cultural value is associated with worthiness (Triandis, Marin, Lisansky, & Betancourt, 1984) and feeling valued (Chochinov, 2002). Moreover,

dignidad is understood "as a concept that includes an individual's sense of self-worth, as well as that individual's experience of others valuing them" (Adames et al., 2014, p. 154). Receiving a dementia diagnosis can be devastating for SGM Latinx. Particularly due to this community's lifelong history with homonegativity, transnegativity, racism, and nativism, the entrance to experiences of ageism and a diagnosis of dementia may increase the patient and their partners' fear of inhumane treatment and loss of agency in any healthcare facility. Healthcare providers are encouraged to demonstrate their sensitivity to such fears. In addition, healthcare providers can have an open conversation with the patient and their family regarding the progression and sequelae of the illness while allowing the patient and family to exercise as much autonomy as they wish in treatment (Adames et al., 2014).

- For optimal treatment, it is critical that healthcare providers remain familiar with the various forms of oppression that a Latinx SGM elder with dementia might experience in their everyday lives and interactions and how these have shaped a healthy distrust of healthcare providers and larger systems of care. *Confianza* (trust), the Latinx cultural value that emphasizes expression of deep, true feelings to only trusted circles of individuals with whom trust has been built, can be understood as a method of survival for those who have faced multiple interlocking systems of oppression. *Confianza* is essential to establish the healthcare provider-patient relationship (Adames et al., 2014). Thus, building relationships with this community may take time and a deliberate effort by healthcare providers to demonstrate that they are deserving of trust.

Establishing Institutional Protocol

The use of a specific protocol for staff interaction with aging Latinx SGM patients that demonstrates sensitivity to disability concerns and dementia-specific experiences may be helpful. Such protocol can include the following:

- Provide solutions to address barriers to care access by providing services that include language interpreters (e.g., Spanish, sign language) available upon patient request at any time.
- Provide specific training about Latinx SGM patients that emphasizes the importance of respecting a patient's identities as an essential component of care. Create and enforce a facility-wide protocol that honors the language, gender pronouns, and gender identity of individuals, regardless of gender expression.
- Display informational pamphlets specific to Latinx SGM patients in multiple places that are easy to find (e.g., throughout the office, in patient rooms, at help desks, and other locations) within a healthcare facility. Pamphlets must be in multiple languages.
- Display information in accessible ways for the differently abled community. For example, information in Braille, sign language, or visual guides should be available throughout facilities.
- Provide accessibility for all differently abled communities at all healthcare office locations, regardless of whether local or state law requires it, e.g., wheelchair access, ramps, available staff to assist in transporting clients, spacious parking for the disabled, and utilize transportation services to help patients without cars access facilities.

Clinical Assessment: Inclusive Protocols for Intake and Preventative Screening

Healthcare agencies are encouraged to implement strategies that capture the most inclusive assessment for Latinx SGM elders. Staff must be trained in dementia assessment, care, illness progression, and the impact of dementia on the behavioral, physical, psychological, and social aspects SGM patients. Intake forms can be designed to contextualize intersecting identities to best inform treatment including the following:

- Include inclusive gender terminology on intake forms, providing transgender and gender nonconforming as options. Leave additional space on intake forms for patients to identify their gender in their own way, outside of binary options.
- Do not assume that any listed gender identity or sexual orientation precludes a history of partners or self-identified categories. Sensitive questions healthcare providers can ask may include "Are your current partner(s) men, women, or other?"
- Create opportunities for patients to describe their relationship status and family structure in their own language. Healthcare providers must also recognize that heteronormative and monogonormative biases exist and work to silence and pathologize diverse family structures. The healthcare provider should commit to an open stance regarding relationship status and focus on working with the patient's loved ones just as they would with any heteronormative, monogamous partnerships. This also includes assessing for a patient's living situation and whether they live with or have children. All families should be viewed as legitimate with diverse constellations.
- SGM Latinx patients require additional considerations and sensitivity in how they are asked about sexual behavior and sexual health. Recognize that SGM elders of the Silent or GI Generation may be more private about their sexual behavior; this may reflect experiences of lifelong harmful attitudes about SGM membership from the dominant culture.
- Avoid asking any questions about sexual behavior, sexual orientation, or gender identity that are not strictly related to the patient's immediate health concerns.
- Assess for dementia-related disorders, and consider how HIV may contribute or exacerbate cognitive decline and other health-related symptoms. Healthcare providers should remain updated with best treatment options for co-occurring HIV and dementia.
- Monitor and assess for risk and signs of physical, sexual, emotional, and financial abuse or violence in patients' lives.
- Become knowledgeable of the increased risk of suicide associated with the progression of dementia, and consider the additional risks the SGM Latinx community faces, such as increased alienation and lack of social support which may exacerbate the threat of suicide. Thus, comprehensive suicide assessment must be included for both the caregiver and patient.

Dementia Diagnosis: Next Steps for Healthcare Providers

When early dementia is detected, healthcare providers are tasked with supplying patients and their families with a plethora of early-stage education about what the patient can expect regarding the progression of disease. The healthcare providers can also provide information about available SGM community resources to bolster social support for both the patient and caregiver. In addition, healthcare providers are encouraged to give caregiver-specific resources to empower and educate caregivers, ensuring they are making connections and staying socially engaged as the disease progresses. Some examples include the following:

- Assess the caregiver's abilities and willingness to provide care through a caregiver profile questionnaire. This questionnaire can help determine whether the caregiver understands the disease, knows where to obtain additional information about the disease, and needs additional support or assistance (Alzheimer's Association, n.d.).
- Become aware that within SGM communities, caregivers are diverse and often are of SGM status as well. Given the existence of diverse family structures, information should be provided both to the caregiver and patient in a way that is nonintrusive and respectful of Latinx cultural values while also ensuring the

caregiver is informed about all available options and resources.

- Give all patients, families, and caregivers information about the benefits of maintaining an ongoing relationship with healthcare providers throughout dementia care. This may be best provided in a pamphlet designed specifically for this community and works best when tailored to unique identity-based concerns. The provider must be ready to explain why this ongoing relationship reaps benefits, as healthy distrust of medical institutions may influence SGM Latinx families' potential inclination to attempt total in-home care without any additional support.

After Diagnosis: Considerations for Creating a Welcoming and Affirming Environment

It is important that once patients, caregivers, and family members begin their relationships with healthcare institutions, healthcare providers, and staff that SGM Latinx patients are provided the most welcoming and affirming environment possible including some of the following:

- Displaying diversity in images that the patient can relate to regarding identity group memberships (e.g., gender, race, sexual orientation, ethnicity) is a simple yet powerful method to establish a welcoming environment that provides subtle messages about the value of the patient's presence in an institution.
- Health providers and institutions can also display images of patients and healthcare providers from diverse SGM communities via online bulletins, emails, and patient portals and on institution websites. One example may include a "Tips for Wellness for SGM-POC" online bulletin.
- Any healthcare facility must ensure that transgender and gender nonconforming patients are allowed to use the restroom that aligns with their gender identity. Unisex restrooms are highly recommended.
- For transgender patients using ongoing hormone replacement therapy (HRT), assurance that this therapy will be provided in a manner consistent with the prevailing standards of care is imperative.
- Note: While these specific recommendations for creating a welcoming and affirming environment are an essential start, they are incomplete without considering how structural inequities exist within healthcare institutions (see *Strong Intersectionality* recommendations).
- For undocumented caregivers, family members, or patients, entrance into palliative care, hospice, or nursing home organizations can be dangerous. This makes alternative, culture-specific, in-home care services (that do not require insurance) all the more necessary for dignified end-of-life support services.

Strong Intersectionality: Structural Policies That Ensure Authentic Living

> *...Let's build a world where I can be, without having to cease being me, where you can be, without having to cease being you...*

Strong intersectionality turns the lens of intersectionality away from intersecting identities within individuals toward interlocking systems of oppression. In this section, recommendations that address systemic inequities experienced by SGM Latinx with dementia are provided.

Addressing Structural Oppression: Ways to Shift Healthcare Culture

Training programs Develop and implement formal training programs throughout the medical field, as well as across disciplines, that specifically discuss the forms of structural oppression impacting aging Latinx SGM while creating dedicated missions to shifting the culture of healthcare to inclusive, culturally competent service delivery.

Nondiscrimination policies Any Area Agencies on Aging (AAA), hospitals, memory clinics, nursing, and long-term care facilities who serve Latinx SGM elders living with dementia must provide policies that focus on protecting the well-being and safety of this community from structural oppression.

- Blanket nondiscrimination policies must be incorporated into institutions' missions and policies. Providing a statement that acknowledges the multiple interlocking forms of oppression and how they uniquely impact individuals who hold multiple minoritized group memberships is affirming of patients' life experiences, and should be explicitly stated on agency websites, pamphlets, and by healthcare providers themselves. These should be presented as nondiscrimination policies regardless of any status. In addition, these policies should include language regarding fair visitation rights for culturally diverse SGM patients within all agencies to show respect for *familismo*, in whatever form it may take in a patient's life.
- Compassionate services must become the norm for SGM Latinx older adults. Staff should be provided specific training about discriminatory behavior experienced in healthcare by SGM People of Color, and alternative behaviors should be taught to professionals to reduce discrimination. Some of these discriminatory behaviors include withholding touch; degrading comments; physical, sexual, and verbal abuse; assault; and discriminatory responses to diverse (consenting) sexual behaviors between older adult patients as well as discriminatory responses to patients' gender expression (e.g., wearing wigs).
- Specific training and policies should be promoted to reduce the pressure older adults feel to hide their SGM identities in fear of discrimination or threats. This training should also address the tripled/quadrupled invisibility this group faces through ageism, heterosexism, and ethnocentrism/racism. In the same vein, healthcare providers should ensure the privacy of SGM patients who want to keep their identity private, who may fear discrimination by their family or community, abuse, or discrimination from staff members and society.

Outreach As a result of the significant barriers to dementia care access, and due to the identified high risk for social isolation and barriers to care addressed in the literature for SGM elders with cognitive impairment and their caregivers (Fredriksen-Goldsen, Jen, et al., 2016; Fredriksen-Goldsen, Kim, et al., 2016), targeted outreach for this community is recommended.

Hiring diverse leaders and staff Additionally, AAA agencies are encouraged to hire staff who identify as SGM individuals, are bilingual (i.e., English/Spanish) and bicultural, and have experience working with individuals living with dementia, as they may be ideal to serve this community (Lambda Legal, 2016; NHCOA, 2016). The image of the healthcare provider must change as the population continues to diversify, and the underrepresentation of diversity in leadership and professional healthcare services must end. This requires conscious efforts by institutions to educate and hire diverse healthcare providers, especially in leadership positions, and to stop discriminatory education and hiring practices.

- In particular, healthcare institutions can reevaluate their requirements for hiring and training procedures for receptionists, as they are the first line of encounter with patients. This sets the tone for how a patient feels about their care, including their sense of safety and dignity. A zero-tolerance policy for discrimination should be mandated for all receptionists, phlebotomists, pharmacy technicians, and "first-line encounter" positions.

Reporting grievances Grievance-reporting processes related to discrimination must be thorough and easy to file in nursing homes, hospice care, and any other in-facility or in-home service.

- While the Health Insurance Portability and Accountability Act (HIPAA) is required to protect the privacy of SGMs, discrimination continues to occur. Filing grievances is time-consuming, and many Latinx families may not know their rights. Therefore, "Know Your Rights" brochures specifically tailored for the SGM community should be available in all settings that an elder Latinx SGM with dementia may encounter and should be provided in understandable language (including easy-to-read Spanish guides). Caregivers should also be provided with "Know Your Rights" information.
- Staff should be available to assist SGM Latinx in filing grievances, and grievances departments should receive oversight related to possible discriminatory practices/dismissal of appeals based on protected statuses.

Transformative Intersectionality: Actions That Honor the Dignity and Right to a Life of Self-Determination

...and where neither you nor I will force one another to be like either me or you...

Transformative intersectionality asks professionals to take the greatest leap – and the most rewarding leap – toward truly transforming the systems of oppression that continue to plague minoritized communities with unacceptable health, education, and opportunity disparities. This starts by taking on policy initiatives and motivating, through local grassroots efforts, changes in legislative agendas and framing of policy issues.

Expanding the Role of Healthcare Providers into SGM Advocacy

Healthcare providers and institutions that provide services to the Latinx SGM population with dementia are encouraged to work toward becoming advocates for their clients by becoming familiar with the laws and accreditation standards that require all healthcare providers to deliver high-quality care regardless of gender identity, sexual orientation, HIV status, age, ethnicity, immigration status, race, disability, and gender. Such advocacy should include the following:

- Coverage of SGM people, people with HIV, and Latinx communities within antidiscrimination and equal opportunity mandates of the Affordable Care Act (ACA).
- Support for the ACA's survival, as well as its expansion of nondiscrimination requirements for employees, employers, and services in healthcare that are currently excluded from nondiscrimination rules within the ACA. Additional antidiscrimination laws should be included and enforced regarding SGM that rely on state agencies such as Medicare and Medicaid.
- Equal access to affordable healthcare insurance should be advocated for by healthcare professionals for same-sex spouses, partners, and their children.

Section 1557 The Department of Health and Human Services (2015) added nondiscrimination laws through Section 1557 in the ACA in 2016. However, Section 1557 is limited to certain groups based on race, color, national origin, sex, age, and disability. Section 1557 also applies to state agencies such as the Children's Health Insurance Program (CHIP), Medicare, and Medicaid (Kaiser Commission on Medicaid and the Uninsured, 2016). Gender identity discrimination is considered a part of sex discrimination (National Center for Transgender Equality, 2017) and includes interpretations prohibiting discrimination based on sexual stereotypes but does not explicitly state protections for sexual minorities (U.S. Department of Health and Human Services, 2016). Enforcement and interpretation varies by state, local laws, and institutions and has been ineffective in preventing discrimination against SGM. In addition, while national origin is a protected status, it does not prohibit policies based on immigration status (Kaiser Commission on Medicaid and the Uninsured, 2016).

- Thus, advocacy for expansion of nondiscrimination rules to explicitly cover sexual orientation, ethnicity, undocumented status, and gender identity protection as a separate category from sex discrimination is required to ensure nondiscrimination is applied to Latinx SGM through Section 1557.

HIPAA Advocating for expansion of rights under the Health Insurance Portability and Accountability Act (HIPAA) to include protecting the privacy of patients regardless of documentation status is necessary for equitable access for Latinx communities, as many families worry that exposure to the healthcare system will endanger undocumented friends or relatives.

- There are many reasons why gender minoritized older adults do not change their gender marker on their birth certificate or other identity documents, including the financial cost, state policies, or documentation status (Porter et al., 2016). Identity documents are also frequently required to gain entrance into retirement communities and other types of senior housing. Healthcare providers can advocate for creating policies that avoid requiring legal gender markers and documentation status in their admission process to late-stage agencies.

Insurance companies A societal-level effort to admonish discriminatory practices by insurance providers that deny or limit coverage for needed care by SGM Latinx patients and families is necessary to allow patients better access to health insurance coverage. Advocacy to include laws that require broader recognition of families of SGM and their legal rights, including nontraditional family structures, is required to protect families' coverage through insurance and rights to decision-making in end-of-life arrangements.

References

Adames, H. Y., Chavez-Dueñas, N. Y., Fuentes, M. A., Salas, S. P., & Perez-Chavez, J. G. (2014). Integration of Latino/a cultural values into palliative health care: A culture centered model. *Palliative & Supportive Care, 12*(2), 149–157. https://doi.org/10.1017/S147895151300028X

Adames, H. Y., Chavez-Dueñas, N. Y., & Organista, K. C. (2016). Skin color matters in Latino/a communities: Identifying, understanding, and addressing Mestizaje racial ideologies in clinical practice. *Professional Psychology: Research and Practice, 47*(1), 46–55. https://doi.org/10.1037/pro0000062

Adames, H. Y., Chavez-Dueñas, N. Y., Sharma, S., & La Roche, M. J. (2018). Intersectionality in psychotherapy: The experiences of an AfroLatinx queer immigrant. *Psychotherapy, 55*(1), 1–7. https://doi.org/10.1037/pst0000152

Addis, S., Davies, M., Greene, G., MacBride-Stewart, S., & Shepherd, M. (2009). The health, social care and housing needs of lesbian, gay, bisexual and transgender older people: A review of the literature. *Health & Social Care in the Community, 17*(6), 647–658.

Administration on Aging. (2015). *A statistical profile of Hispanic older Americans aged 65+*. Retrieved from http://www.aoa.acl.gov/Aging_Statistics/minority_aging/Facts-on-Hispanic-Elderly.aspx

Administration on Community Living. (2017). *A statistical profile of older Hispanic Americans*. Retrieved from https://www.acl.gov/sites/default/files/news%20 2017-03/A_Statistical_Profile_of_Older_Hispanics.pdf

Alzheimer's Association. (2012). Alzheimer's disease facts and figures. *Alzheimer's and Dementia, 8*(2), 8–67.

Alzheimer's Association. (2016a). 2016 Alzheimer's disease facts and figures. *Alzheimer's & Dementia,*

12(4). Retrieved from http://www.alz.org/documents_custom/2016-facts-and-figures.pdf

Alzheimer's Association. (2016b). *Latinos and Alzheimer's*. Retrieved from http://www.alz.org/espanol/about/latinos_and_alzheimers.asp

Alzheimer's Association. (2017). *2017 Alzheimer's disease facts and figures*. Retrieved from https://www.alz.org/documents_custom/2017-facts-and-figures.pdf

Alzheimer's Association. (n.d.). *In brief: For healthcare professionals*. Retrieved from https://www.alz.org/health-care-professionals/documents/InBrief_CarePlanning.pdf

American Psychiatric Association. (2017). *What is gender dysphoria*. Retrieved from https://www.psychiatry.org/patients-families/gender-dysphoria/what-is-gender-dysphoria

American Psychological Association. (n.d.). Lesbian, gay, bisexual and transgender aging. Retrieved from https://www.apa.org/pi/lgbt/resources/aging

Angel, R. J. (2009). Structural and cultural factors in successful aging among older Hispanics. *Family & Community Health, 32*(10), S46.

Angel, R. J., & Angel, J. L. (2015). *Latinos in an aging world: Social, psychological, and economic perspectives*. New York, NY: Routledge.

Araújo, B. Y., & Borrell, L. N. (2006). Understanding the link between discrimination, mental health outcomes, and life chances among Latinos. *Hispanic Journal of Behavioral Sciences, 28*(2), 245–266.

Arce, C. H., Murguia, E., & Frisbie, W. P. (1987). Phenotype and life chances among Chicanos. *Hispanic Journal of Behavioral Sciences, 9*(1), 19–32.

Auldridge, A., Tamar-Mattis, A., Kennedy, S., Ames, E., & Tobin, H. J. (2012). *Improving the lives of transgender older adults: Recommendations for policy and practice*. New York, NY: Services and Advocacy for LGBT Elders & Washington, DC: National Center for Transgender Equality. Retrieved from http://www.lgbtagingcenter.org/resources/pdfs/TransAgingPolicyReportFull.pdf

Betancourt, H., & López, S. R. (1993). The study of culture, ethnicity, and race in american psychology. *American Psychologist, 48*(6), 629–637. https://doi.org/10.1037/0003-066X.48.6.629

Bizumic, B., Duckitt, J., Popadic, D., Dru, V., & Krauss, S. (2009). A cross-cultural investigation into a reconceptualization of ethnocentrism. *European Journal of Social Psychology, 39*(6), 871–899. https://doi.org/10.1002/ejsp.589

Boehmer, U., Bowen, D. J., & Bauer, G. R. (2007). Overweight and obesity in sexual-minority women: Evidence from population-based data. *American Journal of Public Health, 97*(6), 1134–1140.

Bowleg, L. (2013). Once you've blended the cake, you can't take the parts back to the main ingredients: Black gay and bisexual men's descriptions and experiences of intersectionality. *Sex Roles, 68*(11–12), 754–767. https://doi.org/10.1007/s11199-012-0152-4

Bureau of Labor Statistics. (2011). *Labor force characteristics by race and ethnicity*. Retrieved from https://www.bls.gov/opub/reports/race-and-ethnicity/archive/race_ethnicity_2011.pdf

Capielo Rosario, C., Adames, H. Y., Chavez-Dueñas, N. Y., & Renteria, R. (2019). Acculturation profiles of Central Florida Puerto Ricans: Examining the influence of skin color, perceived ethnic-racial discrimination, and neighborhood ethnic-racial composition. *Journal of Cross-Cultural Psychology, 50*(4), 556–576. https://doi.org/10.1177/0022022119835979

Caraci, F., Copani, A., Nicoletti, F., & Drago, F. (2010). Depression and Alzheimer's disease: Neurobiological links and common pharmacological targets. *European Journal of Pharmacology, 626*, 64–71. https://doi.org/10.1016/j.ejphar.2009.10.022

Carteret, M. (2011). *Cultural values of Latino patients and families*. Retrieved from http://www.dimensionsofculture.com/2011/03/cultural-values-of-latino-patients-and-families/

Case, P., Austin, S. B., Hunter, D. J., Manson, J. E., Malspeis, S., Willett, W. C., & Spiegelman, D. (2004). Sexual orientation, health risk factors, and physical functioning in the nurses' health study II. *Journal of Women's Health, 13*(9), 1033–1047. https://doi.org/10.1089/jwh.2004.13.1033

Centers for Disease Control and Prevention [CDC]. (2013). *The state of aging and health in America 2013*. Atlanta, GA: Centers for Disease Control and Prevention, US Dept of Health and Human Services. Retrieved from https://www.cdc.gov/aging/pdf/state-aging-health-in-america-2013.pdf

Centers for Disease Control and Prevention [CDC]. (2015). *Hispanics' health in the United States*. Retrieved from http://www.cdc.gov/media/releases/2015/p0505-hispanic-health.html

Chavez-Dueñas, N. Y., & Adames, H. Y. (in press). Intersectionality awakening model of Womanista: A transnational treatment approach for Latinx women. *Women & Therapy*.

Chavez-Dueñas, N. Y., & Adames, H. Y. (2018). Criminalizing hope: Policing Latino/a immigrant bodies for profit. In S. Weissinger & D. Mack (Eds.), *Law enforcement in the age of black lives matter: Policing black and brown bodies*. Lanham, MD: Lexington Books.

Chavez-Dueñas, N. Y., Adames, H. Y., Perez-Chavez, J., & Salas, S. P. (2019). Healing ethno-racial trauma in latinx immigrant communities: Cultivating hope, resistance, and action. *American Psychologist, 74*(1), 49–62. https://doi.org/10.1037/amp0000289

Chochinov, H. M. (2002). Dignity-conserving care—a new model for palliative care: Helping the patient feel valued. *Journal of the American Medical Association, 287*, 2253–2260.

Choi, S. K., & Meyer, I. H. (2016). *LGBT aging: A review of research findings, deeds, and policy implications*. Los Angeles, CA: The Williams Institute. Retrieved from http://williamsinstitute.law.ucla.edu/wp-content/uploads/LGBT-Aging-A-Review.pdf

Cole, E. R. (2009). Intersectionality and research in psychology. *American Psychologist, 64*(3), 170–180. https://doi.org/10.1037/a0014564

Collins, P. H. (2009). *Black feminist thought: Knowledge, consciousness, and the politics of empowerment* (2nd ed.). New York, NY: Routledge.

Combahee River Collective. (1995). Combahee River Collective statement. In B. Guy-Sheftall (Ed.), *Words of fire: An anthology of African American feminist thought* (pp. 232–240). New York, NY: New Press. (Original work published 1977).

Conron, K. J., Mimiaga, M. J., & Landers, S. J. (2010). A population-based study of sexual orientation identity and gender differences in adult health. *American Journal of Public Health, 100*(10), 1953–1960. https://doi.org/10.2105/AJPH.2009.17416

Cornelius, W. A. (2009). Ambivalent reception: Mass public responses to the "new" Latino immigration to the United States. In M. M. Suarez-Orozco & M. M. Paez (Eds.), *Latinos: Remaking America*. Berkeley, CA: University of California Press.

Crenshaw, K. (1989). Demarginalizing the intersection of race and sex: A black feminist critique of antidiscrimination doctrine, feminist theory and antiracist politics. *University of Chicago Legal Forum, 1989*(1), 139–167.

Crenshaw, K. W. (1991). Mapping the margins: Intersectionality, identity politics, and violence against women of color. *Stanford Law Review, 43*, 1241–1299. https://doi.org/10.2307/1229039

DeNavas-Walt, C., Proctor, B., & Smith, J. C. (2011). *Income, poverty and health insurance coverage in the united states: 2010*. Retrieved from https://www.census.gov/prod/2011pubs/p60-239.pdf

Department of Health and Human Services, Administration on Aging. (2015). *Long-term care ombudsman Program* (OAA, Title VII, Chapter 2, Sections 711/712). Retrieved from http://www.aoa.gov/AoA_programs/Elder_Rights/Ombudsman/index.aspx

Díaz, R. M., Ayala, G., Bein, E., Henne, J., & Marin, B. V. (2001). The impact of homophobia, poverty, and racism on the mental health of gay and bisexual Latino men: Findings from 3 U.S. cities. *American Journal of Public Health, 91*, 927–932. https://doi.org/10.2105/AJPH.91.6.927

Dill, B. T., & Kohlman, M. H. (2011). Intersectionality: A transformative paradigm in feminist theory and social justice. In S. N. Hesse-Biber (Ed.), *Handbook of feminist research: Theory and praxis* (2nd ed., pp. 154–174). Thousand Oaks, CA: Sage.

Dilley, J. A., Simmons, K. W., Boysun, M. J., Pizacani, B. A., & Stark, M. J. (2010). Demonstrating the importance and feasibility of including sexual orientation in public health surveys: Health disparities in the pacific northwest. *American Journal of Public Health, 100*(3), 460–467.

Donahue, P., & McDonald, L. (2005). Gay and lesbian aging: Current perspectives and future directions for social work practice and research. *Families in Society: The Journal of Contemporary Social Services, 86*(3), 359–366.

Espino, R., & Franz, M. M. (2002). Latino phenotypic discrimination revisited: The impact of skin color on occupational status. *Social Science Quarterly, 83*(2), 612–623.

Espinoza, R. (2011). A timetable roundup of recent LGBT research. *Aging, 32*(4), 1–2.

Feldman, J., & Bockting, W. (2003). Transgender health. *Minnesota Medicine, 86*(7), 25–32.

Flores, E., Tschann, J. M., Dimas, J. M., Bachen, E. A., Pasch, L. A., & de Groat, C. L. (2008). Perceived discrimination, perceived stress, and mental and physical health among Mexican-origin adults. *Hispanic Journal of Behavioral Sciences, 30*(4), 401–424.

Fredriksen-Goldsen, K., Cook-Daniels, L., Kim, H., Erosheva, E. A., Emlet, C. A., Hoy-Ellis, C., … Muraco, A. (2014). Physical and mental health of transgender older adults: An at-risk and underserved population. *The Gerontologist, 54*(3), 488–500. https://doi.org/10.1093/geront/gnt021

Fredriksen-Goldsen, K., Kim, H., & Barkan, S. E. (2012). Disability among lesbian, gay, and bisexual adults: Disparities in prevalence and risk. *American Journal of Public Health, 102*(1), e16–e21. https://doi.org/10.2105/AJPH.2011.300379

Fredriksen-Goldsen, K. I. (2011). Resilience and disparities among lesbian, gay, bisexual, and transgender older adults. *Public Policy and Aging Report, 21*(3), 3–7. https://doi.org/10.1093/ppar/21.3.3

Fredriksen-Goldsen, K. I., Emlet, C. A., Kim, H.-J., Muraco, A., Erosheva, E. A., Goldsen, J., & Hoy-Ellis, C. P. (2013). The physical and mental health of lesbian, gay, male and bisexual (LGB) older adults: The role of key health indicators and risk and protective factors. *The Gerontologist, 53*(4), 664–675. Editors' Choice. https://doi.org/10.1093/geront/gns123

Fredriksen-Goldsen, K. I., Jen, S., Bryan, A. E., & Goldsen, J. (2016). Cognitive impairment, Alzheimer's disease, and other dementias in the lives of lesbian, gay, bisexual and transgender (LGBT) older adults and their caregivers: Needs and competencies. *Journal of Applied Gerontology*, 1–25. https://doi.org/10.1177/0733464816672047

Fredriksen-Goldsen, K. I., & Kim, H. J. (2017). The science of conducting research with LGBT older adults-an introduction to aging with pride: National health, aging, and sexuality/gender study (NHAS). *The Gerontologist, 57*(S1), S1–S14.

Fredriksen-Goldsen, K. I., Kim, H. J., Emlet, C. A., Muraco, A., Erosheva, E. A., Hoy-Ellis, C. P., … Petry, H. (2011). *The aging and health report: Disparities and resilience among lesbian, gay, bisexual, and transgender older adults*. Seattle, WA: Institute for Multigenerational Health. Retrieved from http://depts.washington.edu/agepride/wordpress/wp-content/uploads/2012/10/Full-report10-25-12.pdf

Fredriksen-Goldsen, K. I., Kim, H. J., Goldsen, J., Shiu, C., & Emlet, C. A. (2016). *Addressing social, economic, and health disparities of LGBTQ older adults*

& best practices in data collection. Seattle, WA: LGBTQ+ National Aging Research Center, University of Washington. Retrieved from http://age-pride.org/wordpress/wp-content/uploads/2016/05/2016-Disparities-Factsheet-Final-Final.pdf

Fredriksen-Goldsen, K. I., & Muraco, A. (2010). Aging and sexual orientation: A 25-year review of the literature. *Research on Aging, 32*(3), 372–413.

Gates, G. J. (2011). *How many people are lesbian, gay, bisexual and transgender?* Retrieved from http://williamsinstitute.law.ucla.edu/wp-content/uploads/Gates-How-Many-People-LGBT-Apr-2011.pdf

Gates, G. J. (2017). LGBT data collection amid social and demographic shifts of the US LGBT community. *American Journal of Public Health, 107*(8), 1220–1222. https://doi.org/10.2105/AJPH.2017.303927

Goldsen, J., Bryan, A. E. B., Kim, H.-J., Muraco, A., Jen, S., & Fredriksen-Goldsen, K. I. (2017). Who says I do: The changing context of marriage and health and quality of life for LGBT older adults. *The Gerontologist, 57*(S1), S50–S62.

González-Sanders, D. G., & Fortinsky, R. (2012). *Dementia care with black and Latino families: A social work problem-solving approach*. New York, NY: Springer Publishing Company.

Grant, J. M., Mottett, L. A., Tanis, J., Harrison, J., Herman, J. L., & Keisling, M. (2011). *Injustice at every turn: A report of the national transgender discrimination survey*. Washington, DC: National Center for Transgender Equality & National Gay & Lesbian Task Force.

Grzanka, P. R. (2014). Intersectional objectivity. In P. R. Grzanka (Ed.), *Intersectionality: A foundations and frontiers reader* (pp. xi–xxvii). Boulder, CO: Westview Press.

Harper, S. R. (2012). Race without racism: How higher education researchers minimize racist institutional norms. *The Review of Higher Education, 36*(1), 9–29.

History. (2017). *Gay marriage*. Retrieved from http://www.history.com/topics/gay-marriage

Hughes, A. K., Harold, R. D., & Boyer, J. M. (2011). Awareness of LGBT aging issues among aging services network providers. *Journal of Gerontological Social Work, 54*(7), 659–677. https://doi.org/10.1080/01634372.2011.585392

Hudson, W. W., & Ricketts, W. A. (1980). A strategy for the measurement of homophobia. *Journal of homosexuality, 5*(4), 357–372.

Human Rights Campaign. (2017). *HRC & Trans People of Color Coalition release report on violence against the transgender community*. Retrieved from https://www.hrc.org/resources/violence-against-the-transgender-community-in-2019

Institute of Medicine. (2011). *The health of lesbian, gay, bisexual, and transgender people: Building a foundation for better understanding*. Washington, DC: The National Academies Press.

Jerald, M. C., Cole, E. R., Ward, L. M., & Avery, L. R. (2017). Controlling images: How awareness of group stereotypes affects Black women's well-being. *Journal of Counseling Psychology, 64*(5), 487–499. https://doi.org/10.1037/cou0000233

Kaiser Commission on Medicaid and the Uninsured. (2016, July). *Summary of HHS's final rule on nondiscrimination in health programs and activities*. Issue Brief. Washington, D.C. Retrieved from http://files.kff.org/attachment/issue-brief-Summary-of-HHSs-Final-Rule-on-Nondiscrimination-in-Health-Programs-and-Activities

Kim, H., & Fredriksen-Goldsen, K. (2017). Disparities in mental health quality of life between hispanic and non-hispanic white LGB midlife and older adults and the influence of lifetime discrimination, social connectedness, socioeconomic status, and perceived stress. *Research on Aging, 39*(9), 991–1012. https://doi.org/10.1177/0164027516650003

Kim, H. J., Jen, S., & Fredriksen-Goldsen, K. I. (2017). Race/ethnicity and health-related quality of life among LGBT older adults. *The Gerontologist, 57*(S1), S30–S39. https://doi.org/10.1093/geront/gnw172

Kitzinger, C. (2000). Doing feminist conversation analysis. *Feminism & Psychology, 10*(2), 163–193.

Kochhar, R. (2004). *The wealth of hispanic households: 1996 to 2002*. Retrieved from http://www.sas.upenn.edu/~dludden/HispanicWealthREPORT.pdf

Kochhar, R., Fry, R., & Taylor, P. (2011). *Wealth gaps rise to record highs between Whites, Black, Hispanics*. Retrieved from http://www.pewsocialtrends.org/2011/07/26/wealth-gaps-rise-to-record-highs-between-whites-blacks-hispanics/

Krehely, J. (2009). How to close the LGBT health disparities gap: Disparities by race and ethnicity. *Center for American Progress*, 1–5. Retrieved from: https://cdn.americanprogress.org/wp-content/uploads/issues/2009/12/pdf/lgbt_health_disparities_race.pdf

Krogstad, J. M. (2016). *5 facts about Latinos and education*. Retrieved from http://www.pewresearch.org/fact-tank/2016/07/28/5-facts-about-latinos-and-education/

Kronick, R. F., & Hargis, C. H. (1998). *Dropouts: Who drops out and why—and the recommended action*. Springfield, IL: Charles C. Thomas Publishing.

Lambda Legal. (2016). *Creating equal access to quality healthcare for transgender patients: Transgender affirming hospital policies 2013*. Retrieved from http://www.lambdalegal.org/publications/fs_transgender-affirming-hospital-policies

Lewis, J. A., & Neville, H. A. (2015). Construction and initial validation of the Gendered Racial Microaggressions Scale for Black women. *Journal of Journal of Counseling Psychology, 62*(2), 289–302. http://dx.doi.org/10.1037/cou0000062

Lewis, J. A., Williams, M. G., Peppers, E. J., & Gadson, C. A. (2017). Applying intersectionality to explore the relations between gendered racism and health among Black women. *Journal of Counseling Psychology, 64*(5), 475–486. https://doi.org/10.1037/cou0000231

Livingston, G., Minushkin, S., & Cohn, D. (2008). Hispanics and heath care in the United States: Access, information, and knowledge. *Pew Hispanic Center and*

Robert Wood Johnson Foundation, 1–48. Retrieved from: http://www.pewhispanic.org/files/reports/91.pdf

Mackenzie, J. (2009). Working with lesbian and gay people with dementia. *Journal of Dementia Care, 17*(6), 17–19.

Manly, J. J., & Mayeux, R. (2004). Ethnic differences in dementia and Alzheimer's disease. In N. B. Anderson, R. A. Bulatao, & B. Cohen (Eds.), *Critical perspectives on racial and ethnic differences in health in late life*. Washington, DC: National Academies Press.

Marecek, J. (2016). Invited reflection: Intersectionality theory and feminist psychology. *Psychology of Women Quarterly, 40*(2), 177–181. https://doi.org/10.1177/0361684316641090

Martin, L. G., & Soldo, B. J. (1997). *Racial and ethnic differences in the health of older Americans*. Washington DC: National Academies Press.

McGovern, J. (2014). The forgotten: Dementia and the aging LGBT community. *Journal of Gerontological Social Work, 57*, 845–857. https://doi.org/10.1080/01634372.2014.900161

Metlife Mature Market Institute, & The Lesbian and Gay Aging Issues Network of the American Society on Aging. (2010). Out and aging: The MetLife study of lesbian and gay baby boomers. *Journal of GLBT Family Studies, 6*(1), 40–57.

Migration Policy Institute. (2013). *Profile of the unauthorized population: United States*. Washington, DC: Migration Policy Institute. Retrieved from: http://www.migrationpolicy.org/data/unauthorized-immigrant-population/state/US

Montalvo, F. F. (2005). Surviving race: Skin color and the socialization and acculturation of Latinas. *Journal of Ethnic and Cultural Diversity in Social Work, 13*(3), 25–43.

Montalvo, F. F., & Codina, G. E. (2001). Skin color and Latinos in the United States. *Ethnicities, 1*(3), 321–341.

Moradi, B., & Grzanka, P. R. (2017). Using intersectionality responsibly: Toward critical epistemology, structural analysis, and social justice activism. *Journal of Counseling Psychology, 64*(5), 500–513. https://doi.org/10.1037/cou0000203

Morrow, D. E. (2001). Older gays and lesbians: Surviving a generation of hate and violence. In M. E. Swigonski, R. S. Mama & K. Ward (Eds.), *From hate crimes to human rights: A tribute to matthew shepard; from hate crimes to human rights: A tribute to matthew shepard* (pp. 151–169, Chapter xvii, 233 pages). New York, NY: Haworth Press.

Movement, Advancement Project [MAP], SAGE. (2017). *Understanding issues facing LGBT older adults*. Retrieved from https://issuu.com/lgbtagingcenter/docs/understanding_issues_facing_lgbt_ol/8

Mukherjee, S., Molina, L. E., & Adams, G. (2013). "Reasonable suspicion" about tough immigration legislation: Enforcing laws or ethnocentric exclusion? *Cultural Diversity and Ethnic Minority Psychology, 19*(3), 320–331.

National Center for Transgender Equality. (2017). *Know Your Rights: Healthcare*. Retrieved from https://transequality.org/sites/default/files/docs/kyr/KYR-Healthcare-June2017.pdf

National Council of La Raza. (1990). *Immigration reform*. Retrieved from https://www.nclr.org/content/topics/detail/500/

National Hispanic Council on Aging [NHCOA]. (2016). *Status of Hispanic older adults: Insights from the field*. Retrieved from http://www.nhcoa.org/wp-content/uploads/2016/09/2016-NHCOA-Status-of-Hispanic-Older-Adults-report-.pdf

Obergefell v. Hodges, 135 S. Ct. 2071, 576 U.S., 191 L. Ed. 2d 953. (2015). Pew Research Center. (2017). 5 Key Findings about LGBT Americans. Retrieved from https://www.pewresearch.org/fact-tank/2017/06/13/5-key-findings-about-lgbt-americans/

Perreira, K. M., & Telles, E. E. (2014). The color of health: Skin color, ethnoracial classification, and discrimination in the health of latin americans. *Social Science & Medicine, 116*, 241–250. https://doi.org/10.1016/j.socscimed.2014.05.054

Pew Research Center. (2013). A survey of LGBT Americans. Retrieved from https://www.pewsocialtrends.org/2013/06/13/a-survey-of-lgbt-americans/

Porter, K. E., Brennan-Ing, M., Chang, S. C., Dickey, L. M., Singh, A. A., Bower, K. L., & Witten, T. M. (2016). Providing competent and affirming services for transgender and gender nonconforming older adults. *Clinical Gerontologist, 39*(5), 366–388.

Ramos, B., Jaccard, J., & Guilamo-Ramos, V. (2003). Dual ethnicity and depressive symptoms: Implications of being Black and Latino/a in the United States. *Hispanic Journal of Behavioral Sciences, 25*(2), 147–173. https://doi.org/10.1177/0739986303025002002

Santiago-Rivera, A. L., Adames, H. Y., Chavez-Dueñas, N. Y., & Benson-Flórez, G. (2016). The impact of racism on communities of color: Historical contexts and contemporary issues. In A. N. Alvarez, C. T. H. Liang, & H. A. Neville (Eds.), *Cultural, racial, and ethnic psychology book series. The cost of racism for people of color: Contextualizing experiences of discrimination* (pp. 229–245). Washington, DC, US: American Psychological Association. http://dx.doi.org/10.1037/14852-011

Schulden, J. D., Song, B., Barros, A., Mares-DelGrasso, A., Martin, C. W., Ramirez, R., … Heffelfinger, J. D. (2008). Rapid HIV testing in transgender communities by community-based organizations in three cities. *Public Health Reports, 123*(3), 101–114. https://doi.org/10.1177/00333549081230S313

Services & Advocacy for Gay, Lesbian, Bisexual, & Transgender Elders [SAGE], National Hispanic Council on Aging [NHCOA], & Diverse Elders Coalition [DEC]. (2014). *Hispanic LGBT older adult needs assessment*. Retrieved from: http://www.nhcoa.org/wp-content/uploads/2014/02/NHCOA-Hispanic-LGBT-Older-Adult-Needs-Assessment-In-Their-Own-Words.pdf

Shin, R. Q., Welch, J. C., Kaya, A. E., Yeung, J. G., Obana, C., Sharma, R., … Yee, S. (2017). The intersectionality framework and identity intersections in the journal of counseling psychology and the counseling psychologist: A content analysis. *Journal of Counseling Psychology, 64*(5), 458–474. https://doi.org/10.1037/cou0000204

Silverstein, C. (2009). The implications of removing homosexuality from the DSM as a mental disorder. *Archives of Sexual Behavior, 38*(2), 161–163. https://doi.org/10.1007/s10508-008-9442-x

Suarez-Orozco, C., & Suarez-Orozco, M. (2001). *Children of immigration*. Cambridge, MA: Harvard University Press.

Subcomandante Marcos. (2001). *Our word is our weapon: Selected writings*. New York, NY: Seven Stories Press.

Szymanski, D. M., & Carr, E. R. (2008). The roles of gender role conflict and internalized heterosexism in gay and bisexual men's psychological distress: Testing two mediation models. *Psychology of Men & Masculinity, 9*(1), 40–54. https://doi.org/10.1037/1524-92220.9.1.40

Szymanski, D. M., & Moffitt, L. B. (2012). Sexism and heterosexism. In N. A. Fouad, J. A. Carter, & L. M. Subich (Eds.), *APA handbooks in psychology. APA handbook of counseling psychology, Vol. 2. Practice, interventions, and applications* (pp. 361–390). Washington, DC: American Psychological Association.

Taylor, P., Lopez, M. H., Martinez, J., & Velasco, G. (2012). *Language use among Latinos*. Retrieved from http://www.pewhispanic.org/2012/04/04/iv-language-use-among-latinos/

Telzer, E. H., & Vazquez Garcia, H. A. (2009). Skin color and self-perceptions of immigrant and US-born Latinas: The moderating role of racial socialization and ethnic identity. *Hispanic Journal of Behavioral Sciences, 31*(3), 357–374.

Testa, R. J., Sciacca, L. M., Wang, F., Hendricks, M., Goldblum, P., Bradford, J., & Bongar, B. (2012). The effects of violence on transgender people. *Professional Psychology: Research and Practice, 43*(5), 452–459.

Triandis, H. C., Marin, G., Lisansky, J., & Betancourt, H. (1984). Simpatia as a cultural script of Hispanics. *Journal of Personality and Social Psychology, 47*, 1363–1375.

U.S. Census Bureau. (2004). *U.S. interim projections by age, sex, race, and Hispanic origin*. Retrieved from http://www.census.gov/ipc/www.usinterimproj/

U.S. Census Bureau. (2011). *The Hispanic population: Census 2011 brief.* Retrieved from http://www.census.gov/compendia/statab/brief.html

U.S. Department of Health and Human Services. (2012). *Lesbian, gay, bisexual, and transgender health*. Retrieved from: http://www.healthypeople.gov/2020/topicsobjectives2020/overview.aspx?topicid=25

U.S. Department of Health and Human Services. (2016). *Summary: Final rule implementing Section 1557 of the Affordable Care Act*. Retrieved from: https://www.hhs.gov/sites/default/files/2016-06-07-section-1557-final-rule-summary-508.pdf

Vega, W. A., Markides, K. S., Angel, J. L., & Torres-Gil, F. M. (Eds.). (2015). *Challenges of Latino aging in the Americas*. New York, NY: Springer.

Velasquez, L. C. (1993). *Migrant adults' perceptions of schooling, learning, and education*. Doctoral dissertation College of Education, University of Tennessee, Knoxville, TN.

Violence Against Women & Girls. (2015). *Brief on violence against sexual and gender minority women*. Retrieved from http://www.vawgresourceguide.org/sites/default/files/briefs/vawg_resource_guide_sexual_and_gender_minority_women_final.pdf

Violence Against Women & Girls. (2016). *Brief on violence against older women*. Retrieved from http://www.un.org/esa/socdev/documents/ageing/vawg_brief_on_older_women.pdf

Wallace, S. P., Cochran, S. D., Durazo, E. M., & Ford, C. L. (2011). The health of aging lesbian, gay and bisexual adults in California. *Policy Brief (UCLA Center for Health Policy Research), 0*, 1–8.

Wallace, S. P., & Villa, V. M. (2003). Equitable health systems: Cultural and structural issues for Latino elders. *American Journal of Law & Medicine, 29*(2–3), 247–267.

Warner, L. R., & Shields, S. A. (2013). The intersections of sexuality, gender, and race: Identity research at the crossroads. *Sex Roles, 68*(11–12), 803–810. https://doi.org/10.1007/s11199-013-0281-4

Westwood, S. (2015). Dementia, women and sexuality: How the intersection of ageing, gender and sexuality magnify dementia concerns among lesbian and bisexual women. *Dementia*. https://doi.org/10.1177/1471301214564446

Westwood, S., & Price, E. (2016). *Lesbian, gay, bisexual, and trans∗ individuals living with dementia: Concepts, practice and rights*. New York, NY: Routledge.

Williams, M. E., & Freeman, P. E. (2007). Transgender health: Implications for aging and caregiving. *Journal of Gay and Lesbian Social Services, 18*(3–4), 93–108. https://doi.org/10.1300/J041v18n03_06

Witten, T. M. (2014). It's not all darkness: Robustness, resilience, and successful transgender aging. *LGBT health, 1*(1), 24–33.

Witten, T. M. (2015). Elder transgender lesbians: Exploring the intersection of age, lesbian sexual identity, and transgender identity. *Journal of Lesbian Studies, 19*(1), 73–89. https://doi.org/10.1080/10894160.2015.959876

Wu, S., Vega, W., Resendez, J., & Jin, H. (2016). *Latinos and Alzheimer's disease: New numbers behind the crisis projection of the costs for U.S. Latinos living with Alzheimer's disease through 2060*. Retrieved from https://www.ucdmc.ucdavis.edu/latinoaging/news/pdf/Latinos_and_AD_USC_UsA2-Impact-Report.pdf

13 Stress and Coping: Conceptual Models for Understanding Dementia Among Latinos

Julian Montoro-Rodriguez and Dolores Gallagher-Thompson

Abstract

Since the late 1990s, family caregiving theories have contributed significantly to improve our understanding of the dynamic and diverse experiences of caregivers. Some theories have been translated into best practices to deliver a broad range of programs for family caregivers. However, researchers still have not been able to translate most of these theories into interventions for low-income and culturally diverse caregivers. This chapter highlights current knowledge, research, and practice available and, for the most part, based on stress and coping conceptual models, for Latino older adults with dementia and their family caregivers. More specifically, we discuss how existing support services for Latino caregivers are able to address multiple threats associated with their experience of dementia, their cultural family context, and the everyday challenges associated with their low-income status and limited available resources. Several salient theoretical frameworks and innovative efforts that led to development of evidence-based knowledge are examined along with interventions to meet the needs of Latino family dementia caregivers.

J. Montoro-Rodriguez (✉)
University of North Carolina at Charlotte, Charlotte, NC, USA
e-mail: jmontoro@uncc.edu

D. Gallagher-Thompson
Stanford University, School of Medicine, Stanford, CA, USA
e-mail: dolorest@stanford.edu

Introduction

Alzheimer's Disease (AD) and related dementias are conditions characterized by a decline in memory and other mental skills that impact the person's ability to perform daily activities (Alzheimer's Association, 2017). AD is a degenerative brain disease and the most common cause of dementia (Wilson et al., 2012). The recent Alzheimer's Disease Facts and Figures (2017) report describes dementia as a syndrome – a group of symptoms, such as difficulties with memory, language, problem-solving, and other cognitive skills – that has a number of possible causes. These difficulties occur because nerve cells (neurons) in parts of the brain involved in cognitive function have been damaged or destroyed. Dementias are responsible for detrimental health and disability and generate higher levels of health-care cost and depletion of resources for caregivers and people with dementia than for adults with no dementia disorders (Alzheimer's Association, 2017). Furthermore, dementia caregivers report high levels of burden and other negative psychosocial outcomes (depression, anxiety, etc.) while caring for a person with dementia (Mausbach, 2014).

H. Y. Adames, Y. N. Tazeau (eds.), *Caring for Latinxs with Dementia in a Globalized World*,
https://doi.org/10.1007/978-1-0716-0132-7_13

The Alzheimer's Association reports (2017) that an estimated 5.5 million Americans in the United States of America (USA), of all ages, are living with Alzheimer's dementia in 2017 and that, by 2025, the number of people age 65 and older with Alzheimer's dementia is estimated to reach 7.1 million—almost a 35% increase from the 5.3 million age 65 and older affected in 2017 (Hebert, Weuve, Scherr, & Evans, 2013). However, among minorities, older African-Americans are about twice as likely to have Alzheimer's or other dementias as older Whites, and Latinos are about 1.5 times as likely to have Alzheimer's or other dementias as non-Hispanic Whites (Samper-Ternent et al., 2012). The lower socioeconomic status of many older Latinos and the high cost of dementia care, along with the extra financial burdens associated with physical health problems common in older Latinos (e.g., diabetes, heart disease), are associated with an increased demand and need for frequent health services (Alzheimer's Association, 2017). Given that the number of Latinos is also expected to grow in the coming years and reach 20% of older adults needing care by 2050 (Alzheimer's Association, 2017), it becomes a priority to examine the unmet needs of dementia family caregivers among Latinos in the USA. Furthermore, Latinos' own cultural values and beliefs about normative family care can place these caregivers at a disadvantage compared to non-Hispanic White caregivers (Angel & Angel, 2015). In particular, rapidly occurring demographic changes (e.g., increased mobility, smaller family size, less dependence on the extended family, etc.) and changes in patterns of acculturation (greater adaptation to Anglo-American culture) may erode traditional family values (Angel & Angel, 2015; Coon, et al., 2004; Montoro-Rodríguez, Kosloski, Kercher, & Montgomery, 2009). Reviews of the qualitative literature also indicate that Latino caregivers frequently report "feeling caught" in the tension between cultural expectation of family piety and their ability to fulfill caregiving demands (Apesoa-Varano, Gomez, & Hinton, 2015). Latinos often assume multiple family caregiving roles, providing care to older parents and relatives with dementia and, at the same time, caring for their own children and grandchildren. Recent content analysis of qualitative ethnographic interviews of Latino caregivers indicates that regardless of who the care recipient is, caregivers' narratives support similar themes – both positive and negative (Arevalo-Flechas, Martinez, & Flores, 2017). These include duty to family first, sense of satisfaction from caregiving, and willingly sacrificing to be in this role, which provides a sense of purpose. At the same time, however, caregiving is described as physically exhausting, emotionally draining, and financially straining, leading to role challenges, role captivity, legal stressors, fear, and major stress. Earlier quantitative studies comparing Latino to non-Latino caregivers similarly found higher rates of depression and other indices of psychological distress (Hinton, Chambers, Velásquez, Gonzalez, & Haan, 2006; Pinquart & Sorensen, 2005).

Based primarily on stress and coping conceptual models, this chapter will highlight current knowledge, research, and practices available for Latino older adults with dementia and their family caregivers. More specifically, we outline how existing support services and programs for Latino caregivers are able to address multiple threats associated with the experience of dementia, the cultural family context, and the everyday challenges associated with low-income status and limited available resources. Examined in the discussion are the most salient theoretical frameworks and innovative efforts to develop evidence-based knowledge and interventions that contribute to meet the needs of Latino family dementia caregivers.

Theories Associated with Dementia Caregiving Research

During the past three decades, researchers have developed a myriad of social/psychological theories aimed at understanding the experience of family care and its impact on caregivers' well-being. The early theoretical models were developed to examine sources of caregiver stress, as well as how stress and burden affects psychological outcomes, depression, anxiety, physical health, service utilization, and/or institutionalization of care recipients. Development of this first generation of theories began in the 1960s with the work of Thomas Holmes and Richard Rahe. Holmes and Rahe (1967) theorized a direct link between stress and physical and psychological illness. The result was their *Social Readjustment Rating Scale*, which assigned weights to a variety of common stressful life events and used a person's total score as a measure of their risk for illness (Holmes & Rahe, 1967).

This theory was the foundation for the *Stress-Process Model* developed by Leonard Pearlin and colleagues (Pearlin, Mullan, Semple, & Skaff, 1990) and the revised stress model (Folkman, 2008), which examined care demands on the caregiver as stressors and investigated their cumulative effects, as well as explored the role of coping mechanisms to reduce their negative impact. Folkman argued that the coping abilities of the stressed individual were also a predictor of whether the person would be ill/distressed or not (Fig. 13.1). Someone with excellent coping abilities, for example, would not be as severely negatively affected by personal illness as someone with poor coping abilities. Pearlin's pioneering contribution was to apply this Stress-Process Model to a caregiver's situation. This application has led to a variety of caregiver intervention programs, as described in the next section.

Other early models paid particular attention to the *caregiving social content and dyadic interaction* of the caregiver-care recipient relationship (Kahana, Kahana, Johnson, Hammond, & Kercher, 1994), highlighting the importance of both the person and the environment. Later models, including *Caregiver Identity Theory*, articulated by Montgomery and Kosloski (2009, 2013) conceptualize caregiving as a series of identity transitions that result from changes in the caregiving context and in personal norms. Caregiver identity is a set of meanings applied to the self in the care context and is used as a reference point to guide behavior wherein the discrepancy or incongruence between a caregiver's behavior and his or her identity standards is suggested as the major source of caregiver stress (Montgomery, Kwak, & Kosloski, 2016).

Theory-Derived Programs for Caregivers

With the exception of Holmes and Rahe's (1967) work, the theories described in the previous section led to the development of several caregiver intervention programs. The *Holmes-Rahe Stress Inventory* (1967) did not directly lead to any such programs; however, its relevance to the field is in how it provided the foundation for Lazarus and Folkman's (1984) Cognitive Appraisal Theory which, in turn, informed the development of several evidence-based caregiver interventions discussed below.

The first, the *Coping with Caregiving* (CWC) psychoeducational program (Gallagher-Thompson, Ossinalde, & Thompson, 1996), is derived from both Cognitive Appraisal Theory (i.e., Pearlin's *Stress-Process Model*) and Bandura's *Self-Efficacy Theory* (Bandura, 1986). It aims to teach caregivers adaptive coping skills through education, role playing, and home practice using a small group, workshop format. The proposed increase in self-efficacy, resulting from this behaviorally based learning, mediates positive outcomes. In the multiple, randomized controlled trials conducted with this program (or its derivatives), significant reduction in symptoms of depression was common, along with use of

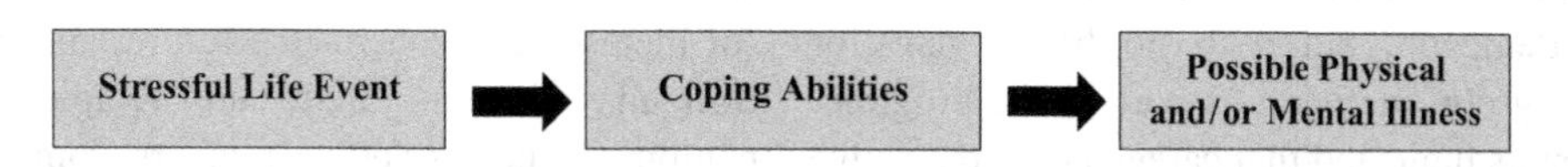

Fig. 13.1 Stress-Process Model. (From Knight and Sayegh (2010). Copyright 2010 by Oxford Academic. Adapted with permission)

more effective cognitive and behavioral coping strategies as compared to emotion-focused coping (see Coon, Keaveny, Valverde, Dadvar, & Gallagher-Thompson, 2012 and Gallagher-Thompson et al., 2012 for comprehensive reviews of this body of work). It is noteworthy that this program has been effectively tailored and translated into Spanish, Chinese, Japanese, and Farsi and then evaluated in several additional studies (Tzuang & Gallagher-Thompson, 2014).

From *Cognitive Appraisal Theory* also came the *Resources for Enhancing Alzheimer's Caregiver Health* (REACH I and REACH II) collaborative research initiatives (Schulz et al., 2003). This body of work began with REACH I, a multisite research program (six sites) sponsored by the National Institute on Aging (NIA) and the National Institute on Nursing Research (NINR). Its main purpose was to test the effectiveness of a variety of interventions unique to each site ranging, for example, from family therapy (with and without technology enhancement) at the Miami, Florida, site to small group workshops at the Palo Alto, California, site. Each site enrolled its own sample and collected its own data. REACH II was an outgrowth of these studies: the most effective features from each successful randomized, controlled trial were incorporated into a new, multicomponent intervention that was administered using standard protocol, across five remaining sites, with the same outcome measures collected at each site. Another unique feature of REACH II was that it enrolled a large, diverse sample of dementia family caregivers – the largest to date in the USA (Belle et al., 2006). The active intervention consisted of components addressing safety, self-care, social support, emotional well-being, and management of problem behaviors. Home visits were conducted (up to nine total) to develop individualized "action plans" for each caregiver. Participation in a telephone-based support group was also offered. The control condition consisted of education about dementia, but with no home visits or support group included as part of the program. The results indicated that caregivers in the active intervention, regardless of ethnicity or race, reported greater significant, positive improvement on an omnibus measure of well-being, from pre- to post-participation as compared to those in the control condition. There was also a trend for greater delay in institutional placement over time – although this was not statistically significant due to the small samples (N's) involved.

The success of REACH II was followed by many "variations on the theme" to make the program more affordable to deliver in community-based settings. One example is the REACHing OUT program (Burgio, Gaugler, & Hilgeman, 2016) which tailored the REACH II for caregivers in the state of Alabama by shortening it and adding local features considered appropriate for that region. Other versions are detailed on the Rosalynn Carter Institute for Caregiving website http://www.rosalynncarter.org. Most recent iterations involve briefer adaptations that are more appropriate to be implemented and tested in community settings. REACH II is listed and described in detail on the National Registry of Evidence-based Programs and Practices website (http://nrepp.samhsa.gov/landing.aspx). REACHing OUT is listed as an evidence-based treatment on the National Alliance for Caregiving (http://www.caregiving.org) and Roselyn Carter Institute for Caregiving websites (http://www.rosalynncarter.org).

A similar program is called Care Partners Reaching Out (CarePRO); it was created as a derivative of REACH II by Coon et al. (2004) who modified REACH II for use in the states of Arizona and Nevada. Coon described the program as follows: "CarePRO aims to enhance the quality of life for both caregivers and their care recipients by teaching caregivers stress management and behavior management skills" (ASU Now, 2013, p. 4). With over five, in-person skills workshops and five telephone calls with an assigned coach, caregivers are taught how to identify sources of frustration, practice mindful breathing, change unhelpful thinking, schedule pleasant activities for themselves, plan for the future, and manage difficult behaviors their care recipients may exhibit. Of the over 600 caregivers from Arizona and Nevada that enrolled in this program since its inception, over 95% reported having increased understanding of dementia and confidence in providing care after completing their 10 sessions. During the launch of this program, over 40 staff members of local Alzheimer's Association chapters in both states were trained and delivered CarePRO in both Spanish and English (Coon, et al., 2016).

The *Stress-Process Model* (Pearlin et al., 1990) has also inspired a wide range of psychosocial programs and educational programs or support interventions for dementia caregivers. Typically, these include some focus on development of more adaptive coping skills to reduce distress. For example, perceived stress may be associated with unhelpful negative thinking about caregiving and the future. In that case, helping caregivers learn how to see their situation from a different perspective can increase their positive coping and reduce stress. Furthermore, the *Stress-Process Model* later led to development of multicomponent interventions that address several factors likely to cause stress, such as low engagement in positive activities and increasing social isolation. These are represented in best practice programs such as *Powerful Tools for Caregivers (PTC)* developed by Schmall and Sturdevant (2000), the *Savvy Caregiver Program* developed by Hepburn, Lewis, Sherman, and Tornatore (2003), and the New York University (NYU) multicomponent counseling program developed by Mittelman, Roth, Haley, and Zarit (2004).

Originally developed to address caregivers of adults with chronic health conditions (Schmall & Sturdevant, 2000), *Powerful Tools for Caregivers (PTC)* is grounded in Self-Efficacy Theory (Bandura, 1986). The premise of PTC is that increased self-efficacy will promote the use of coping behaviors that will, in turn, enhance self-care and well-being among caregivers (Zeiss, Gallagher-Thompson, Lovett, Rose, & McKibbin, 1999). PTC was originally targeted to caregivers of persons with Alzheimer's disease, stroke, and Parkinson's disease (Boise, Congleton, & Shannon, 2005; Kuhn, Fulton, & Edelman, 2003). Research evidence indicates that PTC improves caregivers' self-efficacy, enhances positive views about the caregiver role, increases self-care behaviors, and decreases depression, anger, and guilt (Boise et al., 2005; Kuhn et al., 2003). Recent findings also indicate that this program can be an effective resource for reducing psychological distress and objective burden among spouses caring for disabled partners (Savundranayagam, Montgomery, Kosloski, & Little, 2011).

The *Savvy Caregiver Program* was designed to train family and professional caregivers in the basic knowledge, skills, and attitudes needed to handle the challenges of caring for a family member with Alzheimer's disease and to be an effective caregiver (Hepburn et al., 2003). This intervention has been designated as one of the approved dementia training programs of the Alzheimer's Disease Demonstration Grants to States (ADDGS) program, which is funded through the U.S. Administration on Aging (Stalbaum, Wiener, & Mitchell, 2007). Research evidence demonstrate the program's effectiveness in increasing caregiver skill, knowledge, and confidence as well as reducing caregiver distress (Hepburn et al., 2003; Hepburn, Lewis, Tornatore, Sherman, & Bremer, 2007). The research findings also suggest significant positive outcomes for caregivers who participated in the program, versus those in the control group, with respect to the caregivers' beliefs about caregiving, their reactions to the behavioral symptoms of their care recipient, and their feelings of stress and burden (Hepburn, Tornatore, Center, & Ostwald, 2001; Ostwald, Hepburn, Caron, Burns, & Mantell, 1999).

The *Caregiver Counseling Intervention* developed by Mittelman et al. (2004) at the New York University (NYU) School of Medicine was originally developed for spousal caregivers of persons with Alzheimer's disease or related dementias. This psychosocial intervention provides individual and family counseling along with a support group referral and additional support as needed. Research findings indicate that the NYU intervention reduces negative appraisals of care recipients' behavioral problems, increases positive effects on multiple outcomes for spousal caregivers and persons with dementia, and postpones nursing home placement (Mittelman et al., 2004). The impact of this intervention on depression, caregiver appraisals of behavior problems, and delay in nursing home placement is long-lasting and both clinically and statistically significant. The NYU program has also been found to exert consistent effects in reducing adult child caregivers' negative reactions to disruptive problems (Gaugler, Reese, & Mittelman, 2016). This program has been recognized by the Administration on Aging as an evidence-based protocol.

The third theory discussed in the previous section, *Caregiver Identity Theory*, has also recently been translated into a best practice intervention to manage care planning for caregivers. The *Tailored Caregiver Assessment and Referral (TCARE)* best practice program aims to develop a care plan based on the assessment of caregiver's needs, burden, identity discrepancy, availability and preference for services, and change over time. Early evidence (Montgomery, Kwak, Kosloski, & O'Connell Valuch, 2011) indicates that program promotes the well-being and mental health of caregivers.

In summary, the past two decades have witnessed the development of several theoretical frameworks to understand the conditions under which family caregivers manage stress and maintain their health and positive outlook while caring for their loved ones. As a result, researchers have developed effective interventions to support family dementia caregivers. Nevertheless, the translation from theory to programs, and best practices to support family caregivers, has been limited by practical considerations such as cost, access, and availability of appropriately trained bilingual/bicultural staff. There is also some debate in the field as to the efficacy of these non-pharmacological interventions. The reviews by Gallagher-Thompson et al. (2012) and Olazarán et al. (2010) clearly indicate that psychosocial programs with strong education and skill-building components are successful in improving mental health of distressed caregivers. However, other recent systematic reviews and meta-analyses of psychosocial interventions (e.g., Gaugler & Burgio, 2016) conclude that this body of literature is limited by a lack of high-quality evidence to derive conclusions of efficacy, an absence of clarity about the type of intervention delivered, and variation in study design, outcome measures, and/or screening criteria. Readers are encouraged to draw their own conclusions; however, our opinion is that there are now several evidence-based programs for Latino caregivers – described in detail below – but that the field would benefit from creative thinking about what Latino caregivers need most, and how to provide this to them, over the course of their caregiving journey.

Interventions for Latino Caregivers

Intervention research for Latino family caregivers generally follows the models and practices developed in the caregiving literature over the previous decades and described above. However, the quantity of research studies and evidence-based programs aimed at Latinos and other ethnic and cultural minorities is increasing slowly, compared to their growing needs. In this section, we review the main theoretical models that have influenced past work, as well as the translation efforts to adapt existing programs to make them more relevant and helpful to the Latino community. This section also describes several established support services for Latino caregivers and identifies new and innovative initiatives aimed to increase support for Latino caregivers.

The Alzheimer's Disease Demonstration Grants to States, which was initiated in the early 1990s (Montgomery, Kosloski, Karner, & Schaefer, 2002), helped to establish programs which support family dementia caregivers who were not able to access services because of their geographical and cultural isolation. One of the programs directed at Latino caregivers was the *El Portal Latino Alzheimer's Project* – a dementia-specific outreach and service program targeting Latino caregivers in the Los Angeles County area in California. The project exemplifies an interorganizational, community-based collaboration established to provide dementia care services to a large, urban Latino community in Los Angeles County. Case management was included among the services offered, along with an array of coordinated services ranging from counseling, to referrals, to transportation, to support groups and respite services, both at home and at the adult care center. The key point of the project is that the *Portal Latino Alzheimer's Project* provided ethnic-sensitive services to Latinos using culturally specific outreach and service delivery strategies. Results from the evaluation of service utilization conducted by Aranda, Villa, Trejo, Ramirez, and Ranney (2003) indicated that Latinos reported a reduction in barriers to care and an increase in service utilization. This El Portal Latino Alzheimer's Project was a great model to increase the awareness of Alzheimer's disease as a public health problem. The project was also a way to enhance the dementia care capacity among Latino caregivers, as well as a way of alleviating the negative outcomes of the dementia experience.

Along with El Portal Latino Alzheimer's Project, another attempt to understand how cultural values affect Latino caregivers and other ethnic groups was the *Sociocultural Stress and Coping* model (Aranda & Knight, 1997). The model addressed normative expectations about family care and their impact on caregiving outcomes. The model pointed to "a shared common core to stress and coping in family caregivers of persons with dementia that moves from the stressor of the care recipient's behavior problems to caregivers' appraisals of caregiving as burdensome to poor physical and emotional health outcomes for caregivers" (Knight & Sayegh, 2010, p. 11). In the revised version of this *Stress-Process Model*, the role of cultural values appeared to be group specific, and those values, such as filial piety or *familism*, had a mediating influence on caregiver outcomes, mostly through coping resources such as social support and coping styles rather than through the caregivers' appraisal of burden (Fig. 13.2).

Researchers have proposed an expanded sociocultural model of stress and coping to examine the role of culturally situated factors, such as coping abilities and the caregiver's self-efficacy beliefs, as mediators of negative psychosocial outcomes among Latina and non-Hispanic White caregivers (Montoro-Rodriguez & Gallagher-Thompson, 2009). Their research findings indicate that the effects of ethnicity and other primary stressors (such as the care recipient's behavioral troubles and the caregiver's perceived health) are mediated by coping resources and appraisal of self-efficacy in managing the care, specifically on the level of burden of female caregivers. These findings indicate that both structural and cultural factors appear to affect the caregiving experience.

Additionally, it is important to consider the impact of acculturation on caregiver distress. Acculturation refers to the degree to which someone has adopted the values of the dominant culture of the country to which they have immigrated (Berry, 2003). Acculturation levels vary from community to community, from region to region, and from individual to individual. Programs designed to target specific racial/ethnic minority groups must be mindful of the different levels of acculturation and must modify their program to make it meet the expectation of that groups' collective acculturation level. Angel and Angel (2015) comment on ways to measure acculturation and stressed its complex role in family caregiving.

Many of the caregiver intervention researchers mentioned in the previous section adapted their

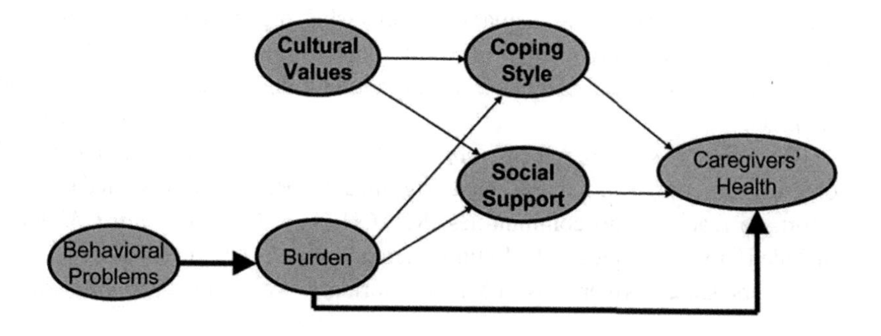

Fig. 13.2 The Revised Sociocultural Stress and Coping Model For Caregivers. (From Knight & Sayegh, (2010). Copyright 2010 by Oxford Academic. Adapted with permission)

Table 13.1 Selected evidence-based intervention programs adapted for Latino caregivers

Theory	Intervention	Cultural adaptations/translations for Hispanic/Latino dementia family caregivers
Folkman and Lazarus: Cognitive Appraisal Theory	*Coping with Caregiving (CWC)/Our Family Journey*	Spanish translations of both available on request; CWC also adapted for Chinese, Farsi, and Japanese Fotonovela: Unidos Podemos! Enfrentando la pérdida de memoria en familia Webnovela: MIRELA – How a Latino family copes
	REACH I *REACH II* *REACHing OUT*	Original REACH materials are in Spanish and English REACHing OUT, aka CALMA (Cuidadores Acompañándose y Luchando para Mejorar y Seguir Adelante) Translation of REACH II, aka CUIDAR (Cuidadores Unidos Inspirados en Dar Amor y Buscar Respuestas) Both of these were specifically adapted for San Diego County (lower literacy)
	CarePRO	Spanish translation Bilingual staff of the Alzheimer's Association trained to implement program in Arizona and Nevada
Pearlin: Stress-/Process Model	*Caregiver Counseling Intervention: NYU program*	Northern Manhattan Hispanic Caregiver Intervention Effectiveness Study, underway in 2017, to examine effectiveness of program among Latino caregivers
	The Savvy Caregiver Program (SCP)	Culturally appropriate Spanish translation Review of program effect on minority caregivers shows positive results
	Powerful Tools for Caregivers (PTC)	Has effectively served Spanish, Korean, Chinese, and Vietnamese caregivers with translated materials
Montgomery: Caregiver Identity Theory	*The Tailored Caregiver Assessment and Referral® (TCARE®)*	TCARE translated into Spanish and adapted for use with Latinos in Louisiana and state of Washington

programs to better meet the needs of Latino communities (see Table 13.1). As previously described, from Lazarus and Folkman's (1984) *Cognitive Appraisal Theory* came *Coping With Caregiving*. A Spanish translation of this program was created to make it more accessible to Latino caregivers, and the program inspired the creation of a *fotonovela* (bilingual picture book) and a *webnovela* (online telenovela) that each showed a Latino caregiving family facing challenging situations and displaying appropriate skills to care for their loved one (described in the next section). *CarePRO* is also available in Spanish and has been delivered by trained bilingual/bicultural staff of local Alzheimer's Association chapters.

The REACH studies, which were also based on cognitive appraisal, likewise made significant efforts to reach Latino communities. REACH I included two samples of Latino dementia caregivers: Cuban-Americans in Miami, Florida, and Mexican-Americans (primarily) in Palo Alto, California. Each site offered a different program: a family-therapy-based approach in Miami (Schulz et al., 2003) and a Cognitive-Behavioral Therapy (CBT) and psychoeducational intervention focusing on coping skills in Palo Alto (the *Coping with Caregiving/CWC* program; Gallagher-Thompson et al., 2003). Based on positive results from these studies and the larger REACH II investigation, the Southern Caregiver Resource Center (SCRC) in San Diego, along with several other community-based organizations and the County of San Diego, developed and implemented Spanish adaptations and translations of REACH (Gallagher-Thompson, Alvarez, et al., 2015). This *REACHing OUT* program, a new and abbreviated version of REACH I, was titled *Cuidadores Acompañándose y Luchando para Mejorar y Seguir Adelante (CALMA)*, and the adapted version of REACH II was named *Cuidadores Unidos Inspirados en*

Dar Amor y Buscar Respuestas (CUIDAR). These were tailored to the needs of lower-literacy individuals and families living in that region (Southern San Diego County, California). These programs included lay Latino community peer educators/counselors or health advocates, known as *promotores/as*, who were considered essential to successful outreach and involvement of the target community. Careful study of the impact of both interventions – CALMA for caregivers experiencing high levels of burden but whose depression was low and CUIDAR for caregivers with high levels of depression – found that *both* programs reduced caregivers' levels of burden and depression (Gallagher-Thompson, Alvarez, et al., 2015). However, since one is much less labor intensive than the other (CALMA is done in a small group format, whereas CUIDAR is done in home visits), the grouping determination suggests a careful assessment at the outset so that caregivers can be referred to the most appropriate program to meet their needs.

These and similar programs have also made efforts to better reach Latino caregivers through targeted outreach, employment of bilingual/bicultural staff, and concentrated training so that fidelity to the "parent" program(s) is maintained. *Powerful Tools for Caregivers* is known to be effective among caregivers from multiple cultural backgrounds and has served Spanish, Korean, Chinese, and Vietnamese caregivers with translated materials. The Northern Manhattan Hispanic Caregiver Intervention Effectiveness Study, underway in 2017 (Luchsinger et al., 2017), examines the effectiveness of the Caregiver Counseling Intervention program among Latino caregivers in New York City.

To help the *Savvy Caregiver Program* better serve Latino caregivers, researchers translated it into a Spanish-language program culturally appropriate to address the needs of the Latino caregivers (Oakes, Hepburn, Ross, Talamantes, & Espino, 2006). Kally et al. (2014) examined the impact of *Savvy Caregiver* among caregivers of Latino, African-American, and Asian/Pacific Islander descent and found that caregivers of all three ethnic groups showed better competence, reduced depression, greater tolerance for care recipients' memory problems, better management of their situation, and improved perception of their situation 6 months after enrollment, with the achieved improvements maintained at 12 months after enrollment in most cases. These encouraging results indicate that the *Savvy Caregiver Program* has been successful at meeting the needs of the minority groups it serves.

Finally, *TCARE*, which is based on *Caregiver Identity Theory*, has been translated into Spanish and adapted for use with Latinos in Louisiana and the state of Washington (Montgomery, 2014). *TCARE* has not been tested to examine whether outcomes for those using the Spanish version are different. However, the feedback from care managers suggests that it works well with Latino caregivers. The measure has been translated into many other languages and is used by researchers in many countries.

Caregiver Interventions Using Information and Communication Technology

Extending the stress and coping conceptual models to the Digital Age has potential because information and communication technologies are becoming effective supportive tools for caregivers. Information technology stresses the role of unified communications. The convergence of audiovisual and telephone networks with computer networks through a single cabling or link system makes possible that large number of people around the world access information via the Internet, many of them through smartphones. This rapid growth, especially in developing countries, has changed everyday life and made information technology a vital resource for many. Dementia is a global phenomenon with increasing incidence and prevalence around the world (Alzheimer Disease International, 2009). Therefore, it is not surprising that caregivers, educators, and health professionals are turning to technology to help dementia patients and their family caregivers to reduce their burden, stress, and depression and to preserve quality of life. Technology has the added benefit of reaching a large number of people and promising permanent access to those resources.

Recent culturally tailored, nontraditional caregiver interventions using a form of information technology include the development of a *fotonovela* and a *webnovela* for Latino dementia caregivers in California. The fotonovela, *Together We Can! Facing Memory Loss as a Family*, is an innovative intervention developed for Latinos at Stanford University (Gallagher-Thompson, Tzuang, et al., 2015). It is a 20-page bilingual "picture book" containing a dramatic storyline wherein Latino actors depict specific challenging scenes designed to illustrate skills for managing stress and difficult behaviors, employing adaptive coping strategies, and asking for help from other family members. This psychoeducational program provides Latino caregivers with accurate information about dementia and illustrates, in a culturally appropriate manner, constructive ways of coping with various stressful situations confronting Latino caregivers. It addresses specific needs by reaching out to Latinos with low-literacy levels and potentially less than accurate knowledge about dementia. Initial results from a study using a sample of Latino caregivers in California found that those in the *fotonovela* condition reported less depression at the conclusion of the program and found it more helpful overall, compared to caregivers in the control condition who received typical Spanish-language educational materials (Gallagher-Thompson, Alvarez, et al., 2015). The *fotonovela* is a culturally tailored program that not only impacts level of depressive symptoms in Latino caregivers by training them with effective coping strategies to deal with stress but also delivers benefits to low-income families, such as helping them to access needed resources. The webnovela, *MIRELA*, is a short, online telenovela in Spanish which is specifically designed for Spanish caregivers to teach Latino caregivers how to cope with dementia caregiving, reduce burden of care, decrease stress, and alleviate depression. MIRELA is based on the *Coping with Caregiving* program (described earlier) as a Cognitive-Behavioral, theory-based psycho-educational intervention that is both entertaining and informative. The randomized controlled trial has just finished, and preliminary analyses of data indicate that MIRELA is significantly more successful in reducing symptoms of depression and caregiving stress than the education-only control condition to which it was compared.

Other technology-based interventions developed for caregivers include the *iCare* program that can be currently found at the *iCarefamily.com* website (iCareFamily, 2010). The *iCare Stress Management e-Training* program is entirely an Internet-based adaptation of the CWC program. The initial evidence for the program indicated that change in perceived stress was significant for the *iCare* condition in comparison to the education-only group condition (Kajiyama et al., 2013). These promising results will need to be replicated and further adapted to increase caregiver's participation and engagement with the program.

The Pew Internet surveys show steady growth in the use of technology by both bilingual and Spanish-dominant Hispanics; the most recent (Brown, López, & Lopez, 2016) notes Latinos' increased dependence on access to the Internet through mobile devices, such as cellphones and tablets. Highlights of technology-based interventions over the last several years show equal expansion in the kinds of communicative and social interactions being developed for persons with dementia in collaboration with their caregivers. Such interventions join telenovelas, now being displayed on tablets and phones (Gallagher-Thompson, Alvarez, et al., 2015), and other telehealth applications (Tirado, 2011).

The research review of interventions targeting Hispanic/Latino caregivers by Llanque and Enriquez (2012) noted the negative impact of caregiving on the caregivers' health although caregivers report less stress and perform greater personal care. The authors caution that Hispanics/Latinos do not represent a cohesive group but instead are a different cultural grouping requiring different and tailored interventions. The flexibility offered by technology can potentially address this diverse cultural group. For example, *StoryCall*, a tablet-based app (*emobile* application) was originally developed on cellphones to support caregivers in any cultural grouping by sharing 60-second stories of how they overcome challenges in daily caregiving such as dispensing medications or soothing agitation. In pilot tests of this research, the caregivers of USA Veterans with dementia were identified as having a desire for maintaining personal privacy and security while also developing or expanding social and community contacts (Davis, Nies, Shehab, & Shenk, 2014). Touch screen technologies are also growing in popularity (Astell et al., 2010; Kerssens et al., 2015; Purves, Phinney, Hulko, Puurveen, & Astell, 2015). Astell et al. (2010) reported on the interactive CIRCA utility, a multimedia touch screen system that contains a wide range of stimuli to prompt reminiscing for adults with dementia. The intention is that people with dementia and caregivers will explore CIRCA together, using the recollections sparked by the media as the basis for conversations. The evaluation of the utility of CIRCA confirmed that people with dementia can use the touch screen system and that the contents prompt them to reminisce. The interactive utility also supports caregivers to interact with people with dementia as more equal participants in the conversation. The results suggest that interacting with the touch screen system is engaging and enjoyable for people with dementia and caregivers alike and provides a supportive interaction environment that positively benefits their relationships. The use of video and multimedia on *emobile* devices allows caregivers to support as well as to go beyond reminiscence (Davis, Shehab, Shenk, & Nies, 2015; García-Betances, Jiménez-Mixco, Arredondo, & Cabrera-Umpiérrez, 2015).

Finally, new interest in wearables that send signals to phones or tablets has spurred the development of interventions to help caregivers address persons with dementia (Mahoney, Coon, & Lozano, 2016), and to share caregiver perspectives (Matthews et al., 2016). Chen and Schulz (2016) offer analysis on how technology can reduce social isolation. Brando, Olmedo, and Solares (2017) review 30 studies that apply technology for diagnosis, for neuropsychological rehabilitation, and for caregiver interventions. The findings include that, while multiple efforts at providing better diagnoses are abundant, the same cannot be said for interventions involving caregivers, who are all too often left out of any analysis of impact on either caregiver or caregiver recipient. In sum, technology is easing its way into the caregiver world with some promising beginnings, but there remains much to do. The programs for caregivers using information technology are also promising, and while they are being tested for caregivers, they are still in the process of being translated and tailored to caregivers of diverse cultural backgrounds.

Community-Based Caregiver Interventions

Another promising area of intervention taking place in communities across the USA refers to the social, environmental, and contextual readiness to accommodate and support caregivers and their families. New community initiatives directed to develop *Age- and Dementia-Friendly Communities* are now gaining popularity. These initiatives are still in the piloting implementation phase, but there are increasing efforts to test their effectiveness in a variety of individual, family, and community outcomes. These initiatives are long-term commitments by organizations from different sectors to agree to a common agenda for social change and propose solutions to current social challenges. The research indicates that collective impact initiatives (as opposed to isolated social change initiatives) are more effective in bringing social change and serve more effectively in providing a unified voice for policy change (Kania & Kramer, 2013). In the United Kingdom, the Alzheimer's Society's Dementia Friends initiative was developed to improve inclusion and quality of life for people with dementia. The goal was to change people's perceptions of dementia and transform the way the community thinks, acts, and talks about the condition. Volunteers become a Dementia Friend as they learned about what it is like to live with dementia and how to do something about it. Dementia Friends are a group of people who have come together to learn more about dementia and the ways they can help, from telling friends about programs to visiting someone living with dementia. *Dementia-friendly communities* encourages communities to sign up and work toward establishing Local Dementia Action Alliances for creating a group of people who have come together to create a dementia-friendly community.

The Dementia Friendly America (DFA) initiative is the USA licensee of the international Dementia Friends program (http://www.dfamerica.org). The DFA has put in motion a movement to more effectively support and serve those across America who are living with dementia and their family and friend care partners. The lead organizations represent all sectors of community and are collectively leveraging their national reach to activate their local affiliates, members, and branches to convene, participate in, and support dementia-friendly community efforts at a local level. In Minnesota, the movement has led to the development of a variety of meaningful opportunities for community engagement and education that include memory cafés, a play in Spanish to increase awareness of dementia, an intergenerational dance hall, book displays at local libraries, a local leaders forum to stimulate community conversation around Alzheimer's disease, and a chorus for individuals with dementia and their caregivers, among other programs (LaPorte, 2016). According to the DFA, collaborative communities are encouraged to progress through four phases as they journey to become dementia friendly: convene-engage-analyze-act. An assessment of the community's needs, community dissemination of priorities, and assessment findings are a key component. The DFA tools have been successfully deployed by dozens of communities across the USA and are now present in more than 80 communities in 25 states to have joined them by the end of 2016. The group's goal is to have dementia-friendly initiatives in all 50 states by the end of 2017. The DFA initiative was acknowledged with the 2015 Public Trust Award presentation at the Leading Age Annual Meeting in Boston, Massachusetts, for its work in fostering dementia friendliness across the country. Certainly, we expect that these initiatives will be culturally tailored interventions.

Summary and Future Research

The theories of family caregiving reviewed in this chapter have been a source of inspiration to develop support services and programs for caregivers and their families. Evidence-based interventions have been developed based on these theories, and some of them have been adapted and translated into culturally appropriate programs for Latino caregivers of elderly patients in the USA and across the world. Although great progress has been made, it is necessary to advance the understanding of the characteristics of caregivers for whom specific programs are/are not effective so that the best and most appropriate recommendations regarding programs be made for each individual. For example, caregivers suffering from significant depression or anxiety may benefit most from a more counseling-oriented approach, whereas those with less stress may benefit from psychoeducational programs, and those with very mild distress may be helped sufficiently with technology-based programs. Alternatives to pharmacological treatments, such as these, need to continue to be developed and tested with diverse caregivers.

In considering the progress made during the past years and looking at what needs to be done to increase support programs for dementia caregivers in the Latino community, we are encouraged by the many efforts directed to develop culturally tailored, evidence-based programs. However, we believe we still face a real challenge to develop evidence-based interventions more specifically directed to Latino caregivers – not just "derivatives" of programs developed primarily for middle-class Caucasians. Such interventions need to identify the main barriers that limit access to social and medical health services and address the main challenges associated with Latinos' lower socioeconomic status, higher level of health problems, and their struggle with their own cultural values in the midst of their process of acculturation.

We propose that research be directed to conceptualize the interactive nature of family caregiving for Latino individuals, their families, and communities, using not only linear models addressing causal relations between stress and caregiving outcomes but also through the lenses of a system-dynamic perspective that builds upon current stress and coping frameworks and the empirical findings in the literature. This approach would illustrate how changes in one factor can have wide-ranging and reciprocal impacts on other factors (Richardson, 2013). The goal in a system-dynamics model is to examine how relationships between endogenous variables (such as supporting programs for adults and caregivers, the availability of family and caregiver resources, etc.) can explain system behavior over time (e.g., caregiver outcomes, such as physical and mental health). From a system perspective, the development of programs to support Latino caregivers and their families will entail the use of a mixed methodology approach wherein researchers working with Latino caregivers and their families and community partners in the broader Latino communities all participate in a discussion to develop a qualitative system-dynamics conceptual model that represents the reciprocal relationships that impact the caregivers' main concerns and specific needs. This conceptual model will highlight the interplay of the existing resources (human and social capital) and the Latinos' aspirations/outcomes, in the context of the larger social and organizational opportunities that contribute to support or neglect such aspirations. From a systemic perspective, the expected system outcomes for caregivers (such as well-being) can be subjected to either reinforcing (positive) or balancing (negative) impacts at any specific time. The core cultural beliefs of Latino family caregivers (such as family orientation) may impact well-being as a reinforcing factor (decreasing stress and depression) and in turn increase the ability of the family to provide quality of care for their members. On the other hand, cultural values (such as the duty to care or the expected caregiving role for women) may impact negatively the well-being of caregivers and in turn diminish their ability to seek the help they need or to maintain or increase their available network of support. A system-dynamics approach that considers the constellation of these reciprocal relationships over time may offer a promising path to better understand the complexity of the caregiving experience among Latino families.

In summary, certain cultural and psychosocial factors can serve either as buffers against stress or can greatly augment the stress that dementia family caregivers experience. Culturally tailored interventions are recommended for use with diverse caregivers in different contexts, such as in-home and community settings, to help in the effective management of stress and depression. Developing community partnerships is encouraged as are collaborations to better engage and initiate change among low-income and underserved dementia caregivers. Research that builds upon conceptual models that incorporate the caregiver's cultural context and specific situational factors is also encouraged. Future research that specifies both moderators and mediators of evidence-based interventions among diverse caregivers will be invaluable to the field. Clearly, this is an expanding area of research, and it is expected that there be an increase in the number of empirical studies showing the effectiveness of various interventions for dementia caregivers among Latinos in the USA.

References

Alzheimer's Association. (2017). Alzheimer's disease facts and figures. *Alzheimer's & Dementia, 13*, 325–373.

Alzheimer's Disease International. (2009). *Caring for people with dementia around the world: Factsheet 5*. Retrieved from http://www.Alz.co.uk/adi/pdf/5caring.pdf

Angel, R. J., & Angel, J. L. (2015). *Latinos in an aging world*. New York, NY: Routledge.

Apesoa-Varano, E. C., Gomez, Y., & Hinton, L. (2015). Dementia informal caregiving in Latinos: What does the qualitative literature tell us? In W. A. Vega, K. S. Markides, J. L. Angel, & F. M. Torres-Gil (Eds.), *Challenges of Latino Aging in the Americas* (pp. 141–169). New York, NY: Springer Publishing.

Aranda, M. P., & Knight, B. G. (1997). The influence of ethnicity and culture on the caregiver stress and coping process: A socio-cultural review and analysis. *The Gerontologist, 37*(3), 342–354.

Aranda, M. P., Villa, V. M., Trejo, L., Ramirez, R., & Ranney, M. (2003). El portal Latino Alzheimer's project: Model program for Latino caregivers of Alzheimer's disease-affected people. *Social Work, 48*(2), 259–271.

Arevalo-Flechas, L. C., Martinez, M., & Flores, B. (2017). Caregiving for grandchildren and dementia caregivers: Not that different among Hispanics. *Innovation and Aging, 1*(Suppl 1), 465.

Astell, A. J., Ellis, M. P., Bernardi, L., Alm, N., Dye, R., Gowans, G., & Campbell, J. (2010). Using a touch screen computer to support relationships between people with dementia and caregivers. *Interacting with Computers, 22*(4), 267–275.

ASU Now. (2013). *ASU research earns prestigious Rosalynn Carter Leadership in Caregiving Award.* Retrieved from: https://www.newswise.com/articles/asu-researcher-earnsprestigious-rosalynn-carter-leadership-in-caregiving-award

Bandura, A. (1986). The explanatory and predictive scope of self-efficacy theory. *Journal of Social and Clinical Psychology, 4*(3), 359–373.

Belle, S. H., Burgio, L., Burns, R., Coon, D., Czaja, S. J., Gallagher-Thompson, D., … Zhang, S. (2006). Enhancing the quality of life of dementia caregivers from different ethnic or racial groups: A randomized, controlled trial. *Annals of Internal Medicine, 145*(10), 727–738.

Berry, J. W. (2003). Conceptual approaches to understanding acculturation. In K. M. Chun, P. B. Organista, & G. Marín (Eds.), *Acculturation: Advances in theory, measurement, and applied research* (pp. 17–38). Washington, DC: American Psychological Association.

Boise, L., Congleton, L., & Shannon, K. (2005). Empowering family caregivers: The powerful tools for caregiving program. *Educational Gerontology, 31*, 573–586.

Brando, E., Olmedo, R., & Solares, C. (2017). The application of technologies in dementia diagnosis and intervention: A literature review. *Gerontechnology, 16*(1), 1–11.

Brown, A., López, G., & Lopez, M. (2016) *Hispanics and mobile access to the internet.* Retrieved from http://www.pewhispanic.org/2016/07/20/3-hispanics-and-mobile-access-to-the-internet/

Burgio, L. G., Gaugler, J. E., & Hilgeman, M. M. (2016). *The spectrum of family caregiving for adults and elders with chronic illness.* New York, NY: Oxford University Press.

Chen, Y.-R., & Schulz, P. J. (2016). The effect of information communication technology interventions on reducing social isolation in the elderly: A systematic review. *Journal of Medical Internet Research, 18*(1), e18. https://doi.org/10.2196/jmir.4596

Coon, D. W., Besst, D. A., Doucet, J. S., Chavez, A., Fenzi, M., Raach, K., … Hirsch, S. (2016). CarePRO: Embedding an evidence-based intervention for caregiver empowerment. *Arizona Geriatrics Society, 22*(2), 9–13.

Coon, D. W., Keaveny, M., Valverde, I. R., Dadvar, S., & Gallagher-Thompson, D. (2012). Evidence-based psychological treatments for distress in family caregivers of older adults. In F. Scogin & A. Shah (Eds.), *Making evidence-based psychological treatments work with older adults* (pp. 225–284). Washington, DC: American Psychological Association. https://doi.org/10.1037/13753-007

Coon, D. W., Rubert, M., Solano, N., Mausbach, B., Kraemer, H., Arguëlles, T., … Gallagher-Thompson, D. (2004). Well-being, appraisal, and coping in Latina and Caucasian female dementia caregivers: Findings from the REACH study. *Aging and Mental Health, 8*(4), 330–345.

Davis, B., Nies, M., Shehab, M., & Shenk, D. (2014). Developing a pilot e-mobile app for dementia caregiver support: Lessons learned. *Online Journal of Nursing Informatics (OJNI), 18*(1).

Davis, B. H., Shehab, M., Shenk, D., & Nies, M. (2015). E-mobile pilot for community-based dementia caregivers identifies desire for security. *Gerontechnology, 13*(3), 332–336.

Folkman, S. (2008). The case for positive emotions in the stress process. *Anxiety, Stress & Coping: An International Journal, 21*(1), 3–14.

Gallagher-Thompson, D., Alvarez, P., Cardenas, V., Tzuang, M., Velazquez, R. E., Buske, K., & Tilburg, L. V. (2015). From ivory tower to real world: Translating an evidence-based intervention for Latino dementia family caregivers into a community settings. In L. Weiss Roberts, D. Reicherter, S. Adelsheim, & S. V. Joshi (Eds.), *Partnerships for mental health* (pp. 105–123). New York, NY: Springer.

Gallagher-Thompson, D., Coon, D. W., Solano, N., Ambler, C., Rabinowitz, Y., & Thompson, L. W. (2003). Change in indices of distress among Latina and Anglo female caregivers of elderly relatives with dementia: Site specific results from the REACH national collaborative study. *The Gerontologist, 43*(4), 580–591.

Gallagher-Thompson, D., Ossinalde, C., & Thompson, L. W. (1996). *Coping with caregiving: A class for family caregivers.* Palo Alto, CA: VA Palo Alto Health Care System. (Note: This refers to a new English language manual for leaders and class participants).

Gallagher-Thompson, D., Tzuang, M., Au, A., Brodaty, H., Charlesworth, G., Gupta, R., … Shyu, Y.-I. (2012). International perspectives on nonpharmacological best practices for dementia family caregivers: A review. *Clinical Gerontologist, 35*(4), 316–355.

Gallagher-Thompson, D., Tzuang, M., Hinton, L., Alvarez, P., Rengifo, J., Valverde, I., & Thompson, L. W. (2015). Effectiveness of a fotonovela for reducing depression and stress in Latino dementia family caregivers. *Alzheimer Disease and Associated Disorders, 29*(2), 146–153.

García-Betances, R. I., Jiménez-Mixco, V., Arredondo, M. T., & Cabrera-Umpiérrez, M. F. (2015). Using virtual reality for cognitive training of the elderly.

American Journal of Alzheimer's Disease and Other Dementias, 30(1), 49–54.

Gaugler, J. E., & Burgio, L. D. (2016). Caregiving for individuals with Alzheimer's disease and related disorders. *The Spectrum of Family Caregiving for Adults and Elders with Chronic Illness*, 15–57.

Gaugler, J. E., Reese, M., & Mittelman, M. S. (2016). Effects of the Minnesota adaptation of the NYU caregiver intervention on primary subjective stress of adult child caregivers of persons with dementia. *The Gerontologist, 56*(3), 461–474.

Hebert, L. E., Weuve, J., Scherr, P. A., & Evans, D. A. (2013). Alzheimer disease in the United States (2010–2050) estimated using the 2010 Census. *Neurology, 80*, 1778–1783.

Hepburn, K., Lewis, M., Tornatore, J., Sherman, C. W., & Bremer, K. L. (2007). The Savvy Caregiver Program: The demonstrated effectiveness of a transportable dementia caregiver psychoeducation program. *Journal of Gerontological Nursing, 33*(3), 30–36.

Hepburn, K. W., Lewis, M., Sherman, C. W., & Tornatore, J. (2003). The Savvy Caregiver Program: Developing and testing a transportable dementia family caregiver training program. *The Gerontologist, 43*(6), 908–915.

Hepburn, K. W., Tornatore, J., Center, B., & Ostwald, S. W. (2001). Dementia family caregiver training: Affecting beliefs about caregiving and caregiver outcomes. *Journal of the American Geriatrics Society, 49*(4), 450–457.

Hinton, L., Chambers, D., Velásquez, A., Gonzalez, H., & Haan, M. N. (2006). Dementia neuropsychiatric symptom severity, help-seeking, and unmet needs in Sacramento Area Latino Study on Aging (SALSA). *Clinical Gerontologist, 29*(4), 1–16.

Holmes, T. H., & Rahe, R. H. (1967). The social readjustment rating scale. *Journal of Psychosomatic Research, 11*, 213–218.

ICareFamily. (2010). *ICare: Stress management training for dementia caregivers*. Video.

Kahana, E., Kahana, B., Johnson, J. R., Hammond, R. J., & Kercher, K. (1994). Developmental challenges and family caregiving: Bridging concepts and research. In E. Kahana, D. E. Biegel, & M. L. Wykle (Eds.), *Family caregiving across the lifespan* (pp. 3–41). Thousand Oaks, CA: Sage.

Kajiyama, B., Thompson, L. W., Eto-Iwase, T., Yamashita, M., Di Mario, J., Tzuang, Y. M., & Gallagher-Thompson, D. (2013). Exploring the effectiveness of an internet-based program for reducing caregiver distress using the iCare Stress Management e-Training Program. *Aging and Mental Health, 17*, 544–554.

Kally, Z., Cote, S. D., Gonzalez, J., Villarruel, M., Cherry, D. L., Howland, S., … Hepburn, L. (2014). The Savvy Caregiver Program: Impact of an evidence-based intervention on the well-being of ethnically diverse caregivers. *Journal of Gerontological Social Work, 57*(6–7), 681–693.

Kania, J. V., & Kramer, M. R. (2013). Embracing emergence: How collective impact addresses complexity. *Stanford Social Innovation Review*. Retrieved from https://ssir.org/articles/entry/embracing_emergence_how_collective_impact_addresses_complexity

Kerssens, C., Kumar, R., Adams, A. E., Knott, C. C., Matalenas, L., Sanford, J. A., & Rogers, W. A. (2015). Personalized technology to support older adults with and without cognitive impairment living at home. *American Journal of Alzheimer's Disease and Other Dementias, 30*(1), 85–97.

Knight, B. G., & Sayegh, P. (2010). Cultural values and caregiving: The updated sociocultural stress and coping model. *The Journals of Gerontology: Series B: Psychological Sciences and Social Sciences, 65*(1), 5–13.

Kuhn, D., Fulton, B., & Edelman, P. (2003). Powerful tools for caregivers: Improving self-care and self-efficacy of family caregivers. *Alzheimer's Care Quarterly, 7*(4), 189–201.

LaPorte, M. (2016). *Building a dementia-friendly movement*. Retrieved from https://changingaging.org/dementia/building-dementia-friendly-movement/

Lazarus, R. S., & Folkman, S. (1984). *Stress, appraisal, and coping*. New York, NY: Springer Publishing Company.

Llanque, S. M., & Enriquez, M. (2012). Interventions for Hispanic caregivers of patients with dementia: A review of the literature. *American Journal of Alzheimer's Disease and Other Dementias, 27*(1), 23–32.

Luchsinger, J. A., Burgio, L., Mittelman, M., Dunner, I., Levine, J. A., Knog, J., … Teresi, J. A. (2017). Northern Manhattan Hispanic Caregiver Intervention Effectiveness Study: Protocol of a pragmatic randomized trial comparing the effectiveness of two established interventions for informal caregivers of persons with dementia. *BMJ Open, 6*, e014082. https://doi.org/10.1136/bmjopen-2016-014082

Mahoney, D. F., Coon, D. W., & Lozano, C. (2016). Latino/Hispanic Alzheimer's caregivers experiencing dementia-related dressing issues: Corroboration of the Preservation of Self model and reactions to a "Smart Dresser" computer-based dressing aid. *Digital Health, 2*. https://doi.org/10.1177/2055207616677129

Matthews, J. T., Campbell, G. B., Hunsaker, A. E., Klinger, J., Mecca, L. P., Hu, L., … Lingler, J. H. (2016). Wearable technology to garner the perspective of dementia family caregivers. *Journal of Gerontological Nursing, 42*(4), 16–22.

Mausbach, B. T. (2014). Caregiving. *The American Journal of Geriatric Psychology, 22*, 743–745.

Mittelman, M. S., Roth, D. L., Haley, W. E., & Zarit, S. H. (2004). Effects of a caregiver intervention on negative caregiver appraisals of behavior problems in patients with Alzheimer's disease: Results of a randomized trial. *Journals of Gerontology – Series B Psychological Sciences and Social Sciences, 59*(1), 27–34. https://doi.org/10.1093/geronb/59.1.P27

Montgomery, R. V. (2014). *Has the use of Tailored Caregiver Assessment and Referral® System impacted the well-being of caregivers in Washington?* Report to the Washington Aging and Long-Term Support Administration. Retrieved from http://leg.wa.gov/

JointCommittees/ADJLEC/Documents/2014-05-19/Tailored_Caregiver_Assess_Ref_System_Impact_Report.pdf

Montgomery, R. V., & Kosloski, K. (2009). Caregiving as a process of changing identity: Implications for caregiver support. *Generations, 33*(1), 47–52.

Montgomery, R. V., & Kosloski, K. (2013). Pathways to a caregiver identity and implications for support services. In R. C. Talley, R. V. Montgomery, R. C. Talley, & R. V. Montgomery (Eds.), *Caregiving across the lifespan: Research, practice, policy* (pp. 131–156). New York, NY: Springer.

Montgomery, R. V., Kosloski, K., Karner, T. X., & Schaefer, J. P. (2002). Initial findings from the evaluation of the Alzheimer's disease demonstration grants to states programs. In R. V. Montgomery & R. V. Montgomery (Eds.), *A new look at community-based respite programs: Utilization, satisfaction, and development* (pp. 5–32). New York, NY: Haworth Press.

Montgomery, R. V., Kwak, J., & Kosloski, K. (2016). Theories guiding support services for family caregivers. In V. L. Bengtson & R. A. Settersten Jr. (Eds.), *Handbook of theories of aging* (pp. 443–462). New York, NY: Springer Publishing Company.

Montgomery, R. V., Kwak, J., Kosloski, K., & O'Connell Valuch, K. (2011). Effects of the TCARE® intervention on caregiver burden and depressive symptoms: Preliminary findings from a randomized controlled study. *The Journals of Gerontology: Series B: Psychological Sciences and Social Sciences, 66*(5), 640–647.

Montoro-Rodriguez, J., & Gallagher-Thompson, D. (2009). The role of resources and appraisals in predicting burden among Latina and non-Hispanic White female caregivers: A test of an expanded socio-cultural model of stress and coping. *Aging & Mental Health, 13*(5), 648–658.

Montoro-Rodríguez, J., Kosloski, K., Kercher, K., & Montgomery, R. V. (2009). The impact of social embarrassment on caregiving distress in a multicultural sample of caregivers. *Journal of Applied Gerontology, 28*(2), 195–217.

Oakes, S. L., Hepburn, K., Ross, J. S., Talamantes, M. A., & Espino, D. V. (2006). Reaching the heart of the caregiver. *Clinical Gerontologist, 30*(2), 37–49.

Olazarán, J., Reisberg, B., Clare, L., Cruz, I., Peña-Casanova, J., del Ser, T., … Muñiz, R. (2010). Nonpharmacological therapies in Alzheimer's disease: A systematic review of efficacy. *Dementia and Geriatric Cognitive Disorders, 30*(2), 161–178. https://doi.org/10.1159/000316119

Ostwald, S. K., Hepburn, K. W., Caron, W., Burns, T., & Mantell, R. (1999). Reducing caregiver burden: A randomized psychoeducational intervention for caregivers of persons with dementia. *The Gerontologist, 39*(3), 299–309.

Pearlin, L. I., Mullan, J. T., Semple, S. J., & Skaff, M. M. (1990). Caregiving and the stress process: An overview of concepts and their measures. *The Gerontologist, 30*(5), 583–594.

Pinquart, M., & Sorensen, S. (2005). Ethnic differences in stressors, resources, and psychological outcomes of family caregiving: A meta-analysis. *The Gerontologist, 45*, 90–106.

Purves, B., Phinney, A., Hulko, W., Puurveen, G., & Astell, A. (2015). Developing CIRCA-BC and exploring the role of the computer as a third participant in conversation. *American Journal of Alzheimer's Disease & Other Dementias, 30*(1), 101–107.

Richardson, G. P. (2013). Concept models in group model building. *System Dynamics Review, 29*, 42–55.

Samper-Ternent, R., Kuo, Y. F., Ray, L. A., Ottenbacher, K. J., Markides, K. S., & Al Snih, S. (2012). Prevalence of health conditions and predictors of mortality in oldest old Mexican Americans and non-Hispanic whites. *Journal of American Medical Directors Association, 13*(3), 254–259. https://doi.org/10.1016/j.jamda.2010.07.010

Savundranayagam, M. Y., Montgomery, R. V., Kosloski, K., & Little, L. T. (2011). Impact of a psychoeducational program on three types of caregiver burden among spouses. *International Journal of Geriatric Psychiatry, 26*(4), 388–396.

Schmall, V. L., & Sturdevant, M. (2000). *The caregiver help book: Powerful tools for caregiving*. Portland, OR: Legacy Health System.

Schulz, R., Burgio, L., Burns, R., Eisdorfer, C., Gallagher-Thompson, D., Gitlin, L. N., & Mahoney, D. F. (2003). Resources for enhancing Alzheimer's caregiver health (REACH): Overview, site-specific outcomes, and future directions. *The Gerontologist, 43*(4), 514–520.

Stalbaum, L., Wiener, J. M., & Mitchell, N. (2007). *Alzheimer's disease demonstration grants to states: Cross-state report on initiatives targeting limited English speaking populations and African American communities.* Retrieved from https://www.alz.org/national/documents/aoagrant_cs_crossstate_07.pdf

Tirado, M. (2011). Role of mobile health in the care of culturally and linguistically diverse U.S. populations. *Perspectives in Health Information Management, 8*, 1–7.

Tzuang, A., & Gallagher-Thompson, D. (2014). Caring for care-givers of a person with dementia. In N. A. Pachana & K. Laidlaw (Eds.), *The oxford handbook of clinical geropsychology* (pp. 797–836). New York, NY: Oxford University Press.

Wilson, R. S., Segawa, E., Boyle, P. A., Anagnos, S. E., Hizel, L. P., & Bennett, D. A. (2012). The natural history of cognitive decline in Alzheimer's disease. *Psychology and Aging, 27*(4), 1008–1017. https://doi.org/10.1037/a0029857

Zeiss, A., Gallagher-Thompson, D., Lovett, S., Rose, J., & McKibbin, C. (1999). Self-efficacy as a mediator of caregiver coping: Development and testing of an assessment model. *Journal of Clinical Geropsychology, 5*(3), 221–230.

Grief, Loss, and Depression in Latino Caregivers and Families Affected by Dementia

14

Dinelia Rosa and Milton A. Fuentes

Abstract

The understanding of the factors associated with the caring for a person with dementia and the related caregiver burden is a key area of study, as the research clearly demonstrates that dementia affects the caregiving relatives as well as their families. Specifically, grief, loss, and depression are common emotions experienced by dementia caregivers. Latino families face greater levels of dementia caregiver burden associated with the cultural values that they subscribe to which may serve as protective factors against dementia caregiver burden or, conversely, place them at a higher risk of dementia caregiver burden. This chapter focuses on grief, loss, and depression in the caregiver(s) and families of Latinos with dementia. Specifically, we discuss Latino cultural experiences and values and illustrate how these may contribute to or ameliorate the caregiver burden. As a backdrop for this analysis, this chapter utilizes General Systems Theory as it relates to Latino families and dementia and outlines treatment options and concerns, including best practices, informal support services, and utilization barriers faced by Latino caregivers.

Both authors contributed equally to the article and are listed alphabetically by first name. We wish to acknowledge the following research assistants from the Clinical and Community Studies Laboratory at Montclair State University for their assistance with the literature review: Richard Avila, Blenda Alexandre, Jojo Hickson, Nour Omran, Drew Weinstein, and Agnieszka Wozniak.

D. Rosa (✉)
Teachers College, Columbia University, New York, NY, USA
e-mail: rosa@tc.columbia.edu

M. A. Fuentes
Montclair State University, Upper Montclair, NJ, USA
e-mail: fuentesm@mail.montclair.edu

Introduction

The focus of this chapter is to address some of the emotional experiences of Latino dementia caregivers including grief, loss, and depression. Highlighted are various cultural aspects such as immigration, acculturation, and cultural values unique to Latinos in the United States of America (USA) such as *familismo* (familism), *marianismo* (marianism), and *machismo* (machismo) that may influence the emotional experiences of Latino dementia caretakers. The first section of this chapter addresses the concepts of caregiving and caregiver burden, emphasizing the role of stress in the caregiver experience, followed by how caregiver burden is understood for many Latino caregivers. This chapter then addresses the experience of loss, grief, and depression among Latino caregivers, highlighting how culturally relevant aspects may contribute to or

H. Y. Adames, Y. N. Tazeau (eds.), *Caring for Latinxs with Dementia in a Globalized World*,
https://doi.org/10.1007/978-1-0716-0132-7_14

ameliorate the stress and care burden. This chapter focuses on interventions that may contribute to the well-being of Latino caretakers, using as a backdrop General Systems Theory and its application to Latino families as it relates to caregiver burden. Interventions relevant to Latino individuals, families, and groups are provided. Finally, this chapter concludes with author observations, remarks, and recommendations for improving the emotional status of Latinos in their caregiving experience.

Latino Caregivers and Caregiver Burden for Dementia

A person who provides care to people who need some degree of ongoing assistance with everyday tasks and provides this care on a regular or daily basis is often referred to as a caregiver (Centers for Disease Control [CDC], 2016). Approximately 15.7 million adult family caregivers care for someone who has Alzheimer's disease (AD) or other dementias (Alzheimer's Association, 2015). This is in addition to formal caregiving through institutions and nursing homes. Latinos have the highest reported prevalence of caregiving at 21%, and together with African-American caregivers, Latinos experience increased burdens from caregiving and spend more time caregiving on average than their White or Asian-American peers (Family Caregiver Alliance, 2015).

Caregivers frequently suffer from depression, exhibit maladaptive coping strategies, and express concerns about their poor quality of life (Galvis & Cerquera-Córdoba, 2016; Molyneux, McCarthy, McEniff, Cryan, & Conroy, 2008; Papastavrou, Kalokerinou, Papacostas, Tsangari, & Sourtzi, 2007; Serrano-Aguilar, Lopez-Bastida, & Yanes-Lopez, 2006). The aforementioned are experiences associated with caregiver burden, which is a multidimensional response to the negative appraisal and perceived stress resulting from taking care of an individual who is ill (Adelman, Tmanova, Delgado, Dion, & Lachs, 2014). This multidimensional response to perceived stress threatens the physical, psychological, emotional, and functional health of caregivers (Carretero, Garces, Rodenas, & Sanjose, 2009; Etters, Goodall, & Harrison, 2008; Parks & Novielli, 2000).

Kim, Chang, Rose, and Kim (2012) examined the multidimensional predictors of caregiver burden in caregivers of individuals with dementia by using nationally representative data. The findings showed that disease-related factors were the most significant predictors, explaining 16% of caregiver burden. These factors were followed by caregiver sociodemographic factors and caregiving-related factors, such as activities of daily living (ADLs), the number of hours of caregiving, use of coping strategies, and caregiver gender. The findings showed that impaired functioning in care recipients predicted caregiver burden and also interacted with demographic and caregiving-related factors (Kim et al., 2012). This study confirmed the findings of a previous study (Etters et al., 2008) in which a comprehensive literature review was conducted, examining the years 1996 through 2006, that identified the factors that influence dementia-related caregiver burden, examined patient and caregiver characteristics associated with caregiver burden, and highlighted evidence-based interventions designed to lessen caregiver burden. The findings showed that dementia caregiving was associated with negative effects on caregiver health. Additionally, factors such as gender, relationship to the patient, culture, and personal characteristics influenced the impact of the caregiving experience (Etters et al., 2008). Other studies have found that spousal caretakers of patients with dementia were far more likely to be diagnosed with depression than spouses of those who did not have dementia (Cooper, Balamurali, & Livingston, 2007; Joling et al., 2010).

Neuroscientific evidence also supports the relationship between stress and negative physical and emotional health. It has been found that elevated stress increases cortisol levels (de Vugt et al., 2005). Furthermore, elevated cortisol levels can predispose caregivers to negative health consequences, leaving caregivers of patients with behavioral and psychological symptoms of dementia (BPSD) at greater risk (de Vugt et al.,

2005). Another cortisol test, hair cortisol concentrations (HCC) proposed as a promising endocrine marker of chronic psychological stress, was used to further explore the biological impact of caregiving burden (Stalder et al., 2014). The results showed elevated HCC in dementia caregivers compared to non-caregiver controls. Caregivers showed a positive association of HCC with self-reported caregiving burden and depression.

More recently, studies have explored other areas related to Latino dementia caregiver burden. A study focusing on resilience among Latino dementia caregivers found that while barriers such as lack of understanding of dementia, mistrust in services, perception of inflexibility in services, and caregiver's belief about their obligations to the caregiver role contributed to the caregiver burden, there were also facilitators to the caregiving role (Macleod, Tantagelo, McCabe, & You, 2017). Factors such as having good communication with the care recipient, having an "expert" (i.e., professional) as point of contact, and having beliefs about the caregiving role that enabled the use of services were all facilitators that reduced the caregiving burden (Macleod et al., 2017). Although studies exploring dementia caregiving burden among Latinos are still somewhat limited, existing studies will be reviewed in subsequent sections of this chapter.

Loss and Grief Among Caregivers

Grief is a normal adaptive response to a broken affectionate bond which causes a loss (Bowlby, 1961, 1969/1982). This adaptive response varies based on the environment and psychological composition of the person grieving (Bowlby, 1961; Parkes, 1972). Grief is part of the normal process of reacting to a loss and can be experienced as a mental, physical, social, or emotional reaction or a combination of all of these (National Institute of Health, 2009). Various theorists have developed different models of grief – all consisting in distinctive stages experienced by the person who grieves (Bowlby, 1961; Kübler-Ross, 1969; Stroebe & Schut, 1999; Worden, 1983). One of the first models about grief was developed by Bowlby (1961), and it stated that stage one of the grief process is characterized by numbness, shock, and disbelief. During the second stage, the person experiences pain, despair, and disorganization. Finally, in stage three, the person feels more hopeful and positive and develops ways of adapting to the new reality of life. With stage three, there is a resolution to reorganize one's life, to start anew, and to build new relationships (Bowlby, 1961). Many other models followed with similar stages including those of Kübler-Ross (1969), Worden (1983), and Stroebe and Schut (1999).

Dementia caregivers experience loss and grief throughout and beyond the caregiving experience. A study exploring family caregiver perceptions of the experience of loss and grief, as it occurs prior to and following the death of a relative with dementia, identified six themes that emerged among caregivers: loss of person and the relationship, loss of hope, pre-death grief, expectancy of death, post-death relief, and caregiving reflections (Collins, Liken, King, & Kokinakis, 1993). A previous study supported similar themes (Austrom & Gendrie, 1990). Loss and grief can become increasingly salient among Latino dementia caregivers due to their personal experiences with immigration. What follows is a discussion of how the experience of immigration, so common among Latinos in the USA, can have a direct impact on the sense of loss and grief of Latino dementia caregivers.

Latino Caregivers: The Immigration Experience of Loss and Grief

Feelings of loss and grief among Latino caregivers in the USA are influenced by their immigration experience. For many Latino caregivers who are also immigrants, the experience of immigration itself entails loss. This loss is experienced at the physical, psychological, and emotional levels (Falicov, 2005, 2014). For immigrants, there is a sense of loss of the motherland, of physical contact with homeland loved ones, and of experiencing direct support from family and friends. Individuals can experience a loss of everything that was known in their homeland environment that helped them maintain a daily rhythm of life's routines such as working a job, going to school, attending worship services, and using their common language. Subsequently, these losses bring about emotional and psychological reactions (Arredondo-Dowd, 1981; Falicov, 2005). Often, immigrants do not have the time or mental space to emotionally process the losses or to talk about their unique immigration narrative/story due to the demands that they face in the new country (Falicov, 2014; Perez, 2016). According to Falicov (2014), "There is no formal structure, no designated 'sacred' place or time, no cultural collective celebration that allows people to come together to mark the migration and to provide a container for the strong emotions everybody is feeling" (Falicov, 2014, p. 79). Therefore, these losses often go unspoken and emotionally unprocessed.

Arredondo-Dowd (1981) used the three stages of grief posited in Bowlby's (1961) grief model to compare the process of losses and grief associated with immigration to the process of grief and mourning precipitated by the death of a loved one. Arredondo-Dowd (1981) stated that, in the immigrant experience, the immigrant first feels a sense of disbelief that he/she has actually left loved ones behind whom they no longer will see on a regular basis (Arredondo-Dowd, 1981). Later on, the immigrant experiences the sense of loss and becomes homesick and makes references to the homeland, now filled with idealization and longing in that referencing. As the immigrant begins the process of adaptation to the new country, new support systems emerge, and there is an acceptance of the new life with potentially increasing identification with the American life (Arredondo-Dowd, 1981). However, unresolved feelings of loss associated with the separation from the motherland and separation from significant family members may surface in unexpected ways when the individual is faced with a new role such as taking care of an ill family member (Vazquez & Rosa, 2011), as in the case of dementia caregivers. The need for accessible support systems, those that were left behind during immigration, becomes salient during these times of caregiving. Often, Latino immigrants who are caregivers for loved ones with dementia are at double risk of caregiver burden, as they experience an overlap of loss and separation feelings (Lorant & Thomas, 2008; Lum & Vanderaa, 2010). This happens first due to the losses and unresolved grief of the immigration experience and second by the experience of loss and grief that occurs with the cognitive, emotional, and physical decline of the family member with dementia.

As the life of the person with dementia changes because of the disease, the caregiver experiences a sense of loss. As compared to how the person with dementia was previously functioning in life, the progressive deterioration of memory, communication, social, and interpersonal skills for the person with dementia, along with their increasing lack of independence, negatively affects the caregiver's sense of connectedness. The caregiver experiences a sense of disconnection between who the person is in present time and with the internalized "object representation" of the person prior to the onset of dementia. According to Bowlby (1961), object representation is the mental representation of an experience with a real person. Thus, there is a realization that the current reality does not match this internalized object representation of who the person with dementia was before the onset of dementia. This process can bring on a sense of emotional, mental, physical, and spiritual loss and grief for the person with dementia, as well as for the caregiver (Vazquez & Rosa, 2011). For Latino immigrants who are also caregivers, the experience of loss and grief for who the loved one was to who they are, as they reach late stages of dementia, can be exacerbated by unresolved feelings from their previous histories of loss and grief related to their immigration experiences, thus adding to the burdened emotional experience. Many times, this can lead to depression.

Latino Caregivers and Depression

Although grief is considered a normal reaction to a loss, many times, prolonged grief can lead to temporary or clinical depression (Chiambretto, Moroni, Guarnerio, Bertolotti, & Prigerson, 2010; Zisook & Shear, 2009). Some studies support the idea that caregivers may experience clinical depression due to their role (Collins et al., 1993; Lindgren, Connelly, & Gaspar, 1999). Other studies have suggested that the symptoms of depression experienced by caregivers and described in the literature are more consistent with "anticipatory grief" or emotional reactions that occur before an impending loss as opposed to clinical depression (Walker & Pomeroy, 1996). However, there is agreement that caregivers are at a higher risk for depression (Chiambretto et al., 2010; Zisook & Shear, 2009).

Latinos caring for persons with dementia have consistently reported higher levels of depression as compared to European-American caregivers (Lawton, Kleban, Moss, Rovine, & Glicksman, 1989; Menselson, Rehkopf, & Kubzansky, 2008). In 2003, Hinton, Haan, Geller, and Mungas explored the frequency and intensity of neuropsychiatric symptoms of the person with dementia and the factors that modify their association with caregiver depression among Latino families. The results indicated that the overall intensity of neuropsychiatric symptoms of the person with dementia was significantly associated with caregiver depression (Hinton, Haan, Geller, & Mungas, Hinton, Haan, Geller, & Mungas, 2003). Cultural values have been shown to influence the expression of emotional issues including loss, grief, and depression (Adames, Chavez-Dueñas, Fuentes, Salas, & Perez-Chavez, 2014). These values are discussed in the next section of this chapter.

Latino Cultural Values Defined

While loss and grief are universal feelings that transcend cultures, the way in which these universal feelings are expressed varies among cultures (Vazquez & Rosa, 2011). The role of cultural values among dementia caregivers has been recently explored among various cultural and ethnic groups (Ajay, Kasthuri, Kiran, & Malhotra, 2017; Bentwich, Dickman, & Oberman, 2017; Lin, Wang, Pai, & Ku, 2017; Liu et al., 2017; Powers & Whitlatch, 2016). Among Latinos, there are specific cultural elements and values that influence the experience and display of their loss and grief (Vazquez & Rosa, 2011). Latino cultural values such as *familismo* (familism), *marianismo* (marianism), and *machismo* (machismo) can certainly inform the understanding of these emotional expressions. Although there are other cultural values of great importance, the aforementioned have specific relevance to the topic of this chapter. What follows is a brief description of the values mentioned and how they may serve as protective factors or as challenges to the caregiver burden that is typically found among Latino dementia caregivers.

Familismo (familism) refers to the cultural value that involves individuals' strong identification with their nuclear and extended families and strong feelings of loyalty, reciprocity, and solidarity among members of the same family. Considered one of the most important cultural values among Latinos, it entails a strong value to close relationships, interdependence, cohesiveness, and cooperation among family members (Santiago-Rivera, Arredondo, & Gallardo-Cooper, 2002). *Familismo* can be a source of support during times of crisis (Santiago-Rivera et al., 2002). It is a strong sense of identification with and loyalty to nuclear and extended family, and it includes a sense of protection of familial honor, respect, and cooperation among family members. Through these values, individuals place their family's needs over their own personal desires and choices.

Familismo is consistent with the collectivistic orientation typical of Latino cultures. A collectivist culture values the needs of a group or a community over the needs of the individual, wherein kinship, family, and community are extremely important (Hui & Triandis, 1986). In these instances, people's tendency to work together to create harmony and group cohesion is extremely valued. In collectivist cultures, most people's social behavior is largely determined by goals, attitudes, and values that are shared with some collectivity (group of persons) (Triandis, 1988).

The collectivistic traditional family welcomes the support of friends or outside members after they have been "legitimized" as part of the family. For example, a close family friend may be considered as a relative and even addressed as an "uncle" or "aunt" who participates as a family member in the caregiving role. Latino families may welcome a caregiver system in which various family members participate in the caretaking, and family members will "take turns" in caregiving based on flexibility and availability. The family members may give emotional support and advice to a burdened caretaker. They also may encourage the affected relative in the help-seeking process from either formal (e.g., primary care or mental health care) or informal sources (e.g., "curanderismo," i.e., an ethnomedical healer) (Ostir, Eschbach, Markides, & Goodwin, 2003).

Familismo may also reduce depression among caregivers. A recent study explored the role of *familismo* as a motivator among Latino caregivers to provide support (Villalobos & Bridges, 2016). The results indicated that *familismo* was predictive of emotional support among caregivers and reduced depression (Villalobos & Bridges, 2016). To some extent, this shared caregiving may reduce the burden placed upon one member only and serve as a buffer to the caregiving burden.

While *familismo* has shown to correlate with lesser burden in USA Latino caregivers, thus denoting the importance of specific cultural contexts influencing dementia caregiving in Latinos (Losada et al., 2006; Villalobos & Bridges, 2016), other studies have found *familismo* to be inversely associated with perceived social support among Latino dementia caregivers (Robinson-Shurgot & Knight, 2005). The same study found no significant effects in caretaker burden and was positively correlated with distress (Robinson-Shurgot & Knight, 2005). Acculturation to Western values of individualism was given as a potential explanation in a previous study with similar results (Knight et al., 2002). *Familism* can also be a source of distress when the family is not able to adequately care for an impaired older adult, there is failure to seek help from outsiders (such as specialized professionals), and there is an expectation of support from family and extended family (Gallagher-Thompson et al., 2003). Further studies are needed in this area to better understand these mixed research findings.

Latino Gender Roles Defined

Marianismo is a Latino gender role that may help explain the increase in Latina dementia caregivers as compared to male Latino caregivers. *Marianismo* was first defined by Stevens (1973a, 1973b) as the devotional following to a belief of female spiritual superiority which teaches that women are semidivine and morally superior to and spiritually stronger than men but also submissive to men. A more contemporary definition states that *marianismo* is a strong or exaggerated sense of traditional femininity, especially in some Latin American cultures, placing increased value on forbearance, self-sacrifice, nurturance, and the limiting of sex to marriage (Gil & Vazquez, 2002).

According to Gil and Vazquez (2002), Latinas with a strong sense of *marianismo* follow traditional mandates that they call the Marianista's Ten Commandments. Among these ten commandments are (1) do not forsake tradition, (2) do not place one's own needs first, (3) do not ask for help, (4) do not discuss personal problems outside the home, and (5) do not change those things which make one unhappy (i.e., that, realistically, one cannot change). A Latina following these culturally traditional *marianista* mandates may feel compelled to take care of the loved one in unconditional, self-sacrificing ways, without "complaining" to others about what is happening with the ill family member. She may also disregard considering self-care, as this would be putting her needs first. In a study exploring how women of Mexican origin conceptualized caregiving as a construct in terms of cultural beliefs, social norms, role functioning, and familial obligations, it was found that the majority of all caregivers had similar views about caregiving as an undertaking by choice, and almost all caregivers engaged in self-sacrificing actions to fulfill the *marianismo* role (Mendez-Luck & Anthony, 2016).

The expectation is that Latinas should take on the primary caregiver responsibility (Borrayo, Goldwaser, Vacha-Haase, & Hepburn, 2007). Although *marianismo* may provide caregiving relief to other family members, it is also likely to place an unequal burden on Latina caregivers. Such expectations are based on the cultural norm that women should be self-sacrificing and nurturing (marianismo) and thus not perceive caregiving as a burden (Borrayo et al., 2007; Zea, Quezada, & Belgrave, 1994). Even when the care is burdensome, the expectation will be that she can and must manage. This outlook may begin to explain why increased symptoms of depression are found among Latina caregivers (Adams, Aranda, Kemp, & Takagi, 2002; Jezzini, 2013; Polich & Gallagher-Thompson, 1997).

Machismo is interrelated with *marianismo* and is associated with the perception of the man as provider and protector of the family (Santiago-Rivera et al., 2002). A summary of the understanding of machismo from various perspectives, including historical, ethical, and contemporary, is described elsewhere (see Adames & Chavez-Dueñas, 2017). *Machismo* is not limited to positive attributes only. It has also been associated with negative behaviors, such as authoritarianism, emotional restriction, and being controlling (Torres, Solberg, & Carlstrom, 2002).

Although Latinas tend to assume the caretaker role more often, at times, Latinos assume the role. However, the influence of men in the caregiver role has not been as carefully considered. In a case study exploring the lived experience of male Latino caregivers of an ill female family member, the men expressed close family bonds and the desire to emotionally protect one another (Broeckelmann, 2015). Additionally, the Latino men found challenges in areas that were foreign to their traditional male role such as having to cook, clean, and do the dishes, among other chores. The findings indicated that male Latino family caregivers can meet role challenges with a sense of responsibility and find increased family strength from the experience (Broeckelmann, 2015). These findings suggest that the male traits of provider and protector of the family may facilitate the caretaker role among Latinos.

Other Key Latino Cultural Elements

Another relevant cultural element among Latino caregivers and their caregiver burden is acculturation. Acculturation, defined as "changes in original cultural patterns that occur as a result of ongoing contact among groups of individuals with different cultures" (Chun, Organista, & Marin, 2003, p. xxxiii), has an impact on the Latino immigrant dementia caregiver. The less acculturated the individual, the stronger the person adheres to traditional cultural values (Chun et al., 2003). This adherence can increase the tendency of a traditional caregiver to care for the loved one following customary and conventional modes. Additionally, less acculturated caregivers confront increased language barriers and limited access to care which are contributing factors to increased depression (Pinquart & Sörensen, Pinquart & Sörensen, 2005). Lower levels of acculturation are associated with limited English language proficiency and increasing barriers to access of mental health care (DuBard & Gizlice, 2008; Kirmayer et al., 2010).

Acculturation has an influence on gender roles, which also affects the Latinos' caregiver burden. Studies suggest that a caretaker's gender is the most important characteristic in determining the caregiving role, with females taking on this role more often as compared to males (Harwood et al., 2000; Torti Jr., Gwyther, Reed, Friedman, & Schulman, 2004). Among less acculturated, traditional Latinos, females are most likely to care for their parents or other family members (Harwood et al., 2000). Similarly, if blood-related females are unavailable, daughters-in-law are more likely to provide such care (Henderson & Gutierrez-Mayka, 1992). Additionally, in the Latino community, family members do not see nursing homes as an alternative due to lack of connection to existing services, language barriers, and strong cultural beliefs that elders are not to be separated from their loved ones when needed most. Instead, daughters are expected to become the primary caregivers (Crist, 2002). Another barrier to Latino caregivers' access to nursing homes and other services available in the community is the stigma associated with mental illness, in part due to a lack of knowledge regarding mental health disorders and treatments among the Latino communities. Studies have described this association between the stigma of mental illness and reduced access to care (Interian et al., 2010; Keyes et al., 2012).

Understanding the Latino culture and how the many cultural elements may support or hinder the well-being of dementia caregivers are fundamental steps for the provision of mental health care and for appropriate interventions for this population. Given that much of the caregiving process occurs within the context of the family system (Family Caregiver Alliance, 2015), the next section of this chapter will consider General Systems Theory and its relationship to Latino families, as well as a discussion of other related and applicable theoretical models. General Systems Theory provides a theoretical understanding consistent with the Latino collectivist worldview orientation. The theory will also be discussed in the context of culturally congruent mental health interventions that are available to support Latino caregivers and Latinos' experience of caregiving burden, grief, and depression.

General Systems Theory as Applied to Latino Caregiving Burden

The construct of family plays a vital role in Latino communities (Arredondo, Gallardo-Cooper, Delgado-Romero, & Zapata, 2014), especially when a family member has dementia (National Alliance for Caregiving and Evercare, 2008). To further understand this intersection, this portion of this chapter provides an overview of General Systems Theory and considers its application to Latino families affected by dementia. General Systems Theory was introduced by biologist Ludwig von Bertalanffy in the late 1920s and later adopted by family therapists to understand the intricacies of familial functioning (Goldenberg & Goldenberg, 2013). Within a General Systems framework, families are acknowledged as complex entities defined by interrelated relationships that are guided by the concepts of organization and wholeness (Goldenberg & Goldenberg, 2013).

A major premise of General Systems Theory is an understanding that the "whole" cannot be simply reduced to the sum of its parts, given that there must be accounting of the multiple and synergistic forces that interact and inform a family's functioning. In other words, in most families, there are typically several members involved, and their interactions inform one another in a dynamic manner so that the by-product of these interactions is multiplicative, rather than summative. These notions are especially true for Latino families where kinship extends beyond the traditional nuclear family to include extended family and close friends (e.g., godparents, in Spanish *padrinos* and *madrinas*) and spiritual leaders (Arredondo et al., 2014).

Additionally, a family's functioning is typically defined and maintained by family rules that are usually unspoken and that help regulate familial transactions and preserve stability, commonly understood as "homeostasis" (Goldenberg & Goldenberg, 2013). As previously outlined, within Latino families, these rules are often informed by relevant cultural values (e.g., collectivism, familismo, and marianismo) as well as other family dynamics (e.g., family secrets). Essentially, these cultural values and dynamics play a key role in the psychological functioning of Latino families. For example, unspoken family rules that are guided by cultural values in more traditional Latino families may include the following: "The family's needs come before the individual's needs"; "As a female, please handle the caretaking duties without protest"; "Papi's (Dad's) dementia is a private matter"; "This matter needs to be handled exclusively by the family."

A critical component of General Systems Theory is the awareness of "subsystems." Subsystems are parts of the greater system that are responsible for particular familial tasks or processes (Goldenberg & Goldenberg, 2013). Essentially, family functions are generally completed by subsystems, which are divisions within the family, such as the spousal, parental, and sibling subsystems. For example, in a Latino family affected by dementia, the two eldest children may be placed in the decision-making role. In Latino families, there may be additional subsystems that assist with familial functions, such as *madrinas* and *padrinos* (i.e., godparents/close friends) or members of the church community (Falicov, 2014). In essence, when a family encounters a challenge such as dementia, it utilizes its subsystems and its related rules to address the challenge and return to a state of "equilibrium," given that a family member with dementia has the power of a disruptive force to the family's previous status quo. Families that are considered "open" in General Systems Theory allow for influences outside of the family to affect its rules, thereby permitting for change and promoting flexibility. Fundamentally, these processes promote "discontinuity" or "incoherence," which eventually allows for a new homeostasis to be achieved at a more advanced level. For example, new information introduced to the family through its openness may lead to better communication or problem-solving. Conversely, families that are deemed "closed" preclude outside forces from informing its infrastructure, possibly leading to familial deterioration or distress (Goldenberg & Goldenberg, 2013). These processes will be informed by the family's cultural values, level of acculturation, and other factors in their background (e.g., support, strengths, previous dysfunction).

Depending on the Latina/o caregiver's background, these processes can play out in various ways. For example, the caregiver's cultural values may guide them to keep the dementia diagnosis a "secret," thus preventing external input and keeping the family system closed to potential support. However, the caregiver's connections to extended systems may mediate this process and allow outside forces to inform the dementia management process. Practitioners who interface with Latino families to address the concerns of dementia will want to consider these dynamics in their formal assessment processes with Latino caregivers and their families in order to ensure their interventions are appropriately planned.

Additionally, families are also a subsystem to a broader system, i.e., a framework embedded in Ecological Systems Theory (Boyd-Franklin, 2006; Bronfenbrenner, 1977; Falicov, 2014). This theory espouses that individuals develop in a series of dynamic and interlocking subsystems, including a microsystem (e.g., families, schools), a mesosystem (i.e., the interaction of microsystems), an exosystem (e.g., neighborhood, school board, local government), and a macrosystem (e.g., laws, cultural norms, federal government). Scholars have also posited that families are sociocultural units that are simultaneously influenced by several cultural factors, such as race and ethnicity (Boyd-Franklin, 2006; Hays, 2008; McGoldrick, Giordano, & Pearce, 2005). Additionally, Falicov's (2014) Multidimensional Ecosystemic Comparative Approach (MECA) aptly illustrates how ecological forces such as poverty, racism, the church, and neighborhoods, along with cultural variables including race, ethnicity, nationality, and language, can all affect the development and functioning of Latino families. As interventionists consider dementia's impact on a family and the related caregivers, there is a need to appreciate the complexity of these significant interweaving forces to ensure a deeper understanding of the family's presenting concerns and background. The cumulative understanding will guide the formulation and treatment process, helping to ensure an optimal outcome for all relevant stakeholders – the patient, the caretaker(s), and the family.

In summary, General Systems Theory and Ecological Systems Theory help to provide an understanding of the context in which Latino families and caregivers exist and function. These frameworks allow interventionists to conceptualize the dimensions that prevent or exacerbate caregiver burden in Latino families. The next section considers interventions that may assist families and caregivers with the challenges of caring for a Latino family member with dementia.

Family and Caregiver-Based Interventions

Dementia is a disease that may affect the entire family, as some family members are often intricately involved in the treatment process (Gonyea, López, & Velásquez, 2016). Specifically, the National Alliance for Caregiving and Evercare (2008) has observed that caregivers are a critical part of the dementia management team. They are typically charged with three major duties: (a) communicating information about the disease to the patient, (b) implementing and monitoring the treatment plan, and (c) communicating progress and concerns to the treating physicians (Albert, 1999). Given the deleterious impact that this role has on caregivers and families, researchers have turned their attention to these entities to learn more about their well-being and to ensure they receive optimal support as they engage in their critical and, at times, taxing functions. This section considers family-based interventions and individual and group-based interventions, as well as other factors that affect the treatment process. Interventionists are encouraged to bear in mind that Latina/os and their families all have their own unique backgrounds informed by the previously discussed cultural values, acculturation level, family interactions, etc. and that consideration of these variables is key to the appropriate conceptualization and planning of treatment.

Latino Family-Based Interventions

While measuring quality of life factors in individuals afflicted by a disorder and those who care for these individuals is common, few studies consider the quality of life of families in aggregate (Ducharme & Geldmacher, 2011). Given the importance of *familismo* in Latino families, this effort to do so holds great appeal. As noted earlier, grief and loss are common occurrences in families managing dementia. According to Walsh and McGoldrick (2004), "Secrecy, myths and taboos surrounding the loss interfere with mastery. When communication is blocked, the unspeakable is more likely to be expressed in dysfunctional symptoms or destructive behaviors" (p. 17).

When possible, it is preferred that interventionists be proactive and engage in preventative efforts with families to assist them in addressing concerns associated with grieving and the caregiving process (Lichtenthal & Kissane, 2008). To assist families and promote effective and sound communication and problem-solving, an adaptation of Kissane's (2000) Family Focused Grief Therapy (FFGT) model may be useful. Having emerged from the palliative care arena, this research-based preventative model may have great utility for Latino families grappling with dementia. The model encourages early intervention; provides families with education on the grieving, death, and dying process; places an emphasis on family strengths; promotes embracing diverse perspectives; facilitates open communication and the establishment of goals; and guides the problem-solving as well as the decision-making processes. The model stresses that interventionists must be mindful of cultural factors, especially the group's cultural values, when adapting and adopting this framework.

Another family-based intervention utilizes a structural family framework to evaluate the role of family functioning on the stress process of caregivers. A representative study involved the families of 181 caregivers, wherein 97 were of Cuban descent (Mitrani et al., 2006). The study revealed that family functioning concerns contributed to the stress and burden levels of caregivers. The authors concluded that family therapists should consider adopting a structural family approach with families affected by dementia. Specifically, interventions should facilitate family cohesion by designing "homework" that engages the caregivers in family activities and providing guidance around conflict resolution. In addition, the "homework" would encourage expressions of affection and levity and would assist in the reduction of anger or negativity toward the individual with dementia. An example of the assessment of family interactional patterns, in case study format, highlights how the patterns can cause or alleviate burden (Mitrani & Czaja, 2000). Since the study focused primarily on Cuban families, adaptions may be necessary for Latino families who originate from other Latin countries.

Individual Interventions: Cognitive-Behavioral Therapy (CBT)

In addition to family-based interventions, another common and effective treatment approach available for Latinos includes Cognitive-Behavioral Therapy (Benuto & O'Donohue, 2015; Pineros-Leano, Liechty, & Piedra, 2017). This therapy provides a combination of strategies, including education, skill-building, cognitive restructuring, and behavior activation (Craske, 2010). Caregivers and families utilize a number of belief systems, informed by cultural values, family rules, and personal dispositions, to guide their behavior. At times, these variables may lead to optimal functioning; at other times, they may lead to maladaptive functioning. Through Cognitive-Behavioral Therapy, caregivers and families are afforded the opportunity to examine their beliefs and worldviews and adjust them accordingly to ensure optimal functioning (Goldenberg & Goldenberg, 2013). What follows are individually based CBT interventions that have been found to be effective with caregivers.

Belle et al. (2006) evaluated the efficacy of a comprehensive, randomized, controlled intervention that was the first to include several ethnic groups (i.e., Euro-Americans, Latinos, and African-Americans). The intervention integrated a cognitive-behavioral approach that included "providing caregivers with education, skills to manage troublesome care recipient behaviors, social support, cognitive strategies for reframing negative emotional responses, and strategies for enhancing healthy behaviors and managing stress" (p. 730). While not made explicit in their implications, from a General Systems perspective, it could be surmised that improved functioning for the caretaker can enhance the family's overall positive functioning (Mitrani et al., 2006).

In examining Latino caregivers, Gallagher-Thompson et al. (2015) evaluated the effectiveness of an innovative approach involving a "fotonovela" to help Latino caregivers learn about the caregiving process while learning effective strategies for managing related stressful situations. As described by the researchers, "This newly created 20-page 'picture book' (in Spanish and English) has a dramatic story line where Latino(a) actors depict specific challenging scenes designed to illustrate key skills for managing difficult behaviors, using adaptive coping strategies, asking for help from other family members, and managing stress" (p. 149).

When compared to the control group who reviewed standard informational pamphlets, the fotonovela "picture book" group reported less depressive symptoms, reported reading and referring to the fotonovela with greater frequency, and found the book more helpful. However, the intervention did not seem to influence how the caregivers responded to behavioral problems. This cost-effective and culturally tailored intervention may have great utility for individuals with low literacy levels. Since most of the participants were of Mexican descent, it is possible that adaptations may be necessary for the applicability to Latino subgroups, e.g., Cubans, Puerto Ricans, etc.

Group Interventions: Cognitive-Behavioral Therapy (CBT)

Another variation of CBT involves shifting the treatment modality from individual-based therapy to group-based therapy. To this end, Gallagher-Thomson, Arean, Rivera, and Thompson (2001) explored the feasibility of providing a group-based, culturally adapted, psychoeducation intervention for Latino caregivers. The eight-week pilot intervention provided cognitive-behavioral skills to 73 Latino caregivers in the San Francisco Bay Area. The intervention focused on the frustration and anger that is typically associated with caregiving practices. After treatment, participants in the intervention group reported fewer depressive symptoms than those participants, while those in the waiting list group reported slightly higher depressive symptoms. Additionally, more individuals in the intervention group reported greater mastery over their angry feelings in the post-interviews than their waiting list counterparts. While the pilot study did not include random assignment or a follow-up component, these preliminary findings with Latino caregivers suggest that psychoeducation in a culturally adapted and supportive environment can help decrease depression and facilitate skill acquisition.

Arango-Lasprilla et al. (2014) examined the effectiveness of a group, cognitive-behavioral, caregiver intervention in Latin America. While the intervention was designed for caregivers in Cali, Colombia, it may have some relevance for first-generation, foreign-born Latinos living in the USA. Utilizing a culturally adapted intervention, the authors provided the experimental group with cognitive strategies (i.e., thought monitoring and restructuring) and behavioral strategies (e.g., relaxation and assertiveness training) which focused on the negative feelings associated with caregiving (e.g., anger, frustration). When compared to the control group who completed an educational program, the intervention group reported greater levels of life satisfaction and lower levels of depression and caregiver burden. However, no differences in stress levels were detected between the groups. The cultural adaptation and the group format make this intervention especially appealing for Latino caregivers, as it considers the unique cultural variables of this particular subgroup and embraces the collectivistic nature of Latinos, in general. Special attention will need to be given to adapting the intervention for Latino caregivers in the USA.

Finally, Gonyea et al. (2016) evaluated a culturally congruent, CBT group intervention for Latino Alzheimer's caregivers. In addition to receiving linguistically compatible groups facilitated by culturally competent therapists, the treatment group ($n = 33$), also known as *Círculo de Cuidado* (Circle of Care), engaged in traditional CBT activities, such as behavior activation, problem-solving, and relaxation exercises, while the control group ($n = 34$) participated primarily in educational activities, including education about Alzheimer's disease, communicating with health-care professionals, and addressing safety concerns. Compared to the control group, participants in the CBT group "reported lower neuropsychiatric symptoms in their relative, less caregiver distress about neuropsychiatric symptoms, a greater sense of caregiver self-efficacy, and less depressive symptoms over time" (p. 292). In addition to addressing the traditional areas of care (e.g., stress, depression, skills), this intervention also focused on neuropsychiatric symptom management (agitation, apathy, depression, and hallucinations) and related disruptive behaviors (wandering, verbal assaults) – symptoms that caregivers often find challenging. The majority of the participants were of Dominican or Puerto Rican descent; thus, adaptations may be necessary for other Latino groups.

Other Treatment Factors

In addition to the formal services described above, some Latino families and caregivers may seek and utilize informal and/or nontraditional support services in times of distress due to both cultural factors (Rogler & Hollingshead, 1985; Rogler, Malgady, & Rodriguez, 1989) and structural inequity concerns (Alegría et al., 2002). As described by Alternative Resource Theory (Rogler & Hollingshead, 1985; Rogler et al., 1989), individuals or ethnic minority groups who are in distress may use informal support and/or familiar social entities to resolve their problems rather than formal social supports (e.g., mental health agencies). For example, some Latinos may address their emotional distress or mental health concerns through close friends (Keefe, Padilla, & Carlos, 1979, spiritual leaders (Kane & Williams, 2000) or indigenous/folk-healing practices (Loera, Muñoz, Nott, & Sandefur, 2009). Again, cultural values such as *familismo* or cultural processes like acculturation may mediate these aforementioned practices with Latino families. For example, Villatoro, Morales, and Mays (2014) found that higher levels of behavioral *familismo* in Latinos were associated with greater use of informal support services. They concluded, "...the underutilization of mental health services...may result from the resourcefulness of supportive familial and social networks that help with the coping of emotional distress" (p. 354).

Conversely, Barrier Theory espouses that institutional and cultural obstacles may prevent Latinos from accessing or utilizing services (Kouyoumdjian, Zamboanga, & Hansen, 2003). Specifically, Alegría et al. (2002) found that some Latinos tend to underutilize specialty mental health services due to linguistic barriers, restricted access to Medicaid specialty services, differences in mental health problems detection, and subpar mental health services. Moreover, Cho, Kim, and Velez-Ortiz (2014) note that gender, education level, nativity status, and insurance eligibility may all serve as barriers to utilization, while Interian, Martinez, Guarnaccia, Vega, and Escobar (2007) note that mental health stigma may also lead to underutilization.

At times, families may need to consider placing their loved ones in a nursing home for more appropriate care. Gaugler, Kane, Kane, and Newcomer (2006) assert that Latino families struggle considerably with this decision, as it is antithetical to many of their cultural values (e.g., *familismo* and collectivism). Nonetheless, institutionalization is a necessary consideration for some families. Gaugler et al. (2006) identified the predictors that led to Latinos being placed a nursing home. Through event history analyses, they considered the context of care, cognitive functioning, caregiver stress and depression, and caregiving resources in 324 Latinos across eight catchment areas in the USA. Interestingly, within the Latino subsample, individuals with cognitive impairment and who lived with their caregivers in Florida were less likely to be placed in a nursing home, while those across the USA who were Medicaid eligible were almost four times more likely to be placed in a nursing home. Additionally, the likelihood of placement was increased if the caregiver that provided more hourly assistance to the care recipient was depressed and employed more adult day services. These findings emphasize the importance of within-group differences (e.g., Cubans vs. other Latinos), regional differences (Florida vs. other states), and cultural values that promote the utilization and support of family and other informal support services (e.g., *familismo*, collectivism).

Closing Remarks

Dementia is a disease that can affect the entire Latino family, especially relatives involved in the direct care of the affected family member, commonly referred to as caregivers. Specifically, family members involved in the management of the disease can be taxed by their responsibilities, thus experiencing dementia caregiver burden. This burden is often associated with grief, loss, and depression. Additionally, the caregiving process is often informed by a number of factors including the Latino family's cultural values, the family's acculturation history, and the family interactional patterns. To address caregiver burden, there are both effective preventative and treatment interventions available to assist caregivers and families affected by dementia and the related grief, loss, and depression. Interventionists are encouraged to consider the following recommendations as they contemplate the most appropriate course of action for managing the grief, loss, and depression often associated with caregiver burden:

- The term Latino is an umbrella term used to describe a very heterogeneous group, encompassing people from more than 20 countries in Latin America with different colonization histories, migration patterns, sociopolitical histories, etc. Interventionists must consider the Latino caregiver and family's overall ethnic background and appropriately adapt the treatment approach to meet the specific needs of the respective participants.
- The acculturation process, cultural values, and familial interaction patterns play key roles in the psychological functioning and distress of caregivers and their families. Interventionists need to accurately assess these factors as they provide their services.
- While there are several interventions available to Latino caregivers and their families, a "one-size-fits-all" approach is not advised when addressing caregiver burden and the related stress, grief, and depression. Interventions must consider personal, familial, and cultural factors and adapt their service plan accordingly.

References

Adames, H. Y., & Chavez-Dueñas, N. Y. (2017). *Cultural foundations and interventions in Latino/a mental health: History, theory, and within group differences*. New York, NY: Routledge.

Adames, H. Y., Chavez-Dueñas, N. Y., Fuentes, M. A., Salas, S. P., & Perez-Chavez, J. G. (2014). Integration of Latino/a cultural values into palliative health care: A culture centered model. *Palliative and Supportive Care, 12*(2), 149–157. https://doi.org/10.1017/S147895151300028X

Adams, B., Aranda, M. P., Kemp, B., & Takagi, K. (2002). Ethnic and gender differences in distress among Anglo American, African American, Japanese American, and Mexican American spousal caregivers of persons with dementia. *Journal of Clinical Geropsychology, 8*(4), 279–301.

Adelman, R., Tmanova, L. L., Delgado, D., Dion, S., & Lachs, M. S. (2014). Caregiver burden a clinical review. *JAMA, 311*(10), 1052–1059. http://jamanetwork.com/

Ajay, S., Kasthuri, A., Kiran, P., & Malhotra, R. (2017). Association of impairments of older persons with caregiver burden among family caregivers: Findings from rural South India. *Archives of Gerontology and Geriatrics, 68*, 143–148.

Albert, S. M. (1999). The caregiver as part of the dementia management team. *Disease Management and Health Outcomes, 5*(6), 329–337.

Alegría, M., Canino, G., Ríos, R., Vera, M., Calderón, J., Rusch, D., & Ortega, A. (2002). Inequalities in use of specialty mental health services among Latinos, African Americans, and non-Latino Whites. *Psychiatric Services, 53*(12), 1547–1555.

Arango-Lasprilla, J. C., Panyavin, I., Merchán, E. H., Perrin, P. B., Arroyo-Anlló, E. M., Snipes, D. J., & Arabia, J. (2014). Evaluation of a group cognitive–behavioral dementia caregiver intervention in Latin America. *American Journal of Alzheimer's Disease & Other Dementias, 29*(6), 548–555. https://doi.org/10.1177/1533317514523668

Arredondo, P., Gallardo-Cooper, M., Delgado-Romero, E., & Zapata, A. L. (2014). *Culturally responsive counseling with Latinas/os*. Alexandria, VA: ACA Press. https://doi.org/10.1002/9781119221609

Arredondo-Dowd, P. (1981). The psychological development and education of immigrant adolescents: A baseline study. *Adolescence, 16*, 175–186.

Association of Alzheimer's Disease. (2015). *Serving Hispanic families: Home and community based services for people with dementia and their caregivers – A toolkit for the Aging Network*. Retrieved from http://www.alz.org/national/documents/aoagrant_kits_his.pdf

Austrom, M., & Gendrie, H. (1990). Death of the personality: The grief responses of the Alzheimer's family caregiver. *American Journal of Alzheimer's Care and Related Disorders Research, 5*(2), 16–27.

Belle, S. H., Burgio, L., Burns, R., Coon, D., Czaja, S. J., Gallagher-Thompson, D., & Zhang, S. (2006). Enhancing the quality of life of dementia caregivers from different ethnic or racial groups: A randomized, controlled trial. *Annals of Internal Medicine, 145*, 727–738.

Bentwich, M. E., Dickman, N., & Oberman, A. (2017). Dignity and autonomy in the care for patients with dementia: Differences among formal caretakers of varied cultural backgrounds and their meaning. *Archives of Gerontology and Geriatrics, 70*, 19–27.

Benuto, L. T., & O'Donohue, W. (2015). Is culturally sensitive cognitive behavioral therapy an empirically supported treatment?: The case for Hispanics. *International Journal of Psychology & Psychological Therapy, 15*(3), 405–421.

Borrayo, E. A., Goldwaser, G., Vacha-Haase, T., & Hepburn, K. W. (2007). An Inquiry into Latino caregivers' experience caring for older adults with Alzheimer's disease and related dementias. *Journal of Applied Gerontology, 26*(5), 486–505. https://doi.org/10.1177/0733464807305551

Bowlby, J. (1961). Processes of mourning. *International Journal of Psychoanalysis, 1961*(42), 317–339.

Bowlby, J. (1969/1982). Attachment and loss: Vol. I: Attachment. New York: Basic Books.

Boyd-Franklin, N. (2006). *Black families in therapy: Understanding the African American experience* (2nd ed.). New York, NY: Gilford Press.

Broeckelmann, N. (2015). Case study of Hispanic male caregivers of cancer patients. *Electronic Theses and Dissertations*. Retrieved from https://repository.tcu.edu/handle/116099117/10290

Bronfenbrenner, U. (1977). Toward an experimental ecology of human development. *American Psychologist, 32*, 515–531.

Carretero, S., Garces, J., Rodenas, F., & Sanjose, V. (2009). The informal caregiver's burden of dependent people: Theory and empirical review. *Archives of Gerontology and Geriatrics, 49*(1), 74–79.

Center for Control Disease. (2016). *Care giving: A public health priority.* Retrieved from https://www.cdc.gov/aging/caregiving/index.htm

Chiambretto, P., Moroni, L., Guarnerio, C., Bertolotti, G., & Prigerson, H. G. (2010). Prolonged grief and depression in caregivers of patients in vegetative state. *Brain Injury, 24*, 581–588.

Cho, H., Kim, I., & Velez-Ortiz, D. (2014). Factors associated with mental health service use among Latino and Asian Americans. *Community Mental Health Journal, 50*(8), 960–967. https://doi.org/10.1007/s10597-014-9719-6

Chun, M., Organista, P. B., & Marin, G. (2003). *Acculturation: Advances in theory, measurement, and applied research.* Washington, D.C.: American Psychological Association.

Collins, C., Liken, M., King, S., & Kokinakis, C. (1993). Loss and grief among family caregivers of relatives with dementia. *Qualitative Health Research, 3*(2), 236–253. https://doi.org/10.1177/104973239300300206

Cooper, C., Balamurali, T. B. S., & Livingston, G. (2007). A systematic review of the prevalence and covariates of anxiety in caregivers of people with dementia. *International Psychogeriatrics, 19*(2), 175–195. https://doi.org/10.1017/S1041610206004297

Craske, M. G. (2010). *Cognitive-behavioral therapy.* Washington, DC: American Psychological Association.

Crist, J. D. (2002). Mexican American elders' use of skilled home care nursing services. *Public Health Nursing, 19*(5), 366–376. https://doi.org/10.1046/j.1525-446.2002.19506.x

de Vugt, M. E., Nicolson, N. A., Aalten, P., Lousberg, R., Jolle, J., & Verhey, F. R. J. (2005). Behavioral problems in dementia patients and salivary cortisol patterns in caregivers. *The Journal of Neuropsychiatry, 17*(2), 201–207. Retrieved from http://neuro.psychiatryonline.org/doi/full/10.1176/jnp.17.2.201

DuBard, C. A., & Gizlice, Z. (2008). Language spoken and differences in health status, access to care, and receipt of preventive services among U. S. Hispanics. *American Journal of Public Health, 98*(11), 2021–2028. https://doi.org/10.2105/AJPH.2007.119008

Ducharme, J. K., & Geldmacher, D. S. (2011). Family quality of life in dementia: A qualitative approach to family-identified care priorities. *Quality of Life Research, 20*(8), 1331–1335.

Etters, L., Goodall, D., & Harrison, B. E. (2008). Caregiver burden among dementia patient caregivers: A review of the literature. *Journal of the American Association of Nurse Practitioners, 9*(8), 423–428.

Falicov, C. J. (2014). *Latino families in therapy.* New York: Guilford Press.

Falicov, J. F. (2005). Ambiguous loss: Risk and resilience in Latino immigrant families. In M. M. Suarez-Orozco, C. Suarez-Orozco, & D. Baolian Qin (Eds.), *The new immigration: An interdisciplinary reader* (pp. 197–206). New York: Taylor & Francis. https://doi.org/10.4324/9780203621028

Family Caregiver Alliance. (2015). *Caregiver statistics, demographics.* Retrieved from https://www.caregiver.org/caregiver-statistics-demographics

Gallagher-Thompson, D., Coon, D. W., Solano, N., Ambler, C., Rabinowitz, Y., Thompson, L. W. (2003). Change in indices of distress among Latino and Anglo female caregivers of elderly relatives With dementia: Site-specific results from the REACH national collaborative study. *The Gerontologist, 43*(4), 580–591.

Gallagher-Thompson, D., Tzuang, M., Hinton, L., Alvarez, P., Rengifo, J., Valverde, I., … Thompson, L. W. (2015). Effectiveness of a fotonovela for reducing depression and Stress in Latino dementia family caregivers. *Alzheimer Disease and Associated Disorders, 29*, 146–153.

Gallagher-Thomson, D. G., Arean, P., Rivera, P., & Thompson, L. (2001). A psychoeducation intervention to reduce distress in Hispanic caregivers: Results of a pilot study. *Clinical Gerontologist, 23*(1/2), 17–32.

Galvis, M. J., & Cerquera-Córdoba, A. M. (2016). Relationship between depression and burden in caregivers of Alzheimer disease patients. *Psicología desde el Caribe. Universidad del Norte, 33*(2), 190–205. Retrieved from http://rcientificas.uninorte.edu.co/index.php/psicologia/article/viewFile/6307/8696

Gaugler, J. E., Kane, R. L., Kane, R. A., & Newcomer, R. (2006). Predictors of institutionalization in Latinos with dementia. *Journal of Cross-Cultural Gerontology, 21*(3–4), 139–155. https://doi.org/10.1007/s10823-006-9029-8

Gil, R. M., & Vazquez, C. I. (2002). *The Maria paradox: How Latinas can merge old world traditions with new world self-esteem.* Indiana: Scenery Press.

Goldenberg, H., & Goldenberg, I. (2013). *Family therapy: An overview.* Belmont, CA: Brooks/Cole, Cengage Learning.

Gonyea, J. G., López, L. M., & Velásquez, E. H. (2016). The effectiveness of a culturally sensitive cognitive behavioral group intervention for Latino Alzheimer's caregivers. *Gerontologist, 56*(2), 292–302. https://doi.org/10.1093/geront/gnu045

Harwood, D. G., Barker, W. W., Ownby, R. L., Bravo, M., Aguero, H., & Duara, R. (2000). Predictors of positive and negative appraisal among Cuban American caregivers of Alzheimer's disease patients. *International Journal of Geriatric Psychiatry, 15*, 481–487. https://doi.org/10.1002/1099-1166(200006)15:6<481::AID-GPS984>3.0.CO;2-J

Hays, P. A. (2008). *Addressing cultural complexities in practice: Assessment, diagnosis, and therapy* (2nd ed.). Washington, DC: American Psychological Association.

Henderson, J. N., & Gutierrez-Mayka, M. (1992). Ethnocultural themes in caregiving to Alzheimer's disease patients in Hispanic families. *Clinical Gerontologist, 11*, 59–74.

Hinton, L., Haan, M., Geller, S., & Mungas, D. (2003). Neuropsychiatric symptoms in Latino elders with dementia or cognitive impairment without dementia and factors that modify their association with caregiver depression. *Gerontologist, 43*(5), 669–677.

Hui, C. H., & Triandis, H. C. (1986). Individualism-collectivism: A study of cross-cultural researcher. *Journal of Cross-Cultural Psychology, 17*, 225–248. https://doi.org/10.1177/0022002186017002006 1986

Interian, A., Ang, A., Gara, M. A., Link, B. G., Rodriguez, M. A., & Vega, W. A. (2010). Stigma and depression treatment utilization among Latinos: Utility of four stigma measures. *Psychiatric Services, 61*(4), 373–379. https://doi.org/10.1176/ps.2010.61.4.373

Interian, A., Martinez, I. E., Guarnaccia, P. J., Vega, W. A., & Escobar, J. I. (2007). A qualitative analysis of the perception of stigma among Latinos receiving antidepressants. *Psychiatric Services, 58*, 1591–1594.

Jezzini, A. T. (2013). Acculturation, marianismo gender role, and ambivalent sexism in predicting depression in Latinas. *Electronic Theses and Dissertations, 317*. http://digitalcommons.du.edu/etd/317

Joling, K. J., van Hout, H. P. J., Schellevis, F. G., van der Horst, H. E., Scheltens, P., & Knol, D. L. (2010). Incidence of depression and anxiety in the spouses of patients with dementia: A naturalistic cohort study of recorded morbidity with a 6-year follow-up. *The American Journal of Geriatric Psychiatry, 18*(2), 146–153. https://doi.org/10.1097/JGP.0b013e3181bf9f0f

Kane, M. N., & Williams, M. (2000). Perceptions of South Florida Hispanic and Anglo Catholics: From whom would they seek help? *Journal of Religion and Health, 39*(2), 107–122.

Keefe, S., Padilla, A., & Carlos, M. (1979). The Mexican-American extended family as an emotional support system. *Human Organization, 38*, 144–152.

Keyes, K. M., Martins, S. S., Hatzenbuehler, M. L., Blanco, C., Bates, M., & Hasin, D. S. (2012). Mental health service utilization for psychiatric disorders among Latinos living in the United States: The role of ethnic subgroup, ethnic identity, and language/social preferences. *Social Psychiatry and Psychiatric Epidemiology, 47*(3), 383–394. Retrieved from https://link.springer.com/article/10.1007/s00127-010-0323-y

Kim, H., Chang, M., Rose, K., & Kim, S. (2012). Predictors of caregiver burden in caregivers of individuals with dementia. *Journal of Advanced Nursing, 68*(4), 846–855. https://doi.org/10.1111/j.1365-2648.2011.05787.x

Kirmayer, L. J., Narasiah, L., Munoz, M., Rashid, M., Ryder, A. G., Guzder, J., … Pottie, K. (2010). Common mental health problems in immigrants and refugees: General approach in primary care. *Canadian Medical Association Journal, 183*(12), 1–9. https://doi.org/10.1503/cmaj.090292

Kissane, D. W. (2000). A model of family-centered intervention during palliative care and bereavement: Focused family grief therapy. In L. Baider, C. L. Cooper, & A. K. Nour (Eds.), *Cancer and the family* (pp. 176–197). New York: John Wiley.

Knight, B. G., Robinson, G. S., Flynn, C. V., Miae, L., Kayoko, C., & Kim, J. H. (2002). Cross cultural issues in caregiving for persons with dementia: Do familism values reduce burden and distress? *Ageing International, 27*(3), 70–94.

Kouyoumdjian, H., Zamboanga, B. L., & Hansen, D. J. (2003). Barriers to community mental health services for Latinos. *Clinical Psychology: Science and Practice, 10*, 394–422.

Kübler-Ross, E. (1969). *On death and dying*. New York: The Macmillan Company.

Lawton, M. P., Kleban, M. H., Moss, M., Rovine, M., & Glicksman, A. (1989). Measuring caregiving appraisal. *Journal of Gerontology, 44*, 61–71. https://doi.org/10.1093/geronj/44.3.P61

Lichtenthal, W. G., & Kissane, D. W. (2008). The management of family conflict in palliative care. *Progress in Palliative Care, 16*(1), 39–45.

Lin, C., Wang, J. D., Pai, M. C., & Ku, L. E. (2017). Measuring burden in dementia caregivers: Confirmatory factor analysis for short forms of the Zarit Burden Interview. *Archives of Gerontology and Psychiatrics, 68*, 8–13. https://doi.org/10.1016/j.archger.2016.08.005

Lindgren, C. L., Connelly, C. T., & Gaspar, H. L. (1999). Grief in spouse and children caregivers of dementia patients. *Western Journal of Nursing Research, 21*(4), 521–537. https://doi.org/10.1177/01939459922044018

Liu, S., Li, C., Shi, Z., Wang, X., Zhou, Y., Liu, S., … Ji, Y. (2017). Caregiver burden and prevalence of depression, anxiety and sleep disturbances in Alzheimer's disease caregivers in China. *Journal of Clinical Nursing, 26*(9–10), 1291–1300. https://doi.org/10.1111/jocn.13601

Loera, S., Muñoz, L. M., Nott, E., & Sandefur, B. K. (2009). Call the curandero: Improving mental health services for Mexican immigrants. *Praxis, 9*, 16–24.

Lorant, V., & Thomas, I. (2008). Contextual factors and immigrants' health status: Double jeopardy. *Health Place, 14*(4), 678–692. https://doi.org/10.1016/j.healthplace.2007.10.012

Losada, A., Robinson Shurgot, G., Knight, B. G., Marquez, M., Montorio, I., Izal, M., & Ruiz, M. A. (2006). Cross-cultural study comparing the association of familism with burden and depressive symptoms in two samples of Hispanic dementia caregivers. *Aging and Mental Health, 10*(1), 69–76. https://doi.org/10.1080/13607860500307647

Lum, T. Y., & Vanderaa, J. P. (2010). Health disparities among immigrant and non-immigrant elders: The association of acculturation and education. *Journal of Immigrant and Minority Health, 12*(5), 743–753. https://doi.org/10.1007/s10903-008-9225-4

Macleod, A., Tantagelo, G., McCabe, M., & You, E. (2017). "There isn't an easy way of finding the help that's available." Barriers and facilitators of service use among dementia family caregivers: A qualitative study. *International Psychogeriatric, 29*(5), 765–776. https://doi.org/10.1017/S1041610216002532

McGoldrick, M., Giordano, J., & Pearce, J. K. (2005). *Ethnicity and family therapy*. New York: Guilford Press.

Mendez-Luck, C. A., & Anthony, K. P. (2016). Marianismo and caregiving role beliefs among U.S.-born and immigrant Mexican women. *Journal of Gerontology, Series B, 1*(5), 926–935. https://doi.org/10.1093/geronb/gbv083

Menselson, T., Rehkopf, D. H., & Kubzansky, L. D. (2008). Depression among Latinos in the United States: A meta-analytic review. *Journal of Consulting and Clinical Psychology., 76*(3), 355–366. https://doi.org/10.1037/0022-006X.76.3.355

Mitrani, V. B., & Czaja, S. J. (2000). Family-based therapy for dementia caregivers: Clinical observations. *Aging & Mental Health, 4*(3), 200–209.

Mitrani, V. B., Lewis, J., Feaster, D. J., Czaja, S. J., Eisdorfer, C., Schulz, R., & Szapocznik, J. (2006). The role of family functioning in the stress process of dementia of caregivers: Structural family framework. *The Gerontologist, 46*(1), 97–105.

Molyneux, G. J., McCarthy, G. M., McEniff, S., Cryan, M., & Conroy, R. M. (2008). Prevalence and predictors of care burden and depression in the care of patients referred to an old age psychiatric service. *International Psychogeriatrics, 20*(6), 1193–1202.

National Alliance for Caregiving and Evercare. (2008). *Evercare study of Hispanic family caregiving in the U.S.: Findings from a national study*. Retrieved from http://www.caregiving.org/data/Hispanic_Caregiver_Study_web_ENG_FINAL_11_04_08.pdf

National Institute of Health. (2009). *Coping with grief*. Retrieved from https://newsinhealth.nih.gov/2009/november/feature1.htm

Ostir, G. V., Eschbach, K., Markides, K. S., & Goodwin, J. S. (2003). Neighbourhood composition and depressive symptoms among older Mexican Americans. *Journal of Epidemiology and Community Health, 57*(12), 987–992.

Papastavrou, E., Kalokerinou, A., Papacostas, S. S., Tsangari, H., & Sourtzi, P. (2007). Caring for a relative with dementia: Family caregiver burden. *Journal of Advanced Nursing, 58*(5), 446–457.

Parkes, C. M. (1972). *Bereavement; studies of grief in adult life*. New York: International Universities Press.

Parks, S. M., & Novielli, K. D. (2000). A practical guide to caring for caregivers. *American Family Physician, 62*(12), 2613–2620.

Perez, R. M. (2016). Lifelong ambiguous loss: The case of Cuban American exiles. *Journal of Family Theory & Review, 8*(3), 324–340.

Pineros-Leano, M., Liechty, J. M., & Piedra, L. M. (2017). Review article: Latino immigrants, depressive symptoms and cognitive behavioral therapy: A systematic review. *Journal of Affective Disorders, 208*, 567–576. https://doi.org/10.1016/j.jad.2016.10.025

Pinquart, M., & Sörensen, S. (2005). Ethnic differences in stressors, resources, and psychological outcomes of family caregiving: A meta-analysis. *Gerontologist, 45*(1), 90–106. https://doi.org/10.1093/geront/45.1.90

Polich, T. M., & Gallagher-Thompson, D. (1997). Preliminary study investigating psychological distress among female Hispanic caregivers. *Journal of Clinical Gerontology, 3*, 1–15.

Powers, S. M., & Whitlatch, C. J. (2016). Measuring cultural justifications for caregiving in African American and White caregivers. *Dementia, 15*(4), 629–645. https://doi.org/10.1177/1471301214532112

Robinson-Shurgot, G., & Knight, B. G. (2005). Influence of neuroticism, ethnicity, familism, and social support on perceived burden in dementia caregivers: Pilot test of the transactional stress and social support model. *Journal of Gerontology, Series B, 60*(6), 331–334. https://doi.org/10.1093/geronb/60.6.P331

Rogler, L. H., & Hollingshead, A. B. (1985). *Trapped: Puerto Rican families and schizophrenia* (3rd ed.). Maplewood, NJ: Waterfront Press.

Rogler, L. H., Malgady, R. G., & Rodriguez, O. (1989). *Hispanics and mental health: A framework for research*. Malabar, FL: Robert E. Krierger Publishing.

Santiago-Rivera, A., Arredondo, P., & Gallardo-Cooper, M. (2002). *Counseling Latinos and la familia: A practical guide*. Thousand Oaks, CA: Sage Publications.

Serrano-Aguilar, P. G., Lopez-Bastida, J., & Yanes-Lopez, V. (2006). Impact on health-related quality of life and perceived burden of informal caregivers of individuals with Alzheimer's disease. *Neuroepidemiology, 27*(3), 136–142.

Stalder, T., Tietze, A., Steudte, S., Alexander, N., Dettenborn, L., & Kirschbaum, C. (2014). Elevated hair cortisol levels in chronically stressed dementia caregivers. *Psychoneuroendocrinology, 47*, 26–30. https://doi.org/10.1016/j.psyneuen.2014.04.021

Stevens, E. P. (1973a). Marianismo: The other face of Machismo. In A. Pescatello (Ed.), *Female and male in Latin America* (pp. 89–101). Pittsburgh, PA: Pittsburgh University Press.

Stevens, E. P. (1973b). Machismo and Marianism. *Society, 10*(6), 57–63. https://doi.org/10.1007/BF02695282

Stroebe, M., & Schut, H. (1999). The dual process model of coping with bereavement: Rationale and description. *Death Studies, 23*(3), 197–224. https://doi.org/10.1080/074811899201046

Torres, J. B., Solberg, V. S. H., & Carlstrom, A. H. (2002). The myth of sameness among Latino men and their machismo. *American Journal of Orthopsychiatry, 72*(2), 163–181. https://doi.org/10.1037/0002-9432.72.2.163

Torti, F. M., Jr., Gwyther, L. P., Reed, S., Friedman, J. Y., & Schulman, K. A. (2004). A multinational review of recent trends and reports in dementia caregiver burden. *Alzheimer Disease & Associated Disorders, 18*(2), 99–109.

Triandis, H. (1988). Collectivism v. Individualism: A reconceptualisation of a basic concept in cross-cultural social psychology. *Cross-Cultural Studies of Personality, Attitudes and Cognition, Chapter 3*, 60–95. Retrieved from https://link.springer.com/chapter/10.1007/978-1-349-08120-2_3

Vazquez, C. I., & Rosa, D. (2011). *Grief therapy with Latinos: Integrating culture for clinicians*. New York, NY: Springer Publishing Company.

Villalobos, B. T., & Bridges, A. J. (2016). Testing an attribution model of caregiving in a Latino sample: The role of familismo and the caregiver-care recipient relationship. *Journal of Transcultural Nursing, 27*(4), 322–332. https://doi.org/10.1177/1043659615590476

Villatoro, A. P., Morales, E. S., & Mays, V. M. (2014). Family culture in mental health help-seeking and utilization in a nationally representative sample of Latinos in the United States: The NLAAS. *American Journal of Orthopsychiatry, 84*(4), 353–363. https://doi.org/10.1037/h0099844

Walker, R. J., & Pomeroy, E. C. (1996). Depression or grief? The experience of caregivers of people with dementia. *Health Social Work, 21*(4), 247–254. https://doi.org/10.1093/hsw/21.4.247

Walsh, F., & McGoldrick, M. (2004). *Living beyond loss: Death in the family*. New York, NY: W.W. Norton.

Worden, J. W. (1983). *Grief counseling and grief therapy* (1st ed.). London: Tavistock.

Zea, M. C., Quezada, T., & Belgrave, F. Z. (1994). Latino cultural values: Their role in adjustment to disability. *Journal of Social Behavior & Personality, 9*, 185–200.

Zisook, S., & Shear, K. (2009). Grief and bereavement: What psychiatrists need to know. *World Psychiatry, 8*(2), 67–74.

Part IV

International Perspectives: The Americas and Beyond

Behavioral and Psychosocial Treatments of Dementia in Mexico

15

Silvia Mejía-Arango, Mariana López-Ortega, and Laura Barba-Ramírez

Abstract

The increase of the elder population in Mexico has led to a growth in individuals with chronic diseases among which the dementias stand out. Approximately 800,000 people in Mexico have some form of dementia, a figure that will triple in the next 30 years. The accelerated process of aging of the population that has occurred in the past years has led to the need by the health system to prioritize the timely diagnosis of degenerative diseases including dementia, leaving the intervention in the background and oriented exclusively to the family as an instrument for the application of treatments. In a context of great social inequalities and economic constraints, the family must cope with the multiple manifestations of dementia related to the progressive deterioration of cognitive functions, within which the behavioral and psychiatric symptoms occupy a predominant place because of their negative effect on the patient's health and on the mental health of the caregiver. The lack of access to nursing homes and related facilities, in-home registered nurses and home health aides, and other home- and community-based services in Mexico leaves the management of behavioral and psychiatric symptoms of dementia in the hands of families and caregivers who accomplish the caring process in highly disadvantaged contexts.

S. Mejía-Arango (✉)
El Colegio de la Frontera Norte, Department of Population Studies, Tijuana, Mexico
e-mail: smejia@colef.mx

M. López-Ortega
National Institutes of Health, Mexico, National Institute of Geriatrics, Cuernavaca, Mexico
e-mail: marianalopezortega@googlemail.com

L. Barba-Ramírez
National Council to Prevent Discrimination, Mexico; El Colegio de la Frontera Norte, Tijuana, Mexico
e-mail: laura.barba@outlook.com

Introduction

Currently, there are 13 million adults 60 years of age or older living in Mexico (INEGI, 2014). While the Mexican life expectancy at birth is 72 years for men and 78 years for women, healthy life expectancy for both sexes is reduced to 66 years, suggesting that Mexican elders live in poor health conditions during their last years of life. Diseases that frequently present during aging and that are main causes of death include dementia, diabetes, heart disease, and stroke (Gutiérrez Robledo & Arrieta Cruz, 2014). Reported prevalence rates for dementia range between 6% and 8% (Mejia-Arango & Gutierrez, 2011). Age- and sex-specific rates indicate increasing prevalence with age and higher prevalence in women; the group with the highest educational level had the lowest prevalence. Hypertension and diabetes

H. Y. Adames, Y. N. Tazeau (eds.), *Caring for Latinxs with Dementia in a Globalized World*,
https://doi.org/10.1007/978-1-0716-0132-7_15

were the two vascular risk factors along with depression that demonstrated an increased risk of dementia (Mejia-Arango & Gutierrez, 2011).

The projections are that there will be 3.5 million elderly people affected by dementia by 2050 in Mexico; this will have a major impact on the overall health-care system of the country (Gutiérrez Robledo & Arrieta Cruz, 2014). Multimorbidity, social disadvantage, and limited access to health-care systems characterize the elder population living in Mexico. Regarding dementia care in Mexico, low levels of awareness and training of health-care staff contribute to low rates of diagnosis, and for those who are diagnosed, the lack of professional knowledge about treatment and care options may also deny people access to post-diagnostic care, treatment, and support (Prince, Comas-Herrera, Knapp, Guerchet, & Karagiannidou, 2016).

When we consider dementia, the first manifestations that come to mind are those symptoms related to cognitive deterioration including loss of memory, inability to recognize places and faces, and language difficulties. But the bigger challenge is often the Behavioral and Psychological Symptoms of Dementia (BPSD). The International Psychogeriatrics Association defined these as "symptoms of disturbed perception, thought content, mood, or behavior that frequently occur in patients with dementia" (Finkel, e Silva, Cohen, Miller, & Sartorius, 1997). These symptoms include depression, psychosis, and aggression, wandering or walking/getting lost, agitation, apathy, and emotional distress. Although BPSD are present in 90% of the cases of dementia at any given point in the duration of their illness, the frequency of BPSD vary among the different types of dementia and over the course of the disease. While some symptoms can be more often recognized in some types of dementia such as the frontotemporal type, the clinical presentation has wide variation (Kar, 2009; Savva et al., 2009). However, BPSD are a major source of significant distress to individuals suffering dementia. Patients subjectively experience changes in emotions, perceptions, thought content, and motor function against a backdrop of impaired cognitive state and ineffective communication which undermine the individual's usual psychological capacities to adequately respond to everyday demands. BPSD can be understood as ineffective attempts of the dementia patient to cope with environmental and physiological stress factors. The purpose of this chapter is to describe the conditions that characterize Mexican elders with dementia in Mexico and their caring system. To accomplish this goal, we present evidence from different studies based on the Mexican population and some theoretical background that provides for an understanding of the results of the studies.

Prevalence of BPSD in Latinos

The most prevalent BPSD are apathy, depression, irritability, agitation, and anxiety. Of the most prevalent symptoms, apathy is often the first reported by caregivers. However, the coexistence of symptoms is a predominant feature in dementia. During the course of the disease, 50% of patients may present at least four symptoms simultaneously (Frisoni et al., 1999). Consequences of BPSD are disturbed sleep, fatigue, increased risk of falls, accidental injuries, inadequate nutrition which produce high levels of burden for the caregivers, and increased use of health-care resources in patients with BPSD ending, most of the time, in institutionalization. The relationship between BPSD and level of cognitive impairment is not always linear, while apathy increases with cognitive decline. Higher prevalence rates of BPSD are observed in the middle stages of dementia, with a reduction during the last years (Lövheim, Sandman, Kallin, Karlsson, & Gustafson, 2006).

Much research on BPSD has been undertaken in the developed world, while the strongest evidence of BPSD in low- and middle-income countries come from the 10/66 Dementia Research Group (Prince, 2000). The study gathered information from 14 sites in Latin America, including Mexico, along with four in India, two in China, one in Taiwan, and one in Africa. Data on BPSD was collected through an interview with the caregiver (Ferri, Ames, & Prince, 2004). Nearly 80%

of the individuals with dementia were reported by their caregivers to have BPSD. Latin Americans ranked between Chinese participants who presented the fewest symptoms and those from India who presented the highest symptoms. However, Latin Americans had the highest prevalence rate in each of the three most frequent psychiatric syndromes (i.e., schizophreniform/paranoid psychosis, depression, anxiety), with depression being the highest (51%). When compared to non-Hispanic Whites, Latinos with dementia living in the USA have also shown higher rates of neuropsychiatric symptoms, with depression being the most common (Hinton, Tomaszewski Farias, & Wegelin, 2008), and a strong association between the presence of neuropsychiatric symptoms in the patient and depression in the caregiver has been reported (Hinton, Chambers, Velásquez, Gonzalez, & Haan, 2006; Hinton, Haan, Geller, & Mungas, 2003).

These results are compelling evidence of the social and health conditions in which aging, dementia, and caregiving are taking place among Hispanics living in their own countries or as an ethnic minority group living in the USA, both of which have aged under economic hardship conditions and health disparities that have a cumulative effect on their physical, cognitive, psychological, and social functioning, thus increasing the burden of disease. As the 10/66 study highlights, the experience of dementia in the developing world, especially in Latin America, is distressing for a high number of affected people (Ferri et al., 2004).

Data from the Mexican Health and Aging Study

In this section, we present data from a group of Mexican elders 60 years of age and older that participated in the Mexican Health and Aging Study (MHAS), a prospective four-wave panel study of a nationally representative cohort of older Mexican adults (Wong et al., 2015). The individuals included in the following analysis participated in wave three (2012) of the study through a proxy interview due to limitations (e.g., health problems, sensory and motor difficulties, or language impairment), which prevented them from answering the questions on their own. As part of the assessment, proxy informants (usually a spouse or an adult child) provided information about the participant's cognitive and functional decline, neuropsychiatric symptoms, and medical history. Based on established cut points for cognitive impairment (Jorm, 2000; Mejía-Arango, Wong & Michaels-Obregón, 2015) and functional decline (Mejia-Arango & Gutierrez, 2011), 1010 individuals were classified as normal (71%) or demented (29%).

Elders with dementia were significantly older compared to those not demented, 64% were female, and 63% did not have a partner mainly due to widowhood. Low educational attainment is a common characteristic among Mexican elders, and in this study's results, 32% and 38% of the subjects in the normal and dementia group, respectively, never attended school. The presence of other medical diseases is also a prevailing condition in this population. The sum of different comorbidities such as hypertension, diabetes, heart attack, pulmonary disease, and stroke is also frequent among both groups but is significantly higher in the elders with dementia, two-thirds of which have one or more comorbidities. Access to a formal health service existed for approximately half of the elders, and although it was higher in those classified with dementia, the other half of the elders depended on out-of-pocket payments or on a less resourceful health service such as *Seguro Popular*.[1] Exacerbating the situation, results showed that only one-third of the participants had a retirement pension, additional evidence of the economic disadvantage of this population (see Table 15.1).

Cognitive and behavioral symptoms were common between both groups; however, almost 91% of the elders with dementia had at least one symptom compared to 59% in the normal group. Analyzing each of the individual behavioral or psychiatric symptoms (see Fig. 15.1), they were all significantly higher in the dementia group: wandering (40.8% vs. 3.6%), hallucinations (43% vs. 6.4%), angry or hostile (58% vs. 36%),

[1] Seguro Popular is a public health insurance that covers a wide range of services without co-payments for its affiliates.

Table 15.1 General characteristics of elders with normal cognition and dementia in Mexico

	Normal	Dementia	
	$N = 716$	$N = 294$	P value
Age, mean (SD)	72.9 (9.3)	81.2 (8.9)	0.0001
Sex			
Male	53.2	36.1	0.0001
Female	46.8	63.9	
Marital status			
Single	4.3	7.1	0.0001
Married	57.5	34.7	
Divorced	9.8	9.9	
Widowed	28.4	48.3	
Education, mean (SD)	3.8 (4.1)	3.2 (3.8)	0.0001
0	32.5	38	0.168
1–6	51.1	49	
7 or more	16.4	13	
Comorbidities			
None	45.6	37.7	0.012
1–2	48.9	52.7	
3–5	5.5	9.6	
Access to formal health service	50.4	58.2	0.023
Has a retirement pension	24.7	30.1	0.17
Number of psychiatric and behavioral symptoms	1.1 (1.2)	2.7 (1.9)	0.0001
None	41.3	9.3	0.0001
1–2	46.5	43.4	
3–4	10.6	25.3	
5–7	1.6	22.1	

Source: Wong et al. (2015)

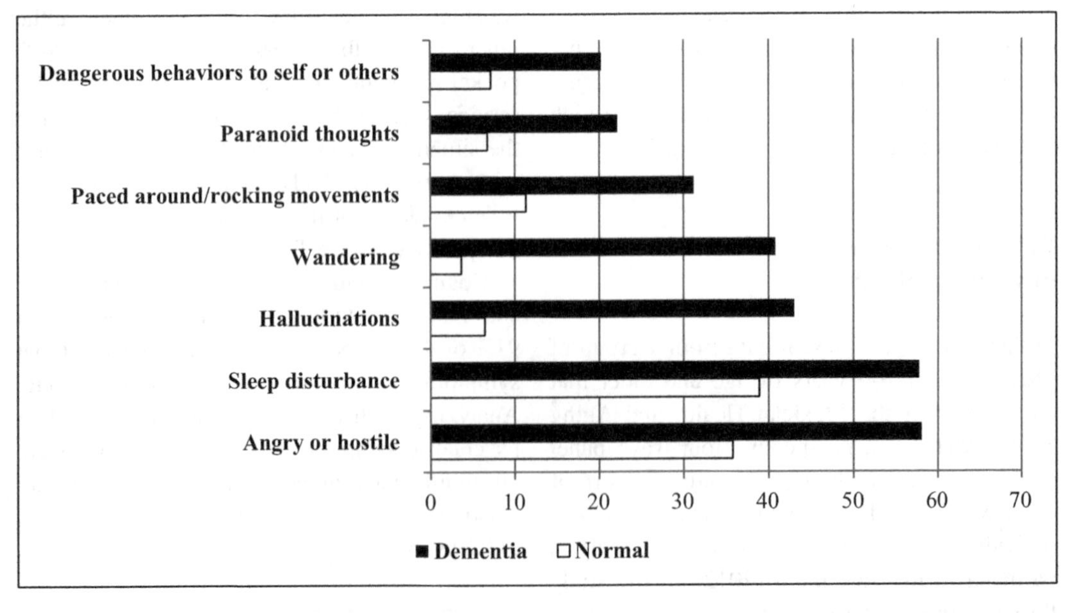

Fig. 15.1 Percentage of behavioral and psychological symptoms in subjects with normal cognition and dementia Wong et al. (2015)

sleep disturbance (58% vs. 39%), dangerous behaviors to self or others (20.2% vs. 7.1%), pacing/rocking movements (31.2% vs. 11.3%), and paranoid thoughts (22.2% vs. 6.7%). After adjusting for sex and age, logistic regression analysis showed that wandering (OR = 14.67, 95%, CI 8.46–25.43), hallucinations (OR = 4.5, 95%, CI 2.81–7.19), pacing/rocking movements (OR = 1.61, 95%, CI 1.01–2.56), and angry/hostile (OR = 1.88, 95%, CI 1.30–2.73) were the symptoms highly associated with the presence of dementia. The pattern of BPSD found indicated that psychiatric symptoms were also present in subjects without dementia, which may lead families to think that these are a normal part of aging, thus further reducing a family's search for health care. On the other hand, wandering, hallucinations, and pacing, all of which were highly prevalent among dementia patients, have been described as embarrassing behaviors that bring shame to the family (Gallagher-Thompson et al., 2003). The associated stigma they have can cause families to mask the symptoms, increasing the possibility of socially isolating the patient.

In summary, Mexican individuals with dementia in this study have a high prevalence of BPSD (91%). Wandering, hallucinations, pacing, and angry behaviors were the most common symptoms. This group had a profile of high health and economic vulnerability consisting of the following: (1) presence of at least one comorbidity in 60% of the group, (2) low educational attainment, (3) lack of a retirement pension in 70% of the group, and (4) no access to formal health service in 40% of the group.

BPSD Interventions and Their Applicability to Mexican Culture

Treatment for BPSD has focused on pharmacological interventions; however, during the last several years, there has been a surge toward psychosocial–environmental interventions for management of behavioral symptoms. Current guidelines recommend non-pharmacological interventions as first-line treatment followed by the least harmful medication for the shortest time possible (Azermai et al., 2012; Gauthier et al., 2010). To understand the basis of non-pharmacological interventions, Cohen-Mansfield (2001) proposed three theoretical models conceptualizing inappropriate behaviors in dementia, including (1) the unmet needs model refers to BPSD as an expression of discomfort and distress in the dementia patient due to their biological, social, and psychological needs that are not apparent to the caregiver or that they do not feel able to fulfill; (2) the *learning and behavioral models* assume that a different learning experience is needed to change the behavior, based on reinforcement contingencies; and (3) the *environmental vulnerability/reduced stress threshold model* assumes that the dementia process results in greater vulnerability to the environment and a lower threshold at which stimuli affects behavior. The different models are complementary. Environmental vulnerability may increase susceptibility to environmental antecedents and consequences or may produce an unmet need when normal levels of stimulation are perceived as overstimulation.

Of the theoretical models presented, non-pharmacological interventions are recommended considering that they address the psychosocial–environmental underlying reason(s) for the behavior; they avoid adverse side effects, drug–drug interactions, and limited efficacy of drugs; and they do not mask the patient's actual need by eliminating the behavior that serves as a signal (when medication is efficacious). Specific non-pharmacological interventions reported in a systematic review (Livingston, Johnston, Katona, Paton, & Lyketsos, 2005) include reminiscence therapy, simulated presence therapy, validation therapy, acupuncture, aromatherapy, light therapy, massage/touch, music therapy, Snoezelen multisensory stimulation, transcutaneous electrical nerve stimulation, behavior management techniques, and other psychosocial interventions such as animal-assisted therapy and exercise, concluding that behavior management therapies appear to have lasting effectiveness for the management of BPSD. The matter in question, however, is the application of these models and interventions to the understanding and management of behavioral problems in Mexican elders with dementia. As discussed in the following text, several conditions of the population such as low availability of formal supportive services, reliance on family caregivers, high emphasis on collectivism (vs. individualism), low levels of formal education, strong cultural beliefs, and use of religious coping characterize the sociocultural context in which Mexican families understand and manage their relative's disease. Given these sociocultural conditions, it is unlikely that the behavioral symptoms of the patient with dementia will be addressed as an expression of unmet needs or explained as a high vulnerability to the environment or that the patient requires a new learning experience to be solved. Moreover, specialized interventions guided by so many different approaches are incongruent with Mexican cultural context. A broader, social contextual framework with culturally sensitive explanations and interventions that necessarily goes beyond the Cohen models is needed to acknowledge the complexity of the caregiving experience and to assure a more comprehensive understanding and management of the patient's condition.

The Dementia Care System in Mexico

In countries with sound health and social care systems, formal care is provided in the form of nursing homes and related facilities, by in-home registered nurses and home health aides, and other home- and community-based services. These care services are paid services and are provided by health-care institutions or trained individuals. Although private, formal care is available in most parts of the country, much of it relies on out-of-pocket payments and therefore is accessible to only a small percentage of the population.

Formal Support

As in many low- and middle-income countries, in Mexico, the lack of long-term care policies and inadequate health-care services, combined with low levels of health insurance, reduce access to health care. The lack of long-term care policies implies that social and health services, as well as strategies of formal public support of older adults, are very few and are divided among different institutions that provide them. Regarding health insurance, the percentage of older adults in Mexico that are insured has increased, mainly after the implementation of *Seguro Popular*, an insurance mechanism offered to those previously not insured (i.e., primarily the poor, those in the informal market, and the self-employed). Data from the Mexican Health and Aging Study (MHAS) indicate that while in 2001 49% of men and 45% of individuals ages 50 years and older reported no health insurance, only 17% of men and 14% or women were not insured in 2012 (Wong et al., 2015). However, issues of quality of care and inadequate services still make effective access to quality services for older adults an important challenge in the country. In addition, only a small proportion of older adults have a retirement pension. By 2013, only 26.1% of those 60 years and older received a pension, with this percentage approximately doubling for men than for women, with 35% and 18.5%, respectively (INEGI, 2014). Finally, while private services in the form of home-care services or long-term care institutions are available, these do not include specialized dementia care, nor personnel that is trained adequately to provide such specialized services. In addition, given their high costs, private services remain available only for a very small percentage of the total population of older adults in Mexico.

Informal Care

Without formal health or social welfare provision, services for older people in low- and middle-income countries rely heavily on informal care from families. In Mexico, the family is the main source of economic and instrumental support for older adults (Angel, Angel, López-Ortega, Robledo, & Wallace, 2016). Even when and where services are available, families frequently remain unaware of them and are left to manage the increasing complexity of dementia care without formal support. Individuals with dementia become increasingly dependent on informal caregivers, such as their spouses, their children, other relatives, or friends. The most common types of networks for older people in Latin American countries are the *family-dependent support network* and the *locally integrated support network* (Thiyagarajan, Prince, & Webber, 2014) which share the commonality that network members do not look for professional help or are unaware of it.

Based on Wenger's (1991) description, the *family-dependent support network* is essentially structured around nearby kin ties, close family relationships, and few friends and neighbors. A shared household with children, or very near but separate households, is the base for this network. High standards of care are based on long-term reciprocity, although increasing dependency demands tend to focus on one person, commonly a daughter. The *locally integrated support network* includes close relationships with local family, friends, and neighbors. It is based on long-term residence and active community involvement that have served to build up well-established reciprocal support relationships. In the study by the 10/66 Dementia Research Group, no differences were observed between Mexican rural and urban areas regarding network type. Approximately 40–45% of the participants lived in these two types of social support networks, while 10–15% lived in locally self-contained, wider community-focused or private networks (Thiyagarajan et al., 2014).

Caregiver

Caregiver characteristics have also been studied by the 10/66 Dementia Research Group (Prince, 2004). In Latin American countries, most caregivers are women (84%), mainly daughters (34%) and wives (26%), ages 40 and older. Nearly all lived with the person with dementia in large households, together with the extended family. Caring was associated with economic disadvantage, as many of the caregivers that were employed (35%) had to cut back on their paid work and frequently employed caregivers to undertake their tasks. This last finding highlights an important characteristic of the caring system in Mexico: nonpaid caregiving is mostly done by women (López-Ortega, 2014; Montes de Oca, 1999; Nigenda, López-Ortega, Matarazzo, & Juárez-Ramírez, 2007; Robles Silva, 2001). The latest data for 2014 indicate that of the total gross domestic product (GDP) spent on health (5.7%), nonpaid health care accounted for 20.5% of the total; the public sector represented 40.7%, and the private sector represented 38.8%. Moreover, when analyzed by type of service, the data show that the 20.5% of nonpaid health work was almost equal to the amount of hospital services with 21.3% of total health expenditures and higher than expected expenditures on primary care (16.5%; INEGI, 2015). While this refers to all household health care, as aging progresses, the proportion of total care for older adults is expected to equally increase.

Besides the economic disadvantage that caregiving represents, several studies have noted the additional burden that dementia symptoms add for caregivers (Gibbons et al., 2014; McDowell, Hill, Nilsson, & Kozma, 2002). Informal care can negatively affect caregiver's mental and physical health, particularly if care is provided frequently or for a prolonged period of time (Sawatzky & Fowler-Kerry, 2003; Schulz & Sherwood, 2008). The number of hours per day spent with a person with dementia, as well as the number of hours spent assisting with activities of daily living (ADLs), has been reported to be the highest among caregivers in Latin American countries compared to other low- and middle-income countries such as India, China, and Nigeria. The psychological impact was also higher among caregivers from these countries as measured by the level of burden and presence of psychiatric morbidity (Prince, 2004). Other characteristics that increase the risk of burden among caregivers include overload, poor relationship with the patient, adverse life events, gender, high level of neuroticism, role captivity, and low levels of confidence (Campbell et al., 2008).

Management of BPSD increases a caregiver's burden from a variety of expected sources such as the physical and emotional strain of providing care often with inadequate support from other family members. Although some studies on filial obligation in Mexico have shown increasing preference for shared responsibilities for caregiving among family members (Robles & Pérez, 2012), others find an increasing divergence between academic discourse on the topic and the reality regarding ideal filial obligation and caregiving. While most individuals declare that children ideally should care for their parents, in reality, they are increasingly less able to do so due to other responsibilities (López-Ortega & Gutiérrez-Robledo, 2015).

Worldwide, psychological interventions with caregivers involve training them to use behavioral management techniques based on different models and approaches such as the progressive lowered stress threshold model (Smith, Gerdner, Hall, & Buckwalter, 2004), psychoeducational interventions with caregivers (Beinart, Weinman, Wade, & Brady, 2012), and supportive counseling and family-based therapy (Mitrani & Czaja, 2000). The results are mixed: those reporting effects on the patient showed improvements at 6 months in mood and ideational disturbance, lowered rate of institutionalization, and a decrease in agitation, as well as improvement in the mental health of the caregivers (Livingston et al., 2005). However, behavioral management techniques centered on individual behaviors were in general more successful in having long-lasting effect over the BPSD. These different types of interventions are all time-consuming, require effort and motivation both from the patient and the caregiver, and also require extensive education for all concerned, conditions that are not easily met in Latin American countries where, as previously mentioned, caregiving is embedded in the life experience of families.

Quality of Care

Quality of care offered by caregivers to elders is another issue to be considered when studying caregiving of Latinos with dementia. Abuse, mistreatment, and social and family violence are widespread conditions that affect many people around the world. The presence of cognitive impairment in the elder population has been associated with the increased likelihood of abuse and social exclusion (Rodríguez, Sotolongo, Gladys, & Calvo, 2012). This points to the social invisibility of this group, the unequal distribution of wealth and services, and the abuse of elders, predominantly of those who live in poverty. In a study in four cities in Mexico, Ruelas González and Salgado de Snyder (2009) found that the prevalence of elder abuses was 23% toward women and 9% toward men. In their study, not having a spouse, having more medical conditions, and depression were the factors associated with elder abuse which, in most cases, the abuse is not reported for fear of retaliation and further abuse. Based on reports from the National Center on Elder Abuse, one in two people with dementia in the USA experience some form of abuse by others (Wiglesworth et al., 2010). Results from the 10/66 study in Mexico (Zúñiga-Santamaría, Sosa-Ortiz, Alonso-Vilatela, Acosta-Castillo, & Casas-Martínez, 2010) find that 26% of the elders with mild to moderate dementia experience abuse. Emotional violence, followed by neglect, was the most common types of abuse. Some of the factors reported to be associated with elder abuse are poor health condition, depression, poverty, lack of family support, social isolation, and a complicated emotional bond between the abuser and the abused (Corsi, 2001).

Cultural Context

The importance that cultural background and social context has on the way that informal care takes place in Mexico is undeniable. Strong cultural values such as familism prioritize reliance on family members rather than on institutions for instrumental, emotional, and material support (Campos, Ullman, Aguilera, & Dunkel Schetter, 2014). This strong involvement in the care of a person with dementia by their family members is done as a duty and as an expression of respect and commitment for the elder. However, several challenges to maintaining this pattern of support are, on the one hand, a shift from extended to nuclear families, rising migration rates, and the empowerment of Mexican women (Nobles, 2013) all of which have an impact on the number of family members available for caregiving. On the other hand, there exists high economic vulnerability for families.

Cultural beliefs concerning the causes and treatment of mental illness can also act as barriers to treatment compliance, not only by the patient but also by caregivers. Family caregivers draw on their cultural and personal resources to understand the nature and meaning of illness. Latinos use family issues as a cause of mental illness (Jimenez, Bartels, Cardenas, Dhaliwal, & Alegría, 2012) attributing symptoms to events that already happened and thus ignoring formal mental health interventions.

Low levels of education among the Mexican older adult population (Wong et al., 2015) also limit treatment engagement, thus preventing dementia symptoms, particularly BPSD, to be addressed and managed correctly. Lack of information about dementia as a progressive, degenerative medical condition in individuals with low education is worsened by the mistaken perception of dementia either as part of "normal aging" or the result of factors explained through cultural beliefs (Gallagher-Thompson, Talamantes, Ramirez, & Valverde, 1996).

Health and Psychological Background

Multiple social, psychological, and biological factors determine the level of health of a person at any point in time. Ultimately, it's the interplay of psychological and sociocultural factors with biological factors that configures the consequences of any illness and allows a comprehensive understanding and management of a person's illness. Among other psychological factors, the role of depression is crucial for poor health outcomes in the elder. The prevalence of depressive symptoms in the Mexican elder population has been estimated between 12% and 50%, and the risk factors include being female, having no paid job, being a caregiver, high levels of poverty, low education, and widowhood (Ávila-Funes, Melano-Carranza, Payette, & Amieva, 2007; Bojorquez-Chapela, Villalobos-Daniel, Manrique-Espinoza, Tellez-Rojo, & Salinas-Rodríguez, 2009).

Several studies have reported a positive association between depression and dementia. When assessing the risk of dementia in diabetic elders, our group (Mejia-Arango & Zuniga, 2011) found that the coexistence of diabetes and depression had almost a fourfold increase on the risk for dementia compared to the increased risk of diabetes alone (RR, 2.71, 95%; CI 1.73–4.24). Factors such as obesity, lack of physical activity, bad habits of self-care, hypercholesterolemia, inflammatory effects, and, in general, the glucometabolic imbalance present in diabetes have all been considered possible mechanisms that mediate this relationship (Wilson et al., 2010).

Specifically for the association between depression and dementia, the hypotheses that have been raised to explain this relationship include (a) depression as a consequence of dementia, (b) depression and dementia as independent conditions that share risk factors and tend to be comorbid, and (c) depression as risk factor for developing dementia (Bennett & Thomas, 2014; Byers & Yaffe, 2011). Mexican elders with both conditions in our study did not show a higher association of obesity, smoking, or a sedentary style. Neither was depression a secondary response to dementia since, at baseline, participants were not demented; however, comorbidity with other cardiovascular risk factors such as hypertension and heart disease was present in the group. This information is evidence of another characteristic that Mexican elders have and that needs to be considered when defining what caring for Latinos with dementia means (i.e., multicomorbidity). Multicomorbidity of the population involves not only the coexistence of different medical conditions but also psychological conditions such as depression that have a negative impact on quality of life. Vulnerability in this group is an expression of different factors including social, economic, health, and psychological aspects.

Recommendations for Practitioners

The ways that dementia is understood and dealt with are not the same in every country. Actually, differences in how people acknowledge the condition may be present among regions within a country (e.g., urban vs. rural). Dementia among Mexicans may be seen as a normal part of aging or as an expression of madness. In both cases, the search for a diagnosis is reduced, thus limiting the possibility of achieving quality of life and care for dementia patients and their caregivers. The need for a fairly consistent understanding of dementia as a medical condition caused by several diseases should be addressed through educational programs with the community and implemented by the health system at national and local levels. For behavioral and psychological symptoms of dementia, education is particularly important because some of these are also present in elders without dementia which gives the family a sense of normality. Stress or excessive worrying could be misinterpreted as the cause of behavioral problems, again limiting the opportunity for adequate management of the disease.

On the other hand, considering the influence that cultural background has on how dementia is approached by the family, it is important that health professionals become aware that culture and ethnicity guide and affect the behavior of those who have to manage, on a daily basis, the problems experienced by a patient with dementia. Insensitivity of health practitioners to cultural beliefs has shown to reduce the search for help (Gallagher-Thompson et al., 2003) and can lead to a sense of mistrust in the health-care system. Educating health practitioners to understand the need for culturally sensitive interventions will allow them to help caregivers identify maladaptive, culturally based assumptions and incorporate the recommendations needed to manage behavioral and psychological symptomatology.

Some examples of culturally sensitive intervention programs with dementia patients come from studies done with Latino families in the USA such as *Círculo de Cuidado* (Circle of Care), a Spanish language, culturally sensitive, targeted cognitive-behavioral therapy (CBT) group intervention for Latinos with Alzheimer's disease and related disorders (ADRD) and caregivers coping with their relatives' neuropsychiatric symptoms (Gonyea, López, & Velásquez, 2016). The groups have shown the need to incorporate several factors in the way that health practitioners help the patient with dementia and their caregivers:

- Individualized or person-centered interventions, focused on behaviors and tailored to the specific concerns of each caregiver.
- Guidance for caregivers to better identify and manage their own distress.
- Use of problem-solving exercises.
- Reinforcement of caregivers' sense of mastery of new skills.
- Removal of linguistic barriers and allowing participants to better articulate the nature of their caregiving experience within their own cultural framework and addressing a critical need "to be heard".
- Engaging caregivers and care recipients in pleasant activities and improving communication.
- Use of relaxation techniques or exercises.
- Incorporating personal connections (*personalismo*) in every stage.

Conclusion

The rapidly aging population and increasing old-age-related diseases such as Alzheimer's disease and other types of dementia represent a high impact upon countries like Mexico. Long-term care systems in different countries include economic incentives to financially support caregivers, and these have been supported through labor legislation. These incentives include a family caregivers' ability to leave their job or have reduced working hours. Similarly, these incentives allow caregivers—particularly those with greater care needs such as those caring for family members with dementia—to continue, at least partially, with their personal activities or to have respite from their caregiving tasks. The aforementioned types of strategies are completely lacking in Mexico. Further research is needed to extend our knowledge regarding who is caring for older adults with dementia, what type of activities they are performing, and at what intensity level are they being performed. Knowing this information can help generate strategies to support these older adults and their primary caregivers.

No specialized diagnosis and treatment for Alzheimer's disease and other dementias or formal, publicly available long-term care services exist in Mexico, and the availability of specific treatment of BPSD symptoms of dementia is also almost nonexistent. It is critically necessary for the health sector to recognize the importance of BPSD and to generate strategies for training primary care practitioners to appreciate their importance and to develop programs to address the treatment concerns of this population. In addition, there is an urgent need for health and social development sectors to consider the high burden that is currently placed on caregivers. Relying on informal care from families in Mexico needs to be reassessed against the backdrop of the social and demographic changes that have led to increasing opportunities for women to carry out activities outside the home, their greater access to formal education, and better opportunities at employment, combined with changes in marriage, divorce, and fertility rates that contribute to increasingly smaller family networks, reduced numbers of potential caregivers, and ultimately to what has been named the "crisis of care" (Arriagada, 2010).

The accelerated process of aging of the Mexican population that has occurred in the past years has led to the need by the health system to prioritize the timely diagnosis of degenerative diseases including dementia, but has left the intervention as an afterthought, and has remained exclusively reliant on the family as an instrument for the application of treatments. As dementia contributes greatly to disability and dependence in older adults, there is a large need to work toward guaranteeing care and support strategies for those with dementia and their family members. As previously discussed, the timely diagnosis and treatment of symptoms of Alzheimer's disease and other dementias in Mexico is still scarce given that no specific national-level policies are currently in place. In 2014, in response to this situation, and to better improve the well-being of people with dementia and related conditions in Mexico, the National Institute of Geriatrics of Mexico's Ministry of Health, together with civil society organizations, has generated a National Alzheimer and Other Dementias Plan (Gutiérrez Robledo & Arrieta Cruz, 2014). The Plan follows the World Health Organization and Alzheimer's Disease International report on Dementia (World Health Organization, 2012), as well as the directives issued during the G8 World Leaders' Summit in 2013 (Global Action against Dementia, 2013). It establishes four guidelines: (1) awareness of the problem, (2) de-stigmatization, (3) promoting the well-being of those with dementia and their families, and (4) prioritizing care and diverse treatments for this condition. Specifically, of the many objectives, numbers eight and nine aim to improve the quality of care delivered by care workers and health professionals in care homes, seek to improve the appropriate antipsychotic prescription to treat challenging behavior, and offer innovative therapies (Gutiérrez Robledo & Arrieta Cruz, 2014). With the implementation of specific strategies that track with the Plan, the expectation is that in the near future, Mexico will provide timely detection of Alzheimer's disease and other dementias, as well as provide appropriate treatments according to the needs of each older adult, their diagnosis, and progression of the condition, including interventions for biological and psychological symptoms of dementia.

References

Angel, J. L., Angel, R. J., López-Ortega, M., Robledo, L. M. G., & Wallace, R. B. (2016). Institutional context of family eldercare in Mexico and the United States. *Journal of Cross-Cultural Gerontology, 31*(3), 327–336.

Arriagada, I. (2010). La crisis del cuidado en Chile. *Revista de Ciencias Sociales, (23)*27, 58–67.

Ávila-Funes, J. A., Melano-Carranza, E., Payette, H., & Amieva, H. (2007). Síntomas depresivos como factor de riesgo de dependencia en adultos mayores. *Salud Pública de México, 49*(5), 367–375.

Azermai, M., Petrovic, M., Elseviers, M. M., Bourgeois, J., Van Bortel, L. M., & Vander Stichele, R. H. (2012). Systematic appraisal of dementia guidelines for the management of behavioural and psychological symptoms. *Ageing Research Reviews, 11*(1), 78–86.

Beinart, N., Weinman, J., Wade, D., & Brady, R. (2012). Caregiver burden and psychoeducational interventions in Alzheimer's disease: A review. *Dementia and Geriatric Cognitive Disorders extra, 2*(1), 638–648.

Bennett, S., & Thomas, A. J. (2014). Depression and dementia: Cause, consequence or coincidence? *Maturitas, 79*, 184–190.

Bojorquez-Chapela, I., Villalobos-Daniel, V. E., Manrique-Espinoza, B. S., Tellez-Rojo, M. M., & Salinas-Rodríguez, A. (2009). Depressive symptoms among poor older adults in Mexico: Prevalence and associated factors. *Revista Panamericana de Salud Publica, 26*(1), 70–77.

Byers, A. L., & Yaffe, K. (2011). Depression and risk of developing dementia. *Nature Reviews Neurology, 7*, 323–331.

Campbell, P., Wright, J., Oyebode, J., Job, D., Crome, P., Bentham, P., & Lendon, C. (2008). Determinants of burden in those who care for someone with dementia. *International Journal of Geriatric Psychiatry, 23*(10), 1078–1085.

Campos, B., Ullman, J. B., Aguilera, A., & Dunkel Schetter, C. (2014). Familism and psychological health: The intervening role of closeness and social support. *Cultural Diversity and Ethnic Minority Psychology, 20*(2), 191.

Cohen-Mansfield, J. (2001). Nonpharmacologic interventions for inappropriate behaviors in dementia: A review, summary, and critique. *The American Journal of Geriatric Psychiatry, 9*(4), 361–381.

Corsi, J. (2001). *Violencia Familiar: Una Mirada interdisciplinaria sobre un grave problema social*. Argentina: Paidos.

Ferri, C. P., Ames, D., & Prince, M. (2004). Behavioral and psychological symptoms of dementia in developing countries. *International Psychogeriatrics, 16*(4), 441–459.

Finkel, S. I., e Silva, J. C., Cohen, G., Miller, S., & Sartorius, N. (1997). Behavioral and psychological signs and symptoms of dementia: A consensus statement on current knowledge and implications for research and treatment. *International Psychogeriatrics, 8*(S3), 497–500.

Frisoni, G., Rozzini, L., Gozzetti, A., Binetti, G., Zanetti, O., Bianchetti, A., & Cummings, J. (1999). Behavioral syndromes in Alzheimer's disease: Description and correlates. *Dementia and Geriatric Cognitive Disorders, 10*(2), 130–138.

Gallagher-Thompson, D., Haley, W., Guy, D., Rupert, M., Argüelles, T., Zeiss, L. M., & Ory, M. (2003). Tailoring psychological interventions for ethnically diverse dementia caregivers. *Clinical Psychology: Science and Practice, 10*(4), 423–438.

Gallagher-Thompson, D., Talamantes, M., Ramirez, R., & Valverde, I. (1996). Service delivery issues and recommendations for working with Mexican American family caregivers. *Ethnicity and the Dementias*, 137–152.

Gauthier, S., Cummings, J., Ballard, C., Brodaty, H., Grossberg, G., Robert, P., & Lyketsos, C. (2010). Management of behavioral problems in Alzheimer's disease. *International Psychogeriatrics, 22*(03), 346–372.

Gibbons, C., Creese, J., Tran, M., Brazil, K., Chambers, L., Weaver, B., & Bédard, M. (2014). The psychological and health consequences of caring for a spouse with dementia: A critical comparison of husbands and wives. *Journal of Women & Aging, 26*(1), 3–21.

Global Action against Dementia. (2013). G8 Dementia Summit, UK. Retrieved from https://www.gov.uk/government/uploads/system/uploads/attachment_data/file/265869/2901668_G8_DementiaSummitDeclaration_acc.pdf.

Gonyea, J. G., López, L. M., & Velásquez, E. H. (2016). The effectiveness of a culturally sensitive cognitive behavioral group intervention for Latino Alzheimer's caregivers. *The Gerontologist, 56*(2), 292–302.

Gutiérrez Robledo LM, & Arrieta Cruz I (coords.) (2014). *Plan de Acción Alzheimer y otras Demencias*. México. 2014. México: Instituto Nacional de Geriatría/Secretaría de Salud.

Hinton, L., Chambers, D., Velásquez, A., Gonzalez, H., & Haan, M. (2006). Dementia neuropsychiatric symptom severity, help-seeking patterns, and family caregiver unmet needs in the Sacramento Area Latino Study on Aging (SALSA). *Clinical Gerontologist, 29*(4), 1–15.

Hinton, L., Haan, M., Geller, S., & Mungas, D. (2003). Neuropsychiatric symptoms in Latino elders with dementia or cognitive impairment without dementia and factors that modify their association with caregiver depression. *The Gerontologist, 43*(5), 669–677.

Hinton, L., Tomaszewski Farias, S., & Wegelin, J. (2008). Neuropsychiatric symptoms are associated with disability in cognitively impaired Latino elderly with and without dementia: Results from the Sacramento Area Latino study on Aging. *International Journal of Geriatric Psychiatry, 23*(1), 102–108.

INEGI. (2014). *Estadísticas a propósito del día internacional de las personas de edad. Datos Nacionales*. México: Instituto Nacional de Geografía y Estadística.

INEGI. (2015). *Sistema de cuentas nacionales de México. Cuenta satélite del sector salud de México 2013. Preliminar. Año base 2008*. México: Instituto Nacional de Estadística y Geografía.

Jimenez, D. E., Bartels, S. J., Cardenas, V., Dhaliwal, S. S., & Alegría, M. (2012). Cultural beliefs and mental health treatment preferences of ethnically diverse older adult consumers in primary care. *The American Journal of Geriatric Psychiatry, 20*(6), 533–542.

Jorm, A. F. (2000). Does old age reduce the risk of anxiety and depression? A review of epidemiological studies across the adult life span. *Psychological Medicine, 30*(01), 11–22.

Kar, N. (2009). Behavioral and psychological symptoms of dementia and their management. *Indian Journal of Psychiatry, 51*(5), 77.

Livingston, G., Johnston, K., Katona, C., Paton, J., & Lyketsos, C. G. (2005). Systematic review of psychological approaches to the management of neuropsychiatric symptoms of dementia. *American Journal of Psychiatry, 162*(11), 1996–2021.

López-Ortega, M. (2014). The family household and informal old age care in Mexico. *International Journal of Sociology of the Family, 40*(2), 235–246.

López-Ortega, M., & Gutiérrez-Robledo, L. M. (2015). Percepciones y valores en torno a los cuidados de las personas adultas mayores. In: Luis Miguel Gutiérrez Robledo y Liliana Giraldo (coord.) *Realidades y expectativas frente a la nueva vejez.* Encuesta Nacional de Envejecimiento. Ciudad de México: Instituto de Investigaciones Jurídicas, Universidad Nacional Autónoma de México, pp. 113–133.

Lövheim, H., Sandman, P. O., Kallin, K., Karlsson, S., & Gustafson, Y. (2006). Relationship between antipsychotic drug use and behavioral and psychological symptoms of dementia in old people with cognitive impairment living in geriatric care. *International Psychogeriatrics, 18*(04), 713–726.

McDowell, I., Hill, G., Nilsson, T. H., & Kozma, A. (2002). Patterns and health effects of caring for people with dementia. *The Gerontologist, 42*(5), 643–652.

Mejia-Arango, S., & Gutierrez, L. M. (2011). Prevalence and incidence rates of dementia and cognitive impairment no dementia in the Mexican population: Data from the Mexican Health and Aging Study. *Journal of Aging and Health, 23*(7), 1050–1074.

Mejía-Arango, S., Wong, R., & Michaels-Obregón, A. (2015). Normative and standardized data for cognitive measures in the Mexican Health and Aging Study. *Salud Pública de México, 57*, s90–s96.

Mejia-Arango, S., & Zuniga, C. (2011). Diabetes mellitus como factor de riesgo de demencia en la población adulta mayor mexicana. *Revista de Neurologia, 53*(7), 397.

Mitrani, V., & Czaja, S. (2000). Family-based therapy for dementia caregivers: Clinical observations. *Aging & Mental Health, 4*, 200–209.

Montes de Oca, V. (1999). Diferencias de género en el sistema de apoyo a la población envejecida en México, en Papeles de Población, Centro de Investigación y Estudios Avanzados de la Población, UAEM, Nueva época Año 5, No. 19, enero-marzo, p. 149–172.

Nigenda, G., López-Ortega, M., Matarazzo, C., & Juárez-Ramírez, C. (2007). La atención de los enfermos y discapacitados en el hogar. Retos para el sistema de salud mexicano. *Salud Pública de México, 49*(4), 286–294.

Nobles, J. (2013). Migration and father absence: Shifting family structure in Mexico. *Demography, 50*(4), 1303–1314.

Prince, M. (2000). Dementia in developing countries. A consensus statement from the 10/66 Dementia Research Group. *International Journal of Geriatric Psychiatry, 15*(1), 14–20.

Prince, M. (2004). Care arrangements for people with dementia in developing countries. *International Journal of Geriatric Psychiatry, 19*(2), 170–177.

Prince, M., Comas-Herrera, A., Knapp, M., Guerchet, M., & Karagiannidou, M. (2016). World Alzheimer report 2016. Improving healthcare for people living with dementia: Coverage, quality and costs now and in the future.

Robles, L. & Pérez, A. C. (2012). Expectativas sobre la obligación filial: Comparación de dos generaciones en México. Revista Latinoamericana de Ciencias Sociales, *Niñez y Juventud, 10*(1), 527–540.

Robles Silva, L. (2001). El fenómeno de las cuidadoras: Un efecto invisible del envejecimiento. *Estudios Demográficos y Urbanos, 16*(3), 61–584.

Rodríguez L, Sotolongo O, Gladys L, & Calvo M. (2012). Violencia sobre personas de la tercera edad con demencia Policlínico Cristóbal Labra Lisa.2010. *Rev Haban Cienc Méd* [Internet]. *11*(5), 709–726.

Ruelas González, M. G., & Salgado de Snyder, V. N. (2009). Factores asociados con el auto-reporte de maltrato en adultos mayores de México. *Rev Chil Salud Publica, 13*(2), 90–99.

Savva, G. M., Zaccai, J., Matthews, F. E., Davidson, J. E., McKeith, I., & Brayne, C. (2009). Prevalence, correlates and course of behavioural and psychological symptoms of dementia in the population. *The British Journal of Psychiatry, 194*(3), 212–219.

Sawatzky, J. E., & Fowler-Kerry, S. (2003). Impact of caregiving: Listening to the voice of informal caregivers. *Journal of Psychiatric and Mental Health Nursing, 10*(3), 277–286.

Schulz, R., & Sherwood, P. R. (2008). Physical and mental health effects of family caregiving. *Journal of Social Work Education, 44*(3), 105–113.

Smith, M., Gerdner, L. A., Hall, G. R., & Buckwalter, K. C. (2004). History, development, and future of the progressively lowered stress threshold: A conceptual model for dementia care. *Journal of the American Geriatrics Society, 52*(10), 1755–1760.

Thiyagarajan, J., Prince, M., & Webber, M. (2014). Social support network typologies and health outcomes of older people in low and middle income countries–A 10/66 Dementia Research Group population-based study. *International Review of Psychiatry, 26*(4), 476–485.

Wenger, G. C. (1991). A network typology: From theory to practice. *Journal of Aging Studies, 5*(2), 147–162.

Wiglesworth, A., Mosqueda, L., Mulnard, R., Liao, S., Gibbs, L., & Fitzgerald, W. (2010). Screening for abuse and neglect of people with dementia. *Journal of the American Geriatrics Society, 58*(3), 493–500.

Wilson, R. S., Hoganson, G. M., Rajan, K. B., Barnes, L. L., De Leon, C. M., & Evans, D. A. (2010). Temporal course of depressive symptoms during the development of Alzheimer disease. *Neurology, 75*(1), 21–26.

Wong, R., Michaels-Obregón, A., Palloni, A., Gutiérrez-Robledo, L. M., González-González, C., López-Ortega, M., … Mendoza-Alvarado, L. R. (2015). Progression of aging in Mexico: the Mexican Health and Aging Study (MHAS). *Salud Pública de México, 57*, S79–S89.

World Health Organization. (2012). *Dementia: A Public Health Priority*. Washington, DC. Retrieved from http://apps.who.int/iris/bitstream/10665/98377/1/9789275318256_spa.pdf.

Zúñiga-Santamaría, T., Sosa-Ortiz, A. L., Alonso-Vilatela, M. E., Acosta-Castillo, I., & Casas-Martínez, M. D. L. L. (2010). Dependencia y maltrato en el anciano con demencia. *Persona y Bioética, 14*, 1.

Dementia Care in Guatemala, Central America

16

Paola Alejandra Andrade Calderón and Tedd Judd

Abstract

Guatemala is a country that is very diverse linguistically, culturally, educationally, and economically. Half the country's population is Mayan, speaks 23 different languages, and lives mostly in rural impoverishment with limited formal education. Most of the remainder of the population is Spanish-speaking and varies in income and education. Dementia care in the country likewise varies widely. Specialized dementia care is in its early stages of development. For example, medical understanding of dementia, its subtypes, and its evaluation and treatment are limited even within the medical professions, and such services reach only small parts of the population, predominantly urban Spanish speakers with strong financial means. Throughout Guatemala, at legal, institutional, and professional levels both public and private, there is considerable interest in modernizing and destigmatizing dementia care and offering greater support to families. The most promising directions in this regard appear to be in developing: dementia awareness and services within the newly developing social services for the elderly; dementia clinics and medical services within the large hospitals; dementia day health and residential care units; and specialized skill in the private sector, especially in neurology, gerontology, and neuropsychology. At a societal level, the challenge will be to develop such services in a manner that offers medical, behavioral, and social support without further stigmatizing dementia or undermining existing positive values and support that are present and reflected in the population's religious beliefs, alternative healing systems, and family and community cultural systems. Close professional and population ties between Guatemala, Mexico, and the United States (U.S.) are likely to have a considerable influence on these processes.

P. A. Andrade Calderón (✉) · T. Judd
Universidad del Valle de Guatemala,
Guatemala City, Guatemala
e-mail: paandrade@uvg.edu.gt; teddjudd@gmail.com

Introduction

Guatemala is located in Central America, bordered by Mexico and Belize to the north, Honduras and El Salvador to the south, the Pacific Ocean to the west, and the Caribbean to the east. The country is considered to have a 40% indigenous population representing 23 spoken languages and a 60% Latino, Spanish-speaking population. Economically, there are great disparities of wealth and high levels of poverty. The population is primarily young and rural; however, the aging population is urbanizing (INE, 2002), and there are many challenges to dementia care in Guatemala.

H. Y. Adames, Y. N. Tazeau (eds.), *Caring for Latinxs with Dementia in a Globalized World*,
https://doi.org/10.1007/978-1-0716-0132-7_16

Given the great diversity of language, culture, and economic resources in Guatemala, practices in dementia care are also greatly varied. In the population at large, and among both public and private mental health and social services, the dementias are generally not widely recognized, nor are they clearly differentiated from other health-related problems of aging. Specialized dementia care is in its beginning stages in Guatemala. In addition, dementia "caregiving" may not even be greatly distinguishable from "just getting on with life." For instance, in our experience, culturally-based attitudes can vary extensively, from great respect for elders, to not recognizing dementia, viewing dementia as a spiritual or mental health problem, infantilization of patients, and its associated stigma, and to a medical model interpretation of the disease process. Concerns related to poverty, lack of transportation, differences in language, and cultural practices along with scarcity of dementia resources often limit families' access to medical, government, and social resources for dementia care.

The number of Guatemalans immigrating to other countries such as the United States, Mexico, and Canada is substantial. Understanding the diversity within Guatemala can assist dementia interventionists from other countries to better comprehend the expectations of this diverse ethnic group. Within Guatemala, there is also a small ex-patriot community of North Americans and other international citizens living in the country who may have been attracted to Guatemala because of its agreeable climate, low cost of living, and other sociocultural factors (Instituto Guatemalteco de Turismo [INGUAT], 2017). Many of these are retirees, and some of these are developing dementia. Residential dementia care based on a North American model is beginning to develop in Guatemala and elsewhere in Central America to serve this population and to attract "medical tourism" from the North because of lower costs. This emerging pattern may also influence the development of local dementia care. On the reverse side of this migration pattern, over a million Guatemalans live in the United States (U.S.) (Pew Research Center, 2012), and it is estimated that over half of these migrants are Mayan.

In this chapter, we aim to foster greater understanding of the ways in which the dementias are viewed and addressed in Guatemala and among Guatemalans so as to help facilitate the development of sensitive and appropriate dementia care throughout the region. To do so, the approach of this chapter is to discuss the broader economic, social, and cultural factors that largely determine the current state of the country's dementia care. While Guatemala differs substantially from its Central American neighboring countries, our hope is that having an awareness and knowledge of the diversity within Guatemala can encourage readers to go beyond this chapter to develop an interest and familiarity with similarly diverse Latin American countries and their approaches to dementia care. Professionals concerned with dementia care need to be aware of this diversity and need to be able to understand the situation of each family they assist in order to be able to assist effectively.

This chapter is based not only on a review of literature given that there is very little scientific literature specifically about dementia in Guatemala. It is also based on direct experience. The first author particularly has visited representative sites of most of the institutions listed in Table 16.1 and has interviewed many of the professionals most directly associated with the development of dementia care in Guatemala.

Demographic Characteristics of Guatemala

Guatemala has a population of 15.6 million inhabitants, of which 51% live in rural areas (National Institute of Statistics, el Instituto Nacional de Estadística [INE], 2016). Guatemala is a multicultural and multilingual country with 40% of the population self-identifying as indigenous and 38.5% as Mayan, and the remaining 1.5% are Garífuna (African-Guatemalan) or Xinca. The K'iche' represent 11.2% of the total population, Q'eqchi' are 9.2%, Kaqchikel are 7.0%, and Mam are 6.1% (INE, 2013). Literacy

Table 16.1 Institutions serving dementia populations in Guatemala

Agency	Status	Coverage	Services provided
Elderly Care			
Ministry of Public Health and Social Assistance (MSPAS)	Public, free	Open to all	Basic health care
Guatemalan Social Security Institute (IGSS)	Public, free care from payroll taxes	Individuals with formal employment, their families, retirees, 17%	Basic health and disability care
Centers for *Comprehensive Medical Care for Pensioners (CAMIP; part of IGSS)*	Public, free care from payroll taxes	IGSS retirees	Geriatric medical clinics, chronic conditions
Social Services Secretariat of the President's wife (SOSEP). National Committee for the Protection of Old Age (CONAPROV)	Public, free	Those in poverty	57 centers (day care) nationwide with transportation, daily food, physical and occupational therapy, primary health-care psychological support
City of Guatemala and other municipalities	Public, free	City residents	Limited health care and preventative care, workshops, senior centers
Elder Residential Care	Private		30% of 70 facilities accept residents with dementia
Dementia Care			
Hospital Roosevelt Clinic of the Elderly (CAM)	Public, free	Open to all	Geriatric medical clinic, dementia clinic 1one day/week
Association Grupo Ermita Alzheimer of Guatemala	Private non-profit NGO, free	Open to all	Dementia medical care, day care, public education
Casa de los Ángeles	Private	Open to all	Living facility designed to care for the elderly with an emphasis on people with Alzheimer's

and written materials in these four languages have only become common in the past decade, and such resources are still uncommon in the indigenous languages that represent smaller numbers of the population. Sixty percent of the population is made up of Ladinos, a term that refers to Spanish speakers who identify with a Westernized and Spanish language culture and who may have a mix of European and indigenous heritage. There is considerable mixing of ethnicity in Guatemala, as well as racism and stigma attached to indigenous identities; thus self-identification may underestimate the indigenous population (Villavicencio, 2017). More than 70% of the population speaks Spanish as their primary language, followed by 29.4% who speak a Mayan language (INE, 2013). In addition, there is considerable bilingualism. Although precise figures are not available, the elderly indigenous populations, especially women, are among those who are most likely to be monolingual in their indigenous language and illiterate and have little to no formal education (INE, 2013).

Guatemala has the largest economy of Central America but some of the greatest economic inequality in all of Latin America. Poverty indices, malnutrition, and maternal-infant mortality are especially high in rural, indigenous zones (World Bank, 2017). Extreme poverty is at 13.33% (1,951,724 people), and general poverty is at 40.38% (5,909,904 people; National Survey of Life Conditions, Encuesta Nacional de Condiciones de Vida [ENCOVI, 2011] (INE, 2011)). According to national surveys, the proportion of the population with a significant injury or illness that did not seek medical care for their medical concern increased from 29.2% in 2000 to 36.2% in 2014. Poverty is the primary reason for this increase. The urban centralization of health services and the lack of adequate public transportation also contribute to the problem (INE, 2011).

Guatemala's Aging Population

The sociodemographic characteristics of Guatemalan society are changing, and the increase in the life expectancy of the population is pushing Guatemala, like the rest of the world, toward facing a demographic transition characterized by a decrease in birth rates and an increase in life expectancy. In Latin America and the Caribbean, the population of adults over 60 years of age tripled from 9 million reported in 1950 (8% of the population) to 41 million in 2000 (10% of the population; United Nations, 2009). In Central America, it is estimated that there are 26.64 million adults over 60 years of age (Alzheimer's Disease International, 2015); and in Guatemala, the National Institute of Statistics (INE, 2013) reported that in 2008 the population aged 65 and over amounted to 584,612 older adults (men and women). There were an estimated 739,518 by 2015 (of a 16,176,133 total population), representing a growth rate of 3.8% compared to 2.34% of the general population. By 2050, there will be approximately 3.7 million people over the age of 60 years; thus, Guatemala will have a much larger aging population in approximately 25 years. In 2015, women accounted for 55.90% of the adult population, and older men represented 48.99%. Of the total number of women, 56% did not have access to education (INE, 2013).

Older adults are located somewhat comparably in urban areas (51%) and rural areas (48%) (INE, 2002). The *departamentos* [districts] with the highest rate of adult population are the districts of Guatemala (which includes the capital, Guatemala City) with 24%—the largest and likely due to the centralization of most services—followed by San Marcos with 7.4% and Huehuetenango with 6.8%. The districts with the lowest rates of older adults are El Progreso with 1.6% (total population of 169,290) and Baja Verapaz with 1.9% (total population of 299,432) (INE, 2002). The distribution of the elderly population has also been affected by poor work structure, shortage of employment opportunities outside of agriculture, and extreme fractionalization of agricultural property (*minifundio*) in rural areas. Some of the highest emigration rates occur in the districts of Baja Verapaz and El Progreso, with rates at 20.5% and 27.7%, respectively (INE, 2002). The reasons for this atypical age distribution of the population are not entirely clear, but it appears that, in these districts, most of the elderly emigrate to the capital or other districts. This implies that a significant number of the population of recently urbanized elderly have become isolated from their familiar countryside, language, and culture. Many individuals were also displaced by the 36 years of civil war that devastated and depopulated many highland Mayan villages—a war that only ended in 1996.

Guatemalans have a life expectancy of 71.72 years (HelpAge International, 2014), and an increase is expected, which can be considered a success in public health policies and socioeconomic development (WHO, 2015). However, this increase in the elderly population is one of the main causes of the increase in the prevalence of chronic and neurodegenerative diseases, such as dementias, which are becoming a current public health problem for Guatemala. It is estimated that there are 47.5 million people with dementia in the world and that each year 7.7 million new cases are recorded (WHO, 2017). It is estimated that 58% of all people with dementia now live in countries classified by the World Bank as low or middle income (Alzheimer's Disease International, 2015). Central America is estimated to have had 1.54 million adults with dementia in 2015, with estimates of 2.97 million for 2030 and 6.88 million for 2050 (Alzheimer's Disease International, 2015). Based on these figures, it is estimated that of the 800,000 older adults currently living in Guatemala, approximately 70,000 to 80,000 could develop some type of dementia. Of this amount, approximately 90% are diagnosed late in the disease process, if at all (INE, 2002). There are many reasons for this situation, beginning with limitations of the health-care system.

Sociological Factors Impacting Mayan and Latino Guatemalan Families Affected by Dementia

In general, it can be said that the primary institution addressing dementia care is the family (Alzheimer Disease International, 2012), and this is especially true in Guatemala. While dementia care in developed countries is increasingly the role of professionals, nursing homes, adult family homes, and specialized dementia care units, these are scarce in Guatemala. Although Guatemalan families are quite diverse in their structure, knowledge, attitudes, and capacity for care, they are especially diverse in their resources and dynamics. The families cannot be stereotyped as having one typical set of values, relationships, or access to information or resources. Larger societal forces are responsible for a great deal of the diversity among families, and so we turn next to those forces.

Guatemala's civil war, although not precisely a race war, to a great extent pitted the largely Latino centralized government and army, often supported by the United States, against a Guatemalan revolutionary movement, which affected many Mayan communities (Asociación de Amigos del País, 2004; Griffith, Stansifer, Horst, & Anderson, 2017). The peace accords of 1996, although still far from fully implemented, included provisions for greater Mayan cultural, linguistic, and political autonomy, along with greater access to the central government and its services. The result is official policies of inclusion, still hindered by the inertia of bureaucracy, habit, racism, and economic inequality.

It is not unreasonable to suspect that the civil war has had consequences for the dementia care of the Guatemalan population. For example, elderly Mayans are often monolingual, illiterate, and poor, with a distrust of centralized institutions, the elderly Maya also have younger family members who may be bilingual, bicultural, educated, and social media savvy. In the best of circumstances, the elders with dementia will be situated in a familiar environment, where they know their community, their neighbors, and the local language(s) and where the community reciprocates in not only knowing them but also in caring for them. Thus, there may be many family members and others to share in caregiver duties, therefore lightening the caregiver burden of any one helping individual family member. The elder's acculturated offspring would be able to negotiate public services for the older adult but would retain the respect and dedication for the elder as associated with the traditional, cultural expectations. In this ideal climate of cultural interdependence, Mayan elders would be most likely to accept help and direction from younger generations, as the younger generations would function as linguistic and cultural brokers to professional dementia services.

In less favorable circumstances, the younger generations will reject the Mayan culture and the care of their elders and migrate to cities or other countries leaving the elderly Maya to fend for themselves. Similarly, when the elderly Maya move or are moved from their original communities, they may experience a sense of abandonment and isolation (Programa de las Naciones Unidas para el Desarrollo PNUD, 2005). Working-class Latino families can experience many similar generational dynamics, but the elders are likely to speak Spanish, be literate, and trust institutions. They will, however, experience many of the same structural barriers to professional care. Middle- and upper-class Guatemalans generally have better access to and trust in professional information and care. Respect for elders from the younger generations is also common. However, elders may be more accustomed to authority and power within the family, and this can present dilemmas when it comes to obtaining dementia care, e.g., for those with anosognosia (the lack of awareness of deficits). In such instances, power struggles concerning family finances, driving, safety, and similar issues can become quite acute.

The INE (2006) has also studied the impact of socioeconomic level, sex, ethnicity, and residence on illness perception and help-seeking by Guatemalans. Women and older people are more likely than others to report health problems. Those in the northwest of the country (a predominantly indigenous area) are the least likely to report health problems on surveys, although they have the poorest health indicators in the nation. This may reflect differences in concepts of health and illness. For example, indigenous responders are less likely to report illness than non-indigenous (24% vs. 21%). This difference is also likely to affect and be affected by access to medications and health services in rural areas. The level of formal education potentially also impacts disease recognition. According to the INE (2006), the health service most frequently consulted is private clinics, especially in urban areas, followed by public health locations and hospitals, clinics in pharmacies, and health locations of the Guatemalan Social Security Institute (IGSS). Although community health centers were opened more than 10 years ago, these are the least consulted. The preference for private clinics demonstrates the dissatisfaction with and inaccessibility of public sector health care.

When the INE survey respondents had a health problem, they reported that they were most likely (51%) to consult a health provider (physician, dentist, psychologist). Other common options are to consult a family member, self-medicate, or do nothing. Indigenous survey respondents are less likely to consult a health provider and equally likely to self-medicate and more likely to consult family, do nothing, or consult a nurse. Respondents in the highest socioeconomic level consult a health professional 70% of the time, while those in the lowest socioeconomic level consult a health professional 16% of the time. Although health care is more costly in the cities, for the poor individual consulting a professional, it takes a much larger proportion of family income to do so. City dwellers report an average 47 minutes to reach a health center, and for rural dwellers, it is 65 minutes to do so. Forty-six percent travel by foot, 29% by bus, and 15% by car. In cities, taxi use is a more common mode of transportation, and in rural areas, it is via bicycle and pickup trucks. These limited transportation options are, of course, much harder on the sick and the poor and make professional health care even more inaccessible.

Spirituality and Alternative and Complementary Medicine

Culture and health are closely related; both greatly influence how resources are used, the nature of the practices, and health behaviors and attitudes. These factors establish the conditions of life and health in a specific geographic region. Culture can determine very distinctive ways of viewing the world, life, health, and illness. Guatemala has been a nominally Catholic country since the Spanish conquest from the 1520s to the 1690s. Traditional Mayan religious beliefs and practices were folded into Catholicism and continue to be practiced and are currently undergoing some revival (United States Department of State, 2016). Evangelical Christianity and Mormonism have made considerable inroads in the past several decades, mostly via missionaries (U.S. Dept. of State, 2016). All of these belief systems have robust traditions of faith healing in Guatemala. Similarly, alternative and complementary medicine in the form of dietary supplements, herbs, and alternative practices are also prominent in Guatemala.

The official health system in Guatemala is based on Western medicine, with an interventionist quality/characteristic. It does not consider the cultural diversity of the population and the population's unique and/or particular knowledge of disease. This current, allopathic system treats people from a singular perspective, i.e., biological, focusing on curing diseases with medications and surgery, without fully considering a patient's other life circumstances, their life conditions, concerns, and spirituality, and excludes alternative, homeopathic, and ethnomedical approaches such as Mayan medicine (Instancia Nacional de Salud, 2004; Instituto de Salud Incluyente, 2010).

The Mayan medical system is built on the three pillars of the Mayan cosmovision of holism, balance, and spirituality. Health is a holistic balance between life forces/energies and between the human, nature, and elements of the greater cosmos. Illness is construed as a lack of that balance. Mayan medicine has traditional practitioners called *comadronas*, *curanderos*, *chayeros*, and *guías espirituales* [midwives, healers, physical medicine doctors, spiritual guides] whose legitimacy rests on their ancestral medical knowledge and the confidence that Mayan families place in them. These practitioners may use medicinal plants, offerings, and ceremonies to restore an individual's balance, and they charge families for these services based on the family's financial means (Icú, 2007). In 2003, the Inclusive Health Model (IHM) [MIS in Spanish] was developed to address public health at the community, family, and individual levels. The MIS is an integrated model that includes the Ministry of Health and Social Welfare working together with Mayan and popular healers (Monzón & Valladares, 2008). The Guatemalan city of Villa Nueva was the first to sign the agreement to implement such a project in June 2017.

These systems may not be invoked in dementia care unless the dementia is perceived as a medical, mental health, or spiritual problem, such as if there are hallucinations or delusions. In Guatemala, our experience is that faith healing is often practiced in parallel with medical healing. For example, squads of evangelists roam public hospitals during visiting hours, loudly praying and singing with whichever patients are willing to participate. Information about patient use of faith healing will sometimes be withheld from medical personnel, especially the pagan practices. However, our direct observations and experiences in the community are that faith healing traditions and practitioners that actively discourage Western medicine are in the minority and that it is less common for faith healing to be used in preference to and in place of medical attention, except when medical attention is not available.

Health-Care and Elder Care Systems in Guatemala

The epidemiological changes and the current characteristics of the health system in Guatemala pose a great challenge to its society. The scarce coverage and poor quality of the different retirement and health-care programs for the elderly is a serious situation due to the lack of infrastructure that can respond to the needs of this population. For example, late diagnosis and the lack of specialized care is a reality in the treatment of elderly people with dementia in Guatemala, despite having the 1997 Protection Act for the Elderly. The law indicates that the State must guarantee and promote the right of those 60 years or older to an adequate standard of living in conditions that offer them education, food, shelter, clothing, geriatric and gerontological health care, recreation, and necessary social services for a purposeful and dignified existence (Congress of the Republic of Guatemala, 1997). This law was intended to produce changes at the level of public scrutiny and to also unify the work, advice, and cooperation of competent public, independent, and private entities. The law was also to develop gerontological programs, as well as to investigate and obtain precise and detailed information of the conditions of life of the elderly population. In fact, it is not a law focused on the individual elderly person with dementia, but exists to coordinate interventions for the timely identification of signs and symptoms that constitute a major neurodegenerative disorder (American Psychiatric Association, 2013) and for the treatment that allows the elder to maintain their best possible quality of life and functionality.

The public institutions that serve dementia populations in Guatemala are outlined in Table 16.1. The main actors of the formal health sector in Guatemala include the Ministry of Public Health and Social Assistance (MSPAS), the Guatemalan Social Security Institute (IGSS), and the municipalities. Also included are universities and entities that offer human services, private organizations, nongovernmental organizations (NGOs), community organizations and cooperation agencies, and professional associations (Health Code, Article 9). The health system is divided into public and private sectors. The public sector consists mainly of the Ministry of Public Health and Social Assistance (MSPAS) and formally provides care to 70% of the population (Pan American Health Organization, 2009). The Guatemalan Social Security Institute (IGSS) provides coverage to less than 17.45% of the population, i.e., those who are formally employed (IGSS, 2007). In theory, these institutions cover over 87% of the population, but many of these services are centralized in the capital of Guatemala City where only 13% of the country's total population is located. In reality, only 48% of the population actually receives services, leaving most rural area residents without adequate health-care access. The Military Health System assists active and retired members of the armed forces and the police, including their families, and serves less than 0.5% of the population. Almost 12% of the population uses the private sector for health-care services (Pan American Health Organization, 2009), including less than 8% who have private insurance. Nongovernmental organizations (NGOs) provide care to about 18% of the population.

With regard to the care of the elderly specifically, programs have been created at the public and private levels, both for profit and non-profit, which seek to support the maintenance of the quality of life and functioning of the elderly population. These include the public program "Golden Years" of the Social Services Secretariat of the President's wife (SOSEP), municipal programs at the country's capital and country's departmental levels, the Center for Comprehensive Medical Care for Pensioners (CAMIP) of the IGSS, clinics in public hospitals, and private associations and centers. The Social Welfare Secretariat of the President's wife (SOSEP) and the National Committee for the Protection of Old Age (CONAPROV) were selected as responsible for promoting, coordinating, conducting, and guiding programs and actions related to the well-being and security of the elderly through the National Program of the Elderly (PRONAM; Congress of the Republic of Guatemala, 1997).

Although the country's previously described Protection Act for the Elderly was approved in 1996, it was not until January 2016 when the formal creation of the National Committee for the Protection of Old Age (CONAPROV) went into effect. Later in April of that year, the National Policy and Plan for Gerontological Care or National Comprehensive Care Plan for the Elderly was formed (SOSEP, 2016). Despite some administrative and legal deficiencies, SOSEP is an attempt to alleviate the lack of care for the elderly. It has proposed, and has as its responsibility, the development of the "My Golden Years" program, which provides comprehensive, daytime, and no direct cost care to a population living in conditions of poverty and extreme poverty. SOSEP currently has 57 centers nationwide and provides free daily food (two snacks and lunch), physical therapy, primary health care (including medical visits, dental care, ophthalmological assessments, daily vital signs, drug control, and minor treatments), psychological support, and occupational therapy. These centers serve a large number of older adults living alone. Two out of ten Guatemalan adults technically qualify for services, given their economic situations, which complicates the challenge of providing sufficient support and care to meet the population's need.

Specific Examples of Guatemalan Elder Care Programs

The Golden Years The *Golden Years* centers do not have specialists who perform the diagnosis of dementia or chart a course of treatment. Individuals with presumed dementia who are in need of such services are referred to public health-care centers where they can be evaluated by a psychiatrist (if the neuropsychiatric symptoms are the primary source of concern) or to a neurologist (if a patient's deficits are mainly related to memory). It is the case that many of the patients and their relatives do not follow through on the initial evaluation due to lack of resources, including time and money to travel to the health system center. Delays in diagnosis and treatment inevitably lead to greater deterioration of the patient's performance of activities of daily living. Fortunately, some of the *Golden Years* centers have a bus service that transports individuals both in the morning and the afternoon. This is the case of the center located in Sansare, El Progreso. The center's administration has indicated that, if this service were not available, patients would not attend given their lack of transportation. In other instances, elders with dementia do not participate in the centers because of mistaken beliefs on behalf of their family members who observe some degree of ability for activities of daily living and thus assume that their elders are sufficiently independent to not need support or assistance.

Municipal Office for the Elderly At the municipal level, the City of Guatemala created the *Municipal Office for the Elderly*, an entity that promotes the social participation of older adults and their connection to a participatory structure of activities. It provides comprehensive care that allows participants to improve their physical and mental conditions. The office addresses health problems, as well as provides supportive techniques that contribute to the improvement of both physical and mental health, such as tai chi, rhythmic gymnastics, and occupational therapy. The program offers psychoeducational workshops that present and explain topics of preventive health and care at the moment of being diagnosed with a chronic, degenerative illness. Three axes guide the approach toward the grouping of older adults: mental health, gender health, and geriatric health. The program seeks to empower older adults by providing psychological treatment, raising awareness of physical health and sexuality, and implementing preventive measures in diseases common with age. Overall, the approach of the municipality of Guatemala is that of prevention by promoting active aging.

General Hospital "San Juan de Dios" and Roosevelt Hospital Although there is no specific program for the population with a diagnosis of dementia, referrals are made to the general hospitals San Juan de Dios and Roosevelt Hospital in cases where the patient requires a specific treatment. In the case of Roosevelt Hospital, the Clinic of the Elderly (CAM) is headed by a geriatrician who directs care and provides for geriatric evaluations. The clinic also initiates pharmacological treatment when appropriate. The clinic is one of the first such projects in a public care hospital dedicated especially to the elderly adult population. When the evaluations confirm a dementia condition, the patient and his or her family are also provided with the clinic's Older Adult School that offers two monthly sessions of psychoeducation services.

Centers for Comprehensive Medical Care for Pensioners (CAMIP)

In view of the need to have a program of care for elderly pensioners, retirees, and beneficiaries, the Guatemalan Social Security Institute (IGSS) proposed the Centers for Comprehensive Medical Care for Pensioners (CAMIP) (NB: not all the population has access to this service.) CAMIP seeks to guarantee better conditions of health for the population previously described, in particular it serves elderly people with diseases such as diabetes, hypertension, and osteopathic/bone problems. Despite the focus on the elderly population, it does not have centers with geriatricians or clinics that treat patients with dementia. In general, the patient who presents with certain changes in cognition and/or behavior is referred to another site with centralized medical specialists.

Association Grupo Ermita Alzheimer of Guatemala Within the difficult reality of the lack of greater care for persons with dementia in Guatemala, there is a private, non-profit association that provides support to the elderly, elders with disabilities, and particularly those with Alzheimer's disease (AD) and their families. Since 1995, the Association Grupo Ermita Alzheimer of Guatemala has reported on the latest research advances in Alzheimer's disease and other dementias. It is the only non-profit center in Guatemala whose work is focused on general, population-based dynamics of dementia treatment. It has the support of Alzheimer Disease International (ADI) and Alzheimer Latin America (ABI), and its President is a member of the Board of Directors of CONAPROV. The program is one that provides care through the intervention of a geriatric physician who determines the diagnosis of dementia, follows up on pharmacological treatment when needed, and recommends physical therapy and supportive services to relatives who have a family member with dementia. All services provided are free of cost to the patient and/or the patient's family. Ermita is a center that provides support during the day; however, both in the Guatemalan capital and in the interior of the country, there are numerous home sites that also provide for the overnight care of older adults, although only some accept patients with dementia symptomatology. Finally, due to the lack of local statistics regarding dementia, Grupo Ermita also has research projects underway to provide data on the prevalence of dementia.

Summary of Health-Care Provision of Dementia Services in Guatemala

The prospect for the care of patients with dementia in Guatemala is not very encouraging because of deficiencies that seem very difficult to change in the short term. Although there are laws that favor specified care, actions are needed to promote and unify the necessary work of public and private institutions. In general, the country is not prepared for the dementia care required of this growing segment of its elderly population, both because hospitals do not have specialists and because there is no geriatric center for elderly care. However, the need for identified care has led to the opening of numerous care homes. Unfortunately, many of these care homes lack the infrastructure and presence of a multidisciplinary team to provide comprehensive care. They function mostly as nursing homes where adults can be "hospitalized" because they can no longer be cared for at home.

According to the directory of geriatric centers and public and private organizations that care for the elderly, there are more than 70 centers, almost 50 of which are in the interior of the country, all of them private, and none with an exclusive focus on care of dementias. The remaining 10% are public. However, up to 30% receive patients with dementia and memory loss, if they have medication to control any presenting behavioral problems. Very few centers comply with the licensing requirements and adherence to the norms of the Ministry of Health to provide their service. Less than five percent of private centers have a geriatrician who provides the required follow-up to pharmacological treatment. However, these are not constituted as dementia care centers. Casa de los Ángeles (2017) is a project that is beginning to operate and is defined as an assisted living facility designed to care for the elderly. Their stated intention is to focus on dementia care and to attract North American residents with the lower cost of care in Guatemala. Presently, it serves foreigners with general neurologic disorders.

Perceptions of the Elderly and of Dementia in Guatemala

Guatemala's greater society's negative perceptions of aging and attitudes regarding dementia add to the deficiencies of the Guatemalan health system regarding the integrative care model that an adult with dementia requires. Undoubtedly, the social and economic inequality that afflicts Guatemalans' day-to-day reality does not stop in its impact on the elderly who, in addition to having a generally low level of formal education and having no access to the labor market, do not have access to basic health services and programs that provide fair pensions and opportunities to generate their own income.

Generally, in our experience, older Guatemalan adults are often excluded, exploited, and, in many cases, considered unproductive so that their quality of life and the respect for their fundamental rights as well as their feelings and aspirations are neglected and disregarded. Unfortunately, a large percentage of this population resorts to public charity for survival and, as a result, often suffers from physical, emotional, and economic abuse, as well as being considered a burden or a hindrance, including by their own relatives. Of course, this situation cannot and should not be representative of the entire Guatemalan population. Cultural diversity underlies variability in the perceptions of aging. For this reason, it is important to note that, more often than not, grandparents in indigenous communities are valued and respected within and outside the community, as their status is seen as synonymous with wisdom, respect, and integrity. The elders play an active part in decision-making, because they are the bearers of the contextual and deeper knowledge of their history, which allows them to guide and advise their community (UNESCO, 2010). This reality is lived especially in the towns of the interior of the Guatemala, while the situation is tempered in some indigenous families that migrate to the capital of the country.

On the other hand, there is still much ignorance about dementia. In the first place, although the term "dementia" has undergone some change, it is still often incorrectly used, leading to much confusion and stigma. Many Guatemalans today believe that dementia is synonymous with "madness" and that individuals who have "dementia" are in psychiatric centers, "tied to their beds," and completely disconnected from reality. Some historical reasons for this perception have been determined. For example, in 1890 in Guatemala City, the first psychiatric hospital in the country opened its doors as the "Asylum for Dementia," where patients were classified as "pensioners, peacefully demented, and furiously insane." It was, undoubtedly, a historical moment for the advancement of mental health in Guatemala since it enabled a specialized care center that sought to improve the quality of treatment of patients with psychiatric conditions (Miranda, 2004). The work was pioneering; however, the terminology used probably influenced the understanding that many people developed for the term "dementia." No doubt many specialists today know the difference between a psychiatric pathology and that which is primarily neurodegenerative. However, not all the population has this information. It is very probable that, at present and due to this perception, a person who is having memory problems consults the neurologist because it is "something of the head," and when the person is "demented or crazy," the individual goes to the psychiatrist. It is telling that even today, Google Translate renders the Spanish *demencia* to the English "insane."

It is important to note that, in spite of advances in science and access to scientific research, many Guatemalan health specialists need to improve their knowledge about the diversity of progressive brain pathologies that impact the functionality of patients. There are other types of dementias apart from Alzheimer's disease (AD), and both the course of the disease and the pharmacological and non-pharmacological approaches to treatment for each are different. However, it seems that the "Alzheimerization" of dementia, i.e., the lack of knowledge about the many other types of dementia apart from Alzheimer's disease, significantly affects the assessment by health specialists in Guatemala, as well as the perception of health deterioration of the elderly patient and its impact on the family. For instance, if there is the presence of short-term memory impairment, according to many specialists, this confirms the presence of Alzheimer's disease (AD). The practitioners may often be correct; however, this assumption precludes them from performing any objective cognitive or neuropsychological assessment that could demonstrate the degree of deterioration and thus confirm if the patient's profile represents AD or some other type of dementia. As suggested throughout this chapter, and as reflected in the nascent attempts of public and private institutions to serve this population, the multidisciplinary work that allows for the construction of a differential diagnosis process and the elaboration of an integrated intervention plan is present in Guatemala only in its earliest of stages. The current, fragmented approach to dementia care also reflects a general, Guatemalan societal piecemeal characteristic of the current work of health specialists.

It is our experience that, regrettably, this same ignorance of the consulted specialists of neurodegenerative diseases is transmitted to the elderly and the population in general. Many of the patients who learn that they do not have Alzheimer's disease but do have dementia are happy and reassured, because they believe that AD is the disease that many people talk about in which "people get lost," "it is 'bad,'" and "there is no cure." They instead believe that dementia (vs. AD) is a condition that can be controlled by medication. It is noteworthy that in many homes and care centers, admission personnel make the distinction by speaking of accepting "healthy, older adults with Alzheimer's Disease" or "persons with dementia." This disinformation about the pathology and its etiology of dementia affects the approach to treatment and therefore the course of the disease. In addition, many Guatemalans still hold the erroneous belief that general aging is synonymous with alterations in cognition, functional impairment, and dependence. Certainly, aging is a multifactorial phenomenon that affects all levels of biological, social, and affective organization. However, that an individual begins with forgetfulness and eventually fully loses memory is not normal, nor inevitable for the aged. This belief also affects the search for care because it is widely expected that all older adults should lose their functioning abilities, especially at the cognitive level, and thus families and patients themselves often do not think to mention such problems to their doctors.

Much of Guatemalan society thinks that the "grandparent who dies healthy and at age 100" is a special case because he or she was always independent and was very active until the last days of his or her life. No doubt, this is likely the kind of healthy aging that most everyone would like to experience. Yet, this is another mistaken belief that leads individuals to delay the search for help until a dementia has advanced to moderate and to severe stages. Unfortunately, it has been our experience that not only is it the family that may hold these beliefs; many physicians also subscribe to this way of thinking, and thus do not initiate the search for options for maintaining and/or improving, if possible, the patient's condition. Once again, the lack of knowledge on the part of the professionals responsible for the decisions that impact the health of the elderly plays an important role in this mistimed approach.

In summary, in Guatemalan society and even among health professionals, there is a limited understanding and acceptance of the medical perspective on Alzheimer's disease and other dementias and evaluation and treatment of these illnesses. This creates a complex and delicate situation for the clinicians who have up-to-date conventional, medical training. Clinicians who are adept at modern evaluation and treatment of the dementias must consider what impact the communication of such diagnoses can have on the patient and the patient's family. Factors including education, urbanization, language, culture, and age will all influence how members of Guatemalan society view and treat their loved ones with dementia. Because accurate dementia education and support services are limited in Guatemala, a diagnosis provided without these key follow-ons can be harmful when it may lead to stigma, isolation, overprotection, exploitation, and even suicide.

It seems the time has come and it is of great necessity to update the perspective held by many Guatemalans of what it is to be an "old" person. Whereas the cultural-historical understanding is that the elderly, particularly those with dementing conditions, are people of little value who represent a burden to society and are unable to care for themselves, health-care professionals can be at the forefront of informing Guatemalan society of medical research findings that can dissuade the population of erroneous and outdated beliefs about the disease process of dementia.

Recommendations for a Guatemalan Integrative Care Approach for Treatment of Dementia

The following is a set of thoughtful considerations that, when applied across the key constituents associated with the care of a patient with dementia and his or her family, can begin to update the Guatemalan population's understanding of dementia. These considerations include the role of families, use of language, explicit consideration of the use of correct medical terminology, help-seeking behaviors of patients and their families, and appropriate use of disclosure.

How might a clinician put all of these considerations together? After making a diagnosis of Alzheimer's disease for an illiterate, village elder who is living alone and residing outside of the village, the clinician can attempt to gain an impression of the patient's family's perspective of the process and outcome. Then, the geriatrician might communicate to the patient and the family: "Don Fulano, you have a disease of your brain in which cells, small parts of your brain, are slowly dying. Your memory and other abilities are slowly getting worse. There is no cure for this disease, but these medicines will help keep you healthy as long as possible. Your daughter should give them to you because you don't remember well enough to take them when you should. It is not safe for you to be using your work tools anymore, so you don't need your machete. But you should keep active with your *cofradia* (community religious/social organization). Your family can help you to remember the meetings and to walk you there. And it is so lonely and not safe for you to be living alone out there anymore. You have been getting lost and having other difficulties. There is a Senior Center in San Juan de Sacatepéquez. Here is note for the social worker there to help you and your family figure out how to keep you safe." The geriatrician may find that the younger and educated family members would be able to comprehend the diagnosis of Alzheimer's disease and access appropriate educational information, including via the Internet.

Role of Families

Professionals must, of course, listen to family members to learn about symptoms and practical problems about their infirm loved one in order to make an accurate diagnosis and offer useful treatment and assistance. In addition, given the great diversity in Guatemalan families, professionals need to be alert to exploring the knowledge, attitudes, skills, and values of family members. Having this information will help clinicians to better understand which family members are likely to play what roles in caregiving, what families may need to know, how families are likely to regard information, what types of care families are likely to be able to offer and implement, and where to direct families to find additional support. Professionals need to be particularly sensitive to understanding whom the various family members are likely to trust and believe. It is not sufficient to presume that families will place full trust in and bestow authority to the interventionist merely because they are a health-care provider.

Use of Language

A best practice for the professional is to work, i.e., speak fluently, in the patient's preferred language. Professionals who speak any of the 23 Mayan languages are relatively rare in Guatemala and even much more difficult to find outside of the country. Linguistically matching patients to professionals is often an administrative challenge. Often, the next best practice is to employ professional interpreters. In North America, the recognized standard of practice when working with Spanish-speaking patients is to employ Spanish-speaking professionals or professional interpreters. However, finding professional interpreters for Mayan languages both within and outside of Guatemala is challenging. Video and phone interpreting, along with active interpreter training efforts, is making professional interpreting increasingly available (National Maya Interpreters' Network, 2017). Spanish-speaking professionals and interpreters who work with Mayans for whom Spanish is a second language need to recognize that they may have difficulty with Spanish, especially with dialects of Spanish from other regions.

Many Mayan families have the expectation that family members who speak Spanish or English will serve as interpreters for others, including interpreting between their family members with suspected dementia and health-care professionals. Such family members rarely have interpreter training and are likely to mix their roles as interpreter and family member. They may answer on behalf of the patient, alter the patient's responses, fail to interpret what is said, add to what is said, and/or interpret poorly. Nevertheless, while use of family members as interpreters is considered poor clinical practice and is even, in most instances, illegal in the United States (Department of Health and Human Services, 2002), it is still a common and even necessary practice in some settings, such as many Guatemalan clinics. Practicalities dictate that professionals be particularly alert to the ethical trade-offs of dealing with such situations so as to optimize both quality and accessibility of care. The Maya Healthcare Toolkit (2017), Wuqu' kawoq (2017), and National Maya Interpreters' Network (2017) can assist providers with this process.

Explicit Consideration of the Use of Correct Medical Terminology

Many Guatemalans and other Latinos associate the word *demencia* [dementia] with psychosis. Many individuals, including physicians, may use "Alzheimer's" for all forms of dementia. It is probably best to avoid the word "dementia," unless the circumstances allow for a full explanation that is likely to be understood and accepted. In this regard, the DSM-5 (American Psychiatric Association, 2013) terms of minor and major neurocognitive disorder may be preferable, even if they are somewhat awkward or unwieldy. It may also be preferable to refer to progressive cerebrovascular disease or small strokes (*derrames, infartos*) rather than "vascular dementia," Lewy body disease rather than "Lewy body dementia," frontotemporal degeneration rather than "frontotemporal dementia," and the like. Such terminology should, of course, be accompanied by explanations. When families are referred to literature, websites, and support groups where the term "dementia" is used, this should then be carefully explained.

Help-Seeking Behaviors of Patients and Their Families

Many Guatemalans may not regard memory loss or behavior changes as medical conditions, or may not seek medical care for such conditions. Primary care providers will need to be proactive in screening for dementia. Even when memory loss or behavior changes are addressed, families may expect that medical attention will be confined to narrow interventions such as solely medications. Families may not expect attention to issues such as safety, competency, advanced directives, caregiving, emotional support, etc. Professional providers will need to be proactive in discussing these important matters.

Appropriate Use of Disclosure Culturally, many Guatemalans will have an expectation that health-care providers will be euphemistic and indirect regarding dementia diagnoses and prognoses. Families may expect that the doctor will be somewhat more frank with the family than with the patient. Families may regard direct communication of the diagnosis with the patient to be potentially damaging to the patient. These expectations likely vary with the age, education, and urbanization of the family members. Nonetheless, providers can address the family's concerns by being alert to signals regarding what the family wants to know, by considering disclosure of diagnosis and prognosis to be a process that may unfold over time, and by considering clinical priorities regarding information disclosure.

Conclusions and Future Directions for Dementia Care in Guatemala

At this point, unfortunately, there is no direct data available concerning the prevalence of dementia due to treatable causes in Guatemala. However, Guatemala has a relatively high prevalence of a number of conditions likely to produce treatable dementia, such as malnutrition, metabolic syndrome, alcoholism, cerebral infections including cysticercosis and dengue fever, hypothyroidism, depression, etc. This suggests that a worthwhile

Guatemalan public health goal would be improved screening for dementia and evaluation for treatable causes. Likewise, the medical work-up for treatable causes of dementia (including vigilance for tropical diseases) should be a priority when evaluating Guatemalans of the diaspora in other countries.

Clearly, there is also work to be done in updating geriatric training programs to include a more comprehensive and modern view of the dementias. In addition, dementia awareness and skill is needed in allied specialties and professions. There is also a need to improve interdisciplinary collaboration. Greater economic resources are needed for care, preventive health, healing, and physical, psychological, emotional, recreational, and mental well-being of those with dementia. It will be necessary to support many existing projects that have demonstrated effectiveness, as well as to plan and develop new projects suitable to Guatemalan socioeconomic and sociopolitical realities. While it is perhaps a priority to first address those with dementia who are isolated and impoverished, such projects need to move beyond individual care to integrative care support approaches that include the family and other caregivers. Ultimately, appropriate dementia care is part of a larger project of improving spaces for elder citizen participation and expression, as well as economic, cultural, and even political integration into Guatemalan society.

References

Alzheimer Disease International [ADI]. (2012). *World Alzheimer report 2012: Overcoming the stigma of dementia*. Retrieved 8/1/17 from https://www.alz.co.uk/research/WorldAlzheimerReport2012.pdf

Alzheimer Disease International [ADI]. (2015). *World Alzheimer report 2015: The global impact of dementia*. Retrieved 8/1/17 from https://www.alz.co.uk/research/world-report-2015

American Psychiatric Association. (2013). *Diagnostic and statistical manual of mental disorders* (5th ed.). Washington, D.C.: Author.

Asociación de Amigos del País. (2004). *Diccionario Histórico Biográfico de Guatemala*. Guatemala. Editorial: Amigos del País, Guatemala.

Casa de los Ángeles. (2017). *Website*. Retrieved 9/27/17 from http://casadelosangeles-alz.com/en

Congress of the Republic of Guatemala. (1997). *Ley de Protección para las Personas de la Tercera Edad Decreto Legislativo No. 80–96*. Retrieved 8/1/17 from http://old.congreso.gob.gt/gt/mostrar_ley.asp?id=880

Department of Health and Human Services. (2002). *Guidance to federal financial assistance recipients regarding Title VI prohibition against national origin discrimination affecting limited English proficient persons*. Retrieved 5/1/07 from http://www.hhs.gov/ocr/lep

Griffith, W. J., Stansifer, C. H., Horst, O. H., & Anderson, T. P. (2017). *Guatemala: Civil War years. Encyclopedia Britannica*. Retrieved 9/29/17 from https://www.britannica.com/place/Guatemala/Civil-war-years

HelpAge International. (2014). *Índice Global de Envejecimiento, AgeWatch 2014. Informe en profundidad*. Retrieved 8/1/17 from http://www.oiss.org/IMG/pdf/ndice_Global_de_Envejecimiento_2014_Informe_en_profundidad_FINAL.pdf

Icú, H. (2007). *Rescate de la medicina Maya e incidencia para su reconocimiento social y político. Estudio de caso Guatemala*. Guatemala: Asociación de Servicios Comunitarios de Salud. Retrieved 8/1/17 from http://www.who.int/social_determinants/resources/csdh_media/mayan_medicine_2007_es.pdf

Instancia Nacional de Salud. (2004). *Una propuesta de salud incluyente. Mediación de la propuesta: hacia un primer nivel de atención en salud incluyente –bases y lineamientos–*. Guatemala: Instancia Nacional de Salud.

Instituto de Salud Incluyente. (2010). *Serie educativa "Por un Modelo Incluyente en Salud"*. Retrieved 8/1/17 from http://www.mspas.gob.gt/images/files/modeloincluyente/ModeloIncluyenteSalud.pdf

Instituto guatemalteco de seguridad social [IGSS]. (2007). *Informe anual de labores 2007*. Retrieved 8/1/17 from http://www.igssgt.org/images/informes/subgerencias/Informe_de_Labores_IGSS_2007.pdf

Instituto Guatemalteco de Turismo [INGUAT]. (2017). *Boletines estadísticos mensuales 2017*. Retrieved 8/1/17 from http://www.inguat.gob.gt/estadisticas/boletines-estadisticos/2017.php

Instituto Nacional de Estadística [INE]. (2002). *Encuesta nacional de condiciones de vida ENCOVI 2000*. Retrieved 8/1/17 from https://www.ine.gob.gt/sistema/uploads/2014/01/15/v8ukWQ78M4VJrnYqyN2o-CumMy1GiBzaf.pdf

Instituto Nacional de Estadística [INE]. (2006). *Encuesta Nacional de Condiciones de Vida ENCOVI-2006*. Retrieved *8/1/17* from https://www.ine.gob.gt/sistema/uploads/2014/01/16/ToW94hMmUnfPw6hGAnGmb2AA7iGw5R8a.pdf

Instituto Nacional de Estadística [INE]. (2011). *Encuesta Nacional de Condiciones de Vida -ENCOVI-2001*. Retrieved *8/1/17* from https://www.ine.gob.gt/sistema/uploads/2014/12/03/qINtWPkxWyP463fpJgnPOQr-jox4JdRBO.pdf

Instituto Nacional de Estadística [INE]. (2013). *Caracterización estadística República de Guatemala 2012*. Retrieved 8/1/17 from https://www.ine.gob.gt/

sistema/uploads/2014/02/26/5eTCcFlHErnaNVeUm m3iabXHaKgXtw0C.pdf

Instituto Nacional de Estadística [INE]. (2016). *Encuesta Nacional de Condiciones de Vida 2014*. Retrieved 8/1/17 from https://www.ine.gob.gt/sistema/uploads/2016/02/03/bWC7f6t7aSbEI4wmuExoNR0oScpSHKyB.pdf

Maya Healthcare Toolkit. (2017). *Maya health toolkit for medical providers*. Bridging Refugee Youth & Children's Services (BRYCS). Retrieved 9/1/17 from http://www.brycs.org/maya-toolkit/

Miranda, H. (2004). *Historia del hospital nacional de salud mental* (tesis de postgrado). Universidad de San Carlos de Guatemala: Citidad de Guatemala, Guatemala.

Monzón, I., & Valladares, R. (2008). *Estudio de costos del Modelo Incluyente en Salud, Informe Final. Guatemala: Proyecto "La construcción social del futuro de la salud"*. Programa de las Naciones Unidas para el Desarrollo, Internacional Development Research Centre. Retrieved 8/1/17 from http://www.saludintegralincluyente.com/ftp/saludintegralincluyente/DOCUMENTOS/PDF/DTN%20paises/Documento-T%C3%A9cnico-Nacional-Guatemala.pdf

National Maya Interpreters' Network. (2017). Retrieved 9/1/17 from http://mayanetwork.breezi.com/

Pan American Health Organization. (2009). *Situación de Salud en las Américas: Indicadores Básicos 2009*. Retrieved 8/1/17 from http://www.paho.org/per/index.php?option=com_content&task=view&id=852&Itemid=558

Pew Research Center. (2012). *Hispanic origin profiles*. Retrieved 9/1/17 from http://www.pewhispanic.org/2012/06/27/country-of-origin-profiles/

Programa de las Naciones Unidas para el Desarrollo [PNUD]. (2005). *Diversidad étnico-cultural y desarrollo humano: La ciudadanía en un Estado plural. Informe Nacional de Desarrollo Humano 2005*. Retrieved 8/1/17 from http://desarrollohumano.org.gt/wp-content/uploads/2016/04/INDH_2005_1.pdf

Secretaría de Obras Sociales de la Esposa del Presidente [SOSEP]. (2016). *Comité Nacional de Protección a la Vejez (CONAPROV)*. Retrieved 8/1/17 from http://www.sosep.gob.gt/?page_id=735

United Nations, Department of Economic and Social Affairs, Population Division. (2009). *World population ageing*. New York: United Nations.

United Nations Educational, Scientific and Cultural Organization [UNESCO]. (2010). *La UNESCO en Guatemala: Una historia de 60 años de cooperación*. Retrieved 8/1/17 from http://unescoguatemala.org/wp-content/uploads/2014/12/LA-UNESCO-EN-GUATEMALA-60-A%2D%2Dos.pdf

United States Department of State. (2016). *International religious freedom report for 2016*. Retrieved 9/23/17 from https://www.state.gov/j/drl/rls/irf/religiousfreedom/index.htm#wrapper

Villavicencio, M. (2017). *Negotiating an indigenous identity in Guatemala City: A survey of Guatemala's population*. Retrieved 10/5/17 from http://anthro-guate.weebly.com/population.html

World Bank. (2017). *Guatemala panorama general*. Retrieved 8/1/17 from http://www.bancomundial.org/es/country/guatemala/overview

World Health Organization. (2015). *Guatemala: WHO statistical profile. Country statistics and global health estimates*. Retrieved 8/1/17 from http://www.who.int/gho/countries/gtm.pdf?ua=1

World Health Organization. (2017). *Dementia*. Retrieved 8/1/17 from http://www.who.int/mediacentre/factsheets/fs362/en/

Wuqu' kawoq. (2017). Retrieved 9/1/17 from www.wuqukawoq.org

Behavioral and Psychosocial Treatments of Dementia in the Caribbean: Cuba, Dominican Republic, and Puerto Rico

17

Ivonne Z. Jiménez-Velázquez,
Juan Llibre-Rodríguez, Daisy Acosta,
Christian E. Schenk-Aldahondo,
and Mackenzie T. Goertz

Abstract

Cuba, the Dominican Republic, and Puerto Rico, otherwise known as the Spanish Insular Caribbean, comprise a subgroup of insular territories that have similar cultural, historical, and colonial backgrounds. The Spanish Insular Caribbean population faces a particularly high risk for dementia, and only steadily emerging, albeit unique behavioral and psychological approaches for treatment. Comprehensive efforts to provide adequate and culturally congruent treatment are complex and characterized by social, economic, political, and public health influences. This chapter begins contextualizing the treatment of dementia by describing the cultural and historical landscape of the Spanish-speaking islands in the Caribbean. Prevalence rates of dementia in these regions are also discussed along with the current strategic, national guidelines for comprehensive care and the systems of healthcare provision in each country. Furthermore, cultural considerations for dementia management are discussed, including a summary of the psychological strengths exemplified within the Cuban, Dominican, and Puerto Rican communities. The chapter concludes with a review of treatment practices and interventions for individuals with dementia and their caregivers.

I. Z. Jiménez-Velázquez (✉)
C. E. Schenk-Aldahondo
Universidad de Puerto Rico, School of Medicine, San Juan, Puerto Rico
e-mail: Ivonne.jimenez1@upr.edu; christian.schenk@upr.edu

J. Llibre-Rodríguez
Universidad de Ciencias Medicas, La Habana, Cuba
e-mail: juan.llibrer@gmail.com

D. Acosta
Universidad Nacional Pedro Henriquez Ureña, Santo Domingo, Dominican Republic
e-mail: daisyacosta1125@gmail.com

M. T. Goertz
Marquette University, Milwaukee, WI, USA
e-mail: Mackenzie.goertz@marquette.edu

Introduction

The Spanish Insular Caribbean refers to those Spanish-speaking countries of the Caribbean region, including Cuba, the Dominican Republic, and Puerto Rico. Together, these populations are experiencing accelerated rates of aging and rapid increases in rates of dementia (Baez & Ibáñez, 2016; Custodio et al., 2017; Ferri & Jacob, 2017; Manes, 2016). This trend mirrors a global rise in neurocognitive disorders in which the worldwide prevalence of dementia is expected to double every

H. Y. Adames, Y. N. Tazeau (eds.), *Caring for Latinxs with Dementia in a Globalized World*,
https://doi.org/10.1007/978-1-0716-0132-7_17

20 years (Prince, Acosta, Castro-Costa, Jackson, & Shaji, 2009). In response, dementia is increasingly being regarded as a public health priority within the Spanish Caribbean, with growing initiatives aimed at its detection, diagnosis, and treatment (Molero, Pino-Ramírez, & Maestre, 2007). At present, dementia remains one of the most serious concerns challenging healthcare services in the Caribbean (Anauati, Galiani, & Weinschelbaum, 2015). While health services are steadily growing, significant advancements are necessary in order to provide culturally responsive, comprehensive care to meet the demands of the rising prevalence rates. Public knowledge about dementia continues to be limited in this region, with families often having little to no awareness of the neurodegenerative disease (Manes, 2016).

The management of dementia is very complex, beginning with its appropriate detection and establishment of accurate diagnosis and severity. Typically, dementia diagnosis requires neuropsychological testing, clinical interview, physical examination, and neuroimaging (Prince et al., 2009), all of which can be difficult to obtain. For example, determining when and where an individual or family member should pursue a diagnostic evaluation may be challenging due to financial burden, lack of information, and/or inability to access appropriate providers. For many individuals in the Spanish Caribbean, meeting with a physician who knows the patient is often the ideal first step, along with the provision of collateral information from a family member. These aspects are often crucial to establish changes from baseline in the patient's cognitive functioning. Thus, establishment of a diagnosis and staging of dementia is the cornerstone of care for this growing population. Moreover, educating the patient and the caregiver is far more effective when the biopsychosocial profile is explored. Typical symptoms (e.g., emotional distress, anxiety, apathy, depression) and behavioral presentations may require consultation with neurology, psychiatry, and psychology services.

The certainty of a dementia diagnosis is frequently controversial and difficult to establish in the early stages of the disease. Therefore, when a physician is asked to confirm if the patient has or may develop dementia, psychoeducation for the patient and the family is warranted. Psychoeducation should focus on the possible etiology, course, and prognosis of the disease, along with ways to manage clinical signs and symptoms, and how it may impact the family system. Clinical experience indicates that there are many places in the Spanish Insular Caribbean wherein general clinicians may be reluctant to establish a diagnostic certainty of dementia for a patient on behalf of their family members, especially at the onset of cognitive deterioration. In part, this reluctance may be attributable to wanting to avoid diagnostic errors which can be potentially confusing to patients who are often of the belief that memory impairment, for example, might be considered a normal aspect of the aging process. However, it is common to see some patients 2–3 years after the initial dementia consult who then demonstrate significant cognitive deterioration that is impacting their functioning.

The incidence and prevalence rates of specific behavioral symptoms vary widely with the type and stage of the dementia, with the presence or lack of other medical conditions, and with institutionalization. Up to 90% of patients with neurocognitive disorders (e.g., dementia, Alzheimer's disease) will develop what is known as behavioral and psychological symptoms of dementia (BPSD) as they progress through the dementia stages (Ferri et al., 2008). These neuropsychiatric symptoms (e.g., emotional distress, anxiety, apathy, depression) represent the largest challenge in daily living for patients and their relatives and are important factors in the increased use of healthcare services and institutionalization (Azermai, 2015; Cerejeira, Lagarto, & Mukaetova-Ladinska, 2012; Kales, Gitlin, & Lyketsos, 2015; Lanctôt et al., 2017; Tible, Riese, Savaskan, & von Gunten, 2017). The nature of dementia care is associated with an intensity of needs, exceeding those required in other health conditions, making care recipients highly dependent on their caregivers to address challenges related to instrumental activities of daily living (IADL) and activities of daily living (ADL; Alzheimer's Association, 2016). The caregiver burden is further exacerbated by families preferring home care for their loved ones as an essential strategy to reduce the cost of long-term care, especially in developing countries. (Black et al., 2013; Chien et al., 2011; Hodgson, Gitlin, Winter, & Hauck, 2013; Jack, O'Brien, Scrutton, Baldry, & Groves, 2015; Kaneda, 2006; OECD, 2011). In addition, families may experience tension associated with the decision to use institutional care, as cultural expectations typically indicate that the elderly person should remain at home rather than residing in a care facility (Neary & Mahoney, 2005). As with any medical condition, clarification of health expectations and goals are at the center of the care for the person with dementia. For patients in the Spanish Caribbean, the role of cultural norms, practices, and values are essential in providing effective, comprehensive, and efficient care.

The goal of this chapter is to describe the current and historical behavioral and psychological treatments for dementia in the Spanish Caribbean. The specific objectives of this chapter are the following: (1) a description of the cultural and historical landscape that contextualizes the current presentation of dementia in the Spanish Caribbean, (2) a review of the available statistics and epidemiology of dementia for the region that are made available through collective and ongoing research initiatives, (3) a presentation of the healthcare systems and strategic guidelines for dementia management that have been implemented within each of the Spanish Caribbean countries, (4) the cultural considerations for dementia management, and (5) a review of treatment practices and supportive interventions for individuals with dementia and their caregivers.

Cultural and Historical Aspects of the Spanish Caribbean

The three island countries of Cuba, the Dominican Republic, and Puerto Rico represent the Spanish Insular Caribbean which are the Spanish-speaking islands in the Western Antilles. These countries have parallel histories but with very different political evolutions. The modern Spanish Insular Caribbean emerged following a separation of the islands colonized by Spain (i.e., the Spanish Greater Antilles) from those that had been colonized by the British (i.e., the Lesser Antilles, Jamaica). The term "West Indies" has also been used to describe their relationship to one another. Moreover, the label of Spanish Caribbean may sometimes be used in a broader context to include the territories of coastal Colombia, Mexico, Panama, and Venezuela. For the purpose of this chapter, the term Spanish Caribbean is used to refer specifically to the collective islands of Cuba, the Dominican Republic, and Puerto Rico.

While each of these countries possesses their own vibrant and unique cultural heritage, there is much that is shared among the islands. Similarities exist within language, folklore, food, art, and social landscape that have grown from the rich legacy of the Indigenous (e.g., Taino, Carib) and African people and the influence of the Spanish European colonizers in the region (Adames & Chavez-Dueñas, 2017). Accordingly, there are many similarities among the islands with regard to how dementia presents, how it is understood, and how it is cared for including the ancestral healing traditions that often permeate interventions of modern, Western medicine. In an effort to help provide a context for these issues, what follows are the cultural highlights and historical aspects of the region.

Racial Landscape

The colonial period in the Spanish Caribbean gave rise to three racial groups: African, Indigenous, and white European. The current racial landscape of the region is a consequence of the colonization acts carried out by white European colonizers, including the exploitation and destruction of the lives of Indigenous and African peoples through the transatlantic slave trade (Chavez-Dueñas, Adames, & Organista, 2014). Understanding the direct implications of racism on those with dementia in the Spanish Caribbean is challenging, in part because the influence of race is typically denied and/or minimized (Adames, Chavez-Dueñas, & Organista, 2016; Borrell, 2005; Chavez-Dueñas et al., 2014).

Another illustration of why it is difficult to examine the role of race is apparent in the lack of data that exists regarding racial demographics in the region. For example, the Cuban government does not collect any data on the racial identity of its citizens (Castellanos & Gloria, 2018). Today, a racialized caste system remains based on skin color and phenotypical characteristics (Adames & Chavez-Dueñas, 2018; Chavez-Dueñas et al., 2014). Within this system, individuals with lighter skin tone and more white European-looking characteristics are often afforded greater access to political, social, and economic resources. In contrast, the experience of Afro-Caribbean individuals is, and has been, that of restricted access to education, healthcare, employment, and economic mobility (Adames & Chavez-Dueñas, 2018; Castellanos & Gloria, 2018; Chavez-Dueñas et al., 2014).

Geographical and Social Landscape

Agricultural and industrial practices have largely shaped the geographical organization of the Spanish Caribbean. The majority of land use (i.e., over 50%) in Cuba and the Dominican Republic is agricultural, compared to Puerto Rico which is only 22% agricultural land use (Central Intelligence Agency [CIA], 2015). Major cities in the region, with some of the highest population densities in the world, include Havana, Cuba, with a population of over 2 million; Santo Domingo of the Dominican Republic, with close to 1 million inhabitants; and the greater metropolitan area of San Juan, Puerto Rico, with approximately 2.5 million inhabitants (CIA, 2015). Residents of the three islands have experienced a large diaspora in the last century with large migrations to the United States of America (USA) and the establishment of well-known communities such as Little Havana in Miami, Florida; Spanish Harlem and Washington Heights in New York City; and various Boston city neighborhoods (Adames & Chavez-Dueñas, 2018; Castellanos & Gloria, 2018). Recently, migration to the USA is accelerating for Puerto Rico more than any other island in the Caribbean (CIA, 2015).

Rates of aging are accelerating in recent years across the Spanish Caribbean islands (Baez & Ibáñez, 2016). To illustrate, Cuba and Puerto Rico both have the largest population of individuals aged 65 and older at 15%, whereas the population of the Dominican Republic is much smaller at only 7% (World Bank, 2017). Life expectancy rates vary, with the age of life expectancy at birth in Cuba at 78.9 years, 81 years in Puerto Rico, and 71.3 years in the Dominican Republic (CIA, 2015). Finally, the overall illiteracy rates for persons over the age of 15 years are 0.2% in Cuba, 8.2% in the Dominican Republic, and 6.7% in Puerto Rico (CIA, 2015).

With regard to nutrition and food practices, common cuisine in the region includes rice; beans; cassava, as well as other starchy tubers; and plantain-based dishes. A combination of meat, rice, and beans is a common meal and often considered to be the basic ingredients of a healthy diet. Access to meat and other animal products may vary due to economic instability on the islands. Frequently, protein-rich grains serve as a substitute for meat. Avocadoes are a popular vegetable rich in fats, and tropical fruits are available year-round. In recent decades, Puerto Rico has seen a surge in fast-food commercial establishments from the U.S. mainland that has influenced a dietary trend of shifting away from traditional nutritional practices.

Incidents of smoking and alcohol consumption are high in the region, in particular amidst thriving tobacco and rum production industries in Cuba and the Dominican Republic. For example, almost 25% of all Cuban adults smoke tobacco, with rates as high as 31% for men and Afro-Cubans (Gorry, 2013). Puerto Ricans also report high rates of smoking, along with other risk factors for cancer including obesity and hepatitis C (American Cancer Society, 2015).

Catholicism is a common religion of the region, with 50% of Cubans identifying as Catholic (Castellanos & Gloria, 2018) and as many as 80% reporting a Catholic religious affiliation in the Dominican Republic (Country Reports, 2016). *Santería*, a blend of African spirituality and Catholic beliefs, is also practiced in the area, particularly among communities in Cuba (Castellanos & Gloria, 2018). Indigenous healing practices such as *espiritísmo* and *curanderismo* are common among Latino communities, including those in the Spanish Caribbean. These practices are grounded in the belief of a spiritual and "invisible world" that influences human behavior (Comas-Díaz, 1981). Examples include belief in syndromes influenced by supernatural forces such as *el mal de ojo* [the evil eye]. In Puerto Rico, *espiritísmo* is practiced widely and is associated with positive psychological outcomes (Torres Rivera, 2005).

Study of Dementia in the Spanish Caribbean

There are few population-based studies regarding the description and frequency of the noncognitive symptomatology of dementia in the Caribbean. The most comprehensive data to date comes from the 10/66 Dementia Research Group, an assemblage of over 30 research clinics across the globe associated with Alzheimer's Disease International (ADI). The name 10/66 refers to the less than one-tenth (i.e., 10%) of all population-based research on dementia that has been directed at the estimated two-thirds (i.e., 66%) of people living with dementia that reside in low- and middle-income countries and territories including Cuba, the Dominican Republic, and Puerto Rico (Prince, 2009). Beginning in 1998, 10/66 researchers have sought to inform the health and social welfare of these individuals and families living with dementia. Readers are encouraged to visit the 10/66 project website at https://www.alz.co.uk/1066/ for a comprehensive review of this ongoing research initiative and for a list of publications to date.

In its first phase, the 10/66 Dementia Research Group conducted cross-sectional catchment area surveys with a total of 21,000 respondents aged 65 and over from 11 countries and territories: Brazil, Cuba, the Dominican Republic, Mexico, Peru, Puerto Rico, and Venezuela in Latin America; China and India in Asia; and Nigeria and South Africa on the African continent (NB: the first three authors of this chapter served as principal investigators of phase one of the project in Cuba, the Dominican Republic, and Puerto Rico; see Ferri et al., 2008; Llibre-Rodríguez et al., 2008). This large-scale study used dementia diagnostic measures and interview protocols that were developed and validated in the 10/66 pilot studies conducted in 26 countries worldwide (see Prince, Acosta, Chiu, Scazufca, & Varghese, 2003). Overall this research aimed to (1) provide a diagnostic algorithm for dementia that was educationally and culturally congruent and (2) gather information on care arrangements for people with dementia, including the impact on caregivers (Prince et al., 2009). Ultimately, the study provided a wide range of data regarding the prevalence of dementia and associated subtypes of: mental disorders broadly, physical health status, anthropometric factors, sociodemographic characteristics, information regarding disability, functional abilities, utilization of health services, care characteristics, and caregiver burden.

A total of 6,963 people aged 65 and over were surveyed in Cuba, the Dominican Republic, and Puerto Rico (Ferri et al., 2008). Prevalence of dementia was relatively high in these three countries, with an estimated 10.8% of respondents meeting diagnostic criteria in Cuba, 12% in the Dominican Republic, and 11.7% in Puerto Rico (Ferri et al., 2008). These findings are elevated compared to worldwide, age-standardized prevalence rates for populations above the age of 60. Regarding the latter, these are estimated at 5–7%, with higher prevalence in Latin America at 8.5% and lower in the four sub-Saharan African regions, where prevalence lies between 2% and 4% (Ferri & Prince, 2010; Prince et al., 2013). Significant increases in dementia within the Spanish Caribbean are anticipated in the near future, with estimates suggesting a 215% increase in prevalence between the years 2010 and 2050, compared to an estimated 100% increase in higher income countries and more than 300% increase in India and China (Prince et al., 2009).

Healthcare Systems and Dementia Service Provision in the Spanish Caribbean

Global dementia research has helped inform existing healthcare systems by providing practical and attainable recommendations for comprehensive dementia care. For example, findings from the 10/66 Dementia Research Project have informed recommendations for "packages of care" that are specific to low- and middle-income countries, including those within the Spanish Caribbean (Ferri & Jacob, 2017; Prince et al., 2009). According to the 10/66 Dementia Research Group, the primary goals for dementia management include (a) early diagnosis; (b) detection and appropriate treatment of BPSD; (c) optimization of physical health, cognition, activity, and well-being; and (d) provision of information and support to caregivers (Prince et al., 2009). Current systems of healthcare delivery in Cuba, the Dominican Republic, and Puerto Rico include varying elements of these management strategies. Given that healthcare systems significantly impact individual health status and characterize dementia management, the following sections provide brief synopses of current healthcare frameworks and dementia response strategies within each of the Spanish Caribbean countries and territory.

Cuba

Cuba addresses the country's dementia care by way of its health system that is based on principles of universal health coverage and access, free medical and dental services, and comprehensive and regionalized healthcare. Specifically, the government assumes fiscal and administrative responsibility for the health of its citizens, and healthcare is free at the point of delivery. There is a high integration of physical and mental healthcare within Cuba's universal system (Basauri, 2008; Clay, 2015). This collaborative focus largely emerged in 1995 following the Pan American Health Organization (PAHO) conference in Havana, Cuba, known as *Reorienting Psychiatry towards Primary Care*, in which the region was called upon to better integrate mental health services within communities and adopt greater prevention efforts (Gorry, 2013). Over the past several decades, mental health has been approached with greater urgency, and general healthcare has shifted to embrace a more holistic emphasis on prevention, treatment, and rehabilitation. For example, all Cubans are recommended to have annual physical and mental health screenings, with the availability of home-visiting nurses to conduct evaluations if the individual is unable to attend in-person appointments (Clay, 2015).

Cuba, along with Costa Rica and Mexico, was among the first of middle- and low-income countries in the world to develop a national strategic plan for dementia care (Alzheimer's Association, 2017; Ames, O'Brien, & Burns, 2017; Bupa, 2015; Custodio et al., 2017; Llibre-Rodríguez et al., 2017). Guidelines for dementia care in Cuba were developed and approved in 2016 by key stakeholders including medical professionals and dementia experts working within diverse settings such as academic institutions, the Cuban Ministry of Public Health (MINSAP), and community clinics (Bosch-Bayard et al., 2016). The guidelines broadly include (a) education regarding the rights of people with cognitive impairment, (b) professional development, (c) research, (d) health promotion, and (e) dementia prevention (Bosch-Bayard et al., 2016). Initiatives under these guidelines have increased early detection, created a national dementia registry, and developed programs to support patients and their families, including memory clinics, day centers, and comprehensive rehabilitation services for cognitive-related problems.

Healthcare in Cuba is organized by means of three levels, including primary, secondary, and tertiary. At the primary care level, family doctors and nurses are responsible for identifying patients with memory or behavioral problems; detecting at-risk patients through control of chronic, noncommunicable diseases; and consulting with specialists in internal medicine, psychiatry, psychology, and geriatrics (De Jesus Llibre Rodriguez, 2013). These services are generally housed within *consultorios*, i.e., interdisciplinary clinics serving small neighborhoods or communities (Gorry, 2013). Also at the primary level, multiservice community polyclinics are structured to serve as referral sites for neighborhood *consultorios* and tend to offer more specialized equipment and care often being staffed by geriatricians, psychiatrists, and staff associated with optometry, audiometry, physiotherapy, and rehabilitation services (Bosch-Bayard et al., 2016; Gorry, 2013). Finally, secondary care in Cuba is provided by larger hospitals, including psychiatric facilities, while tertiary care exists through research institutes where medical specialties are predominantly housed.

The healthcare staff at community clinics (i.e., *consultorios*, polyclinics) provide specialized consultation to older adults and their caregivers. These often include specialists in psychiatry, psychology, social work, nursing, and occupational therapy who offer care to patients with psychological and behavioral symptoms (Gorry, 2013). Moreover, they provide support for patient caregivers. Specialized services include cognitive stimulation and other interventions including music and dance therapy, arts and crafts activities, group therapy, and physical rehabilitation. For example, therapeutic groups known as *Círculos de Abuelos* [Grandparent Gatherings] are supported by family physicians and polyclinics that offer opportunities for regular exercise (Coyula March, 2010). By 2015, over 945,000 older adults across Cuba were participating in *Círculos de Abuelos*, comprising approximately one-third of the total elderly population (Bosch-Bayard et al., 2016). Thus, these group-based interventions play a prominent role in the management of dementia while also fostering a sense of community among those impacted by the condition.

Residential treatment programs such as *Hogares de Ancianos* [nursing homes] are also available options for elders with dementia (Gorry, 2013). However, these typically require the patient to pay a modest fee. These residential treatment programs typically offer caregiving and meals. They also provide social and recreational activities. Another community initiative to support elders with cognitive decline is *the Older Adult University*, an outgrowth of the University of Havana, which provides continuing education opportunities for seniors (Coyula March, 2010). Lastly, community clinics are also known to coordinate the *Caregivers' School* for those taking care of older adults with dementia. These schools offer practical courses to help improve care and quality of life for both the patient and the caregiver, such as proper hygiene, proper nutrition, and how best to engage in stimulating recreational activities (Gorry, 2013).

Dominican Republic

Healthcare in the Dominican Republic has undergone several changes over the past several decades that have been essential in bolstering dementia management efforts. For example, in 2001, a law was passed issuing the Dominican Social Security System (SDSS) which established funding programs for the elderly, disabled, and unemployed. Included in the plan was the Family Health Insurance, (i.e., the *Seguro Familiar de Salud* [SFS]). This health insurance consists of a subsidized basic health plan and a health services plan known as *El Plan de Servicios de Salud* (WHO and Pan American Health Organization, 2008). In 2014, the country adopted a model of primary healthcare based on integrated health service delivery networks. These new initiatives have broadly focused on coordinating, managing, and articulating policies, resources, and structures aimed at goals common to all institutional stakeholders of the country's national health system. At the same time, health insurance coverage has increased steadily among residents of the country, moving from only 43% coverage in 2011 to 65% coverage in 2015 (Pan American Health Organization [PAHO], 2017).

Furthermore, the Dominican Republic has implemented a national strategic plan for comprehensive dementia care. Led by the Department of Mental Health of the Ministry of Health, the National Dementia Plan was implemented in 2017. The objectives of the plan include (a) strengthening the capacity of leadership and governance in the development of national public policies to support people with dementia and their families; (b) developing programs focused on prevention and risk reduction, with an emphasis on public health, gender, and patient rights, as well as access to integral health services that facilitate early diagnosis and treatment; (c) creating programs and developing action plans that can provide psychoeducation to caregivers and families of people with dementia; (d) potentiating the capacities of the human resources in health for attention to the dementias; and (e) promoting and supporting the development of research on dementia that will allow for generating information for decision-makers to take action.

Despite comprehensive efforts to improve healthcare in the Dominican Republic, there continues to be a need for more efficient mental health services in the country. Large-scale efforts to integrate mental health services into primary care are underway (Caplan et al., 2018). However, there continues to be a notable lack of resources within four main areas including an insufficient fiscal budget for mental health services, and shortages in essential medications, treatment facilities, and human resources (Caplan et al., 2018).

At the individual level, access to adequate healthcare is influenced by stigma and limited mental health literacy, in addition to challenges in accessing care due to lack of time off from work and the ability to pay for transportation, medications, and other healthcare-related costs (Caplan et al., 2018). Another accessibility issue is that of the government healthcare system's lack of provision of coverage for psychiatric medications. This suggests that many pharmacological treatments might be more difficult to obtain for individuals with dementia. Moreover, with dementia, cognitive symptoms and awareness of the syndrome are not typically reasons for initiating medical care; rather it is the presence of BPSD that prompts family members to seek help (Caplan et al., 2018). Thus, psychiatrists are commonly the first professionals to be contacted by families.

Puerto Rico

The healthcare system in Puerto Rico is coordinated by private and public services (Gil-Fournier, 2014). The healthcare system is financed in a complex manner by contributions from the U.S. federal government, the Puerto Rican government, private employers, and out-of-pocket payments by individuals. For most of the island's working class, a principal form of insurance is the Government Health Insurance (*Plan de Salud del Gobierno* [PSG]), known locally as *Reforma* (i.e., Health Reform), which serves as a modified, state-run form of Medicaid for those experiencing economic disadvantage. Beginning in 2015, Puerto Rico developed and implemented a strategic Dementia National Plan that includes elements similar to those in Cuba and the Dominican Republic (Departamento de Salud de Puerto Rico, 2015c). In association with the comprehensive plan, the Puerto Rico Health Department began a web-based registry for patients with Alzheimer's disease that has yielded some data from the more than 12,000 entries (Departamento de Salud de Puerto Rico, 2015a, 2015b, 2016; Torres, 2016).

Communities that reside on the island of Puerto Rico are largely shaped by the island's complex history with the USA government and the fact that the island remains, to this day, under colonial status with limited agency to govern and represent its own people. A significant manifestation of Puerto Rico's colonial status is evident in the fact that island-based Puerto Ricans do not receive the same social welfare benefits as do U.S. citizens who reside on the mainland (Acosta-Belén & Santiago, 2006). For example, while virtually all mainland states receive a federal match rate for Medicaid expenditures at 100%, the match rate for Puerto Rico is only 55% (U.S. Department of Health and Human Services, 2015). Moreover, Puerto Ricans on the island are not eligible to receive Supplemental Security Income (SSI), and those who qualify may receive only a small fraction of the stipend available under the Aid to the Aged, Blind, or Disabled federal program as compared to those receiving the stipend on the USA mainland.

Thus, while Puerto Ricans may enjoy benefits of qualifying for Medicaid/Medicare, the conditions under which they are granted access are significantly restricted.

In general, health services are managed through subscriptions of an insurer-based coverage plan and with primary, secondary, and tertiary services accessed through referral to a specialist from primary care providers. It also involves utilization and review mechanisms such as prior authorization. In sum, there is no universal type of healthcare coverage, and integration of services is carried out through screening and disease prevention in primary care physicians' (PCP) offices, along with administrators' and insurers' coordination of mental health benefits. Puerto Rico has also experienced an "Americanization" of its medical care system in ways that can impede the use of long-standing, Indigenous, and ethnocultural healing practices of the community. For example, physicians and other healthcare professionals have been systematically prohibited from collaborating with *curanderos/curanderas* and of endorsing traditional healing practices (Trujillo-Pagan, 2014).

Migration patterns are also an important factor in accessing healthcare. The most recent financial crisis in Puerto Rico that preceded, and was exacerbated by, the devastating impact of Hurricane Maria in 2017 has contributed to rising migration to the U.S. mainland (Capielo Rosario, Adames, Chavez-Dueñas, & Renteria, 2019). Beginning as early as 2010, more Puerto Ricans lived on the mainland than on the island (Krogstad, 2015). Migrants who return to the island or engage in circulatory migration, i.e., those who migrate back and forth between the island and the mainland U.S. (Acevedo, 2004; Capielo Rosario et al., 2019), often face additional barriers to connecting with social welfare benefits (i.e., Medicare/Medicaid). Additionally, returning migrants may experience a shift in their social relationships due to having been away. Thus, migrant individuals with dementia may no longer have the same social connections to support the care they need upon returning to the island.

Finally, the calamitous toll of Hurricane Maria has greatly impacted virtually all aspects of life on Puerto Rico, including management of dementia. The hurricane resulted in numerous disastrous consequences for Puerto Ricans on the island including restricted access to healthcare and other basic needs such as potable water, electricity, stable shelter, and safe roads (Capielo Rosario et al., 2019; Capielo, Schaefer, Ballesteros, Monroig, & Qiu, 2018). Despite a severe lack of support from the USA government, Puerto Ricans have demonstrated great solidarity, creative ingenuity, and amplified resilience in the aftermath of the hurricane's destruction (Capielo et al., 2018).

Spanish Caribbean Cultural Influences in Dementia Management

Among the health systems in the Spanish Caribbean, there is a mix of allopathic medicine, cosmopolitan medicine (i.e., biomedicine), and Indigenous healing practices (Henderson & Gutierrez-Mayka, 1992). These approaches often combine ancestral wisdom with influences from religious and spiritual frameworks. For example, Cuba's healthcare, in particular the mental health specialty, includes significant emphasis on natural and traditional medicine such as homeopathy, floral therapy, as well as Eastern-influenced practices such as Tai Chi and auricular acupressure (Gorry, 2013). It is also common for individuals to engage in health-promoting behaviors at home that have been passed down through generational wisdom, such as the practice of drinking a spoon of castor oil each day, taking herbal teas and supplements, or drinking *Agua de Azahar* [honey and orange floral water] to cope with anxiety.

Communities of the Spanish Caribbean exemplify numerous psychological strengths that support the care of dementia. These strengths emanate from the rich cultural heritage of the region, and, while each country possesses its own unique characterization of these strengths, there is a collective thread that weaves them together (Adames & Chavez-Dueñas, 2017). The following section reviews a selection of common, culturally derived values and psychological strengths that influence dementia care in the Spanish Caribbean region.

The Latino psychological strength of *connectedness to others*, including connection to family (i.e., *familismo*), serves as the cornerstone of dementia care (Adames & Chavez-Dueñas, 2017). This orientation toward connection promotes intergenerational care networks in which the well-being of the individual is intertwined with the well-being of the group. Moreover, connections with family can facilitate cultural pride, provide a buffer against negative coping responses, and reinforce cultural identity (Castellanos & Gloria, 2018). The 10/66 Dementia Research Group recommendations for dementia care call for community solidarity, in which intergenerational networks are essential to enhance awareness of dementia, reduce stigma, and bolster existing care systems for those suffering (Prince et al., 2009). The essence of community solidarity is well aligned with connectedness to others, in addition to the cultural values of engaging with others through respect (i.e., *respeto*), pleasing interactions (i.e., *personalismo*), and care (i.e., *cariño*).

Multigenerational households that are common in the Spanish Caribbean are also more likely to engage in help-seeking behaviors with medical professionals (Prince et al., 2009), often due to younger generations having greater access to health information and having less adherence to stigmatizing beliefs about dementia. Within these multigenerational households (e.g., children living with parents or grandparents), it is typical for elders to be cared for within the context of cultural values that promote honor, respect, and reverence for elders (Adames, Chavez-Dueñas, Fuentes, Salas, & Perez-Chavez, 2014).

The Latino psychological strengths of *determination* and *adaptability* (Adames & Chavez-Dueñas, 2017) are boldly exemplified among dementia patients and caregivers in the Spanish Caribbean region who demonstrate endless drive and courage as they creatively navigate and learn to thrive amidst challenging circumstances including restricted access to healthcare, systemic discrimination, and environmental disasters. Additionally, humor is a central Latino cultural value that can be essential to the care of persons with dementia. Humor, including telling jokes and laughing together, is a central element of connection and well-being among Latino people (Castellanos & Gloria, 2018). The practice of *choteo* (i.e., joking/goofing off) can be a useful strategy to help navigate the challenging circumstances that accompany dementia. Finally, communities in the Spanish Caribbean may cope with the difficulties of dementia through *collective emotional expression*, in which a range of emotions such as rage, sadness, and joy are expressed through rhythm and dance (Adames & Chavez-Dueñas, 2018).

Caregiving in the Spanish-Speaking Islands of the Caribbean

Caregivers play an essential role in the management of dementia. These individuals may be family, community members, or paraprofessionals and may be paid or unpaid for their caretaking role. Caregiver strain is common due to the difficultly of caring for individuals with dementia, including the presentation of BPSD. In the Spanish Caribbean, individuals will commonly use prayer as an effective strategy to reduce stress and anxiety, which, along with faith, offer important values and coping mechanisms for many Latino elders and their family members. The need for caregiver's education of BPSD management is essential for the caregivers' well-being yet also provides benefit to the person with dementia who is receiving care. The caregiver's psychological support needs should be addressed with clinically informed stress management methods and skills for emotional expression (Black et al., 2013; García-López et al., 2008; Gitlin, Marx, Stanley, & Hodgson, 2015; Hodgson et al., 2013; Lin et al., 2013; Miranda-Castillo et al., 2010).

Addressing Behavioral and Psychological Symptoms of Dementia

Identifying and treating behavioral and psychological symptoms of dementia (BPSD) is a key component of dementia management as these symptoms typically have the most profound impact on quality of life for both the patient and caregiver (Prince et al., 2009). BPSD are also known as noncognitive behaviors, challenging behaviors, or neuropsychiatric symptoms of dementia. A majority of dementia patients will experience some degree of BPSD throughout the course of the disease, such as aggression, agitation, disinhibition, affect lability, and apathy (Kales et al., 2015; Lyketsos et al., 2002).

Some of these symptoms, such as agitation, may become so distressing that they provoke help-seeking more rapidly than more covert symptoms, such as apathy and depression. BPSD may involve symptoms that are easy to interpret as intentional, i.e., deliberate misbehavior such as agitation or refusal to abide by social norms. These symptoms contribute to stigma and can result in blame being directed at the caregiver or family. BPSD are significant contributors to caregiver strain and are a common reason for patient institutionalization (Prince et al., 2009). Ultimately, the classification, assessment, and quantification of these symptoms are the first step toward their successful management of BPSD (Shiota & Sugimoto, 2016).

Successful treatment of BPSD involves a combination of non-pharmacological and pharmacological treatments, along with psychoeducation for the caregiver (Abraha et al., 2017; Azermai, 2015; de Oliveira et al., 2015; Livingston et al., 2014; Watchman & Kerr, 2014). The use of non-pharmacological interventions is common, yet there remains a need for further research-based, randomized controlled trials (RCTs) in order to establish a robust evidence base for these treatments. Non-pharmacological treatments are often aimed at cognitive rehabilitation, such as cognitive stimulation therapy, computerized cognitive therapy, and reality orientation therapy. Multisensory, stimuli-based interventions are also common non-pharmacological approaches, including reminiscence therapy, music and dance therapy, and sensory therapy (Prince et al., 2009). Non-pharmacological approaches also include interventions for caregivers and nursing home staff including psychoeducation, caregiver support group therapy, case management, and individual counseling.

For most patients, pharmacological treatments often remain the first line of treatment for many clinicians managing BPSD. Pharmacological treatments can be used to target cognitive impairment, behavioral symptoms (e.g., agitation, aggression), and psychological symptoms (e.g., depression, anxiety, psychosis; Prince et al., 2009). It is important to note that pharmacological management of neuropsychiatric symptoms in dementia has raised concerns due to the potential harmful side effects of some medications, including increased risk of death and cerebrovascular adverse events (Goldman & Holden, 2014; Kales, 2015; Lanctôt et al., 2017; McClam, Marano, Rosenberg, & Lyketsos, 2015; Olazarán et al., 2010; Panza et al., 2015; Rabins et al., 2010; Tsolaki et al., 2011; Zec & Burkett, 2008; Zeisel, Reisberg, Whitehouse, Woods, & Verheul, 2016). In addition to the more severe potential side effects, these medications may cause decreased appetite and stomach upset. Despite the potential side effects, pharmacological treatments are known to improve amnestic cognitive symptoms and may help to stabilize patient functioning in cases of mild and moderate Alzheimer's disease.

The pharmacological management of BPSD is as extensive as the classification of the symptoms. Agitation, depression, misidentification syndromes, vegetative-type symptoms, and psychosis are among the symptoms that are treated with a combination of antidepressants (Banerjee et al., 2011; Porsteinsson et al., 2014; Seitz et al., 2011) and antipsychotics (Farlow & Shamliyan, 2017; Maust et al., 2015). In Puerto Rico, the use of haloperidol is common in many emergency rooms for addressing agitation. In hospitals, and for indigent families in the Dominican Republic, where the accessibility and costs of newer drugs are difficult, very small amounts of haloperidol may be very beneficial where agitation threatens the safety of the patient and that of others. Since there is no regulation for the use of common antipsychotics, efforts toward educating the medical profession toward their appropriate and safe use are the best regulatory mechanism currently available.

Conclusion

Dementia represents an immense challenge for the public health of developing countries, particularly for those in rapidly changing demographic areas such as the Spanish Insular Caribbean. Such changes, coupled with BPSD, are becoming more common and have implications for diagnosis, medical management, and the stress-producing impact on the family, including the financial burden of dementia care costs. Thus, the comprehensive assessment of elderly patients in clinical settings of these world regions is essential for optimal clinical management. This is especially important as centers providing these assessments are scarce throughout the Caribbean and may be available only in select academic centers or geriatric clinics.

While there are numerous limitations and challenges to the accessibility of dementia care in the Caribbean, recent changes in healthcare provision and the implementation of strategic national dementia management such as the "packages of care" in Cuba, the Dominican Republic, and Puerto Rico offer a promising framework in order to move forward a comprehensive care system. It is clear that a collaborative, interdisciplinary team approach for dementia care is the optimal setting for this booming patient population as interdisciplinary teams with clinical and social science backgrounds are able to identify the cases that can benefit not only from medical but also social and psychological interventions. Additionally, because caregivers play a pivotal role in the management of dementia, support for these individuals must be included in comprehensive treatment approaches. Support groups and caregiver schools may serve as platforms for the continuity of this wealth of knowledge of caregiving (Reinhard, Given, Petlick, & Bemis, 2008). Equally important is the implementation of pharmacological and non-pharmacological measures and the education and training in the management of BPSD aimed at primary and secondary care staff and caregivers, all of which would have a substantial impact on reducing stress in the family and on the high costs of patient hospitalization and early institutionalization (Kales et al., 2015; Tible et al., 2017).

Finally, continually attending to the need to incorporate the wisdom, values, and traditional practices of the Spanish Caribbean culture is central to better management of dementia (Freidenberg & Jiménez-Velazquez, 1992). As highlighted in this chapter, Spanish Caribbean communities exemplify various psychological strengths that have allowed individuals to manage and cope with many challenges, including illness and systemic barriers (Adames & Chavez-Dueñas, 2018). Thus, efforts to support dementia care on the Spanish-speaking islands of the Caribbean must take into account each island's history, cultural landscape, and community qualities that exist within this region. Dementia care should consider all these elements in order to provide individuals with the care and support they deserve.

References

Abraha, I., Rimland, J. M., Trotta, F. M., Dell'Aquila, G., Cruz-Jentoft, A., Petrovic, M., ... Cherubini, A. (2017). Systematic review of systematic reviews of non-pharmacological interventions to treat behavioural disturbances in older patients with dementia. The SENATOR-OnTop series. *BMJ Open, 7*(3), e012759. https://doi.org/10.1136/bmjopen-2016-012759

Acevedo, G. (2004). Neither here nor there: Puerto Rican circular migration. *Journal of Immigrant & Refugee Services, 2*(1–2), 69–85.

Acosta-Belén, E., & Santiago, C. E. (2006). *Puerto Ricans in the United States: A contemporary portrait* (p. 84). Boulder, CO: Lynne Rienner Publishers.

Adames, H. Y., & Chavez-Dueñas, N. Y. (2017). *Cultural foundations and interventions in Latino/a mental health: History, theory, and within group differences*. New York, NY: Routledge.

Adames, H. Y., & Chavez-Dueñas, N. Y. (2018). The drums are calling: Race, Nation, and the complex history of Dominicans. In P. Arredondo (Ed.), *Latinx immigrants: Transcending acculturation and xenophobia* (pp. 95–109). New York, NY: Springer. https://doi.org/10.1007/978-3-319-95738-8_6

Adames, H. Y., Chavez-Dueñas, N. Y., Fuentes, M. A., Salas, S. P., & Perez-Chavez, J. G. (2014). Integration of Latino/a cultural values into palliative health-care: A culture centered model. *Journal of Palliative & Supportive Care, 12*(2), 149–157. https://doi.org/10.1017/S147895151300028X

Adames, H. Y., Chavez-Dueñas, N. Y., & Organista, K. C. (2016). Skin color matters in Latino/a communities: Identifying, understanding, and addressing Mestizaje Racial Ideologies in clinical practice. *Professional Psychology: Research and Practice, 47*, 46–55. https://doi.org/10.1037/pro0000062

Alzheimer's Association. (2016). Alzheimer's disease facts and figures. Alzheimer's & Dementia. *The Journal of the Alzheimer's Association, 12*(4), 459–509.

Alzheimer's Association. (2017, August 7). *Global efforts*. Retrieved from https://www.alz.org/advocacy/global-efforts.asp#

American Cancer Society. (2015). *Cancer facts & figures for Hispanics/Latinos 2015–2017*. Atlanta: American Cancer Society.

Ames, D., O'Brien, J. T., & Burns, A. S. (2017). *Dementia* (5th ed.). Boca Raton, FL: CRC Press.

Anauati, M. V., Galiani, S., & Weinschelbaum, F. (2015). The rise of noncommunicable diseases in Latin America and the Caribbean: Challenges for public health policies. *Latin American Economic Review, 24*(1), 11. https://doi.org/10.1007/s40503-015-0025-7

Azermai, M. (2015). Dealing with behavioral and psychological symptoms of dementia: A general overview. *Psychology Research and Behavior Management, 8*, 181–185. https://doi.org/10.2147/PRBM.S44775

Baez, S., & Ibáñez, A. (2016). Dementia in Latin America: An emergent silent tsunami. *Frontiers in Aging Neuroscience, 8*, 253. https://doi.org/10.3389/fnagi.2016.00253

Banerjee, S., Hellier, J., Dewey, M., Romeo, R., Ballard, C., Baldwin, R., Bentham, P., Fox, C., Holmes, C., Katona, C., KNapp, M., Lawton, C., Lindsey, J., Livingston, G., McCrae, N., Moniz-Cook, E., Murray, J., Nurock, S., Orrell, M., O'Brien, J., Poppe, M., Thomas, A., Walwyne, R., Wilson, K., & Burns, A. (2011). Sertraline or mirtazapine for depression in dementia (HTA-SADD): A randomised, multicentre, double-blind, placebo-controlled trial. *The Lancet, 378*(9789), 403–411. Retrieved from https://doi.org/10.1016/S0140-6736(11)60830-1

Basauri, V. A. (2008). Cuba: Mental health and community participation. In J. M. Caldas De Almeida & A. Cohen (Eds.), *Innovative mental health programs in Latin America and the Caribbean* (pp. 62–78). Washington, D.C.: Pan American Health Organization. Retrieved from http://www1.paho.org/hq/dmdocuments/2008/MHPDoc.pdf

Black, B. S., Johnston, D., Rabins, P. V., Morrison, A., Lyketsos, C., & Samus, Q. M. (2013). Unmet needs of community-residing persons with dementia and their informal caregivers: Findings from the maximizing independence at home study. *Journal of the American Geriatrics Society, 61*(12), 2087–2095. Retrieved from http://www.ncbi.nlm.nih.gov/pubmed/24479141

Borrell, L. N. (2005). Racial identity among Hispanics: Implications for health and well-being. *American Journal of Public Health, 95*, 379–381. https://doi.org/10.2105/AJPH.2004.058172

Bosch-Bayard, R. I., Llibre-Rodríguez, J. J., Fernández-Seco, A., Borrego-Calzadilla, C., Carrasco-García, M. R., Zayas-Llerena, T., Moreno-Carbonell, C. R., Reymond-Vasconcelos, A. G. (2016). Cuba's strategy for Alzheimer disease and dementia syndromes. *MEDICC Review, 18*(4), 9–13.

Bupa. (2015, October 7). *The Americas: Leading the way in dementia policy*. Retrieved from https://www.bupa.com/sharedcontent/articles/the-americas-leading-the-way-in-dementia-policy

Capielo, C., Schaefer, A., Ballesteros, J., Monroig, M. M., & Qiu, F. (2018). Puerto Ricans on the U.S. mainland. In P. Arredondo (Ed.), *Latinx immigrants: Transcending acculturation and xenophobia* (pp. 187–210). New York, NY: Springer.

Capielo Rosario, C., Adames, H. Y., Chavez-Dueñas, N. Y., & Renteria, R. (2019). Accultuaiton profiles of Central Florida Puerto Ricans: Examining the influence of skin color, perceived ethnic-racial discrimination, and neighborhood ethnic-racial composition. *Journal of Cross-Cultural Psychology, 50*(4), 556–576. https://doi.org/10.1177/0022022119835979

Caplan, S., Little, T. V., Reyna, P., Sosa Lovera, A., Garces-King, J., Queen, K., & Nahar, R. (2018). Mental health services in the Dominican Republic from the perspective of health care providers. *Global Public Health, 13*(7), 874–898. https://doi.org/10.1080/17441692.2016.1213308

Castellanos, J., & Gloria, A. M. (2018). Cuban Americans: From golden exiles to dusty feet–freedom, hope, endurance, and the American dream. In P. Arredondo (Ed.), *Latinx immigrants: Transcending acculturation and xenophobia* (pp. 75–94). New York, NY: Springer.

Central Intelligence Agency. (2015, October 30). *The world factbook*. Retrieved from https://www.cia.gov/library/publications/the-world-factbook/

Cerejeira, J., Lagarto, L., & Mukaetova-Ladinska, E. B. (2012). Behavioral and psychological symptoms of dementia. *Frontiers in Neurology, 3*, 73. https://doi.org/10.3389/fneur.2012.00073

Chavez-Dueñas, N. Y., Adames, H. Y., & Organista, K. C. (2014). Skin color prejudice and within group racial discrimination: Historical and current impact on Latino/a populations. *Hispanic Journal of Behavioral Sciences, 36*, 3–26. https://doi.org/10.1177/0739986313511306

Chien, L. Y., Chu, H., Guo, J. L., Liao, Y. M., Chang, L. I., Chen, C. H., & Chou, K. R. (2011). Caregiver support groups in patients with dementia: A meta-analysis. *International Journal of Geriatric Psychiatry, 26*(10), 1089–1098. https://doi.org/10.1002/gps.2660

Clay, R. A. (2015). Getting to know Cuba. *Monitor on Psychology, 46*(6), 30. Retrieved from http://www.apa.org/monitor/2015/06/cover-cuba.aspx

Comas-Díaz, L. (1981). Puerto Rican espiritismo and psychotherapy. *American Journal of Orthopsychiatry, 51*(4), 636–645.

Country Reports. (2016). *Dominican Republic facts and culture*. Retrieved from http://www.countryreports.org/country/DominicanRepublic.htm

Coyula March, M. (2010). Havana: Aging in an aging city. *MEDDIC Review, 12*(4), 27–29.

Custodio, N., Wheelock, A., Thumala, D., Slachevsky, A., Cogram, P., Caramelli, P., & Pe, N. (2017). Dementia in Latin America: Epidemiological evidence and implications for public policy. *Frontiers in Aging Neuroscience, 9*, 221. https://doi.org/10.3389/fnagi.2017.00221

De Jesus Llibre Rodriguez, J. (2013). Aging and dementia: Implications for Cuba's research community, public health and society. *MEDICC Review, 15*(4), 54–59.

de Oliveira, A. M., Radanovic, M., De Mello, P. C. H., Buchain, P. C., Vizzotto, A. D. B., Celestino, D. L., … Forlenza, O. V. (2015). Non-pharmacological interventions to reduce behavioral and psychological symptoms of dementia: A systematic review. *BioMed Research International, 2015*, 1. https://doi.org/10.1155/2015/218980

Departamento de Salud de Puerto Rico. (2015a). *Datos del registro de Alzheimer en Puerto Rico.*

Departamento de Salud de Puerto Rico. (2015b). *La Enfermedad del Alzheimer: Detección y Diagnóstico Temprano*. San Juan, PR. Retrieved from http://www.salud.gov.pr/Dept-de-Salud/PublishingImages/Pages/Unidades-Operacionales/Secretaria-Auxiliar-para-la-Promocion-de-Salud/Módulo de Alzheimer PR 2015.pdf

Departamento de Salud de Puerto Rico. (2015c). *Plan de acción para Puerto Rico: Estrategias para mejorar el cuidado y la calidad de vida de las personas con la enfermedad de Alzheimer*. San Juan, PR. Retrieved from http://www.salud.gov.pr/Dept-de-Salud/PublishingImages/Pages/Unidades-Operacionales/Secretaria-Auxiliar-para-la-Promocion-de-Salud/Plan de Acción.pdf

Departamento de Salud de Puerto Rico. (2016). *Informe de la salud en puerto.*

Farlow, M. R., & Shamliyan, T. A. (2017). Benefits and harms of atypical antipsychotics for agitation in adults with dementia. *European Neuropsychopharmacology, 27*(3), 217–231. https://doi.org/10.1016/j.euroneuro.2017.01.002

Ferri, C. P., Acosta, D., Guerra, M., Huang, Y., Jacob, K. S. K. S., Krishnamoorthy, E. S. E. S., & Prince, M. (2008). Prevalence of dementia in Latin America, India, and China: A population-based cross-sectional survey. *The Lancet, 372*(9637), 464–474. https://doi.org/10.1016/S0140-6736(08)61002-8

Ferri, C. P., & Jacob, K. S. (2017). Dementia in low-income and middle-income countries: Different realities mandate tailored solutions. *PLoS Medicine, 14*(3), 2–5. https://doi.org/10.1371/journal.pmed.1002271

Ferri, C. P., & Prince, M. (2010). 10/66 Dementia Research Group: Recently published survey data for seven Latin America sites. *International Psychogeriatrics/IPA, 22*(1), 158–159. https://doi.org/10.1017/S1041610209990901

Freidenberg, J., & Jiménez-Velazquez, I. Z. (1992). Assessing impairment among Hispanic elderly. *Clinical Gerontologist, 11*(3–4), 131–144. https://doi.org/10.1300/J018v11n03_10

García-López, R., Romero-González, J., Perea-Milla, E., Ruiz-García, C., Rivas-Ruiz, F., & De Las Mulas Béjar, M. (2008). Estudio piloto sin grupo control del tratamiento con magnesio y vitamina B6 del síndrome de Gilles de la Tourette en niños. *Medicina Clinica (Barcelona), 131*(18), 689–692.

Gil-Fournier, I. M. (2014). *El Sistema De Salud Pública – Puerto Rico. Oficina Económica y Comercial de la Embajada de España en San Juan*. Retrieved from http://www3.icex.es/icex/cma/contentTypes/common/records/mostrarDocumento/?doc=4740370

Gitlin, L. N., Marx, K., Stanley, I. H., & Hodgson, N. (2015). Translating evidence-based dementia caregiving interventions into practice: State-of-the-science and next steps. *The Gerontologist, 55*(2), 210–226. https://doi.org/10.1093/geront/gnu123

Goldman, J. G., & Holden, S. (2014). Treatment of psychosis and dementia in Parkinson's disease. *Current Treatment Options in Neurology, 16*(3), 281.

Gorry, C. (2013). Community mental health services in Cuba. *MEDICC Review, 15*(4), 11–14. https://doi.org/10.1192/pb.16.10.648

Henderson, J. N., & Gutierrez-Mayka, M. (1992). Ethnocultural themes in caregiving to Alzheimer's disease patients in Hispanic families. *Clinical Gerontologist, 11*(3–4), 59–74. https://doi.org/10.1300/J018v11n03_05

Hodgson, N., Gitlin, L. N., Winter, L., & Hauck, W. W. (2013). Caregiver's perceptions of the relationship of pain to behavioral and psychiatric symptoms in older community residing adults with dementia. *The Clinical Journal of Pain, 30*(5), 1. https://doi.org/10.1097/AJP.0000000000000018

Jack, B. A., O'Brien, M. R., Scrutton, J., Baldry, C. R., & Groves, K. E. (2015). Supporting family carers providing end-of-life home care: A qualitative study on the impact of a hospice at home service. *Journal of Clinical Nursing, 24*(1–2), 131–140. https://doi.org/10.1111/jocn.12695

Kales, H. C. (2015). Common sense: Addressed to geriatric psychiatrists on the subject of behavioral and psychological symptoms of dementia. *The American Journal of Geriatric Psychiatry, 23*(12), 1209–1213. https://doi.org/10.1016/j.jagp.2015.10.001

Kales, H. C., Gitlin, L. N., & Lyketsos, C. G. (2015). Assessment and management of behavioral and psychological symptoms of dementia. *BMJ, 350*, h369–h369. https://doi.org/10.1136/bmj.h369

Kaneda, T. (2006, October 3). *Health care challenges for developing countries with aging populations*. Retrieved from http://www.prb.org/Publications/Articles/2006/HealthCareChallengesforDevelopingCountrieswithAgingPopulations.aspx

Krogstad, J. (2015). *Puerto Ricans leave in record numbers for mainland U.S.* [online] Pew Research Center. Retrieved from http://www.pewresearch.org/fact-tank/2015/10/14/puerto-ricans-leave-in-record-numbers-for-mainland-u-s/

Lanctôt, K. L., Amatniek, J., Ancoli-Israel, S., Arnold, S. E., Ballard, C., Cohen-Mansfield, J., … Boot, B. (2017). Neuropsychiatric signs and symptoms of Alzheimer's disease: New treatment paradigms. *Alzheimer's & Dementia: Translational Research & Clinical Interventions, 3*, 440–449. https://doi.org/10.1016/j.trci.2017.07.001

Lin JS, O'Connor E, Rossom RC, et al. (2013). Screening for Cognitive Impairment in Older Adults: An Evidence Update for the U.S. Preventive Services Task Force [Internet]. Rockville (MD): Agency for Healthcare Research and Quality (US); (Evidence Syntheses, No. 107). Available from: https://www.ncbi.nlm.nih.gov/books/NBK174643/

Livingston, G., Kelly, L., Lewis-Holmes, E., Baio, G., Morris, S., Patel, N., … Cooper, C. (2014). Non-pharmacological interventions for agitation in dementia: Systematic review of randomised controlled trials. *The British Journal of Psychiatry, 205*(6), 436–442. https://doi.org/10.1192/bjp.bp.113.141119

Llibre-Rodríguez, J., Valhuerdi, A., Sanchez, I. I., Reyna, C., Guerra, M. A., Copeland, J. R. M., … Prince, M. J. (2008). The prevalence, correlates and impact of dementia in Cuba: A 10/66 group population-based survey. *Neuroepidemiology, 31*, 243. https://doi.org/10.1159/000165362

Llibre-Rodríguez, J. J., Valhuerdi-Cepero, A., López-Medina, A. M., Noriega-Fernández, L., Porto-Álvarez, R., Guerra-Hernández, M. A., … Marcheco-Teruel, B. (2017). Cuba's aging and Alzheimer longitudinal study. *MEDICC Review, 19*(1), 31–35. https://doi.org/10.1590/medicc.2017.190100008

Lyketsos, C. G., Lopez, O., Jones, B., Fitzpatrick, A. L., Breitner, J., & DeKosky, S. (2002). Prevalence of neuropsychiatric symptoms in dementia and mild cognitive impairment: Results from the cardiovascular health study. *JAMA, 288*(12), 1475–1483. Retrieved from http://jama.jamanetwork.com/article.aspx?doi=10.1001/jama.288.12.1475

Manes, F. (2016). The huge burden of dementia in Latin America. *The Lancet Neurology, 15*(1), 29. https://doi.org/10.1016/S1474-4422(15)00360-9

Maust, D. T., Kim, H. M., Seyfried, L. S., Chiang, C., Kavanagh, J., Schneider, L. S., & Kales, H. C. (2015). Antipsychotics, other psychotropics, and the risk of death in patients with dementia: Number needed to harm. *JAMA Psychiatry, 72*(5), 438–445. https://doi.org/10.1001/jamapsychiatry.2014.3018

McClam, T. D., Marano, C. M., Rosenberg, P. B., & Lyketsos, C. G. (2015). Interventions for neuropsychiatric symptoms in neurocognitive impairment due to Alzheimer's disease. *Harvard Review of Psychiatry, 23*(5), 377–393. https://doi.org/10.1097/HRP.0000000000000097

Miranda-Castillo, C., Woods, B., Galboda, K., Oomman, S., Olojugba, C., & Orrell, M. (2010). Unmet needs, quality of life and support networks of people with dementia living at home. *Health and Quality of Life Outcomes, 8*, 132. https://doi.org/10.1186/1477-7525-8-132

Molero, A. E., Pino-Ramírez, G., & Maestre, G. E. (2007). High prevalence of dementia in a Caribbean population. *Neuroepidemiology, 29*(1–2), 107–112. https://doi.org/10.1159/000109824

Neary, S. R., & Mahoney, D. F. (2005). Dementia caregiving: The experiences of Hispanic/Latino caregivers. *Journal of Transcultural Nursing, 16*(2), 163–170.

OECD. (2011). The impact of caring on family Carers. In *Help wanted? Providing and paying for long-term care* (pp. 85–98). Paris, France: OECD Publishing. Retrieved from www.oecd.org/els/health-systems/47884865.pdf

Olazarán, J., Reisberg, B., Clare, L., Cruz, I., Peña-Casanova, J., del Ser, T., … Muñiz, R. (2010). Nonpharmacological therapies in Alzheimer's disease: A systematic review of efficacy. *Dementia and Geriatric Cognitive Disorders, 30*(2), 161–178. https://doi.org/10.1159/000316119

Pan American Health Organization [PAHO]. (2017). *Dominican republic. Health in the Americas* (Vol. I). Retrieved from http://www.paho.org/salud-en-las-americas-2017/?p=4014

Panza, F., Solfrizzi, V., Seripa, D., Imbimbo, B. P., Santamato, A., Lozupone, M., … Logroscino, G. (2015). Progresses in treating agitation: A major clinical challenge in Alzheimer's disease. *Expert Opinion on Pharmacotherapy, 16*(17), 2581–2588. https://doi.org/10.1517/14656566.2015.1092520

Porsteinsson, A. P., Drye, L. T., Pollock, B. G., Devanand, D. P., Frangakis, C., Ismail, Z., … Lyketsos, C. G. (2014). Effect of citalopram on agitation in Alzheimer disease. *JAMA, 311*(7), 682. Retrieved from http://jama.jamanetwork.com/article.aspx?doi=10.1001/jama.2014.93

Prince, M., Acosta, D., Castro-Costa, E., Jackson, J., & Shaji, K. S. (2009). Packages of care for dementia low- and middle-income countries. *PLoS Medicine, 6*(11), 1–9. https://doi.org/10.1371/journal.pmed.1000176

Prince, M., Acosta, D., Chiu, H., Scazufca, M., & Varghese, M. (2003). Dementia diagnosis in developing countries: A cross-cultural validation study. *Lancet, 361*(9361), 909–917. https://doi.org/10.1016/S0140-6736(03)12772-9

Prince, M., Bryce, R., Albanese, E., Wimo, A., Ribeiro, W., & Ferri, C. P. (2013). The global prevalence of dementia: A systematic review and metaanalysis. *Alzheimer's & Dementia, 9*(1), 63–75.e2. Retrieved from https://doi.org/10.1016/j.jalz.2012.11.007

Prince, M. J. (2009). The 10/66 dementia research group – 10 years on. *Indian Journal of Psychiatry, 51*(Suppl 1), S8–S15.

Rabins, P. V., Blacker, D., Rovner, B. W., Rummans, T., Schneider, L. S., Tariot, P. N., Blass, D. M. (2010). *Practice guideline for the treatment of patients with Alzheimer's disease and other dementias*. Retrieved from http://psychiatryonline.org/pb/assets/raw/sitewide/practice_guidelines/guidelines/alzheimers.pdf

Reinhard, S. C., Given, B., Petlick, N. H., & Bemis, A. (2008). *Supporting family caregivers in providing care. Patient safety and quality: An evidence-based handbook for nurses*. Agency for Healthcare Research and Quality (US). https://doi.org/NBK2665 [book accession]

Seitz, D. P., Adunuri, N., Gill, S. S., Gruneir, A., Herrmann, N., & Rochon, P. (2011). Antidepressants for agitation and psychosis in dementia. *Cochrane Database of Systematic Reviews (Online), 2*(2), 8191. https://doi.org/10.1002/14651858.CD008191.pub2

Shiota, S., & Sugimoto, Y. (2016). Classification of the behavioral and psychological symptoms of dementia and associated factors in inpatients in psychiatric hospitals – With special reference to rehabilitation. *Journal of Alzheimer's Disease & Parkinsonism, 6*(4), 258. https://doi.org/10.4172/2161-0460.1000258

Tible, O. P., Riese, F., Savaskan, E., & von Gunten, A. (2017). Best practice in the management of behavioural and psychological symptoms of dementia. *Therapeutic Advances in Neurological Disorders, 10*(8), 297–309. https://doi.org/10.1177/1756285617712979

Torres Rivera, E. (2005). Espiritismo: The flywheel of the Puerto Rican spiritual traditions. *Interamerican Journal of Psychology, 39*(2), 295–300.

Torres, W. (2016). *Alzheimer's disease registry. XXVI Jornadas Neurológicas – Academia Puertorriqueña de Neurología* (Vol. 362, p. 329). San Juan, PR: Academia Puertorriqueña de Neurología. https://doi.org/10.1056/NEJMra0909142

Trujillo-Pagan, N. E. (2014). *Modern colonization by medical intervention: U.S. medicine in Puerto Rico (Studies in Critical Social Sciences)*. Leiden: BRILL.

Tsolaki, M. N., Kounti, F., Agogiatou, C., Poptsi, E., Bakoglidou, E., Zafeiropoulou, M., ... Vasiloglou, M. (2011). Effectiveness of nonpharmacological approaches in patients with mild cognitive impairment. *Neurodegenerative Diseases, 8*(3), 138–145. Retrieved from http://www.karger.com/doi/10.1159/000320575

U.S. Department of Health & Human Services. (2015). *Puerto Rico*. [online] Retrieved from https://www.medicaid.gov/medicaid/by-state/puerto-rico.html

Watchman, K., & Kerr, D. (2014). Non-pharmacological interventions. In *Intellectual disability and dementia: Research into practice* (1st ed., p. 334). London, UK: Jessica Kingsley Publishers.

World Bank. (2017). *Population ages 65 and above (% of total)*. Retrieved from https://data.worldbank.org/indicator/SP.POP.65UP.TO.ZS

World Health Organization and Pan American Health Organization. (2008). *WHO-AIMS report on mental health systems in Central America and Dominican Republic*. Retrieved from http://www.who.int/mental_health/evidence/WHO-AIMS/en/

Zec, R. F., & Burkett, N. R. (2008). Non-pharmacological and pharmacological treatment of the cognitive and behavioral symptoms of Alzheimer disease. *NeuroRehabilitation, 23*(5), 425–438. Retrieved from http://www.ncbi.nlm.nih.gov/pubmed/18957729

Zeisel, J., Reisberg, B., Whitehouse, P., Woods, R., & Verheul, A. (2016). Ecopsychosocial interventions in cognitive decline and dementia. *American Journal of Alzheimer's Disease & Other Dementiasr, 31*(6), 502–507. https://doi.org/10.1177/1533317516650806

18 Dementia Treatment and Health Disparities Among Puerto Ricans: The Impact of Cultural and Political Histories

Cristalís Capielo Rosario, Amber Schaefer, and Jhokania De Los Santos

Abstract

This chapter provides specific recommendations to health-care providers regarding how to work with Puerto Ricans who have dementia, including how to work with their family members and caregivers. The recommendations, however, do not take place in a vacuum as they draw on Puerto Rico's sociopolitical history. The continuous colonization of Puerto Rico, first by Spain in 1493 and since 1898 by the United States of America (USA), is a predominant factor impacting health-care practices for Puerto Ricans and a constitutive force on the cultural identity of island-based and USA mainland Puerto Ricans. Although USA citizenship allows Puerto Ricans to migrate to the USA without restrictions, the sociopolitical association between the USA and Puerto Rico has not translated into equal humanitarian treatment or advancement for Puerto Ricans on the island or on the USA mainland. On the contrary, colonialism in Puerto Rico has caused significant economic and social disruption. In turn, this disruption is associated with health disparities across Puerto Rican populations and the deterioration of the island's health system. Hence, understanding the experiences of island-based and mainland Puerto Ricans requires an examination of how colonialism affects the lived experiences and health of this population including those impacted by dementia. To this end, this chapter describes dementia and health disparities among island-based and USA mainland Puerto Ricans and summarizes how Puerto Rico's cultural contact with Spain and the USA has important implications for the delivery of health services to Puerto Ricans. This chapter also provides a description of USA mainland Puerto Ricans' sociodemographic and health profiles and a discussion of historical and contemporary colonialism in Puerto Rico.

C. Capielo Rosario (✉) · A. Schaefer
Arizona State University, Tempe, AZ, USA
e-mail: Cristalis.Capielo@asu.edu; aschaef2@asu.edu

J. De Los Santos
University of Georgia, Athens, GA, USA
e-mail: jdelossantos@uga.edu

H. Y. Adames, Y. N. Tazeau (eds.), *Caring for Latinxs with Dementia in a Globalized World*,
https://doi.org/10.1007/978-1-0716-0132-7_18

Introduction

Case Vignette

Mayra, a 67-year-old Puerto Rican native, had been diagnosed with dementia of the Alzheimer's Disease (AD) type 3 years prior in her hometown of San Juan. A widowed mother of three and grandmother of eight, most of Mayra's family members have left the island for the United States of America (USA) mainland, including two of her children and their families. Currently, Mayra lives with her youngest daughter, Yesika, age 29, and Yesika's husband in a low-income neighborhood just outside of San Juan. As her disease continued to progress, Mayra was unable to keep up with the demands of her job, resulting in her being laid off after 12 years of service. Since losing her job 6 months ago, Mayra spends most her time at home and going to physician appointments, accompanied by her daughter. This loss of autonomy and purpose has resulted in Mayra feeling depressed. She has reached out to her family, neighbors, and faith community for support, but she is finding it increasingly difficult to manage her diagnosis.

Yesika, a mother of two, and a full-time concierge and hotel manager, has been struggling to care for her ailing mother, as her mother requires more supervision and assistance with daily tasks. Because of these challenges, the family has been considering the possibility of sending Mayra to the state of Florida, where two of her children, Luis, 34, and Iván, 38, are currently living. While the adult children are hoping to send Mayra to the USA mainland within the next year, Mayra herself has expressed reservations about the possibility of needing to leave her homeland. Having spent all her life on the island of Puerto Rico, Mayra is hesitant about moving to the USA mainland due to the drastic cultural differences and her limited English proficiency. Moreover, she fears losing the small amount of autonomy that she currently has, as she would need to rely fully on her two sons and daughters-in-law for all her needs.

In discussing Mayra's situation, the family is also considering the costs and benefits of having Mayra in the USA as opposed to remaining in Puerto Rico. As would be expected, a central theme of the family's discussion has been access to health care and its associated costs. Mayra has already had to change physicians twice since being diagnosed, as the numbers of practicing physicians on the island dwindled due to continued migration of healthcare professionals to the USA mainland. Her children fear that Mayra will soon lose her current physician, ultimately making it increasingly more difficult to find another who is familiar with her case. On the other hand, they know that Mayra's Medicare benefits will increase if she were to go to the mainland, because the island experienced an 11% payment reduction for Medicaid benefits in 2015. The family remains concerned about the cost of care on the mainland. Even if all three children were to contribute to the associated costs for treatment, Mayra's dementia is progressing at a rate that indicates that she will require a higher level of care soon, e.g., in-home nursing services. With her current rate of disease progression, Mayra's children will have to decide soon, as the demands to care for their mother will continue to increase at an accelerated rate.

This vignette serves as an example of a common experience of Puerto Rican families that have a loved one with Alzheimer's disease (AD) or other dementias.

Behavioral and Psychosocial Treatments of Dementia Utilized on the Island of Puerto Rico

Research has begun to highlight the impact of Alzheimer's disease and other related dementias on Latinxs. According to the Alzheimer's Association (2017), USA Latinxs are 1.5 times more likely to develop dementia than non-Hispanic Whites. Puerto Ricans display more severe AD symptomatology than other Latinxs as well as other ethnic groups (Camacho-Mercado, Figueroa, Acosta, Arnold, & Vega, 2016). Diabetes and metabolic syndrome appear to be risk factors for dementia. Puerto Ricans' high rates of diabetes and metabolic abnormalities place them at greater risk for developing AD and other related dementias (Alzheimer's Association, 2017). In fact, AD is the fourth leading cause of death in Puerto Rico, accounting for 6.4% of all deaths (Rodríguez Ayuso, Geerman, & Pesante, 2012).

A closer look at AD mortality rates among Puerto Ricans reveals significant differences between Puerto Ricans living on the USA mainland and those living on the island. Figueroa, Steenland, MacNeil, Levey, and Vega (2008) demonstrated that mortality associated with AD of Puerto Ricans on the island is higher than that of Puerto Ricans living on the USA mainland. The underlying cause(s) of the different mortality rate and clinical profile of AD patients in Puerto Rico are still unknown; however, a retrospective study conducted by Camacho-Mercado et al. (2016) suggested that higher rates of AD on the island could be attributed to higher rates of poverty and lack of access to health care. These covariates merit closer consideration as risk factors for AD in future studies. Furthermore, it is also important to expand research on the transnational experience of Puerto Ricans and its role in obtaining health care across borders in general. More specifically, expanded research is needed to understand this transnational experience in terms of identifying dementia. For instance, how would physicians in the USA identify and treat dementia in the case of Mayra? Elements of this transnational identity of island-based and USA mainland-based duality can help inform dementia care and treatment.

Alzheimer's Disease and Other Dementias Among Puerto Ricans

Given Puerto Rico's history of medical colonization, political ties, and geographic proximity to the USA, it is no surprise that the tools of medical and mental health-care practice are similar. However, strained economic circumstances and reduced health-care funding have been detrimental to the quality of care in Puerto Rico (Rivera-Hernandez, Leyva, Keohane, & Trivedi, 2016) which, in turn, may influence AD mortality rates among island Puerto Ricans (Camacho-Mercado et al., 2016). To illustrate, almost half of Puerto Ricans on the island have access to government-funded USA entitlement programs, e.g., Medicaid, Medicare, and Social Security benefits (U.S. Department of Health & Human Services, 2015), yet 9.4% of Puerto Ricans aged 18 and older are not covered by any form of insurance (U.S. Census Bureau, 2015). Although most Puerto Ricans on the island have access to health insurance, access to quality health care is poor.

Inaccessibility to health care has been exacerbated by continued colonialism[1] and natural disasters inflicting damage on the island. Specifically, the debt owed by Puerto Rico's Health Insurance Administration (ASES) poses a significant threat to exacerbating a deteriorating health-care system (Vazquez, 2015). More than 40% of the island's debt is due to health care and a lack of federal support for Medicaid in particular. This debt has triggered a surge of developments, including longer wait times for procedures, overcrowded emergency rooms, and attempts to charge patients directly for

[1] Colonialism, defined as the occupation and exploitation of a territory's land and its resources (Fanon, 1965), is further advanced when the colonizer distorts the history and culture of those colonized and when the colonizer portrays the colonized as inferior to the colonizer (David, 2008; Memmi, 1996). Embedded in the histories and experiences of Puerto Rican communities are these colonial dynamics.

health-care services, while a major physician exodus is taking place (Shin, Sharac, Luis, & Rosenbaum, 2015). Despite the fact that the Puerto Rican government must comply with all federal health-care regulations and make all required payments, it is treated differently than any other jurisdiction in the USA. As described in the vignette, Puerto Rico experienced an 11% cut in funding for Medicaid in 2015, thereby increasing health disparities on the island (Alvarez & Goodnough, 2015). Mayra's case serves as an example of the lived experiences of patients on the island with AD and other related dementias and therefore demonstrates the need for more federal support to decrease the health disparities on the island. Given the sociopolitical history that Puerto Rico has with the USA, the federal government must restore equity, eliminate the Medicaid funding cap, and eliminate the barriers to parity in health service delivery and services among Puerto Ricans.

In order to begin a conversation about behavioral and psychosocial treatments for dementia and AD, providers must first understand the research surrounding the development of evidence-based treatments and practices. For instance, behavioral and psychological treatments for dementia and AD patients are vastly underrepresented in the literature in general, with an even greater dearth when considering Latinx patients and their caregivers (Gallagher-Thompson, Haynie, Takagi, Valverde, & Thompson, 2000). This has been previously attributed to difficulty in research participant recruitment and high attrition rates of Latinx patients with AD and their caregivers (Gallagher-Thompson et al., 2004; Gallagher-Thompson, Solano, Coon, & Areán, 2003; Karlawish et al., 2011). Some of the barriers to treatment and to conducting research on the efficacy of treatments within the Latinx community include (a) lack of knowledge about AD, (b) perceptions of memory loss as a normal process of aging, (c) sociopolitical and structural barriers to care, and (d) lack of knowledge of Latinx cultural values on the part of researchers and clinicians (Karlawish et al., 2011). These barriers are especially pronounced when working with Latinx patients and caregivers as they are less likely than non-Latinx Whites to utilize AD treatments and participate in research studies (Karlawish et al., 2011). Moreover, the exodus of practicing physicians and behavioral health workers to the USA mainland (Shin et al., 2015) has left many Puerto Ricans, like Mayra in the vignette, searching for new physicians and/or therapists to manage the behavioral and psychosocial symptoms related to AD. This disruption in health services contradicts dementia care practice recommendations (Meeks et al., 2018). Thus, there is a clear need for further research to inform effective psychological and behavioral treatments on the island. Yet, without substantial support from federally funded programs, culturally appropriate research and treatments for Puerto Rican patients with dementia remain sparse.

While the literature about psychological treatments for AD *patients* is limited, research surrounding the well-being of *caregivers* of patients with AD has been steadily increasing over the last two decades (Aranda, Villa, Trejo, Ramírez, & Ranney, 2003). Within this growing body of literature, support groups and psychoeducation courses appear to be the most widespread forms of psychological support for caregivers of patients with dementia. However, much of this research is confined to the USA mainland, and is not necessarily culturally specific to the unique needs of Latinx caregivers, nor directed at the necessities of Puerto Ricans. Therefore, research specifically relating to Puerto Rican caregivers of dementia patients remains underrepresented (Llanque & Enriquez, 2012). Greater research efforts could be of pivotal assistance to caregivers, such as Mayra's children (on the USA mainland, as well as the island), in order to receive the best and most necessary support so as to care for their loved ones.

According to the Alzheimer's Association (2017), many people living with AD find the behavioral and psychological changes caused by continued brain deterioration to be the most challenging. Of the examined behavioral and psychological implications of AD, depression, anxiety, and irritability are among the most reported side effects in the early stages of the degenerative disease. These early signs of AD often go undiagnosed and are commonly mistaken by patients and their families as normal signs of aging. This is especially true for patients who lack access to screening, as the disease can only be diagnosed with formal medical and neuropsychological testing and neuroimaging (Alzheimer's Association, 2017). As the disease progresses, several symptoms including withdrawal, increased confusion, and changes in sleep patterns begin to appear as the staging of the disease moves to the mild stage. This is typically when caregivers and loved ones will notice the disease, resulting in increased rates of diagnosis (Alzheimer's Association, 2017). From this point, individuals typically live, on average, 4–8 more years after diagnosis, with some individuals living up to 20 years after diagnosis. Figure 18.1 provides the clinical trajectory of dementia and need for intervention.

In the USA, many organizations and medical professionals work with individuals with dementia and their caregivers to navigate and alleviate these symptoms. However, the same cannot be said when considering the island of Puerto Rico. When working with Puerto Ricans on the island, there is a decrease in the number of services available and a clear need to address these concerns in a culturally congruent manner. For instance, Latinx individuals are more likely to seek support from their faith communities, friends, and family before seeking professional health-care services (Sue & Sue, 2012). This is, in large part, attributed to a culture of distrust of professionals who have historically been represented by predominantly White males who demonstrated a lack of multicultural competence (Comas-Diaz, 2006). Therefore, the inclusion of spiritual beliefs, cultural values, and family support throughout the course of treatment should be

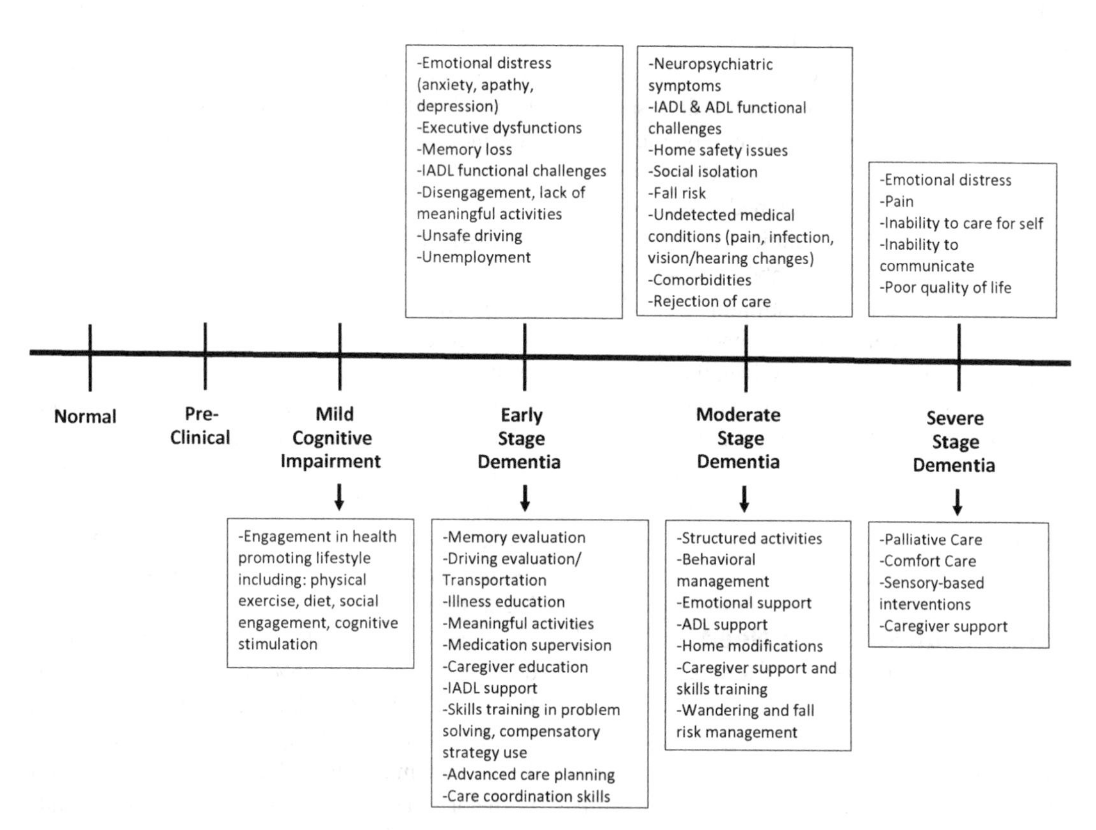

Fig. 18.1 The clinical trajectory of dementia and need for intervention. (From Gitlin, L. N., Hodgson, N. A., & Choi, S. S. W. (2016).Home-based interventions targeting persons with dementia: What is the evidence an where do we go from here? In M. Boltz & J. E. Galvin (Eds.), *Dementia care: An evidence-based approach*. New York, NY: Springer.)

emphasized (Adames & Chavez-Dueñas, 2017). To illustrate, when working with a Puerto Rican patient with dementia, incorporating a spiritual leader in the treatment plan, instead of sole reliance on the traditional European-American model of care, can help blend spiritual practices with psychological treatments. In addition, incorporating culturally appropriate ways of communicating should also be considered key when working with the Puerto Rican community, e.g., use of the Spanish language and associated dialects (Karlawish et al., 2011). These practices can help bridge the cultural divide between a traditional, Western-based model of care with one that is more culturally attuned to the needs of Puerto Rican patients with dementia.

An example of culturally competent psychoeducational program for Puerto Ricans is *Un Café por el Alzheimer* [A Coffee for Alzheimer] which incorporates investigative work from federally funded and private research entities to educate the public and those affected by the disease. This organization is one of very few examples of foundations that provides culturally specific resources to patients with dementia on the island of Puerto Rico. *Un Café por el Alzheimer* provides a forum through social media that connects professionals and community members and encourages community members to use *cuentos* [stories] and *testimonios* [testimonies] to educate others about dementia and to encourage social support among participants. Given the mass exodus of physicians from the island and the decreased federal funding for Medicaid recipients, organizations such as *Un Café por el Alzheimer* are going to become increasingly necessary.

What Health Providers Need to Know About Puerto Ricans

The following section describes demographic information useful in understanding Puerto Ricans' needs and living conditions. It specifically discusses the geographic and sociodemographic differences between mainland- and island-born Puerto Ricans. A broad understanding of Puerto Ricans living in both regions can facilitate the integration of cultural and geographic factors when health providers work with individuals such as Mayra and her family.

Geographical Dispersion

More Puerto Ricans live in the USA (5.5 million) than in Puerto Rico (3.2 million; U.S. Census, 2017). On the USA mainland, Puerto Ricans represent the second largest Latinx group, accounting for 9% of the total USA Latinx population (Brown & Patten, 2014). While Puerto Ricans are dispersed throughout all regions of the USA mainland, according to the U.S. Census (2017), most live in the Northeast (51%) and South (32%). Regarding ethnic identification, 55% of Puerto Rican adults report they are more likely to identify as "Puerto Rican." Twenty-eight percent describe themselves most often as "American" and less than 15% self-reference with the term "Hispanic" or "Latino" (Cohn, Patten, & Lopez, 2014).

Sociodemographic Profile

Puerto Ricans living in the USA are not a monolithic group. Notable demographic differences exist between mainland-born and island-born Puerto Ricans who are residing on the USA mainland. For instance, the median age of those born on the island is 47 years of age, while the median age among those born on the mainland is 22 years of age (Pew Research Center, 2015). Language use also appears to differ between mainland- and island-born Puerto Ricans. While 42% of mainland-born Puerto Ricans are English-dominant, less than 15% of island-born Puerto Ricans are English-dominant (Pew Research Center, 2015). Regarding educational attainment, 18% of mainland-born Puerto Ricans possess a bachelor's degree compared to 15% of island-born Puerto Ricans (Cohn et al., 2014). The profile of Puerto Ricans on the mainland also appears

Table 18.1 Comparison of socioeconomic profile of Puerto Ricans in New York and Florida

Characteristic	New York Population 1,113,123	Florida Population 1,128,225
Poverty level (all families)		
Percent below poverty level	21.5%	16.9%
Employment status (16 years or older)		
Percent unemployed	5.5%	4.0%
Percent employed	51.4%	58.0%
Educational attainment (25 years or older)		
Less than high school	26.7%	14.3%
High school diploma (or equivalent)	30.3%	28.9%
Some college or associate's degree	26.3%	35.9%
Bachelor's degree or higher	10.9%	14.4%
Language use (5 years or older)		
English only	39.9%	30.4%
Language other than English	60.1%	69.6%
Speaks English less than "very well"	16.5%	19.7%

Source: U.S. Census Bureau, 2017 American Community Survey, 1-year estimates

Table 18.2 Socioeconomic profile of main Latinx groups in the USA mainland

	Latinx group		
Characteristic	Cuban	Mexican	Puerto Rican
Income (past 12 months)			
Median household income	$50,142	$49,439	$44,731
Poverty level (all families)			
Percent below poverty level	12.2%	17.8%	19.3%
Employment status (16 years or older)			
Percent unemployed	2.9%	4.0%	5.0%
Percent employed	60.2%	63.3%	57.6%
Educational attainment (25 years or older)			
Less than high school	18.8%	36.2%	20.2%
High school diploma (or equivalent)	29.9%	28.6%	29.8%
Some college or associate's degree	23.6%	23.4%	30.5%
Bachelor's degree or higher	18.1%	8.5%	13.0%
Language use (5 years or older)			
English only	21.9%	28.4%	40.8%
Language other than English	78.1%	71.6%	59.2%
Percent who Speaks Ensligh less than "very well"	39.4%	28.7%	17.2%

Source: U.S. Census Bureau, 2017 American Community Survey, 1-year estimates

to differ by USA region. For example, the median yearly household income for Puerto Ricans in New York is $38,505, while in Florida, it is $46,735 (U.S. Census, 2017). Table 18.1 describes other significant socioeconomic advantages Florida Puerto Ricans have over those in New York. The share of Puerto Ricans who live in poverty, 27%, is higher than the rate for the general USA population (16%) and that of the overall Latinx population (25%). The rate of Puerto Rican home ownership (38%)—one of the strongest indicators of economic security—is lower than the rate for both the Latinx (45%) and the USA population (64%) as a whole. Table 18.2 further compares the socioeconomic profile of the three main Latinx groups on the USA mainland.

General Health Profile

Interest in the health status of Puerto Ricans was sparked by studies documenting counterintuitive nativity differences in health within Latinxs. For example, foreign-born individuals were believed to face challenges in adapting to a new and different society and therefore would show higher levels of psychological distress and behavioral dysfunction than their USA-born counterparts from similar ethnic backgrounds (Fuligni & Perreira, 2009). However, findings from systematic studies of representative samples of immigrants illustrated that immigrants had similar or even better psychological health than USA-born Latinxs (Fuligni & Perreira, 2009). This is known as the immigrant health paradox (Vega & Sribney, 2011). While this phenomenon has been observed among immigrants of Mexican and Central American descent, researchers have not been able to find consistent support for the healthy immigrant paradox in first-generation Puerto Ricans (Capielo, Delgado-Romero, & Stewart, 2015; Sánchez et al., 2014).

Data on Puerto Rican health outcomes portray a discouraging profile. For example, a 2015 report by the American Cancer Society found that cancer—the leading cause of death among USA Latinxs—was highest among mainland Puerto Ricans as well as island-based Puerto Ricans, with prostate cancer accounting for 18.3% of deaths in men and breast cancer accounting for 18.9% of all cancer deaths in women (Tortolero-Luna et al., 2013). Cancer rates appear to be partially correlated with high rates of smoking, hepatitis C, and obesity among mainland and island Puerto Ricans (American Cancer Society, 2017).

Puerto Ricans also appear to be disproportionately afflicted with cardiovascular disease, the second leading cause of death among USA Latinxs. Diabetes is another prevalent chronic disease in Puerto Rico, with an estimated 14.9% of individuals over the age of 18 diagnosed (Rodríguez-Vigil, Rodríguez-Chacón, & Valcarcel, 2016). According to Colón-Ramos et al. (2017), the death rate from diabetes is twice that of their USA counterparts and higher than any other racial or ethnic group in the USA. A study published by researchers in *The Hispanic Community Health Study/Study of Latinos* (Heiss et al., 2014) found that the overall prevalence of metabolic syndrome—a risk factor for cardiovascular disease and diabetes—was highest among Puerto Ricans, with Puerto Rican women having a higher risk than Puerto Rican men (40% vs. 32%).

Puerto Ricans are also disproportionately afflicted by mental health disorders. Puerto Ricans are consistently at a higher risk for substance abuse, suicidality, and mood disorders (Ai, Appel, Huang, & Lee, 2012; Alegría et al., 2008; Alegría, Canino, Stinson, & Grant, 2006; González, Tarraf, Whitfield, & Vega, 2010; Sánchez et al., 2014). The prevalence of mental health disabilities (Rivera & Burgos, 2010) and psychiatric disorders (Alegría et al., 2006; Alegría, Mulvaney-Day, Woo, Torres, et al., 2007) is also higher among USA Puerto Ricans compared to Mexicans, Cubans, and island-based Puerto Ricans.

Taken together, mainland Puerto Rican adults consistently report poorer health status and higher rates of chronic illness and psychological distress compared to non-Hispanics and other Latinx groups (National Center for Health Statistics, 2017). These indicators suggest that Puerto Ricans are at a disadvantage compared to the USA population and other Latinx groups irrespective of their documentation and citizenship status. It has been hypothesized that health disparities, particularly higher rates of mood disorders among Puerto Ricans, may be explained by the unique sociopolitical status of the island of Puerto Rico (Alegría et al., 2008; Sánchez et al., 2014) and its history of colonization (Capielo Rosario, Schaefer, Ballesteros, Rentería, & Davis, 2019). While the association between colonialism and Puerto Rican mental health has received very little attention, research has begun to uncover the influence of colonialism on depression symptomatology among mainland Puerto Ricans. In Capielo Rosario, Schaefer, et al. (2019), researchers found a positive association between colonial mentality, i.e., internalized inferiority toward one's own ethnic culture while perceiving the colonizer's culture as superior (David & Okazaki, 2006), and depression symptoms which, in turn, were mediated by acculturative stress.

These findings underscore the need to integrate Puerto Rico's colonial history in training and research in order to understand the health of Puerto Ricans and how Puerto Ricans respond to health-care interventions.

The Political Matters: Dementia Care of Puerto Ricans

While much of the literature has focused on the symptomatology and treatment of dementia, very little has been written on how Puerto Rico's political state has impacted dementia care and treatment. In order to understand why there is such a high need for health and social services in Puerto Rico, as they pertain to individuals with dementia, it is important to first comprehend the historical context that has culminated in Puerto Rico's present state. The following section is to provide health-care providers with an overview of how

Puerto Rico's unique sociopolitical status has impacted various facets of life for Puerto Ricans, including access to health care and political representation.

As it is the case for many Latinx populations, African and Indigenous influences as well as European colonialism are also threaded into the fabric of Puerto Rican culture and identity (Rodríguez-Silva, 2012). However, persistent USA colonialism in Puerto Rico since 1898 distinguishes Puerto Ricans from other Latinx groups (Capielo, Schaefer, Ballesteros, Monroig, & Qiu, 2018; Duany, 2002). After American forces defeated Spain at the end of the Spanish-American War in 1898, the USA and Spanish governments signed the Treaty of Paris. As part of the treaty, Spain ceded Puerto Rico to the USA, converting the island into an unincorporated territory unable to continue a path toward statehood (Capielo et al., 2018; Whalen, 2005). In 1917, through the Jones Act, USA citizenship was imposed on anyone born in Puerto Rico in 1898 or after (Whalen & Vázquez-Hernández, 2008). USA citizenship, however, has not translated into "equal treatment or protections" for island Puerto Ricans (Capielo et al., 2018, p. 190). For instance, while residents of the island can freely travel between the USA mainland and the island, they have no political representation in the USA Congress and are not able to vote in presidential or congressional USA elections (Capielo Rosario, Adames, Chavez-Dueñas, & Rentería, 2019). In 1952, Puerto Ricans were allowed to vote for their own governor and establish a constitution; however, the Puerto Rican constitution is subject to the constitution of the USA (Torruella, 2017).

More than a century of USA colonialism has relegated Puerto Ricans living on the island to second-class status. Puerto Rico has no means to enact trade agreements or make military decisions (Lugo-Lugo, 2006). Puerto Ricans are also not allowed representation at the United Nations, the World Bank, or the International Monetary Fund (Lugo-Lugo, 2006). These restrictive participation policies have greatly limited the island's ability to address its ongoing debt and current financial crisis (Torruella, 2017). Colonialism is also reflected in the limited access island Puerto Ricans have to social programs (e.g., Social Security, Medicare, Medicaid) that are available to non-Puerto Rican USA citizens (Acosta-Belén & Santiago, 2006). Although island Puerto Ricans pay state and payroll taxes (Capielo et al., 2018; Torruella, 2017), they receive far fewer benefits than USA mainland residents (Levis-Peralta et al., 2016). For example, island Puerto Ricans cannot participate in the Affordable Care Act (ACA) exchange program (Severino Pietri, 2017). Instead, they receive a grant from the USA federal government to help cover health services, but at a significant lower rate than the states and the District of Columbia (Severino Pietri, 2017). In 2015, Puerto Ricans received less than 35% in cost coverage per member, per month as compared to Medicaid recipients living in the USA (Levis-Peralta et al., 2016).

Despite USA hegemony of Puerto Rico, and perhaps in response to USA colonialism, Puerto Ricans have maintained their cultural identity by their use of the Spanish language, adherence to musical traditions rooted in their Taíno and African ancestry, and preference for relationships based on interdependence (Duany, 2002; Rodríguez-Silva, 2012). Research also illustrates that island Puerto Ricans have a strong level of familiarization and attachment to White American cultural practices, cultural values, and identity (Capielo Rosario, Lance, Delgado-Romero, & Domenech Rodríguez, 2019). The following section traces the island's history of colonialism and migration, both needed to contextualize the implications for dementia treatment of Puerto Ricans.

Puerto Ricans and Colonialism

Nota Bene: The following summary is advantaged by the cited works of sociologist Nicole Elise Trujillo-Pagan, who through comparative-historical and discourse analysis methodologies has detailed how USA colonial policies in Puerto Rico have led to health inequalities in order to advance USA political and economic goals.

Puerto Rico has been a colony for more than 500 years. Its colonization began in 1493 with the arrival of Spanish conquistadors (Capielo Rosario, Schaefer, Ballesteros, Rentería, & Davis, 2019). As a Spanish colony, Puerto Rico experienced great mortality among its native people (Taínos) and Black slaves, along with devastating exploitation of its natural resources (Chavez-Dueñas, Adames, & Organista, 2014). It has been suggested that toward the end of the Spanish rule, the island was treated more as a province and less as a colony in order to appease those living on the island and to prevent independence movements such as those in other Latin American countries (Torruella, 2017). In the early 1800s, the island was allowed to establish trading relationships with other nations, including the USA (Cintrón Aguilú, 2014). However, by the late 1890s, Spain began to curtail trade with the USA by imposing tariffs and eventually ended all trade in 1897 (Vélez, 2017). Spanish interference in trading between Puerto Rico and the USA is believed to have also contributed to the USA invasion of Puerto Rico in 1898 at the end of the Spanish-American War (Cintrón Aguilú, 2014).

The Treaty of Paris, the peace treaty accorded between Spain and the USA at the end of the Spanish-American War, helped to shape Puerto Ricans' distinct political status and subsequent migration patterns to the USA (Whalen, 2005). The treaty forced Spain to cede Puerto Rico to the USA (Baker, 2002; Whalen & Vázquez-Hernández, 2008). When American forces invaded the island in 1898, General Nelson Miles asserted that the goals of the USA were:

> to bring to this beautiful island the greatest degree of liberty… to bring protection, not only to you, but also to your property, promoting your prosperity and bestowing upon you the guarantees and the blessings of the liberal institutions of our government. (Berbusse, 1966, p. 79; as cited in Trujillo-Pagan, 2014, p. 71)

The USA government echoed objectives of promoting island prosperity and self-governance (Trujillo-Pagan, 2014). However, the interventions that followed were antithetical to these goals and only solidified the island's colonial status, this time at the hand of the USA government (Trujillo-Pagan, 2014). Contradictions between the stated goals of the USA and the actual treatment of Puerto Ricans were apparent in all aspects of island living. Colonial policies that followed the devastating 1899 Hurricane San Ciriaco further weakened economic recovery while strengthening USA control over the island (Baker, 2002; Trujillo-Pagan, 2014). Challenged by the discontent among Puerto Ricans and growing demands for independence, a year after Hurricane San Ciriaco, the Foraker Act—also known as the Organic Act of 1900—established a limited civil government on the island (Torruella, 2017). This civil government had limited representation, and the USA refused demands for economic recovery efforts. As impoverished conditions took hold on the island, the USA declared that the Puerto Ricans' purported inferiority was the reason the island could not prosper (Trujillo-Pagan, 2014).

Colonization of the island also included mandatory public health interventions. The dominant doctrine at the time, held by both the affluent class of the island and the USA government, was that Puerto Ricans were "degenerated" by sickness and poverty and thus Puerto Ricans were incapable of self-government (Go, 2008). *Jíbaros* (Puerto Ricans that reside in rural or mountainous regions of the island) were often described as lazy and of low moral character and in need of medical interventions to regenerate *jíbaros* (Trujillo-Pagan, 2013). Early interventions included sanitation regulations, compulsory smallpox vaccination, and eradication of hookworm. The sanitation and medical campaign also mandated physicians and municipalities to collect demographic information, including the individual's skin color, political affiliation, and location (Willrich, 2011). During the hookworm campaign, Puerto Ricans were labeled as "scientifically different" and of weak physical constitution as many were stricken with iron deficiency anemia (Trujillo-Pagan, 2013). However, instead of developing nutrition interventions, the hookworm campaign expanded the strategies used during an earlier sanitation campaign, which quarantined people and allowed the USA to safeguard its trade interests by monitoring ports.

The institutions created to carry out such interventions deeply affected the landscape of medical and public health practice of the island, which continues to reverberate today.

For instance, the Superior Board of Health (SBOH) created in 1899 was done so without the input of Puerto Ricans (Trujillo-Pagan, 2014) yet had jurisdiction over all the medical and public health interventions previously described, as well as the ones that followed. Besides leading all public health efforts, it also established social policy. For instance, the SBOH mandated physicians to obtain medical licensure and accreditation from the USA and prohibited physicians from forming apprenticeships with *curanderos/curanderas* or *parteras* (i.e., midwives also known as *comadronas*), even though there were insufficient numbers of physicians to attend to all the medical needs in Puerto Rico (Trujillo-Pagan, 2014).

Efforts by USA colonizers to "Americanize" medical and public health practices on the island received strong opposition from Puerto Ricans (Trujillo-Pagan, 2014). Local physicians frequently challenged the theory that hookworm, and not malnutrition, was the cause of iron-deficient anemia. Puerto Rican women continued their practice of using *comadronas* and *parteras* despite facing violence or jail time. Many Puerto Ricans refused the compulsory "imposition of sanitation and unnecessary vaccination" (Trujillo-Pagan, 2014, p. 221).

Today, health delivery in Puerto Rico reflects a bifurcation between traditional healing practices and the historical, compulsory interventions and institutions created under USA control. For example, SBOH was the predecessor for the island's contemporary *Departamento de Salud* (Trujillo-Pagan, 2014). Current public health assessments and procedures through this entity parallel public health policies on the USA mainland. Nonetheless, remaining widely accepted and practiced on the island are culturally based healing practices such as *espiritísmo*, which reflect the belief in a spiritual and invisible world that influences human behavior and communication with deceased family members to ameliorate distress (Kelley & Fitzsimons, 2000).

Puerto Rican Migration and Its Impact on Health

Differentiated citizenship and economic and social policies have contributed to creating an atmosphere for cyclical poverty in Puerto Rico and out-migration from the island to the USA mainland (Baker, 2002; Vélez, 2017). Puerto Rican migrants' patterns can be described in three manners: (1) the "one-way migrants" who move permanently to the USA mainland, (2) the "return migrants" who after many years return from the USA mainland back to the island and reestablish residence on the island, and (3) the "circular migrants" who migrate back and forth between the island and the USA mainland, spending extensive periods of residence in both places (Acevedo, 2004). Although Puerto Rican migration is technically classified as "internal migration" by the U.S. Census (Mayol-García, 2019), Puerto Rican migration is best understood from a transnational and colonial perspective. While the former acknowledges the multiple and significant cultural boundaries that island Puerto Ricans cross when they migrate to the USA mainland (Duany, 2002), the latter describes the migration that results from economic and social interference caused by historical and contemporary colonialism in Puerto Rico (Vélez, 2017). Because both transnational and colonial migrations clarify important aspects of Puerto Rican migration; Puerto Ricans are increasingly called transnational colonial migrants (e.g., Duany, 2010; Silver & Velez, 2017). The magnitude of the intersection between transnationalism and colonialism demonstrates a compelling need for understanding how Puerto Rican migratory processes influence the health of this population.

Although Puerto Ricans exhibit a high rate of mobility similar to that of other ethnic groups, the factors associated with migration and mobility are different for Puerto Ricans (Rivera-Batiz & Santiago, 1996; Vélez, 2017). There is a substantial difference between pre-migration expectations and post-migration lived experiences among Puerto Ricans. Puerto Ricans may be more optimistic about their post-migration upward mobility due to their status as USA citizens (Silver & Velez, 2017). However, Puerto Ricans may subsequently experience heightened awareness of changes in social status and greater risks for psychopathology in the context of unmet or thwarted expectations for upward social mobility (Burgos & Rivera, 2012). Changes in social status and psychological stress reactions associated with a marginalized social position may heighten psychological vulnerability to major depression episodes (MDE; Camacho-Mercado et al., 2016).

Findings from cross-sectional and longitudinal research suggest that downward social mobility is associated with poor coping skills, psychiatric hospitalizations, and functional impairment (Das-Munshi, Leavey, Stansfeld, & Prince, 2011; Nicklett & Burgard, 2009; Tiffin, Pearce, & Parker, 2005; Timms, 1998). Upward social mobility has been associated with reduced mental health risk (Timms, 1996). Alcántara, Chen, and Alegría (2014) determined that Puerto Ricans who perceived themselves as belonging to a lower social status in the USA, as compared to their perceived social status on the island, exhibited higher chances of having an MDE, relative to those who did not perceive any differences in social status related to migration. In addition, Puerto Ricans showed a pronounced difference in likelihood for an MDE compared to migrants from Mexico and Cuba. Such research indicates that social status may serve as a proxy for negative affect and physiological reactivity that are associated with enduring perceptions of relatively low social status (Alcántara et al., 2014). In addition, acculturation appears to be another migration-related phenomenon linked to depression symptoms among Latinxs (Alegría et al., 2007; Torres, 2010; Torres & Rollock, 2007). For Puerto Ricans in particular, while the association between acculturation and depression seems equivocal, the stress associated with acculturation, e.g., pressure to abandon heritage culture in favor of the dominant culture, appears to be an important predictor of depression symptoms (Capielo et al., 2015; Capielo Rosario & Dillon, 2019).

Moreover, depression is a frequent comorbid condition in patients with dementia (Modrego & Ferrández, 2004), with the rate of MDE ranging from 23% to 55% (Zubenko et al., 2003). Additionally, some longitudinal studies of elderly subjects concluded that depressive symptoms might be associated with the risk of developing dementia disorders. A longitudinal study of elderly individuals found that those with depression at baseline were more than twice as likely to develop dementia of the Alzheimer's type (AD) compared to those without depression (Modrego & Ferrández, 2004). Furthermore, patients with dementia and depression appear to respond well to interventions that aim to reduce depression symptoms (Alexopoulos, Meyers, Young, Mattis, & Kakuma, 1993). Therefore, it is important for health providers to understand how migration and migration-related factors may place Puerto Ricans at risk for developing depression symptoms and how depression may, in turn, affect the development, progression, and treatment of dementia in this population.

Puerto Rican Migration and Economics

Puerto Rican migration is inextricably connected to the island's colonial history and the economic consequences of colonization (McGreevey, 2018). Catalysts for migration between 1945 and the early 1970s, a period known as the Great Puerto Rican Migration (Duany, 2002; Vélez, 2017), included the significant reduction in the agriculture industry, coupled with the limited success of corporate tax elimination for American companies that relocated to the island (Vélez, 2017). During this period, Puerto Rican migration was mostly concentrated in the Northeast USA (Baker, 2002). Upon their arrival, Puerto Ricans faced racial, housing, and employment discrimination and segregation (Burgos & Rivera, 2012). These limiting dynamics led to the overrepresentation of Puerto Ricans in low-wage and light manufacturing jobs, creating a cycle of unemployment and poverty (Baker, 2002; Burgos & Rivera, 2012; Whalen & Vázquez-Hernández, 2008).

While Puerto Ricans on the USA mainland were facing dire living and economic conditions, Puerto Rico was undergoing a reformation of economic policies that sought to increase investments and employment on the island. In 1976, the USA Congress approved Section 936 of the Internal Revenue System, which gave tax incentives to USA companies that moved to Puerto Rico and reinvested some of the financial gains back on the island (Feliciano & Green, 2017). Section 936, known in Puerto Rico as *Industrias* 936, was associated with a significant increase of the island's gross domestic product (GDP; Merling, Cashman, Johnston, & Weisbrot, 2017). As Puerto Rico's economy improved, Puerto Rico experienced a decrease in migration to the USA mainland (Baker, 2002; Whalen & Vázquez-Hernández, 2008). However, after the law was accused of being unfair for USA taxpayers, in 1996, Section 936 was revoked, and all tax incentives were phased out within the next 10 years (Cabán, 2017; Collins, Bosworth, & Soto-Class, 2006). The end of *Industrias* 936 coupled with the approval of new international trade agreements between the USA, Mexico, Canada, and China (see Merling et al., 2017 for a review) sent the Puerto Rican economy once again into recession. As the economy of the island faltered, migration to the USA mainland accelerated. However, new relocation patterns post-migration began to emerge.

Disillusioned with poor economic and living conditions in the northern region of the USA, Puerto Ricans saw in Florida better housing, employment opportunities, and a lower cost of living. By 1990, Florida had surpassed New Jersey with the second largest concentration of mainland Puerto Ricans, followed only by New York (Duany & Silver, 2010). According to statistics provided by the government of Puerto Rico, 761,000 island Puerto Ricans have migrated to the USA mainland since the mid-2000s (Commonwealth of Puerto Rico, 2017). In 2015 alone, for every 8 Puerto Ricans that returned to the island, 36 left (Commonwealth of Puerto Rico, 2017). Thirty-eight percent of those who migrated chose the southern region of the USA as their new destination, with Florida leading the way (Pew Research Center, 2015).

Puerto Rican migration is only expected to increase. A long history of failed economic policies has led the island to accrue over $100 billion dollars in debt, creating an economic and social crisis on the island. In an attempt to manage the debt, as perhaps yet another illustration of the island's colonial status, the U.S. Congress passed the Puerto Rico Oversight, Management, and Economic Stability Act (PROMESA) (Cabán, 2017). PROMESA established an unelected board, known in Puerto Rico as *La Junta de Control Fiscal*, to oversee the island's finances and reduce the island's debt via austerity measures (Cabán, 2017; Merling et al., 2017). Such measures have further stagnated the island's economy and contributed to the deterioration of Puerto Rico's public services, education, and health systems (Cabán, 2017; Merling et al., 2017; Sparks & Superville, 2017). This bill supersedes any previous Puerto Rican laws, particularly those inconsistent with PROMESA's objective of restructuring or managing the debt (PROMESA, 2016). Puerto Rico's governor is only allowed to be a non-voting member of the board (PROMESA, 2016). To further curtail spending and meet objectives imposed by PROMESA's board, Puerto Rico's government has begun to place a lifetime limit on the medical services for Puerto Ricans who receive services through Medicaid. This new austerity effort will negatively impact the medical services received by approximately one-half of Puerto Rico's population (Severino Pietri, 2017).

Puerto Rico's economic and migration crisis was further exacerbated by the devastation caused by Hurricane María of 2017, which impacted the island of Puerto Rico as the first category 4 hurricane on the island in over 85 years. This natural disaster wreaked substantial damage including the destruction of the power grid and many roads. Reaching parts of the island to provide aid and essential services proved difficult. For months, much of the island remained without power or potable water making life difficult for many vulnerable populations, including the elderly. Many families had no other choice but to migrate to the USA mainland. It is estimated that more than 130,000 Puerto Ricans migrated to the USA

Table 18.3 Promising recommendations when working with Puerto Rican dementia patients and their caregivers

Stage	Intervention
Mild cognitive impairment	Assess how patient and family understand illness (e.g., What according to the family/patient is causing the symptoms? Are these perceived as problematic?) Assess for mental and chronic health conditions (e.g., depression, diabetes) that may further exacerbate symptoms related to dementia and integrate related treatments to dementia care Assess endorsement of cultural values such as *familismo* (desire to keep family together) and *fatalismo/aguante* (believing that one must endure illness) by patient and caregivers and how these values may inform working alliance with physicians and staff and future care planning Evaluate access to health care and/or potential barriers to continuation of care (e.g., periods of circular migration between Puerto Rico and the USA for both patient and physician, change in socioeconomic status) Establish a plan for continuation of services and treatment for Puerto Rican dementia patients in the island, in the event that their current physician would leave to the USA mainland Integrate and promote culturally aligned lifestyle changes: physical exercise (e.g., dancing), diet (e.g., modify Puerto Rican foods to follow low-fat, low-sodium standards), and social engagement (e.g., community and/or church activities) Integrate cultural and/or spiritual healing practices (e.g., prayer, connecting with folk healers, or obtaining support from spiritual leaders) Use formal and informal conversations to educate patient and support system about dementia (e.g., support and psychoeducation groups for patients and caregivers) Connect dementia patients and their caregivers to programs like *Café por el Alzheimer*, and make such resources available and accessible through various mediums (e.g., social media, pamphlets, flyers) Facilitate support groups that encourage community members to use *cuentos* [stories] and *testimonios* [testimonies] to educate others about dementia and that encourage social support among participants
Early stage	Assess continuation of care (e.g., Has the patient moved temporarily to Puerto Rico or to live with family members in other regions of the USA? If so, has care continued in new place of residence? If the patient moved temporarily to Puerto Rico, were there any significant changes in health coverage?) Caregiver support and skills training (e.g., integrating psychoeducation and support groups) Continued formal and informal illness education for patient and support system while also promoting strength-based coping strategies the family and community may already be using/practicing (e.g., prayer) Continued assessment of patient's and caregivers' perceptions of medical treatment and disease etiology. Given the history of medical systems in Puerto Rico, when the patient or caregiver may not agree with the treatment course, or not trust it, and the patient or caregiver may not feel comfortable sharing concerns with physicians or care providers. Validation of these concerns and distrust may be integral to the working alliance and the course of treatment Make counseling services available to address the patients and the family's fears and anxieties around the progression of the disease
Moderate stage	Continued assessment of continuation of care and/or barriers to continued care Continued caregiver/family support and skills training Integrate emotional and/or spiritual support for the patient and the caregivers
Severe stage	Continue to integrate emotional and/or spiritual support for the patient and the caregivers Explore families' belief around palliative care and incorporate such humane treatment into the palliative care

mainland in the aftermath of Hurricane María (Sutter, 2018). Multiple news organizations reported on the arrival of elderly Puerto Ricans to the USA, and among the main reasons noted for the elderly Puerto Ricans' migration was the need to receive medical and social services they were unable to receive in Puerto Rico after the hurricane (González, 2017). Emerging data on the psychological and social ramifications of Hurricane María's effects indicate that Puerto Ricans who were directly (Scaramutti, Salas-Wright, Vos, & Schwartz, 2019) and indirectly (Capielo Rosario, Abreu, Gonzalez, & Cárdenas, in press) exposed to the Hurricane María's devastation experienced significant traumatic stress symptoms. Although less is known about the specific psychological consequences on elderly Puerto Ricans, it is known that elderly survivors of natural disasters experience a wide range of symptoms including posttraumatic stress disorder and depression (Kessler et al., 2008; Solomon & Green, 1992) and a great need for supportive social services (Fernandez, Byard, Lin, Benson, & Barbera, 2002). It stands to reason that health interventions with elderly Puerto Ricans who suffer from dementia require medical treatments that also incorporate psychological interventions and other supportive services such as housing and financial assistance.

Given the sociopolitical history of Puerto Rico, it is crucial that health-care providers understand how policies and laws, as well as migration patterns, impact the provision and delivery (or lack thereof) of interventions and treatment of dementia. The following recommendations describe ways in which providers can work with Puerto Ricans with dementia, based on the sociopolitical context offered in this chapter (see Table 18.3).

Conclusion

To advance social justice in the work associated with Puerto Rican clients in general, and patients with AD or dementia in particular, a call to action goes out for researchers, interventionists, and all health-care providers: that these professionals seek, with purpose, to gain knowledge and specialized training for work with Puerto Ricans and that they participate in and learn about Puerto Rican communities and may be able to recognize and reinforce the strengths of Puerto Rican patients' cultural values and, most importantly, strive to recognize and work through their own biases and assumptions. An understanding of the transcultural and transnational experiences associated with the different types of Puerto Rican migration patterns provides the backdrop against which culturally congruent intervention frameworks can be constructed. A call to action also demands that providers understand Puerto Rico's political history and its implications on dementia care. Having knowledge of the long-standing history that the USA has with Puerto Rico is key to formulating interventions that can contextualize some of the cultural and systemic barriers to treatment that a patient may likely experience. The recommendations provided in this chapter aim to assist providers in carrying out interventions and treatments that are culturally congruent to Puerto Rican communities with individuals living with AD and dementia.

References

Acevedo, G. (2004). Neither here nor there: Puerto Rican circular migration. *Journal of Immigrant & Refugee Services, 2*(1–2), 69–85. https://doi.org/10.1300/J191v02n01_05

Acosta-Belén, E., & Santiago, C. E. (2006). Puerto Ricans in the United States: A contemporary portrait. Boulder, CO: Lynne Rienner Publishers.

Adames, H. Y., & Chavez-Dueñas, N. Y. (2017). *Cultural foundations and interventions in Latino/a mental health: History, theory and within group differences*. New York, NY: Routledge.

Ai, A. L., Appel, H. B., Huang, B., & Lee, K. (2012). Overall health and healthcare utilization among Latino American women in the United States. *Journal of Women's Health, 21*(8), 878–885. https://doi.org/10.1089/jwh.2011.3431

Alcántara, C., Chen, C. N., & Alegría, M. (2014). Do post-migration perceptions of social mobility matter for Latino immigrant health? *Social Science & Medicine, 101*, 94–106. https://doi.org/10.1016/j.socscimed.2013.11.024

Alegría, M., Canino, G., Shrout, P. E., Woo, M., Duan, N., Vila, D., … Meng, X. L. (2008). Prevalence of mental illness in immigrant and non-immigrant US Latino groups. *American Journal of Psychiatry, 165*(3), 359–369. https://doi.org/10.1176/appi.ajp.2007.07040704

Alegría, M., Canino, G., Stinson, F. S., & Grant, B. F. (2006). Nativity and DSM-IV psychiatric disorders among Puerto Ricans, Cuban Americans, and non-Latino Whites in the United States: Results from the National Epidemiologic Survey on Alcohol and Related Conditions. *Journal of Clinical Psychiatry, 67*(1), 56–65. https://doi.org/10.4088/JCP.v67n0109

Alegría, M., Mulvaney-Day, N., Woo, M., Torres, M., et al. (2007). Correlates of past-year mental health services use among Latinos: Results from National Latino and Asian American study. *American Journal of Public Health, 97*(1), 76–83. https://doi.org/10.2105/AJPH.2006.087197

Alexopoulos, G. S., Meyers, B. S., Young, R. C., Mattis, S., & Kakuma, T. (1993). The course of geriatric depression with "reversible dementia": A controlled study. *American Journal of Psychiatry, 150*(11), 1693–1693. https://doi.org/10.1176/ajp.150.11.1693

Alvarez, L., & Goodnough, A. (2015). Puerto Ricans brace for crisis in health care. *The New York Times*. Retrieved from https://www.nytimes.com/2015/08/03/us/health-providers-brace-for-more-cuts-to-medicare-in-puerto-rico.html?searchResultPosition=1

Alzheimer's Association. (2017). *10 Early signs and symptoms of Alzheimer's*. Retrieved from https://www.alz.org/alzheimers_disease_10_signs_of_alzheimers.asp

American Cancer Society. (2017). *Cancer facts & figures for Hispanics/Latinos 2015–2017*. Atlanta: American Cancer Society.

Aranda, M. P., Villa, V. M., Trejo, L., Ramírez, R., & Ranney, M. (2003). El Portal Latino Alzheimer's project: Model program for Latino caregivers of people with Alzheimer's disease. *Social Work, 48*(2), 259–271. https://doi.org/10.1093/sw/48.2.259

Baker, S. S. (2002). *Understanding mainland Puerto Rican poverty*. Philadelphia, PA: Temple University Press.

Berbusse, E. J. (1966). *The United States in Puerto Rico, 1898–1900*. Chapel Hill, NC: University of North Carolina Press.

Brown, A., & Patten, E. (2014). *Statistical portrait of Hispanics in the United States, 2012: Table 9*. Washington, D.C.: Pew Research Center.

Burgos, G., & Rivera, F. I. (2012). Residential segregation, socio-economic status, and disability: A multilevel study of Puerto Ricans in the United States. *CENTRO: Journal of the Center for Puerto Rican Studies, 24*(2), 14–47.

Cabán, P. (2017). Puerto Rico and PROMESA: Reaffirming colonialism. *New Politics Journal, 14*(3), 120–125.

Camacho-Mercado, C. L., Figueroa, R., Acosta, H., Arnold, S. E., & Vega, I. E. (2016). Profiling of Alzheimer's disease patients in Puerto Rico: A comparison of two distinct socioeconomic areas. *SAGE Open Medicine, 4*(1), 1–8. https://doi.org/10.1177/2050312115627826

Capielo, C., Delgado-Romero, E. A., & Stewart, A. E. (2015). A focus on an emerging Latina/o population: The role of psychological acculturation, acculturative stress, and coping on depression symptoms among Central Florida Puerto Ricans. *Journal of Latina/o Psychology, 3*(4), 209–223. https://doi.org/10.1037/lat0000039

Capielo, C., Schaefer, A., Ballesteros, J., Monroig, M. M., & Qiu, F. (2018). Puerto Ricans on the U.S. mainland. In P. Arredondo (Ed.), *Latinx immigrants: Transcending acculturation and xenophobia* (pp. 187–210). New York, NY: Springer.

Capielo Rosario, C., Abreu, R., Gonzalez, K., & Cárdenas, E. (in press). "That day no one spoke": Florida Puerto Ricans reaction to Hurricane María. *The Counseling Psychologist*.

Capielo Rosario, C., Adames, H. Y., Chavez-Dueñas, N. Y., & Rentería, R. (2019). Acculturation Profiles of Central Florida Puerto Ricans: Examining the influence of perceived ethnic-racial discrimination and neighborhood ethnic-racial composition. *Journal of Cross-Cultural Psychology, 50*(4), 556–576. https://doi.org/10.1177/0022022119835979

Capielo Rosario, C., & Dillon, F. (2019). Ni de aquí, ni de allá: Puerto Rican acculturation-acculturative stress profiles and depression. *Cultural Diversity and Ethnic Minority Psychology*. Advance online publication. https://doi.org/10.1037/cdp0000272

Capielo Rosario, C., Lance, C. E., Delgado-Romero, E. A., & Domenech Rodríguez, M. M. (2019). Acculturated and Acultura'os: Testing bidimensional acculturation across Central Florida and Island Puerto Ricans. *Cultural Diversity & Ethnic Minority Psychology, 25*(2), 152–169. https://doi.org/10.1037/cdp0000221

Capielo Rosario, C., Schaefer, A., Ballesteros, J., Rentería, R., & Davis, E. J. R. (2019). A caballo regalao no se le mira el colmillo: Colonial mentality and Puerto Rican depression. *Journal of Counseling Psychology, 66*, 396–408. https://doi.org/10.1037/cou0000347

Chavez-Dueñas, N. Y., Adames, H. Y., & Organista, K. C. (2014). Skin color prejudice and within-group racial discrimination: Historical and current impact on Latino/a populations. *Hispanic Journal of Behavioral Sciences, 36*(1), 3–26. https://doi.org/10.1177/0739986313511306

Cintrón Aguilú, A. (2014). *Puerto Rico's economy under Spanish rule*. Enciclopedia de Puerto Rico. Retrieved from https://enciclopediapr.org/en/encyclopedia/economy-under-spanish-rule/

Cohn, D. V., Patten, E., & Lopez, M. H. (2014). *Puerto Rican population declines on island, grows on US mainland*. Washington, D.C.: Pew Research Center.

Collins, S. M., Bosworth, B. P., & Soto-Class, M. A. (2006). *The economy of Puerto Rico: Restoring growth*. New York, NY: Brookings Institution Press.

Colón-Ramos, U., Rodríguez-Ayuso, I., Gebrekristos, H. T., Roess, A., Pérez, C. M., & Simonsen, L. (2017). Transnational mortality comparisons between archipelago and mainland Puerto Ricans. *Journal of Immigrant and Minority Health, 19*(5), 1009–1017. https://doi.org/10.1007/s10903-016-0448-5

Comas-Diaz, L. (2006). Latino healing: The integration of ethnic psychology into psychotherapy. *Psychotherapy: Theory, Research, Practice, Training, 43*(4), 436–453. https://doi.org/10.1037/0033-3204.43.4.436

Commonwealth of Puerto Rico. (2017). *Informe de Progreso de Reto Demográfico*. Retrieved from https://jp.pr.gov/Portals/0/Reto%20Demogr%C3%A1fico/Documentos/Informe%20Progreso%20de%20Reto%20Demografico%20Jun%2030%202017(editado%20con%20firmas%20y%20carta).pdf?ver=2017-08-08-110017-593

Das-Munshi, J., Leavey, G., Stansfeld, S. A., & Prince, M. J. (2011). Migration, social mobility and common mental disorders: Critical review of the literature and meta-analysis. *Ethnicity & Health, 17*(1–2), 17–53. https://doi.org/10.1080/13557858.2011.632816

David, E. J. R. (2008). A colonial mentality model of depression for Filipino Americans. *Cultural Diversity & Ethnic Minority Psychology, 14*(2), 118–127. https://doi.org/10.1037/1099-9809.14.2.118.

David, E. J. R., & Okazaki, S. (2006). The Colonial Mentality Scale (CMS) for Filipino Americans: Scale construction and psychological implications. *Journal of Counseling Psychology, 53*(2), 241–252.

Duany, J. (2002). Mobile livelihoods: The sociocultural practices of circular migrants between Puerto Rico and the United States. *International Migration Review, 36*(2), 355–388. https://www.jstor.org/stable/4149457

Duany, J. (2010). The Orlando Ricans: Overlapping identity discourses among middle-class Puerto Rican immigrants. *CENTRO: Journal of the Center for Puerto Rican Studies, 22*(1), 84–115.

Duany, J., & Silver, P. (2010). The "Puerto Ricanization" of Florida: Historical background and current status: Introduction. *CENTRO: Journal of the Center for Puerto Rican Studies, 22*(1), 4–32.

Fanon, F. (1965). *Studies in a dying colonialism*. New York, NY: Monthly Review Press.

Feliciano, Z. M., & Green, A. (2017). *US Multinationals in Puerto Rico and the Repeal of Section 936 Tax Exemption for US Corporations* (No. w23681). National Bureau of Economic Research. Retrieved from http://www.nber.org/papers/w23681

Fernandez, L. S., Byard, D., Lin, C. C., Benson, S., & Barbera, J. A. (2002). Frail elderly as disaster victims: Emergency management strategies. *Prehospital and Disaster Medicine, 17*(2), 67–74. https://doi.org/10.1017/S1049023X00000200

Figueroa, R., Steenland, K., MacNeil, J. R., Levey, A. I., & Vega, I. E. (2008). Geographical differences in the occurrence of Alzheimer's disease mortality: United States versus Puerto Rico. *American Journal of Alzheimer's Disease & Other Dementias, 23*(5), 462–469. https://doi.org/10.1177/1533317508321909

Fuligni, A. J., & Perreira, K. M. (2009). Immigration and adaptation. In F. A. Villarruel, G. Carlo, J. M. Grau, M. Azmitia, N. J. Cabrera, & T. J. Chahin (Eds.), *Handbook of US Latino psychology: Developmental and community-based perspectives* (pp. 99–114). Thousand Oaks, CA: Sage.

Gallagher-Thompson, D., Haynie, D., Takagi, K., Valverde, I., & Thompson, L. W. (2000). Impact of an Alzheimer's disease education program: Focus on Hispanic families. *Gerontology & Geriatrics Education, 20*(3), 25–40. https://doi.org/10.1300/J021v20n03_03

Gallagher-Thompson, D., Singer, L. S., Depp, C., Mausbach, B. T., Cardenas, V., & Coon, D. W. (2004). Effective recruitment strategies for Latino and Caucasian dementia family caregivers in intervention research. *American Journal of Geriatric Psychiatry, 12*(5), 484–490. https://doi.org/10.1176/appi.ajgp.12.5.484

Gallagher-Thompson, D., Solano, N., Coon, D., & Areán, P. (2003). Recruitment and retention of Latino dementia family caregivers in intervention research: Issues to face, lessons to learn. *The Gerontologist, 43*(1), 45–51. https://doi.org/10.1093/geront/43.1.45

Go, J. (2008). *American empire and the politics of meaning: Elite political cultures in the Philippines and Puerto Rico during U.S. Colonialism*. Durham, NC: Duke University Press.

González, D. (2017). *Thousands of Puerto Ricans flee hurricane-ravaged island for mainland, including Arizona*. AZ Central. Retrieved from https://www.azcentral.com/story/news/local/phoenix/2017/10/31/thousands-puerto-ricans-flee-hurricane-ravaged-island-mainland-including-arizona/801305001/

González, H. M., Tarraf, W., Whitfield, K. E., & Vega, W. A. (2010). The epidemiology of major depression and ethnicity in the United States. *Journal of Psychiatric Research, 44*(15), 1043–1051. https://doi.org/10.1016/j.jpsychires.2010.03.017

Heiss, G., Snyder, M. L., Teng, Y., Schneiderman, N., Llabre, M. M., Cowie, C., … Loehr, L. (2014). Prevalence of metabolic syndrome among Hispanics/Latinos of diverse background: The Hispanic Community Health Study/Study of Latinos. *Diabetes Care, 37*(8), 2391–2399. https://doi.org/10.2337/dc13-2505

Karlawish, J., Barg, F. K., Augsburger, D., Beaver, J., Ferguson, A., & Nunez, J. (2011). What Latino Puerto Ricans and non-Latinos say when they talk about Alzheimer's disease. *Alzheimer's & Dementia, 7*(2), 161–170. https://doi.org/10.1016/j.jalz.2010.03.015

Kelley, M. L., & Fitzsimons, V. M. (2000). *Understanding cultural diversity: Culture, curriculum, and community in nursing*. Sudbury, MA: Jones and Bartlett.

Kessler, R. C., Galea, S., Gruber, M. J., Sampson, N. A., Ursano, R. J., & Wessely, S. (2008). Trends in mental illness and suicidality after Hurricane Katrina. *Molecular Psychiatry, 13*(4), 374–384. https://doi.org/10.1038/sj.mp.4002119

Levis-Peralta, M., Llompart, A., Rodriguez, M. C., Sanchez, M. C., Sosa-Pascual, A., Matousek, S. B., & Day, R. (2016). *Description of the state health care environment: Puerto Rico state health innovation plan*. Rio Piedras: Puerto Rico Department of Health.

Llanque, S. M., & Enriquez, M. (2012). Interventions for Hispanic caregivers of patients with dementia: A review of the literature. *American Journal of Alzheimer's Disease & Other Dementias, 27*(1), 23–32. https://doi.org/10.1177/1533317512439794

Lugo-Lugo, C. R. (2006). U.S. congress and the invisibility of coloniality: The case of puerto rico's political status revisited. *CENTRO: Journal of the Center for Puerto Rican Studies, 18*(2), 125–145.

Mayol-Garcia, Y. H. (2019). Migration, living arrangements, and poverty among Puerto Rican-origin children: Puerto Rico and the United States. Paper presented at the Population Association of America Annual Meeting. Austin, TX.

McGreevey, R. C. (2018). *Borderline citizens: The United States, Puerto Rico, and the politics of colonial migration*. Ithaca: Cornell University Press.

Meeks, S., Fazio, S., Pace, D., Kallmyer, B. A., Maslow, K., & Zimmerman, S. (2018). Alzheimer's association dementia care practice recommendations. *The Gerontologist, 58*(S1), S1–S9. https://doi.org/10.1093/geront/gnx182

Memmi, A. (1996). *Retrato del colonizado*. Buenos Aires, Argentina: Ediciones La Flor.

Merling, L., Cashman, K., Johnston, J., & Weisbrot, M. (2017). *Life after debt in Puerto Rico: How many more lost decades*. Center for Economic and Policy Research. Retrieved from http://cepr.net/images/stories/reports/puerto-rico-2017-07.pdf

Modrego, P. J., & Ferrández, J. (2004). Depression in patients with mild cognitive impairment increases the risk of developing dementia of Alzheimer type: A prospective cohort study. *Archives of Neurology, 61*(8), 1290–1293. https://doi.org/10.1001/archneur.61.8.1290

National Center for Health Statistics. (2017). *Health, United States, 2016, with chartbook on long-term trends in health* (No. 2017). Government Printing Office.

Nicklett, E. J., & Burgard, S. A. (2009). Downward social mobility and major depressive episodes among Latino and Asian-American immigrants to the United States. *American Journal of Epidemiology, 170*(6), 793–801. https://doi.org/10.1093/aje/kwp192

Pew Research Center. (2015). *Hispanics of Puerto Rican origin in the United States, 2013*. Retrieved from http://www.pewhispanic.org/2015/09/15/hispanics-of-puerto-rican-origin-in-the-united-states-2013/ Puerto Rico Oversight, Management, and Economic Stability Act of 2016, S.2328 (114th Congress, 2015-2016). https://www.pewresearch.org/hispanic/2015/09/15/hispanics-of-puerto-rican-origin-in-the-united-states-2013/

Puerto Rico Oversight, Management, and Economic Stability Act. S.2328 — 114th Congress (2015-2016).

Rivera, F. I., & Burgos, G. (2010). The health status of Puerto Ricans in Florida. *CENTRO: Journal of the Center for Puerto Rican Studies, 22*(1), 199–217.

Rivera-Batiz, F., & Santiago, C. E. (1996). *Island paradox: Puerto Rico in the 1990s*. New York, NY: Russell Sage Foundation.

Rivera-Hernandez, M., Leyva, B., Keohane, L. M., & Trivedi, A. N. (2016). Quality of care for white and Hispanic Medicare advantage enrollees in the United States and Puerto Rico. *JAMA Internal Medicine, 176*(6), 787–794. https://doi.org/10.1001/jamainternmed.2016.0267

Rodríguez Ayuso, I. R., Geerman, K., & Pesante F. (2012). *Puerto Rico community health assessment: Secondary data profile*. Retrieved from http://www.salud.gov.pr/Estadisticas-Registros-y-Publicaciones/Publicaciones/Evaluacion%20de%20la%20Salud%20de%20la%20Comunidad%20Puertorriqueña%20Perfil%20de%20Datos%20Secundarios%20(Inglés).pdf

Rodríguez-Silva, I. (2012). *Silencing race: Disentangling blackness, colonialism, and national identities in Puerto Rico*. New York, NY: Palgrave Macmillan.

Rodríguez-Vigil, E., Rodríguez-Chacón, M., & Valcarcel, J. J. R. (2016). Correlation of global risk assessment with cardiovascular complications in patients with diabetes mellitus living in Puerto Rico. *BMJ Open Diabetes Research and Care, 4*(1), 1–7. https://doi.org/10.1136/bmjdrc-2016-000279

Sánchez, M., Cardemil, E., Adams, S. T., Calista, J. L., Connell, J., DePalo, A., … Rivera, I. (2014). Brave new world: Mental health experiences of Puerto Ricans, immigrant Latinos, and Brazilians in Massachusetts. *Cultural Diversity and Ethnic Minority Psychology, 20*(1), 16–26. https://doi.org/10.1037/a0034093

Scaramutti, C., Salas-Wright, C., Vos, S., & Schwartz, S. (2019). The mental health impact of Hurricane Maria on Puerto Ricans in Puerto Rico and Florida. *Disaster Medicine and Public Health Preparedness, 13*(1), 24–27. https://doi.org/10.1017/dmp.2018.151

Severino Pietri, K. (2017). *Understanding Puerto Rico's health care crisis*. Center for Puerto Rican Studies. Retrieved from https://centropr.hunter.cuny.edu/events-news/puerto-riconews/health-care/understanding-puerto-rico%E2%80%99s-health-care-crisis

Shin, P., Sharac, J., Luis, M. N., & Rosenbaum, S. (2015). *Puerto Rico's community health centers in a time of crisis (Policy Brief No. 43)*. Retrieved from https://publichealth.gwu.edu/sites/default/files/downloads/GGRCHN/Policy%20Research%20Brief%2043.pdf

Silver, P., & Velez, W. (2017). "Let me go check out Florida": Rethinking Puerto Rican diaspora. *CENTRO: Journal of the Center for Puerto Rican Studies, 29*(3), 98–125.

Solomon, S. D., & Green, B. L. (1992). Mental health effects of natural and human made disasters. *PTSD Research Quarterly, 3*, 1–8.

Sparks, S. D., & Superville, D. R. (2017). Re-opening Puerto Rico's schools takes a back seat to Island's basic needs. *Education Week*. Retrieved from https://www.edweek.org/ew/articles/2017/09/27/re-opening-puerto-ricos-schools-takes-a-back.html

Sue, D. W., & Sue, D. (2012). *Counseling the culturally diverse: Theory and practice* (6th ed.). Hoboken, NJ: John Wiley & Sons.

Sutter, J. D. (2018). 130,000 left Puerto Rico after Hurricane Maria, Census Bureau says. *CNN News*. Retrieved from https://www.cnn.com/2018/12/19/health/sutter-puerto-rico-census-update/index.html

Tiffin, P. A., Pearce, M. S., & Parker, L. (2005). Social mobility over the lifecourse and self reported mental health at age 50: Prospective cohort study. *Journal of Epidemiology & Community Health, 59*(10), 870–872. https://doi.org/10.1136/jech.2005.035246

Timms, D. W. (1996). Social mobility and mental health in a Swedish cohort. *Social Psychiatry and Psychiatric Epidemiology, 31*(1), 38–48.

Timms, D. W. (1998). Gender, social mobility and psychiatric diagnoses. *Social Science & Medicine, 46*(9), 1235–1247.

Torres, L. (2010). Predicting levels of Latino depression: Acculturation, acculturative stress, and coping. *Cultural Diversity and Ethnic Minority Psychology, 16*(2), 256–263. https://doi.org/10.1037/a0017357

Torres, L., & Rollock, D. (2007). Acculturation and depression among Hispanics: The moderating effect of intercultural competence. *Cultural Diversity and Ethnic Minority Psychology, 13*(1), 10–17. https://doi.org/10.1037/1099-9809.13.1.10

Torruella, J. R. (2017). To be or not to be: Puerto Ricans and their illusory U.S. citizenship. *CENTRO: Journal of the Center for Puerto Rican Studies, 29*(1), 108–135.

Tortolero-Luna, G., Zavala-Zegarra, D., Pérez-Ríos, N., Torres-Cintrón, C. R., Ortíz-Ortíz, K. J., Traverso-Ortíz, M., ... Ramos-Cordero, M. (2013). *Cancer in Puerto Rico, 2006–2010*. Puerto Rico Central Cancer Registry. Retrieved from http://www.rcpr.org/Portals/0/Reporte%20Anual/Cancer%20in%20Puerto%20Rico%202006-2010.pdf

Trujillo-Pagan, N. E. (2013). Worms as a hook for colonizing Puerto Rico. *Social History of Medicine, 26*(4), 611–632. https://doi.org/10.1093/shm/hkt001

Trujillo-Pagan, N. E. (2014). *Modern colonization by medical intervention U.S. medicine in Puerto Rico (studies in critical social sciences)*. Boston, MA: BRILL.

U.S. Census Bureau. (2015). *U.S. Census Bureau: American Community Survey 5-year estimates*. Retrieved from https://www.census.gov/quickfacts/PR

U.S. Census Bureau. (2017). *American Community Survey, 1-year estimates*. Retrieved from https://factfinder.census.gov/faces/tableservices/jsf/pages/productview.xhtml?pid=ACS_17_1YR_S0201&prodType=table

U.S. Department of Health & Human Services. (2015). *Puerto Rico*. [online] Retrieved from https://www.medicaid.gov/medicaid/by-state/puerto-rico.html

Vazquez, S. (2015). *ASES Abona $44 millones a deuda con proveedores*. Metro. Retrieved from http://www.metro.pr/noticias/ases-abona-44-millones-a-deuda-con-proveedores/pGXoiD!3OWQVIbcYIPr2

Vega, W. A., & Sribney, W. M. (2011). Understanding the Hispanic health paradox through a multi-generation lens: A focus on behaviour disorders. In G. Carlo, L. J. Crockett, & M. A. Carranza (Eds.), *Health disparities in youth and families: Research and applications* (pp. 151–168). New York, NY: Springer Science Business Media.

Vélez, W. (2017). A new framework for understanding Puerto Ricans' migration patterns and incorporation. *Centro Journal, 29*, 126–153.

Whalen, C. T. (2005). *Puerto Rican diaspora: Historical perspectives*. Philadelphia, PA: Temple University Press.

Whalen, C. T., & Vázquez-Hernández, V. (2008). *The Puerto Rican Diaspora: Historical Perspectives*. Philadelphia, PA: Temple University Press.

Willrich, M. (2011). *Pox: An American history*. New York, NY: Penguin.

Zubenko, G. S., Zubenko, W. N., McPherson, S., Spoor, E., Marin, D. B., Farlow, M. R., ... Sunderland, T. (2003). A collaborative study of the emergence and clinical features of the major depressive syndrome of Alzheimer's disease. *American Journal of Psychiatry, 160*(5), 857–866. https://doi.org/10.1176/appi.ajp.160.5.857

Dementia Diagnosis, Treatment, and Care in Colombia, South America

19

Yakeel T. Quiroz, Paula Ospina-Lopera, Valeria L. Torres, Joshua T. Fuller, Amanda Saldarriaga, Francisco Piedrahita, Alexander Navarro, and Francisco Lopera

Abstract

Dementia, characterized by a gradual onset and progressive decline of cognitive function that impacts daily life, is the only top ten cause of death with no cure or means of prevention. Research with a Colombian extended family with a genetic mutation causing Alzheimer's disease (AD) has helped scholars reconceptualize Alzheimer's as a continuum that begins several years before symptom onset and has helped set the stage for a clinical trial seeking to find out whether the disease may be preventable. While dementia research ventures in Colombia have expanded at an exponential pace in the past decade, access to quality medical and psychosocial care for dementia remains limited. In this chapter, the current state of dementia in Colombia, South America, is explored, where it is estimated that more than 200,000 individuals are diagnosed with AD. The chapter also describes existing behavioral and psychosocial interventions for patients and caregivers in Colombia including educational workshops, support groups, cinema forums, and the use of social media to communicate the latest developments in dementia research. Special attention is paid to Colombian governmental dementia policy, research, socioeconomic factors impacting diagnosis and access to care, and current treatment developments.

Y. T. Quiroz (✉)
Massachusetts General Hospital, Harvard Medical School, Boston, MA, USA
e-mail: yquiroz@mgh.harvard.edu

P. Ospina-Lopera · A. Saldarriaga · F. Piedrahita
A. Navarro · F. Lopera
Grupo de Neurociencias de Antioquia, Medellín, Colombia
e-mail: paulaospinalopera@gmail.com; amandaluciasaldarriaga@gmail.com; fpiedrahitapiedrahita@gmail.com; alexander.navarro@gna.org.co; francisco.lopera@gna.org.co

V. L. Torres
Massachusetts General Hospital, Harvard Medical School, Boston, MA, USA;
Florida Atlantic University, Boca Raton, FL, USA
e-mail: vtorres2015@fau.edu

J. T. Fuller
Massachusetts General Hospital, Harvard Medical School, Boston, MA, USA;
Boston University, Boston, MA, USA
e-mail: jtfuller@bu.edu

Introduction

Dementia, characterized by a gradual onset and progressive decline of cognitive function that impacts daily life, is the only top ten cause of

H. Y. Adames, Y. N. Tazeau (eds.), *Caring for Latinxs with Dementia in a Globalized World*,
https://doi.org/10.1007/978-1-0716-0132-7_19

death with no cure or means of prevention (Alzheimer's Association, 2017). The most common forms of dementia are neurodegenerative/neurocognitive disorders (NCDs), which include Alzheimer's disease (AD), Parkinson's disease, frontotemporal lobar degeneration, Lewy body disease, and Huntington's disease (American Psychiatric Association, 2013). Cerebrovascular disorders, such as small vessel disease and multi-infarcts, may also cause vascular dementia (American Psychiatric Association, 2013).

The prevalence of dementia has increased globally, partially explained by individuals living longer compared to previous decades. As such, people with dementia exist in all countries. In this chapter, we explore the current state of dementia in Colombia, South America, where it is estimated that 256,000 individuals are diagnosed with AD (ADI/BUPA, 2013). Notably, the ADI/BUPA (2013) report is almost certainly an underestimation of the number of Colombians living with dementia, as it did not include those diagnosed with non-AD dementias. Other studies have found that the prevalence of dementia is approximately 1.3% for Colombians at age 50; this number rises to 3.04% for individuals 70 years of age and older (Pradilla, Vesga, Boris, & León-Sarmiento, 2003). Certain regions within Colombia appear to have a higher prevalence of dementia, with the rate of dementia being 2.2% in the southwest and 1.9% in the east (Pradilla et al., 2003). A study in Manizales, Colombia, found a prevalence rate of dementia at 6% among older adults (Díaz Cabezas, Marulanda Mejía, & Martínez Arias, 2013). One study stands out, reporting a rate of dementia of 23.6% among Colombians older than the age of 60 (Goodling, Amaya, Parra, & Ríos, 2006).

The monetary impact of dementia in Colombia is substantial. Including the costs incurred by caregivers, the costs for each patient with dementia in Colombia throughout the course of the disease – 8 years on average – are estimated to be more than 99 million Colombian pesos or approximately US $33,500 (Prada, Takeuchi, & Ariza, 2014). Accordingly, the national annual cost of dementia care in Colombia is estimated to be over 100 billion Colombian pesos (Prada et al., 2014). The goal of this chapter is to describe existing behavioral and psychosocial interventions for patients and caregivers in Colombia. Special attention is paid to Colombian governmental dementia policy, research, socioeconomic factors impacting diagnosis and access to care, and current treatment developments.

The "Paisa" Mutation and Familial Alzheimer's Disease

In Colombia, Nobel Prize literature laureate Gabriel García Márquez arguably produced one of the earliest characterizations of dementia. A native Colombian, García Márquez's novel, *One Hundred Years of Solitude*, tells the story of a family in a small village struggling with an insomnia plague, which resulted in their inability to remember the "name and notion of things" (García Márquez, 1970; Rascovsky, Growdon, Pardo, Grossman, & Miller, 2009). In brilliant fashion, García Márquez characterized what it is now known as semantic dementia, before the medical community would formally recognize the condition in 1975 (Rascovsky et al., 2009).

Nearly two decades after García Márquez wrote of dementia in his works, Francisco Lopera, M.D., and his colleagues at *El Grupo de Neurociencias de Antioquia* (*GNA*) identified the largest known population affected with a single genetic mutation for early-onset, autosomal dominant AD (Cornejo, Lopera, Uribe, & Salinas, 1987; Lopera et al., 1994). This kindred population, which has been followed for more than 20 years, includes approximately 5000 members, of which approximately 1800 are carriers of the presenilin-1 (*PSEN1*) E280A mutation (also called the "Paisa[1]" mutation). These *PSEN1* mutation carriers have been extensively characterized (Acosta-Baena et al., 2011; Aguirre-Acevedo et al., 2016; Quiroz et al., 2010; Quiroz et al., 2015). For instance, the work with these families has helped scholars and clinicians reconceptualize AD as a continuum, which begins sev-

[1]The term "*Paisa*" refers to people who are originally from the northwest region of Colombia in the Andes mountain range. The region is formed by the *departamentos* [states] of Antioquia, Caldas, Risaralda, and Quindio. The main cities are Medellin, Pereira, Manizales, and Armenia.

eral years before the onset of clinical symptoms and is followed by mild cognitive impairment (MCI; cognitive problems that do not impact functioning) and dementia.

New disease-modifying treatments are being developed around the world. While treatment at almost any stage of AD is desirable, logically a disease-modifying treatment should be initiated as early as possible to allow the patient to have the best possible functioning for the longest possible time. The ideal time for pharmacological intervention is prior to the onset of clinical symptoms, allowing the onset of the disease to be delayed or at times prevented entirely. Recently, the Banner Alzheimer's Institute in Phoenix, Arizona, in the USA and the GNA launched the Alzheimer's Prevention Initiative Colombia Clinical Trial (Reiman et al., 2012). This initiative seeks to evaluate a range of disease-modifying AD treatments in presymptomatic *PSEN1* mutation carriers within 15 years of their estimated age of clinical onset, thus providing access to some of the most promising investigational treatments in development for this group of individuals in Colombia. The current API Colombia Clinical Trial seeks to test the efficacy of a new drug, *crenezumab* – an "anti-amyloid" antibody medication – in 300 Colombians who belong to the large kindred population with the *PSEN1* mutation (Tariot & Reiman, 2016). Research with the "Paisa" mutation has also facilitated the study and characterization of other dementias in Colombia, including frontotemporal lobar degeneration, and CADASIL (cerebral autosomal dominant arteriopathy with subcortical infarcts and leukoencephalopathy). While dementia research ventures in Colombia have expanded at an exponential pace in the past decade, access to quality medical and psychosocial care for dementia remains sparse, with healthcare facilities largely centralized in urban areas.

National Colombian Policies for Dementia Patients

As of 2017, 30 nations have developed "national Alzheimer's disease plans," a government policy tailored by a nation to address the burdens of dementia based on specific demographic and cultural needs (Alzheimer's Disease International, 2017). Only four Latin-American countries (Argentina, Costa Rica, Cuba, Mexico) and one Latin-American commonwealth of the USA (Puerto Rico) have formally adopted national Alzheimer's disease plans (Alzheimer's Disease International, 2017). Chile is in the process of implementing its plan; however, the plan is yet to be formally adopted (Gajardo & Abusleme, 2016). Colombia does not presently have a national Alzheimer's plan, nor does the country have a less formal "Alzheimer's Disease Strategy" (Alzheimer's Disease International, 2017).

Unfortunately, as described in Vargas et al. (2014), Colombia and other Latin-American countries face an immense challenge as local governments place a greater emphasis on combatting issues like illiteracy rather than the public health burden posed by terminal diseases such as dementia. As such, access to trained geriatricians is limited throughout the country and is almost non-existent in rural or less developed areas of Colombia (Gómez, Curcio, & Duque, 2009). The number of home care agencies in Colombia is also small, and long-term care facilities for the treatment of those with advanced dementia are accessible only for patients and families that have financial resources to cover the elevated costs (Gómez et al., 2009).

In more recent years, Colombia has witnessed the development of memory clinics where multidisciplinary groups of doctors and healthcare workers gather and work with patients to facilitate treatment and diagnosis of neurocognitive disorders (Medina-Salcedo, 2010). Memory clinics are also facilitating the development of new epidemiological studies allowing for further characterization of dementia in Colombian populations. However, access to the memory clinics is often limited by the *Plan Obligatorio de Salud de Colombia* [Mandatory Health Plan of Colombia] because these clinics are usually not considered part of the plan for basic level of care, even though it is recognized that receiving services provided by these clinics have potential benefits for patients (Medina-Salcedo, 2010).

El Grupo de Neurociencias de Antioquia (GNA) has been a pioneer in providing services to patients with memory problems and their families. The GNA's team has integrated interdisciplinary research, teaching, clinical, and social services for dementia patients and families in Colombia. Since 2014, the GNA has partnered with the *Alzheimer's Prevention Initiative* to conduct a Colombian registry of AD, as well as to increase public awareness of the disease in the country (Rios-Romenets et al., 2017). Another memory clinic, *Intellectus*, was founded in 2011 in Bogotá as part of the *Hospital Universitario San Ignacio* [San Ignacio Hospital]. *Intellectus* is associated with the *Pontificia Universidad Javeriana* [Javeriana University] and the *Instituto de Envejecimiento de la Facultad de Medicina* [Institute of Aging in the Department of Medicine]. The clinic provides services for those with dementia and related disorders. In addition, the *Fundación Acción Familiar Alzheimer Colombia* [Colombian Foundation for AD Action] is a nongovernmental organization whose mission is "to offer information, training, orientation, and help to those who care for someone affected by Alzheimer's disease" (Fundación Acción Familiar Alzheimer Colombia, 2008, p. 1). As the burden of dementia increases in Colombia, centralized coordination of resources and education about dementia is essential. A national plan that addresses modifiable dementia risk factors would be a strong foundation upon which further improvements in dementia diagnosis, treatment, and care can best be designed, assessed, and improved.

Sociocultural Factors Impacting Dementia Diagnosis

The consideration of sociocultural factors that impact individuals with dementia in Colombia is of great consequence. Goodling et al. (2006) conducted a study in Neiva, Colombia, where they investigated socioeconomic factors impacting the risk of developing dementia. Notably, 81% of participants diagnosed with dementia had 0–3 years of education, and 82.9% were from the lowest socioeconomic status (SES) bracket (Goodling et al., 2006). Cerebrovascular risk factors (e.g., hypertension, diabetes) were the most common health risk factors in their sample. Additionally, the authors reported that most cases (59.9%) were explained by a degenerative dementia, followed by vascular (23%) and mixed etiologies (13.8%).

Socioeconomic factors in Colombia, especially low levels of formal education and high levels of illiteracy, also play an important role in the delay patients may experience in receiving a clinical evaluation and a prompt diagnosis of dementia. Limited awareness of the disease and social stigma of dementia are further obstacles for early detection, care, and treatment of dementia throughout Latin America, including Colombia (Manes, 2016).

Pharmacological Treatments of Dementia

In the United States of America (USA), the Food and Drug Administration (FDA) has approved several pharmacological treatments for AD including three cholinesterase inhibitors (CI; donepezil/Aricept, galantamine, and rivastigmine) and two forms of memantine (Namenda, Namzaric (donepezil + memantine; Mayo Foundation for Medical Education and Research MFMER, 2017)). Further information about these pharmacological treatments for dementia can be found at the U.S. Department of Health and Human Services, National Institute on Aging's Alzheimer's Disease Medications Fact Sheet (https://www.nia.nih.gov/health/how-alzheimers-disease-treated). In Colombia, the national *Plan de Beneficios en Salud* (PBS) [Plan of Health Benefits] outlines what "treatments, medications, and [medical] supplies the general population has the right to access" (El Ministerio de Salud y Protección Social, 2017, para. 1). The Resolution 6408 of El Ministerio de Salud y Protección Nacional [Ministry of Health and National Protection] states that generic brands of donepezil, galantamine, rivastigmine, and memantine are available for the treatment of dementia through the PBS. Despite the availability of these medications, it is estimated that only 16.5% of dementia patients in Colombia are receiving some form of pharmacological treatment (Prada et al., 2014).

Beyond the medications that target underlying changes in cognition due to dementia, neuropsychiatric symptoms that commonly accompany dementia, such as changes in personality and mood, are also the target of pharmacological treatment through the use of antidepressants, anxiolytics, and antipsychotics (Lyketsos et al., 2002). While awaiting a successful pharmacological advancement that can target the underlying disease mechanism of AD (and other dementias), of equal prominence in the treatment of dementia is the need for non-pharmacological, behavioral, and psychosocial interventions, all of which are important therapeutic tools for patients, families, and caregivers.

Non-pharmacological Treatments of Dementia in Colombia

While pharmacological treatments of dementia seek to treat the cognitive and behavioral manifestations of dementia, non-pharmacological treatments aim to maintain patient autonomy and the highest quality of life possible during the inevitable dementia-related cognitive decline. Non-pharmacological or psychosocial treatments for dementia often employ an interdisciplinary approach and involve patients and their caregivers and families throughout the course of the disease.

As the disease progresses, dementia patients may experience medical complications [e.g., motor symptoms, gastrointestinal (GI) problems] that can be ameliorated by the implementation of behavioral interventions. For instance, exercises such as weight training and water aerobics are important for maintaining or improving motor function, which may lead to long-term gains including higher quality of life, greater patient autonomy, and improved relationships between patients and their family (Heyn, Abreu, & Ottenbacher, 2004). Similarly, several GI symptoms can be significantly improved by changing a patient's diet (Harris, 2001; Ibarzo, Suñer, Martí, & Parrilla, 2015; Prado, 2000). Another common experience in patients with dementia is a decline in dental health (Ghezzi & Ship, 2000; Gitto, Moroni, Terezhalmy, & Sandu, 2001; Kocaelli, Yaltirik, Yargic, & Özbas, 2002; Rutkauskas, 1997). Behavioral interventions to improve dental health of these patients are aimed at maintaining oral functions, (e.g., chewing and swallowing) and preserving the utterance of speech, all of which help patients stay socially involved.

Beyond maintaining physical health and autonomy, recent psychosocial treatments for dementia patients have focused on cognitive interventions and offering enriched educational and cultural opportunities for older adults (IV Congreso Iberoamericano de Universidades para mayores CIPUAM, 2011). Participation in educational opportunities promotes cognitive and social growth, as well as functional independence in older adults who are at risk for developing dementia because of their advanced age (Herrera & del Carmen, 2009; Hincapié, 2015; IV Congreso Iberoamericano de Universidades para mayores CIPUAM, 2011). In Medellín, Colombia, there are two academic institutions now welcoming senior citizens in their college-level classes. One of these institutions, *La Universidad de Antioquia* [University of Antioquia], has been offering classes to older adults for over 10 years now through the *El Programa de Aulas Universitarias Para Mayores* [Aulas University classes for older adults program]; more than 480 older adults have participated in these classes (Herrera & del Carmen, 2009). The *Tecnológico de Antioquia* [Technological University of Antioquia] also has a similar program (Hincapié, 2013). Inspired by these educational programs, the city of Medellín has designed "community spaces" throughout the city that promote intergenerational learning in topics such as literature, culture, technology, and gerontology. These spaces also host cognitive and exercise activities in which people of all ages, including those with dementia, can participate (Hincapié, 2013). While the management of dementia symptoms is of utmost concern for the medical community and family members, programs to assist caregivers in their duties and emotional burden are equally important given the direct impact that caregivers have on the quality of life for people with dementia.

Caregivers of Dementia Patients in Colombia

It is a cultural expectation among Colombians for family members to take care of other family members who are chronically ill. Thus, understanding the experiences of Colombian caregivers and how this role impacts their physical and mental health is a top priority (Ávila-Toscano, García-Cuadrado, & Gaitá n-Ruiz, 2010; Cerquera Cordoba, Pabón Poches, & Uribe Báez, 2012; Méndez, Giraldo, Aguirre-Acevedo, & Lopera, 2010). By understanding Colombian caregiving perspectives, investigators hope to design psychosocial and educational interventions that will reduce the risk of physical and mental disorders in caregivers (Cerquera Córdoba, Poches, & Katherine, 2015). Some studies conducted in this area have focused on formal caregivers in nursing settings (De la Cuesta, 2004). However, most of the Colombian patients with dementia remain in their homes where their caregiver is a family member, friend, or neighbor who usually does not receive monetary compensation for the caregiving work (National Alliance for Caregiving, 1997).

Caregiver Burden

Caregiver burden is the psychological distress that often accompanies caring for someone with dementia. As defined by Roig and Torres (1998), "caregiver burden" is a mental state resulting from the physical tasks, emotional demands, and restricted ability to socialize that is a consequence of caring for a chronically ill person. Protective factors exist that can help ameliorate aspects of caregiver burden. For example, caregiver self-care is vital to providing a high quality of life to a patient with dementia or another chronic illness. Thus, when considering psychosocial and educational interventions for caregivers, it is important to examine how caregivers effectively handle the stress and frustrations that their caregiving role creates.

Studies conducted in Colombia have not been conclusive regarding the nature of caregiver burden in informal caregivers. However, caregiver burden in Colombians has been linked to the patient's independence in daily functioning and level of cognitive impairment (Ocampo et al., 2007). Other factors impacting the level of caregiver's burden include self-perception of burden, patient's mood, and the degree to which the family unit is functional (Ocampo et al., 2007). Other studies examining the relationship between caregiver burden and family functioning are inconsistent (Dueñas et al., 2006) and may be explained, at least in part, by the different methodologies used in the studies, as well as the different demographic characteristics of the study samples. In addition, caregivers' personal background, including socioeconomic status, formal education, age, occupation, and relationship with the patient, seem to contribute to the burden experienced by an informal caregiver of a dementia patient in Colombia (Cerquera Córdoba, Granados Latorre, & Buitrago Mariño, 2012).

Caregiver Anxiety and Depression

Being a caregiver increases the risk for developing depressive symptoms, anxiety, and other mental and physical disorders. Cerquera Córdoba et al. (2012) reviewed the presence of depressive symptoms in the informal caregivers of relatives with dementia. Their review emphasizes the importance of treating the patient and attending to the needs of the informal caregiver (Pérez, 2008 cited by Cerquera Córdoba et al., 2012, p. 364). Similar to other empirical studies, Cerquera Córdoba et al. (2012) have suggested that depressive symptoms in caregivers are more pronounced in the initial years of care when fears and concerns related to personal performance, patient outcomes, and the changes in everyday life are most salient. Ultimately, the concerns and worries that can cause these depressive symptoms are reduced as caregivers develop a support system and learn to cope with the challenges of caring for a dementia patient (Ávila-Toscano et al., 2010).

Resilience as a Protective Factor for the Caregiver

The capacity to face adverse, traumatic, or risky situations and benefit from the experience without developing a mental or physical illness are the main components of resilience (Carretero, 2010, cited by Cerquera & Pabón, 2015, p. 76). Given the difficult nature of caregiving and how providing care to a dementia patient is wrought with complex situations, improving resilience in dementia caregivers is a major target of psychosocial interventions. A study on burden and resilience in caregivers of dementia patients in Bucaramanga, Colombia, illustrates some of the realities of caregiving in the country (Cerquera & Pabón, 2015). Specifically, 100 informal caregivers of patients with possible AD were part of the *Investigación Calidad en la Tercera Edad* [Study on Quality Investigation in the Elderly] at the Universidad Pontificia Bolivariana. The study was also interested in examining how caregivers handle stressful family situations and legal matters. Participants in the study (a) were 18 years of age or older, (b) were related to the patient, (c) lived in the same home as the patient, (d) were not receiving any financial compensation for their work, and (e) cared for the patient for at least 3 months and for 8 hours per day. The results revealed that 48.5% of caregivers in the sample reported severe burden and 20.8% reported mild-to-moderate burden. Overall, 79.3% of caregivers in the study reported some type of burden. Religion was the most frequently used coping strategy for the burden of their role (63.4%), and higher resilience was related to fewer symptoms of depression. Those individuals with higher levels of resilience also reported stronger emotional support systems, more social connections, and more frequent participation in leisure activities.

Based on the aforementioned results of Cerquera and Pabón (2015), cognitive-behavioral interventions for caregiver resilience were developed. Mori (2008; described by Cerquera & Pabón, 2015, p. 77) developed one such resilience intervention program in Perú. This program's objectives are to improve self-care strategies and, by proxy, improve the quality of care provided to patients with dementia. The researchers suggest that psychological interventions are most effective if the interventions provide information, emotional support, and training (Cerquera & Pabón, 2015).

Psychosocial and Educational Interventions for Dementia Patients

Nonprofit organizations in Colombia have developed programs for promoting healthy habits and support systems among patients and families. As previously described, the GNA is a hub of dementia research and care in Colombia. For the past decade, the group has hosted cognitive stimulation workshops for healthy older adults from the community, as well as for patients with MCI and mild dementia. These workshops teach participants cognitive exercises to maintain cognitive function for as long as possible. Although research in this area is in its initial stage, one study conducted by GNA researchers demonstrated a relative improvement in cognitive functioning in patients with amnestic MCI following their participation in 2 months of cognitive stimulation at 3 hours per week across a total of 24 sessions (Velilla-Jiménez, Soto-Ramírez, & Pineda-Salazar, 2010). In addition, the researchers noted a reduction in self-reported memory concerns of the seven participants after the stimulation sessions (Velilla-Jiménez et al., 2010). As discussed by these authors and others, cognitive stimulation is usually conducted in a group setting; thus, it is possible that social interaction also plays a significant role in improving cognitive function in these patients (Rozo, Rodríguez, Montenegro, & Dorado, 2016). Another study conducted by the GNA reported that older adults who participate in cognitive stimulation programs improve or maintain their level of functioning in everyday activities, as well as their processing speed and selective attention (Valencia et al., 2008).

GNA's Social Plan for Colombian Families Affected by Dementia

Part of *El Plan Social* [Social Plan] from the GNA, which seeks to provide support to more than 700 families affected by AD, is to develop an outreach program that encourages community engagement in the care of patients with dementia (Ospina-Lopera et al., 2015). The outreach program includes (a) support groups for patients with MCI and early dementia, (b) caregiving skills training, (c) educational workshops for children and young adults, (d) social media as a tool to increase awareness, and (e) community movie forums. Each initiative is briefly discussed below.

GNA Support Groups for Colombian Patients with Mild Cognitive Impairment (MCI) and Early Dementia

Support groups are often considered to be one of the most important and effective initiatives to alleviate the burden of dementia (De la Cuesta, 2004). Support group leaders (e.g., psychologists, social workers, other trained mental health staff or volunteers) facilitate the interaction between dementia patients, caregivers, and family members. During the ten sessions (at 2 hours per session), groups highlight the shared and unique experiences of patients and caretakers. As an educational activity, the support groups serve to provide training on symptom management. Furthermore, patients who present with dementias defined primarily by declines in non-memory domains (e.g., Parkinson's disease, ataxias, Huntington's disease) are encouraged to make choices related to their own course of care. Ultimately, support groups promote the well-being of caregivers and patients through the exploration of the emotional experiences and specific needs of each patient group.

GNA Caregiving Skills Training

The GNA also offers a more formal training on basic patient care skills of activities of daily living (e.g., feeding, bathing) and safe patient handling, which is designed for nursing staff and informal caregivers. Each training cycle provides ten, 4-hour sessions. These sessions also teach skills to cope with caregiver burden, depression, and anxiety.

GNA Educational Workshops for Children and Young Adults

Children and young relatives of people with dementia are also a target audience of the GNA workshops given the important role they can play in the care of an older family member with dementia (Montorio, Fernández, López, & Sánchez, 1998). These workshops have been designed with three main goals including (a) increasing awareness of dementia, (b) allowing children and youth to voice their concerns about the disease and caregiving experience, and (c) empowering youth to help their families cope with the stresses of caring for their relative with dementia. The typical structure of these workshops is five, 4-hour sessions and a yearly reunion for those who have completed the workshops. Outside the city of Medellín, the workshops take place during the weekend, for approximately 8–10 hours. These workshops also serve as an avenue to disseminate information about cutting-edge dementia research and care.

GNA's Focus on Social Media as a Tool to Increase Awareness

Using social media outlets such as Facebook and Twitter, the GNA has established constant communication with patients and families suffering from dementia, as well as members of the community at large. Social media has allowed the GNA to share new research findings and to promote GNA activities related to the caregiver training. The GNA has also used social media to increase awareness of dementia and AD, as well as brain health and healthy lifestyles.

GNA Community Movie Forum

The GNA hosts movie forums every 3 months to allow communities to discuss films on dementia and caregiving. Movie forums provide patients and families with the opportunity to meet local experts in the field and to learn more about Alzheimer's disease and dementia. This experience also allows the community to better understand the feelings and struggles of caring for a dementia patient and increases the community's awareness of brain disease.

Conclusion

Despite limited epidemiological information of dementia in Colombia, the nation is considered a pioneer in AD research, in part because of the work with the world's largest extended family with early-onset, autosomal dominant AD. Unfortunately, Colombia does not yet have a national dementia plan; however, nongovernmental organizations such as *El Grupo de Neurociencias de Antioquia* (GNA), *Intellectus*, and the *Fundación Acción Familiar Alzheimer* are leaders in providing resources and support to dementia patients and their families. Nonetheless, caregiver burden is a serious concern in Colombia, and there continues to be a need for investigators to develop self-care and resilience training programs. Future research should include (a) conducting large epidemiological dementia studies in Colombia, including representation of individuals from rural and urban areas, and (b) systematizing and assessing the efficacy of non-pharmacological interventions for dementia that have been developed in Colombia to date.

References

Acosta-Baena, N., Sepulveda-Falla, D., Lopera-Gómez, C. M., Jaramillo-Elorza, M. C., Moreno, S., Aguirre-Acevedo, D. C., … Lopera, F. (2011). Pre-dementia clinical stages in presenilin 1 E280A familial early-onset Alzheimer's disease: A retrospective cohort study. *Lancet Neurology, 10*(3), 213–220. https://doi.org/10.1016/S1474-4422(10)70323-9

ADI/BUPA. (2013). *Dementia in the Americas: Current and future cost and prevalence of Alzheimer's disease and other dementias*. Retrieved from https://www.alz.co.uk/research/world-report-2013

Aguirre-Acevedo, D. C., Lopera, F., Henao, E., Tirado, V., Muñoz, C., Giraldo, M., … Langbaum, J. B. (2016). Cognitive decline in a Colombian kindred with autosomal dominant Alzheimer disease: A retrospective cohort study. *JAMA Neurology, 73*(4), 431–438.

Alzheimer's Association. (2017). 2017 Alzheimer's disease facts and figures. *Alzheimer's & Dementia, 13*(4), 325–373.

Alzheimer's Disease International [ADI]. (2017). *National Alzheimer's and dementia plans*. Retrieved from: https://www.alz.co.uk/dementia-plans/national-plans. Accessed July 21, 2017.

American Psychiatric Association. (2013). *Diagnostic and statistical manual of mental disorders* (5th ed.). Arlington: American Psychiatric Publishing.

Ávila-Toscano, J. H., García-Cuadrado, J. M., & Gaitá n-Ruiz, J. (2010). Habilidades para el cuidado y depresión en cuidadores de pacientes con demencia. *Revista Colombiana de Psicología, 19*(1), 71–84.

Carretero, R. (2010). Resiliencia. Una revisión positiva para la prevención e intervención desde los servicios sociales. *Nómadas. Revista Crítica de Ciencias Sociales y Jurídicas., 27*(3), 1–13.

Cerquera, A., Pabón, D. (2015). *Programa de intervención Psicológica para cuidadores informales de pacientes con Demencia tipo Alzheimer en Bucaramanga y su Área Metropolitana. Experiencias significativas en Psicología y salud mental.* Retrieved from Colegio Colombiano de Psicólogos: https://issuu.com/colpsic/docs/experiencias_significativas/1?e=18058890/35327536

Cerquera Córdoba, A. M., Granados Latorre, F. J., & Buitrago Mariño, A. M. (2012). Sobrecarga en cuidadores de pacientes con demencia tipo Alzheimer. *Psychologia. Avances de la disciplina, 6*(1), 35–45.

Cerquera Cordoba, A. M., Pabón Poches, D. K., & Uribe Báez, D. M. (2012). Nivel de depresión experimentada por una muestra de cuidadores informales de pacientes con demencia tipo Alzheimer. *Psicología desde el Caribe, 29*(2), 360–384.

Cerquera Córdoba, A. M., Poches, P., & Katherine, D. (2015). Modelo de intervención psicológica en resiliencia para cuidadores informales de pacientes con Alzheimer. *Diversitas: Perspectivas en Psicología, 11*(2), 181–192.

Cornejo, W., Lopera, F., Uribe, C., & Salinas, M. (1987). Descripción de una familia con demencia presenil tipo Alzheimer. *Acta Médica Colombiana, 12*(2), 55–61.

De la Cuesta, C. (2004). Cuidado artesanal. *La invención ante la adversidad. Medellín: Facultad de Enfermería, Universidad de Antioquia, 38.*

Díaz Cabezas, R., Marulanda Mejía, F., & Martínez Arias, M. H. (2013). Prevalence of cognitive impairment and dementia in people older 65 years in a Colombian urban population. *Acta Neurológica Colombiana, 29*(3), 141–151.

Dueñas, E., Martínez, M. A., Morales, B., Muñoz, C., Viáfara, A. S., & Herrera, J. A. (2006). Síndrome del cuidador de adultos mayores discapacitados y sus implicaciones psicosociales. *Colombia Médica, 37*(2), 31–38.

El Ministerio de Salud y Protección Social. (2017). *Plan de beneficios en salud.* Retrieved from https://www.minsalud.gov.co/salud/POS/Paginas/plan-obligatorio-de-salud-pos.aspx

Fundación Acción Familiar Alzheimer Colombia. (2008). Retrieved August 03, 2017, from http://www.alzheimercolombia.org/

Gajardo, J., & Abusleme, M. T. (2016). Plan nacional de demencias: antecedentes globales y síntesis de la estrategia chilena. *Revista Médica Clínica las Condes, 27*(3), 286–296.

García Márquez, G. (1970). *One hundred years of solitude*, trans. *Gregory Rabassa*. New York: Avon.

Ghezzi, E. M., & Ship, J. A. (2000). Dementia and oral health. *Oral Surgery, Oral Medicine, Oral Pathology, 69*(1), 2–5.

Gitto, C. A., Moroni, M. J., Terezhalmy, G. T., & Sandu, S. (2001). The patient with Alzheimer's disease. *Quintessence International, 32*(3), 221–231.

Gómez, F., Curcio, C. L., & Duque, G. (2009). Health care for older persons in Colombia: A country profile. *Journal of the American Geriatrics Society, 57*(9), 1692–1696.

Goodling, M., Amaya, E., Parra, M., & Ríos, A. (2006). Prevalencia de las demencias en el municipio de Neiva 2003–2005. *Acta Neurológica Colombiana, 22*(3), 243–248.

Harris, N. G. (2001). *Nutrición y Dietoterápia de Krause*. Décima Edición: McGraw-Hill. Interamericana editores, S:A de C.A. 319, 320.

Herrera, Z., & del Carmen, M. (2009). La experiencia del Aula Universitaria de Mayores: enseñanza-aprendizaje de cuidado y autocuidado. Medellín, Colombia. *Investigación y Educación en Enfermería, 27*(2), 244–252.

Heyn, P., Abreu, B. C., & Ottenbacher, K. J. (2004). The effects of exercise training on elderly persons with cognitive impairment and dementia: A meta-analysis. *Archives of Physical Medicine and Rehabilitation, 85*(10), 1694–1704.

Hincapié, N. (2013). Alfabetización Digital como Cátedra Universitaria para Adultos Mayores. Congreso Iberoamericano de Programas Universitarios con Adultos Mayores PUMA 2013. ISBN: 978-607-495-246-9, Universidad de la Habana, Cuba.

Hincapié, N. (2015). Relación del modelo Gerontagógico con la práctica de los programas educativos para adultos mayores en Medellín Colombia. Congreso Iberoamericano de Programas Universitarios para adultos Mayores CIPUAM 2015. 9822-34902-1- I – SM.DOCX Olhar de Profesoor, Universidad Estadual de Ponta Grossa. Curitiba, Brasil.

Ibarzo, M. A., Suñer, R., Martí, A., & Parrilla, P. (2015). Manual de alimentación del paciente neurológico. *Sociedad Española de Enfermería Neurológica.*

IV Congreso Iberoamericano de Universidades para mayores CIPUAM. (2011). *Aprendizaje a lo largo de la vida, envejecimiento activo y cooperación internacional en los programas universitarios para mayores.* España: Alicante.

Kocaelli, H., Yaltirik, M., Yargic, L. I., & Özbas, H. (2002). Alzheimer's disease and dental management. *Oral Surgery, Oral Medicine, Oral Pathology, Oral Radiology, and Endodontology, 93*(5), 521–524.

Lopera, F., Arcos, M., Madrigal, L., Kosik, K., Cornejo, W., & Ossa, J. (1994). Demencia tipo Alzheimer con agregación familiar enAntioquia, Colombia. *Acta Neurológica Colombiana, 10*(4), 173–187.

Lyketsos, C. G., Lopez, O., Jones, B., Fitzpatrick, A. L., Breitner, J., & DeKosky, S. (2002). Prevalence of neuropsychiatric symptoms in dementia and mild cognitive impairment: Results from the cardiovascular health study. *JAMA, 288*(12), 1475–1483.

Manes, F. (2016). The huge burden of dementia in Latin America. *The Lancet Neurology, 15*(1), 29.

Mayo Foundation for Medical Education and Research [MFMER]. (2017). *Alzheimer's: Drugs help manage symptoms.* Mayo Clinic. Retrieved from http://www.mayoclinic.org/diseases-conditions/alzheimers-disease/in-depth/alzheimers/art-20048103

Medina-Salcedo, J. M. (2010). Las clínicas de memoria en el diagnóstico y tratamiento de las demencias. *Acta Neurológica Colombiana, 26*(3, supl. 1), 3–3.

Méndez, L., Giraldo, O., Aguirre-Acevedo, D., & Lopera, F. (2010). Relación entre ansiedad, depresión, estrés y sobrecarga en cuidadores familiares de personas con demencia tipo Alzheimer por mutación e280a en presenilina 1. *Revista chilena de Neuropsicología, 5*(2), 137–145.

Montorio, I., Fernández, M., López, A., & Sánchez, M. (1998). La Entrevista de Carga del Cuidador. Utilidad y validez del concepto de carga. *Anales de Psicología, 14*(2), 229–248.

Mori, M. (2008). Una propuesta metodológica para la intervención comunitaria. *Liberabit, 14*, 81–90.

National Alliance for Caregiving. (1997). *Family caregiving in the US – findings from a national survey (1997).* Bethesda, MD: National Alliance for Caregiving.

Ocampo, J. M., Herrera, J. A., Torres, P., Rodríguez, J. A., Loboa, L., & García, C. A. (2007). Sobrecarga asociada con el cuidado de ancianos dependientes. *Colombia Médica, 38*(1), 40–46.

Ospina-Lopera, P., Gutiérrez-Tamayo, A. L., Serna-Guzmán, C., Uribe-Pérez, C. M., Alzate-Echeverri, D., & Rivera-Otálvaro, M. T. (2015). Estudio Poblacional: Caracterización socio-familiar de los afectados con Enfermedad de Alzheimer Familiar Precoz (EAFP). *Integración Académica en Psicología, 3*(8), 94–107.

Pérez, M. (2008). Las intervenciones dirigidas a los cuidadores de adultos mayores con enfermedad de Alzheimer. *Revista Habanera de Ciencias Médicas, 7*(3), 1–11.

Prada, S. I., Takeuchi, Y., & Ariza, Y. (2014). Costo monetario del tratamiento de la enfermedad de Alzheimer en Colombia. *Acta Neurológica Colombiana, 30*(4), 247–255. Retrieved from http://www.scielo.org.co/scielo.php?script=sci_arttext&pid=S0120-87482014000400004&lng=en&tlng=es

Pradilla, A., Vesga, A., Boris, E., & León-Sarmiento, F. E. (2003). National neuroepidemiological study in Colombia (EPINEURO). *Revista Panamericana de Salud Pública, 14*(2), 104–111.

Prado, M. E. (2000). *Cuidados Paliativos en el anciano y en enfermedad terminal.* Retrieved from www.angelfire.com/pe/VvaSerena/anciano.html

Quiroz, Y. T., Budson, A. E., Celone, K., Ruiz, A., Newmark, R., Castrillón, G., ... Stern, C. E. (2010). Hippocampal hyperactivation in presymptomatic familial Alzheimer's disease. *Annals of Neurology, 68*(6), 865–875.

Quiroz, Y. T., Schultz, A. P., Chen, K., Protas, H. D., Brickhouse, M., Fleisher, A. S., ... Shah, A. R. (2015). Brain imaging and blood biomarker abnormalities in children with autosomal dominant Alzheimer disease: A cross-sectional study. *JAMA Neurology, 72*(8), 912–919.

Rascovsky, K., Growdon, M. E., Pardo, I. R., Grossman, S., & Miller, B. L. (2009). 'The quicksand of forgetfulness': Semantic dementia in One Hundred Years of Solitude. *Brain, 132*(9), 2609–2616.

Reiman, E. M., Quiroz, Y. T., Fleisher, A. S., Chen, K., Velez-Pardo, C., Jimenez-Del-Rio, M., ... Lopera, F. (2012). Brain imaging and fluid biomarker analysis in young adults at genetic risk for autosomal dominant Alzheimer's disease in the presenilin 1 E280A kindred: A case-control study. *The Lancet Neurology, 11*(12), 1048–1056. https://doi.org/10.1016/S1474-4422(12)70228-4

Rios-Romenets, S., Lopez, H., Lopez, L., Hincapie, L., Saldarriaga, A., Madrigal, L., ... Ramirez, L. (2017). The Colombian Alzheimer's prevention initiative (API) registry. *Alzheimer's & Dementia, 13*(5), 602–605.

Roig, M. V., & Torres, M. C. A. (1998). La sobrecarga en los cuidadores principales de enfermos de Alzheimer. *Anales de Psicología, 14*(2), 215.

Rozo, V., Rodríguez, O., Montenegro, Z., & Dorado, C. (2016). Efecto de la implementación de un programa de estimulación cognitiva en una población de adultos mayores institucionalizados en la ciudad de Bogotá. *Revista Chilena de Neuropsicología, 11*(1), 12–18.

Rutkauskas, J. S. (1997). The dental clinics of North America. In *Clinical decision-making in geriatric dentistry*. Philadelphia: Saunders.

Tariot, P. N., & Reiman, E. M. (2016). The Alzheimer's prevention initiative. *Alzheimer's & Dementia: The Journal of the Alzheimer's Association, 12*(7), P327.

Valencia, C., López-Alzate, E., Tirado, V., Zea-Herrera, M. D., Lopera, F., Rupprecht, R., & Oswald, W. D. (2008). Efectos cognitivos de un entrenamiento combinado de memoria y psicomotricidad en adultos mayores. *Revista de Neurologia, 46*(8), 465–471.

Vargas, E. A., Gallardo, Á. M. R., Manrrique, G. G., Murcia-Paredes, L. M., Riaño, M. C. A., & Dneuropsy, G. (2014). Prevalence of dementia in Colombian populations. *Dementia & Neuropsychologia, 8*(4), 323–329.

Velilla-Jiménez, L. M., Soto-Ramírez, E., & Pineda-Salazar, D. (2010). Efectos de un programa de estimulación cognitiva en la memoria operativa de pacientes con deterioro cognitivo leve amnésico. *Revista Chilena de Neuropsicología, 5*(3), 185–198.

Behavioral and Psychosocial Treatments of Dementia in Spain

20

Juan P. Serrano Selva and Margaret Gatz

Abstract

By 2050, Spain is projected to be home to the largest amount of elders in the world, due to a combination of high life expectancy, low birth rates, and the baby boomer generation entering their elder years. Consequently, in the next decades, Spain will have one of the highest dependency rates in the world. The country has nearly 800,000 people with dementia, with more than half in a situation of dependency. A conservative estimate suggests that the cost of each dependent is approximately 30,000 euros per year. When extended to all Spaniards with dementia, the cost reaches 24,000 million euros a year between direct costs for medical and professional care and indirect costs to caregivers and relatives who must alter their working life. Meanwhile, the dependency subsidy suffers from administrative blockages and budget cuts. The situation brings enormous challenge, affecting the patient, the caregiver, and the nature of care. Effective intervention for patients with Alzheimer's disease and other dementias increasingly includes the presence of psychologists specialized in the care of older adults. Psychologists work in both public and private settings including hospital units (memory units, dementia units, psychiatric units, etc.), day centers, residential centers, and universities and research centers. There is increasing use and evidence of the effectiveness of psychosocial, or non-pharmacological, interventions to address cognitive symptoms, such as altered memory or language, and especially psychological and behavioral symptoms associated with dementia, such as depression, anxiety, delusional disorders, altered behavior, etc., with advantages over psychotropic drugs with respect to greater effectiveness and fewer side effects.

J. P. Serrano Selva (✉)
Universidad de Castilla-La Mancha, Albacete, Spain
e-mail: JuanPedro.Serrano@uclm.es

M. Gatz
University of Southern California,
Los Angeles, CA, USA
e-mail: gatz@usc.edu

Introduction

To understand dementia care in Spain requires understanding the demographic and socioeconomic contexts, the philosophy that drives systems of care, and the evolution of the role of family within the Spanish culture. In the following sections, we begin with key contextual factors in Spain, in recent years to the present, including the aging of the population, the socioeconomic situation of the country, the immigration, and the extent of public funding for care. We then consider the epidemiology of dementia in

H. Y. Adames, Y. N. Tazeau (eds.), *Caring for Latinxs with Dementia in a Globalized World*,
https://doi.org/10.1007/978-1-0716-0132-7_20

Spain, specifically the already elevated number of dementia cases and the projected increase in prevalence of dementia. Considering this background, we describe the usual care protocol for those with dementia including types of services, living situation, and support for caregivers. We lastly consider the impact on dementia care of Spain's decentralized system of autonomous communities, as well as the evolution of family structure and cultural values about care.

Spain has 17 autonomous communities and 2 autonomous cities, Ceuta and Melilla, which are islands that are part of the African continent. This structure was established less than 40 years ago by the 1978 Constitution, after General Francisco Franco's regime ended. Creation of autonomous communities did not take place immediately, as the balance of federal and regional powers for many of the autonomous communities is still being negotiated into the twenty-first century. The autonomous communities differ in population, population density, urbanicity, gross regional product, unemployment rate, age structure, and history (Government of Spain, 2017).

Demographic Context

Rates of dementia in any country are strongly determined by population demography and the aging process (Harper, 2014). Spain has undergone greater shifts in its demographics over the last hundred years compared to most countries. Life expectancy at birth doubled between 1910 and 2010, increasing by more than 40 years. Currently, the life expectancy for women is 85.4 years and 79.9 years for men (INE [National Statistics Institute], 2016b), which is the highest life expectancy in Europe and among the highest in the world. Life expectancy for women is the second highest in the world, only surpassed by Japanese women, who live approximately 87 years (WHO, 2014).

According to the figures from the Continuous Population Census (INE), since January 2016, there were 8,657,705 older adults (65 years and over), constituting 18.4% of the total population (46,557,008) in Spain. The proportion of octogenarians continues to grow at a faster rate, now accounting for 6% of the total population. However, we should also highlight two major, current factors related to population patterns including (a) the low birth rate and (b) the ongoing economic crisis which started in 2008 and continues today (INE, 2016b).

The current birth rate in Spain is 12 births per 1000 persons in the population per year (Andrés de Llano et al., 2015), which is equivalent to 1.3 children per woman (INE, 2012). Concerning the economic crisis, the unemployment rate currently stands at 18.75% (INE, 2017), which is a key socioeconomic indicator. The impact of the 2008 national economic recession has not only affected the birth rate, but it has also had a major impact on the migratory flows that would normally compensate for the low birth rate. Since the onset of the economic crisis, the number of immigrants entering Spain has fallen dramatically. In addition, a significant number of immigrants established in Spain have decided to return to their home countries, especially those from Latin America. The increasing number of qualified young Spanish people emigrating to other countries further exacerbates the situation, which will ultimately affect Spain's work production network (INE, 2016a).

In light of these trends, the total population of Spain is projected to fall from over 46 million to just over 41 million over the next 50 years. By way of context, the current population of California is 39.5 million and is expected to exceed 46 million within the next 20 years (State of California, 2017). The Spanish population is becoming both smaller and older. According to projections by the INE (2016c), in 2066, there will be more than 14 million older adults in Spain, representing 34.6% of the predicted population of 41,068,643 inhabitants. Similarly, the Spanish Alzheimer's Association (CEAFA) forecasts that, of the total population in 2050, 31% will be aged 65 years or over (CEAFA, 2013). However, if we adjust figures to consider migratory factors, assuming a population of 41.5

million, this percentage would rise to 38% of total population.

In short, these forecasts indicate a highly complex scenario with one of the highest life expectancies in the world in conjunction with one of the lowest birth rates. The combination of these two factors, together with a notably declining population, especially among young adults, is resulting in a rapidly aging society. Age is one of the main risk factors for the emergence of neurodegenerative diseases; thus, the prevalence of these diseases will undoubtedly escalate. Additionally, this situation will ultimately lead to a lower number of persons of working age, many of whom will be unemployed, without sufficient resources to maintain themselves (OECD, 2015) and without the possibility to help the social system. These individuals will have to bear the costs of an increasingly aging population with higher levels of dependence (Prince, Prina, & Guerchet, 2013).

Prevalence of Dementia in Spain

Although different sources provide somewhat disparate estimates based on different age cutoffs, there does seem to be evidence that Spain is one of the countries with the highest percentage of dementia patients among individuals aged over 65 in the world, according to a report from the OECD (Health at a Glance, 2013). This study claims that 6.3% of Spaniards over the age of 60 have some degree of dementia. Only France (6.5%) and Italy (6.4%) present a higher rate. The mean rate in OECD countries is eight tenths of a point lower than in Spain (5.5%). The rate in a number of Northern European countries is higher than the mean, such as Sweden (6.3%) and Norway (6.2%), as well as in other leading economic powers such as the United States, the United Kingdom, and Japan (all three with 6.1%) and Germany (5.8%). Prevalence rates reflect the age structure of the population (previously discussed in the prior section of this chapter), as well as how long individuals survive after receiving a dementia diagnosis.

A wide-ranging report by PricewaterhouseCoopers (2012), the firm of economic consultants, established a figure between 500,000 and 800,000 dementia patients in Spain, with a prevalence of 10% among those older than 65 years, 25% among persons above 85 years and 50% in individuals older than 90 years, with 150,000 new cases per year. A meta-analysis conducted in 2011 suggests a prevalence of between 5.2% and 16.3% for those aged 65 and older, with a rate of 22% for men and 30% for women over 85 years old (Jurczynska, Eimil, López de Silanes & Luque, 2011). Broadly speaking, due to their longer life expectancy, the prevalence is higher in women, with the rate increasing with age (De Pedro-Cuesta et al., 2009), in accordance with evidence reported in world literature on dementia (Prince et al., 2013).

The distribution of types of dementia in Spain is not strikingly different from most comparison countries. The prevalence of Alzheimer's disease (AD) in Spain is around 6% in persons over 70 years of age, accounting for 70% of dementia cases (Gascón-Bayarri et al., 2007). Vascular dementia (VD) accounts for 12.5–27% of dementias (Casado & Calatayud, 2009; Gascón-Bayarri et al., 2007; Molinuevo & Peña-Casanova, 2009), while the prevalence of dementia with Lewy bodies in persons over age 70 years in Spain is about 1%. A study conducted with adults over age 70 in El Prat de Llobregat (Barcelona) showed a prevalence of 0.3%. The incidence of frontotemporal lobar degeneration (FTLD) cannot be estimated due to a lack of data (Gascón-Bayarri et al., 2007).

Disparity in prevalence of dementia between geographical regions of Spain is noticeable. While rates of dementia among those aged 65 and older are fairly similar across autonomous regions, the population prevalence varies with the proportion of the population who are aged 65 and older. Five autonomous communities in the northwest of Spain (including Galicia and the Basque Country) have the highest proportion of those aged 65 and older, therefore indicating a higher prevalence of dementia. Regions with the lowest proportion of those aged 65 and older (which turns out to be approximately half that of regions with the highest proportion) include the south of Spain, such as Ceuta and Melilla (CEAFA, 2013).

Costs Associated with Care and Treatment of Dementias

The cost of caring for a person with Alzheimer's disease in Spain is estimated to vary between 27,000 and 37,000 euros per year, with costs rising as the disease progresses (Prince et al., 2013; Turró-Garriga et al., 2010). This translates into a total cost of 24,000 million euros per year, which is estimated to rise to 33,000 million in 2030 and 56,000 million by 2050 as the number of persons with dementia increases. Care and treatment for those with dementia is covered by the Territorial Council of the System for Autonomy and Dependent Care. At the same time, it is estimated that families absorb 87% of the costs of care (Castañeira, Rodríguez, & Nunes, 2009). Accordingly, dementia has been defined as the "indirect cost disease" due to the burden of these costs on families (Coduras et al., 2010; Domínguez, Castro & López-Alemany, 2002).

Territorial Council of the System for Autonomy and Dependent Care

In late 2006, a new System for Promotion of Personal Autonomy and Assistance for Persons in a Situation of Dependency (SAAD *in Spanish*) was established in Spain by Law 39/2006 of 14 December, known as the Dependency Act (BOE, 2006). The Dependency Act recognized the universal right to social services for all Spanish citizens who are unable to care for themselves. A goal of the Dependency Act was to provide benefits to facilitate individuals' maintaining personal autonomy in their usual daily setting for as long as possible, thereby leading to the development of care for dependent persons from the provision of a public service to its current configuration as a form of integrated protection. The Territorial Council of the System for Autonomy and Dependent Care was designed to cooperate with the implementation of the new system. Public services are provided via social service agencies by the respective autonomous communities.

The current population of dependent persons in Spain is 2,141,404, of whom 1,462,292 are over the age of 65. Women account for 72.3% of this population, while men account for 27.7%. An estimated 88% of dependent persons aged 65 and older have a diagnosis of dementia (IMSERSO, 2005b).

Levels of Dependence

In Law 39/2006, dependence is classified into the three grades (moderate, severe, and major) according to the activities with which a person has difficulty and how much assistance is required. Grade I: Moderate dependence is a category by which a person requires help to perform various activities of daily living, at least once a day, or needs intermittent or limited support for personal autonomy. Grade II: Severe dependence is classified for persons that require help to perform various activities of daily living two or three times a day, but do not require the permanent support of a caregiver or do not have extensive needs for personal autonomy. Grade III: Major dependence is classified for persons requiring help to perform activities of daily living several times a day and, due to their total loss of physical, mental, intellectual, or sensory autonomy, who need the indispensable and continuous support of another person or need generalized support for personal autonomy.

According to the National Association of Social Service Directors and Managers (2017), the current distribution of dependent persons is as follows: of 1,213,873 dependent persons, Grade I is assigned to 394,212 persons (32.5%), Grade II to 455,741 persons (37.5%), and Grade III to 363,920 persons (30.0%). Of the total number of persons assessed as dependent, the system provides the benefits or services indicated in the law to 71% (865,564 persons). The remaining

29% (348,309 persons) are still waiting to receive the services to which they are entitled, a process that has been called the "dependency limbo" (Oliva, Peña, & García, 2015a; Peña-Longobardo, Oliva-Moreno, & Garcia-Armesto, 2016). More than 122,000 persons are entitled to services who are not attended to have either Grade I or Grade II dependence (35% of those waiting). Some of this is explained by the accumulated impact of the recession, which motivated Royal Decree 20/2012 and which cut back on benefits and created delays for applicants (Oliva, Peña & García, 2015b; Peña-Longobardo et al., 2016). The shortfall reached 3773.9 million euros in December 2016, owing to adjustments made in the General State Budget.

Benefits and Funding

The benefits established in Law 39/2006 to which dependent persons have access include services to support autonomy and monetary benefits. The original design of SAAD was to contract for services; however, in 2015, monetary benefits for family care accounted for 39.0% of all benefits (IMSERSO, 2015). Peña-Longobardo et al. (2016) relate this shifting of the composition of benefits to the economic crisis. Services are largely delivered in the home through agency contracts, designed to meet the needs of people who have difficulty performing activities of daily living (ADLs). The range of services encompasses home help with personal care and household needs, telecare services, day or night centers, and residential care. Monetary benefits may be of several types:

Monetary Benefit Linked to a Service This benefit payment, which is received on a regular basis, is only granted when it is not possible to access a public or subsidized attention and care service, with the amount depending on the degree and level of dependency and on the beneficiary's financial situation.

Cash Benefit to Family for Care A monetary benefit for family care was established for access on an exceptional basis, and the Royal Decree 20/2012 tightened the conditions for entitlement to these benefits. The Territorial Council of the System for Autonomy and Dependent Care establishes the conditions of access to this benefit in accordance with the person's recognized level of dependence and financial situation. The council is also charged with providing information and promoting non-professional caregivers who sign up for training programs.

Monetary Benefit for Personal Care The objective is to contribute to the hiring of a personal assistant, for a set number of hours, to aid in performing activities of daily living. Funding originates from a combination of the General State Administration, the autonomous communities, and co-payments. The *General State Administration* must determine the minimum level of protection guaranteed for each beneficiary. A Royal Decree establishes the State's contribution to this level of care. The minimum amounts are determined in accordance with the level of dependence of the person attended to and are calculated based on a coefficient which penalizes economic benefits for care in the family setting if this benefit is part of the individual attention program. The *autonomous communities* each year contribute an amount at least equal to the contribution made by the General State Administration in their region. Each autonomous community may establish other funding to increase the amounts provided for in Law 39/2006, which is considered additional funding. The recipients of dependency benefits are also required to participate in the funding through co-payments.

Total SAAD funding, in 2015, amounted to a cost of 7126 million euros in benefits and services. Indeed, the last quarter of 2016 boasted the largest expenditure yet for dependent care. Ten years after the implementation of the System for Promotion of Personal Autonomy and Care for Dependent Persons, figures show that 2.6% of the Spanish population needs support to perform activities of daily living, that is, 1,213,873 persons were assessed as dependent as of 31 December 2016. To this number, we could add

103,238 potential dependent persons pending assessment at that date (National Association of Directors and Managers in Social Services, 2017).

Usual Care Situation for Those with Dementia

Different modalities of health care are necessary to permit adequate, comprehensive, multidisciplinary, and coordinated attention to the needs of persons with dementia. The Spanish National Health System recognizes two basic levels of care: primary care and specialized care. A third level known as social and health care, which is unevenly implemented across the different autonomous communities, provides resources for inpatient, outpatient, and home care. Together, these levels permit comprehensive, continuous care until death.

(a) Primary care: This is a person's first point of contact with the health service system in which a team of professionals work on the prevention, detection, and diagnosis of disease, provide treatment, as well as monitor the treatment of illness, manage resources, and accompany patients until the end of their lives.
(b) Specialized care: This second level of care provides specific, specialized care in the diagnosis, treatment, and rehabilitation for health problems, which, due to their characteristics, cannot be solved in primary care.
(c) Social and health care (ASS): This third level, based on a model of comprehensive, multidisciplinary attention to older adults, guarantees the care of dependent persons with chronic disease and those nearing the end of life.

The Primary Health Team is a key service throughout the dementia process. It is responsible for early detection of dementia, diagnosing, requesting and interpreting basic complementary examinations, starting treatment, and, if necessary, referring patients to the Specialized Dementia Care Team. The Primary Care Team is then responsible for continuous care and coordination of resources, monitoring, and managing cases (Coll de Tuero & López-Pousa, 2011). The Specialized Dementia Care Team is a multidisciplinary team in a particular service area, which offers a comprehensive, specialized approach to care of dementia patients. This multidisciplinary team is comprised of (1) an expert in treating dementias (i.e., neurologists, geriatricians, or psychiatrists), (2) psychologists/neuropsychologists, (3) nurses, (4) social workers, and (4) administrative staff.

Dementia patients who have been assessed by the State Administration as meeting one of the levels of dependency may access the services provided for by Law 39/2006. The social services available to promote the personal autonomy and care of dependent persons include telecare service, home help service, day care and night care centers, and residential care services. Table 20.1 provides brief descriptions of these services.

Non-pharmacological Therapies

Non-pharmacological therapies may be defined as "any theoretically based, nonchemical, focused and replicable intervention, conducted with the patient or the caregiver, which potentially provided some relevant benefit" (Olazarán et al., 2010, p. 162). These interventions are available to various extents in different communities. Currently, a plan exists for developing a set of State Referral Centers. One of the centers, launched by the Institute for the Elderly and Social Services (IMSERSO) and located in Salamanca, specializes in Alzheimer's disease and other dementias in order to promote the best care for people with Alzheimer's and their families (IMSERSO, 2015). The following are among the pioneering non-pharmacological interventions:

Table 20.1 Social services to promote personal autonomy and care of dependent persons

Service	Description
Telecare	This service provides care for its beneficiaries by means of information and communication technologies, with the support of the required human resources. There is immediate support in situations of emergency or insecurity, loneliness, or isolation
Home help	This service provides assistance in the homes of dependent persons to attend to their needs related to activities of daily living. The service is provided by accredited organizations and companies
Day care and night care centers	This service provides comprehensive care for dependent persons during the day or night. The aim is to enhance or maintain their level of personal autonomy and support families and caregivers
Residential care	This service is provided in residential care centers with the service depending on the type and level of dependence of the patient. The service may be permanent when the care center becomes the patient's main residence or temporary when the service is provided for a limited period for convalescence, vacation periods, and weekends or if the routine, non-professional caregiver is ill or off work. Residential care service is provided by public agencies in their own centers or in subsidized centers

(a) *Creative/Therapeutic Dance*: A creative/therapeutic dance program, currently delivered by the State Referral Center for Alzheimer's Disease and Other Dementias in Salamanca, is based on theories of dance/movement therapy (Wengrower & Chaikin, 2008).

(b) *Psychomotor Stimulation*: Psychomotor re-education is a comprehensive, integrated activity designed to facilitate and enhance the physical, perceptual, psychological, and social development of an individual through body awareness and physical activities (Ferrer, Ortiz, & Avila-Castells, 2013). The Center provides guidelines for implementation.

(c) *Snoezelen Multisensory Stimulation*: The "snoezelen," or multisensory room, was developed in Holland in the 1970s (Klages, Zecevic, Orange, & Hobson, 2011). The word combines the ideas of sniffing and tranquility. The rooms deliver controlled multisensory stimulation, using light, sound, color, scents, textures, etc.

(d) *Dog-Assisted Intervention*: The Institute for the Elderly and Social Services (IMSERSO, 2007) has a research program that uses dog-assisted interventions with people with dementia which involves a social worker and animal-handler team (Animal Assisted Intervention International, 2016). By involving dogs in social work and care, socialization, ambulation, and a sense of well-being are promoted. There are both individual and group formats.

(e) *Music Therapy*: Since 2012, a range of music programs have been implemented, including creative music therapy which features improvisation, group participation in music to promote socialization, stimulation music aimed to preserve cognitive function, and the involvement of family in such music programs (Federación Mundial de Musicoterapia, 2011; Thaut, 2005).

(f) *Integral Cognitive Activation Program for Dementias (ICAPD)*: The ICAPD is a comprehensive cognitive stimulation program that collaborates with a research project at the University of Salamanca; the program's goals aim to maintain existing patient cognitive functions at the highest possible level while preventing behavioral problems and serving as a base for implementing psychoeducational support programs (García Meilán & Carro Ramos, 2014). ICAPD consists of a first phase of preliminary assessment, a 6-month intervention in which patients perform specific tasks and exercises, and a final assessment to monitor and evaluate the program's effectiveness. There is an extensive manual to guide implementation (García Meilán & Carro Ramos, 2014).

(g) *GRADIOR Program*: This computer-based neuropsychological software program acts as a rehabilitation system which provides training programs with the goal of recovering higher cognitive functions (attention, memory, orientation, etc.) in persons with cognitive deficits or deterioration (González-Palau, Franco, Toribio, Losada, Parra, & Bamidis, 2013). A touch screen allows the user to interact directly with the computer, which helps the system become more intuitive and accessible. The program permits the design of fully customized training for each user, taking into account their unique deficits and capabilities. The software collects data continuously, reviews it periodically, and facilitates information on the patient's progress.

(h) *Cognitive Psychostimulation*: Broadly speaking, cognitive stimulation encompasses techniques designed to maintain or enhance cognitive and functional capacities, behavior, and affect. The activities aim to stimulate and train different cognitive domains (e.g., perception, attention, reasoning, abstraction, memory, language, orientation processes, and praxis). Olazarán and Muñiz (2004) developed the program.

(i) *Reminiscence*: Reminiscence therapy is used to stimulate episodic or autobiographical memory – especially the emotional aspects of memory (Butler, 1963). Photos, music, recordings, old newspaper articles, domestic objects, informal chats, and other resources are used to stimulate emotional memories. This technique facilitates interpersonal relationships and communication in order to increase the patient's sensation of well-being and self-esteem.

(j) *Robot Therapy:* This therapy uses a robot called PARO, developed to look like a baby seal. PARO was developed in Japan and is equipped with artificial intelligence and multiple sensors which permit the robot to behave and interact with users as if it were a real animal (Shibata & Wada, 2011). Therapeutic robots offer an alternative to animal-assisted therapy by avoiding the complications of real animals while generating similar effects. The goals of robot therapy encompass (a) psychological needs, such as relaxation and motivation, (b) physiological improvements in vital signs, and (c) social stimulation, facilitating communication between patients and caregivers.

(k) *Reality Orientation Therapy*: This is a set of techniques to improve orientation in order to reduce confusion and disconnection from the environment (Spector et al., 2003). These techniques include verbal instructions to orient the individual to time, place, person, and situation, delivered informally during the day or in a group format, while also being offered as activities, such as discussion of current news.

(l) *Wii Therapy*: This therapy utilizes video games as a play-based, alternative form of cognitive stimulation (Padala et al., 2012). Using the Wii video console and brain training games, this therapy aims to generate forms of play which stimulate higher cognitive functions while promoting social interaction.

These diverse, non-pharmacological interventions are designed to accommodate moderate to severe cases of dementia, as well as to promote healthy brain aging in those without dementia or with mild cognitive impairment, while assisting those with neurocognitive disorders, and which can be re-purposed for dementia patients. The extent of empirical validation is currently mixed, but recognition of the advantages of non-pharmacological therapies is growing, especially as psychotropic drugs' limitations and adverse effects are concerns (Cobos & Rodriguez, 2012).

Caregivers

In Spain, care is defined in the preamble to the Dependency Act, Article II, as family members, health-care personnel, or both as providers of care to dependent persons (Rogero-García, 2009). Non-professional care services are those provided to dependent persons in the home by family members or other individuals such as friends and neighbors. Non-professional caregivers receive no financial payment for the work (Feldberg et al., 2011). Non-professional care is also called informal support. In Spain, 83% of family caregivers are female (43% are daughters, 22% are wives, and 7.5% are daughters-in-law). The mean age of an informal caregiver is 52 years old. Out of the total informal caregiver population, 77% are married and 20% of those married over age 65; 17% share their care work with other family responsibilities. In 80% of informal care cases, the work is unremunerated, and 60% share residence with the person they care for (Domingo, Sierra, Valero, & Castiñeira, 2015). Although a formal system exists in Spain to support those requiring dependent care and the caregivers, such services are not sufficiently widespread to cover the actual needs of this population. Thus, services such as psychotherapy or psychoeducation are often not within reach of those with fewer economic resources (Mateos, Franco & Sánchez, 2010).

Formal care is also referred to as professional care. Formal care includes tasks carried out by specialized professionals who are trained to provide care services beyond those that persons can do for themselves or others (Rodríguez, 2005). There are two types of formal care: those that are delivered by institutions and those that are paid for by the family. Formal caregivers represent a group of professionals who perform high-risk work in hard working conditions, receive low salaries, and experience long, laborious shifts (Delgado et al., 2014). In this respect, families are involved in caregiving both as informal caregivers and by paying for professional care.

In a nationally representative survey conducted by the Spanish National Statistics Institute (INE) in 2008 of dependent adults aged 65 and older who were living at home, 47.5% were receiving only informal care, another 9.8% received a combination of informal and formal care, 4.9% received formal care only, and the balance had no care (Rodríguez, 2013). Those who received only formal care tended to be women with low income residing in a capital location. Until Law 39/2006, home care services were scarce, with somewhat more possibility of receipt of public services occurring in certain municipalities. The 2008 INE survey also indicated an increase in dependent older adults living alone, which put further pressure on informal care systems (Rodríguez, 2013).

Support for Caregivers: Sources and Types of Support

Law 39/2006 established the System for Autonomy and Dependent Care (SAAD) which stipulates a catalogue of services to support informal caregivers. These include day care centers, home help, and temporary stays in residential care centers known as family respite services (Zambrano & Guerra, 2012). Family respite programs provide temporary accommodation for dependent persons, thus giving informal caregivers time to rest and perform tasks other than those related to caring for the family member (Institute for the Elderly and Social Services, IMSERSO, 2007). There is considerable variability among the different autonomous communities with respect to availability of different family respite options.

Other resources for caregivers include mutual help groups that provide help, advice, protection, and companionship for caregivers during the process of change in their lives, including services that help caregivers adjust to the limitations imposed by dependence and its associated problems. These groups meet periodically and are not necessarily directed by professionals (Torres, Ballesteros & Sánchez, 2008). Public and private entities in many towns provide training courses and workshops for caregivers. These workshops target dependent persons' family caregivers. IMSERSO provides an online listing of these training courses and workshops (http://www.imserso.es/imserso_01/innovacion_y_apoyo_tecnico/formacion_especializada/actividades_2017/index.htm).

In addition, La Caixa-Spanish Red Cross hosts a caregiver hotline. As part of the "One

Table 20.2 Organizations that offer advocacy and information for families

Organization	Description and corresponding website
Spanish Confederation for Associations of Relatives of People with Alzheimer's Disease and Other Dementias (CEAFA)	A national, nongovernmental organization, whose aim is to place Alzheimer's disease on the political agenda and pursue the necessary social commitment and raising of awareness so as to be able to represent and defend the interests of all those who live with Alzheimer's disease Website: www.ceafa.es
Spanish Alzheimer's Foundation (FAE)	The aim of this organization is to improve the quality of life of the patient, caregiver, and family. This includes a large variety of activities ranging from providing information (help lines, interviews with caregivers and families, leaflets, guides, conferences, symposia, etc.) to representing families before the social and health-care authorities. Website: www.alzfae.org
Alzheimer's Families Association (AFAL)	This association, founded in Madrid in 1989, has more than 4000 members. It forms part of the federal structure of the associations for families affected by Alzheimer's disease which exist across Spain. Its work is dedicated to information, training, advice, and support for families and persons close to patients. Website: www.afal.es
Alzheimer's Federation of the Community of Madrid (FAFAL)	This non-profit organization founded in 1999 is an organization that represents diverse associations in Madrid. It collaborates and cooperates with different public and private organizations working in the field of Alzheimer's disease and other dementias within the social and health-care services sectors Website: www.fafal.org
Pasqual Maragall Foundation	This foundation for research on Alzheimer's disease was founded as a response to a Spanish politician's publicly stated commitment on 20 October 2007, when he announced he was diagnosed with the illness Website: www.fpmaragall.org
Queen Sofía Foundation	This is a private, non-profit cultural and social care foundation with a presence across Spain. The Queen Sofía Foundation Alzheimer's center is a social and health-care project which created a model care complex that approaches the disease from three perspectives: research, training, and care services for Alzheimer's disease sufferers. It boasts a Center for Research on Neurological Disease, managed by the Carlos III Health Institute, as well as a Training Center and a Care Center, both managed by the Madrid Regional Ministry for Social Services. Website: http://www.fundacionreinasofia.es
CIEN Foundation	This foundation was established with the aim of creating a network of centers to support, promote, and coordinate research in all the fields of basic clinical and epidemiological neurology. The country's Center for Biomedical Research in Neurodegenerative Diseases is charged with the management of the Center. At the same time, it is also tasked with the management of the Alzheimer's Project Research Unit which occurred when the Queen Sofía Foundation ceded the building to the CIEN Foundation Website: http://www.fundacioncien.es/
Castilla-León Regional Federation of Alzheimer's Families (AFACAYLE)	This federation comprises 28 Alzheimer's Families Associations in the Castilla-León autonomous community. Website: https://www.afacayle.es/
The María Wolff Foundation	This organization promotes clinical research in Alzheimer's disease-type dementias. It focuses primarily on non-pharmacological interventions and therapies administered through social and health-care facilities such as day and residential care centers. The foundation's aim is to assure that patients, their direct caregivers, and professionals have access to the best techniques available, i.e., recognized at an international level, to enhance the quality of life of persons with dementia. Website: http://www.mariawolff.org/

(continued)

Table 20.2 (continued)

Organization	Description and corresponding website
The ACE Foundation	This Catalan Institute of Applied Neurosciences is a private organization devoted to the diagnosis, treatment, research, and support for persons with dementia, and especially those who suffer Alzheimer's disease, their families, and caregivers. The foundation's activities include services (diagnosis, treatment, and personalized monitoring), research (genetics, clinical trials, neuroimaging, etc.), and dissemination to scientific and lay audiences Website: http://www.fundacioace.com/quienes-somos/
Galicia Alzheimer's Federation (FAGAL)	This is a federation of families affected by Alzheimer's disease and other dementias. This non-profit entity comprises 13 associations in Galicia, accounting for more than 5000 families. The federation provides multiple services through 260 professionals from the social and health-care sectors. Website: www.fagal.org

Note: Other associations from other autonomous communities include Andalusia Alzheimer's Federation (FEAFA), Asturias Alzheimer's Federation (AFA), Canary Islands Alzheimer's Federation, Castilla-León Alzheimer's Federation, Madrid Alzheimer's Federation (AFAL), Murcia Alzheimer's Federation in Cieza (ACIFAD), Aragon Alzheimer's Federation, Balearic Islands Alzheimer's Federation, Catalonia Alzheimer's Federation, Basque Country Alzheimer's Federation, and Extremadura Alzheimer's Federation. In Valencia, Asociacion Familiares Alzheimer de Valencia (AFAV) and Federación Valenciana de Asociaciones de Familiares de Enfermos de Alzheimer (FEVAFA) provide an umbrella network bringing together all the associations in the province. In Catalonia, specifically in Barcelona, there exists the Alzheimer's Families Association (FAFAC); in Lleida, there is the Alzheimer's Families Association (AFALL). Murcia has a federation and an association: Federacion de Asociaciones de Familiares y Enfermos de Alzheimer de la Región de Murcia (FFEDARM) and Asociación de familiares de enfermos de Alzheimer de la Región de Murcia (AFAMUR). In Andalusia, Seville is home to the Santa Elena Alzheimer's Association. In Galicia, there is another association, Asociación de Familiares de enfermos de Alzheimer y otras demencias de Galicia (AFAGA)

Caregiver, Two Lives" program, the caregiver hotline provides information and helps address questions that informal caregivers might have about dementia; it also provides emotional support and strategies for coping with the difficulties associated with care. Internet resources for caregivers are also available. Plentiful information, resources, and guides for caregivers are accessible through a number of online links, including the official society of psychologists in the Valencian community (http://www.inclusio.gva.es/web/dependencia/manuales-profesionales). There are also a number of organizations, both national and regional, that offer advocacy, information for families, and sometimes direct services. These are listed in Table 20.2.

International Associations that Collaborate with Spain

Spain values international collaboration through its membership in the European Union and other international bodies. It encourages Alzheimer's disease researchers to publish in international journals and to work with scholars in other countries. As such, Spain has joined key international associations that support people with Alzheimer's disease and their families. Some of these organizations are listed in Table 20.3.

Regional and Cultural Considerations

There is no national plan or strategy in Spain that would provide a coordinated multi-professional solution to Alzheimer's disease, such as legal and ethical measures to protect persons affected by the disease and their caregivers. Rather, dementia patients' needs were addressed within the Dependency Act, which has a different mandate and is not limited to older adults. For now, the source of services to patients is Law 39/2006. In 2017, the Spanish Ministry of Social Services and Equality proposed the elaboration and enactment of the National Plan for Alzheimer's Disease and other Dementias as one of the priorities of this parliamentary term. The State Referral Center (CRE) for Attention to Persons with Alzheimer's Disease and Other Dementias,

Table 20.3 Key international associations that support people with Alzheimer's disease

Organization	Description and corresponding website
Alzheimer's Disease International (ADI)	The ADI is the worldwide federation of Alzheimer's disease associations supporting people with dementia and their families. The Spanish organization the *Spanish Confederation for Associations of Relatives of People with Alzheimer's Disease and Other Dementias* (CEAFA) appears as a member in the federation's strategic plan. Website: www.ceafa.es
Alzheimer's Europe (AE)	Over time, the associations have grown considerably, both in the size of each organization and in terms of the number of national associations which now exist. As the organizations grew, so did their work mandate which extended to include campaigns for particular issues as well as policy work. As of November 2016, Alzheimer's Europe has 39 member associations from 34 countries. Spain is part of this organization through the membership of the Spanish Alzheimer's Foundation (FAE). Website http://www.alzfae.org/
Alzheimer's Iberoamérica (AIB)	Spain collaborates with this organization through the *Spanish Confederation for Associations of Relatives of People with Alzheimer's Disease and Other Dementias* (CEAFA). Members include Spain, Portugal, and regions in the Americas where Spanish or Portuguese are predominant languages. Its objectives include to serve as a platform to connect members, unify criteria in public policies, advise and support members on matters of interest, promote scientific and social research related to dementias, and promote international cooperation between organizations with the same or similar aims. Website: http://alzheimeriberoamerica.org/
European Association Working for Carers (Eurocarers)	Spain belongs to this organization. Eurocarers was established to advance the issue of informal care at both national and European Union (EU) levels by (A) carrying out and supporting research on issues that concern carers in order to help build the evidence for sound advocacy, communication, and – ultimately – policy development; (B) advocating the interests of carers with a focus on their health, pensions and social security, social inclusion, and employment; (C) encouraging and facilitating the development of representative and sustainable carers' organizations in all EU states; and (D) promoting the development of inclusive and patient-centered care systems which fully recognize the role, contribution, and added value of caregivers Website: http://www.eurocarers.org
Confederation of Family Organizations in the European Community (COFACE)	This is a pluralistic network of civil society associations representing the interests of all families. With more than 60 member organizations in 23 Member States of the European Union, COFACE Families Europe represents more than 25 million families in Europe. In June 2016, the General Assembly agreed on a new name: COFACE Families Europe. Website: http://www.coface-eu.org/

located in Salamanca, held a meeting on 28 April 2017 of the State Dementia Group (GED), led by IMSERSO, to create the National Alzheimer's Plan to define relevant social and health-care policies in Spain.

The country's economic crisis of 2008 gave rise to the term "the dependence limbo" (Oliva, Peña & García, 2015a), which refers to individuals who, although their rights as dependents have been officially recognized, have still received no benefits or financial subsidy. Individuals who are actually receiving either economic benefits or services are known as the *benefit-receiving population*. Technically, the *dependence limbo* or wait-listed population should be the result of the difference between persons with rights to benefits and those actually receiving benefits. However, as noted earlier, contrary to the initial principle of providing in-home services, monetary benefits to families became the more common practice during the first years of the System for Autonomy and Independent Care (Oliva, Peña & García, 2015b).

When examined, the distribution of benefits across the autonomous communities is such that there are major differences between regions (Jimenez-Martín, Vilaplana & Andrea, 2016). According to these statistics, in Aragon, the Balearic Islands, and Navarre, monetary payments account for 60–70% of total benefits, an amount much higher than the national average. By contrast, in Castilla-La Mancha, La Rioja,

Andalusia, and Galicia, this figure is approximately 35–40%; in Madrid, the proportion is 22%. Home care in Galicia accounts for 28%, while in the Balearic Islands, the proportion is only around 1%. For residential care, the percentage ranges from 7% to 8% (in Ceuta and Melilla and Castilla-León) and 26% (in Cantabria). In addition, in some regions such as Andalusia, telecare represents almost 30%.

It is worth underscoring the regional differences not only in the receipt of needs applications and rulings issued but also in the awarding of benefits. Correa and Jiménez-Aguilera (2015) highlight the imbalances between regions regarding the receipt of applications. These differences clearly reflect a heterogenous system of dependent care at the national level, with actual application of the system at different degrees or levels across the autonomous communities. The benefit models are dissimilar, resulting in different degrees of commitment and treatment measures regarding the level, scope, and quality of the system. It should be noted that, despite some exceptions, the autonomous communities with the lowest gross domestic product (GDP) (such as in Ceuta and Melilla, Extremadura, Andalusia, and Murcia) consistently occupy the top of the rankings, revealing a relationship between more generous coverage and low per capita income.

Regarding individual co-payment, as indicated by Oliva (2014), in some regions, it is determined by the economic situation of the beneficiary, while other autonomous communities opt for a combination of information based on income and net worth, which, due to the absence of a common legal framework, has resulted in multiple co-payment formulas. At the regional level, there is uneven representation between the development of the main dimensions of the system (i.e., number of applications, rulings, persons entitled to benefits, and beneficiaries) and the offer and actual distribution of services. This reflects the difference in resources (e.g., underfunding in certain regions) and the diverse preferences across the national territory.

In sum, the systems of attention to dependence are radically different from the original concept, and across different regions of the country, arguably requiring different solutions. Finally, regarding the future of this system, it is of some concern whether it will be able to meet the needs of all the dependent persons on the waiting list, the future demands of an aging population, as well as the predicted numbers of moderately dependent persons entering the system (Oliva, Peña, & García, 2015a).

Caregiving and the Family

The tradition of care in the family assumes that women, essentially the mothers or daughters of the person cared for, are those who take on the role of main caregiver in most cases (Bover-Bover, 2006; Crespo & López, 2008; García-Calvente, Mateo-Rodríguez, & Maroto-Navarro, 2004; Vaquiro & Stiepovich, 2010). In Spain, this is true for 84% of the cases of informal care (IMSERSO, 2005a). The sociodemographic profile of the female caregiver is that of a person who is usually unemployed with a low level of education, responsible for running the home, and an immediate family member who lives with the dependent person (García-Calvente et al., 2004; Vaquiro & Stiepovich, 2010).

The health-care needs of dependent individuals are covered firstly by the family (informal care) and secondly by the health system and social services (formal care). Informal care, in some studies (Rapp et al. 2011), represented more than 80% of total care. Traditionally, and especially in Mediterranean countries, the family has taken a leading role in the care of the elderly and the sick (IMSERSO, 2014; Rogero-García, 2009). The family provides support for carrying out activities of daily living (ADLs) and instrumental activities of daily living (IADLs) to people in situations of dependency. However, the distribution of care work is usually shared unequally within the family. One member often assumes most of the burden, usually a woman, who then becomes the primary caregiver. The profile of the primary caregiver is a female (76.3%), middle-aged (between 45 and 64 years of age), with a family relationship with the patient (e.g., wife or daughter), who lives with the dependent person, holds a low socioeconomic status, and has no paid work (Rivera, Casal & Currais, 2009). This appears as a pattern that is repeated for the primary caregiver of patients with Alzheimer's disease and other dementias throughout the country (World Health Organization & Alzheimer's Disease International, 2012).

In a recent study, Ruiz, Gonzalez, and Mainar (2015) analyzed the main demographic and socioeconomic conditions of the primary caregivers of those with Alzheimer's disease and dementia sufferers, along with the caregiver's employment situation. Ruiz, Gonzalez, and Mainar used empirical analysis of the data obtained from surveys of 694 primary caregivers of persons with Alzheimer's disease and persons with dementia through the Andalusian Associations of Relatives of Alzheimer's Disease Patients. The results showed that the employment rate of working-age caregivers is much lower than that of the general population, especially in older women with low levels of education who live with the patient. The data revealed that caregiving in the home represents the main restriction of access to the labor market. In other words, living with the patient is an additional handicap, and this handicap is even greater for women. In addition, this limitation is significantly greater when care is provided in the home.

Conclusion

Life expectancy in Spain is one of the highest in the world. In addition, the country also has one of the fastest aging populations. By contrast, the birth rate is among the lowest in Europe. Consequently, in years to come, the country will have a large aging population with a high percentage of persons over the age of 80 years, which, in turn, will lead to an elevated prevalence of dementia. The country is also witnessing marked migratory flows of immigrants returning to their country of origin, spurred by the 2008 economic recession which represents a constant drain of young people leaving to live abroad. These two factors will lead to the creation of a population pyramid with an extremely high percentage of older adults and an excessively small working-age population. Together, this represents major social, economic, and political problems. It will be difficult to sustain the two key pillars of the welfare state, namely, the pension system and care for dependent persons, which will have a direct impact on the resources devoted to persons and families afflicted by dementia.

Additional factors are now effecting a change in the profile of the caregiver in Spain. Both the changes in family structure and the generalized integration of women in the labor market have brought about a change in who the caregiver is for the person with dementia. Extended families are disappearing and are being replaced by smaller nuclear families. The concept of the family as a permanent institution has become blurred, associated with high levels of separation and divorce. The number of different family models is increasing (e.g., people living alone, single-parent families, civil unions, etc.). Above all, there have been deep-seated changes in the social conditions of women (active integration into the world of remunerated work and social institutions) and their role in the family. The geographical mobility of different family members has also increased, creating distance between direct family members and weakening the networks of family solidarity. This situation is exacerbated by the fact that Spain has no well-established national strategy or plan for Alzheimer's disease. The plan has started to function timorously only in 2017 and will need years before it is totally implemented. In short, Spain's social, political, and economic situations make it difficult to develop effective strategies to offer the resources and responses required to cope with what is an increasingly concrete and difficult problem, namely, that Spain has one of the highest dementia rates in the world.

References

Andrés de Llano, J. M., Alberola, S., Garmendia, J. R., Quiñones, C., Cancho, R., & Ramalle-Gómara, E. (2015). Evolución de la natalidad en España. Análisis de la tendencia de los nacimientos entre 1941 y 2010. *Anales de Pediatría, 82*(1), e1–e6.. Retrieved from. https://doi.org/10.1016/j.anpedi.2014.03.018

Animal Assisted Intervention International. (2016). *Animal assisted intervention*. Retrieved from http://www.aai-int.org/aai/animal-assisted-intervention/

BOE. (2006). Promotion of Personal Autonomy and Assistance for Persons in a Situation of Dependency Act. (Act 39/2006 of 15th December). Retrieved from https://www.boe.es/boe/dias/2006/12/15/pdfs/A44142-44156.pdf

Bover-Bover, A. (2006). El impacto de cuidar en el bienestar percibido por mujeres y v arones de mediana edad: una perspectiva de género. *Enfermería Clínica, 16*(2), 69–76.

Butler, R. N. (1963). The life review: An interpretation of reminiscence in the aged. *Psychiatry, 26*, 65–76.

Casado, I., & Calatayud, T. (2009). Epidemiología y factores de riesgo. En: Molinuevo JL, Peña-Casanova, J., (eds.). *Guía oficial para la práctica clínica en demencias: conceptos, criterios y recomendaciones*. Barcelona, Spain: Prous Science, SAU. Thomson Reuters. Guías oficiales de la Sociedad Española de Neurología, N° 8, 23–50.

Castiñeira, B. R., Rodríguez, B. C., & Nunes, L. C. (2009). Provisión de cuidados informales y enfermedad de Alzheimer: valoración económica y estudio de la variabilidad del tiempo. *Hacienda Pública Española, 189*, 107–130.

CEAFA. (2013). *Alzheimer – Una cuestión de estado*. Confederación Española de Alzhiemer (CEAFA).

Cobos, F. J. M., & Rodríguez, M. M. M. (2012). A review of psychological intervention in Alzheimer's disease. *International Journal of Psychology & Psychological Therapy, 12*, 373–388.

Coduras, A., Rabasa, I., Frank, A., Bermejo-Pareja, F., Lopez-Pousa, S., Lopez-Arrieta, J. M., et al. (2010). Prospective one-year cost-of-illness study in a cohort of patients with dementia of Alzheimer's disease type in Spain: The ECO study. *Journal of Alzheimers's Disease, 19*(2), 601–615.

Coll de Tuero, G., & López-Pousa, S. (2011). Informe sobre atenció primària. *Atencion Primaria, 43*(11), 585–594.

Correa, M., & Jiménez-Aguilera, J. D. D. (2015). "Sombras y sombras en la aplicación de la ley de dependencia," Gaceta Sanitaria, disponible en. https://doi.org/10.1016/j.gaceta.2015.09.001, acceso 10-11-15.

Crespo López, M., & López Martínez, J. (2008). Cuidadoras y cuidadores: el efecto del género en el cuidado no profesional de los mayores. *IMSERSO, 35*, 1–33.

De Pedro-Cuesta, J., Virués-Ortega, J., Vega, S., Seijo-Martínez, M., Saz, P., Rodríguez, F., et al. (2009). Prevalence of dementia and major dementia subtypes in Spanish populations: A reanalysis of dementia prevalence surveys, 1990-2008. *BioMed Central Neurology, 19*, 9–55.

Delgado, E., Suarez, O., de Dios, R., Valdespino, I., Sousa, Y., & Braña, G. (2014). Características y factores relacionados con sobrecarga en una muestra de cuidadores principales de pacientes ancianos con demencia. *SEMERGEN-Medicina de Familia, 40*(2), 57–64. https://doi.org/10.1016/j.semerg.2013.04.006

Domínguez, A., & López, J. M. (2002). La enfermedad de los costes indirectos. *Revista Española de Economía de La Salud, 1*, 52–54.

Domingo, E. P., Sierra, M. G., Valero, M. M., & Castiñeira, M. P.-O. (2015). El Libro Blanco del párkinson en España: Aproximación, análisis y propuesta de futuro. Madrid: Real Patronato sobre Discapacidad (Ministerio de Sanidad, Servicios Sociales e Igualdad) y Federación Española de Párkinson. Retrieved from http://www.fedesparkinson.org/libro_blanco.pdf

Federación Mundial de Musicoterapia. (2011). Defining music therapy 2011. Retrieved from https://sobremusicoterapia.wordpress.com/tag/federacion-mundial-de-musicoterapia/

Feldberg, C., Tartaglini, M. F., Clemente, M. A., Petracca, G., Cáceres, F., & Stefani, D. (2011). Vulnerabilidad psicosocial del cuidador familiar: Creencias acerca del estado de salud del paciente neurológico y el sentimiento de sobrecarga. *Neurología Argentina, 3*(1), 11–17.

Ferrer, S. I., Ortiz, C. A., & Avila-Castells, P. (2013). Impat of a psychomotor re-education guide on the quality of life of patients with Alzheimer's disease. *Revista Medica de Chile, 141*, 735–742.

García-Calvente, M., Mateo-Rodríguez, I., & Maroto-Navarro, G. (2004). El impacto de cuidar en la salud y lacalidad de vida de las mujeres. *Gaceta Sanitaria, 18*(2), 83–92.

García Meilán, J.J. & Carro Ramos, J. (2014). *Programa de Actuación Cognitiva Integral en Demencias (PACID)*. Instituto de Neurociencias de Castilla y León (INCYL) Universidad de Salamanca. Retrieved from https://www.alzheimeruniversal.eu/2014/07/29/programa-de-actuacion-cognitiva-integral-en-demencias-pacid/

Gascón-Bayarri, J., Reñé, R., Del Barrio, J. L., De Pedro-Cuesta, J., Ramón, J. M., Manubens, J. M., et al. (2007). Prevalence of dementia subtypes in El Prat de Llobregat, Catalonia, Spain: The PRATICON study. *Neuroepidemiology, 28*(4), 224–234.

González-Palau, F., Franco, M., Toribio, J. M., Losada, R., Parra, E., & Bamidis, P. (2013). Designing a computer-based rehabilitation solution for older adults: The importance of testing usability. *PsychNology Journal, 11*, 119–136.

Government of Spain. (2017). *General information, autonomous communities*. Retrieved from http://en.administracion.gob.es/pag_Home/espanaAdmon/comoSeOrganizaEstado/ComunidadesAutonomas.html

Harper, S. (2014). Economic and social implications of aging societies. *Science, 346*(6209), 587–591. https://doi.org/10.1126/science.1254405

IMSERSO. (2005a). *Libro Blanco sobre la atención a las personas en situación de dependencia en España*. Madrid, Spain: IMSERSO.

IMSERSO. (2005b). *Atención a las personas en situación de dependencia en España*. Libro Blanco: Ministerio de Trabajo y Asuntos Sociales. Instituto de Mayores y Servicios Sociales.

IMSERSO. (2007). *El apoyo a los cuidadores de familiares mayores dependientes en el hogar: desarrollo del programa "Cómo mantener su bienestar."* Ministerios de Trabajo y Servicios Sociales. Madrid, Spain.

IMSERSO. (2014). *Informe 2012. Las personas mayores en España*. Madrid, Spain: Instituto de Mayores y Servicios Sociales.

IMSERSO. (2015). *CRE Alzheimer Salamanca. Non-pharmacological therapies*. Retrieved from http://www.crealzheimer.es/crealzheimer_01/terapias_no_farmacologicas/index.htm

INE. (2012). Proyecciones de población 2012 [Notas de Prensa]. Madrid.

INE. (2016a). *Cifras de Población a 1 de enero de 2016 – Estadística de Migraciones 2016* [Notas de Prensa]. Retrieved June 30, 2016, from http://www.ine.es/prensa/np980.pdf

INE. (2016b). Avance de la explotación estadística del Padrón a 1 de enero de 2016. *Datos provisionales*. [Nota de Prensa]. Retrieved from http://www.ine.es/prensa/np966.pdf

INE. (2016c). *Proyecciones de Población 2016-2066*. [Nota de Prensa]. Madrid.

INE. (2017). *Encuesta de Población activa (EPA)*. [Nota de Prensa]. Retrieved from http://www.ine.es/daco/daco42/daco4211/epa0217.pdf

Jimenez-Martín, S., Vilaplana, C., & Andrea, A. (2016). Observatorio de la Dependencia (febrero 2016). *Studies on the Spanish Economy eee2016–05*. Fedea.

Jurczynska, C., Eimil, M., López de Silanes, C., & Llanero, M. (2011). Impacto Social de la Enfermedad de Alzheimer y otras Demencias. Feen (Fundación Española de Enfermedades Neurológicas).

Klages, K., Zecevic, A., Orange, J. B., & Hobson, S. (2011). Potential of Snoezelen room multisensory stimulation to improve balance in individuals with dementia: A feasibility randomized controlled trial. *Clinical Rehabilitation, 25*, 607–616.

Mateos, R., Franco, M., & Sánchez, M. (2010). Care for dementia in Spain: The need for a nationwide strategy. *International Journal of Geriatric Psychiatry, 25*(9), 881–884.

Molinuevo, J. L., & Peña-Casanova, J. (2009). Grupo de estudio de neurología de la conducta y demencias. Guía oficial para la práctica clínica en demencias: conceptos, criterios y recomendaciones. Barcelona: Sociedad Española de Neurología (SEN). Guía N° 8.

National Association of Directors and Managers in Social Services. (February, 2017). Observatory Opinions. XVII Opinion of the observatory of the law 9/2006 of promotion of the personal autonomy and attention to the persons in situation of dependence. Retrieved from: http://www.directoressociales.com/documentos/dictamenes-observatorio.html

OECD. (2013). *Health at a Glance*. OECD indicators. Retrieved from http://www.oecd-ilibrary.org/docserver/download/8113161e.pdf?expires=1504696762&id=id&accname=guest&checksum=F5266D7DD1C8FB779958D0218EDCD488

OECD. (2015). *Informe De Diagnóstico De La Estrategia De Competencias De La Ocde - España*. OECD. Retrieved from http://skills.oecd.org/developskills/documents/Spain_Diagnostic_Report_Espagnol.pdf

Olazarán, J., & Muñiz, R. (2004). Possible and recommended cognitive stimulation. In J. M. M. Lage & T of Being (Eds.), *Alzheimer's: The necessary pragmatics*. Madrid, Spain: Aula Médica.

Olazarán, J., Reisberg, B., Clare, L., Peña-Casanova, J., Del Ser, T., et al. (2010). Non-pharmacological therapies in Alzheimer's disease: a systematic review of efficacy. *Dementia and Geriatric Cognitive Disorder, 30*, 161–178. https://doi.org/10.1159/000316119

Oliva J. (2014). Sistema de autonomía personal y atención a la dependencia: análisis y líneas de avance en tres dimensiones. Actas de la Dependencia N. 12, *Fundación Caser*, disponible en: http://www.fundacioncaser.org/sites/default/files/j.oliva_analisisylineas_web.pdf

Oliva, J., Peña, L., & García, S. (2015a). *Looking back to move forward: Spanish System for Promotion of Personal Autonomy and Assistance for persons in a situation of dependency (Part I)*. Disponible en: http://www.hspm.org/countries/spain25062012/countrypage.aspx, acceso 2-10-15.

Oliva, J., Peña, L., & García, S. (2015b). *Looking back to move forward: Spanish System for Promotion of Personal Autonomy and Assistance for persons in a situation of dependency (Part II)*. Disponible en: http://www.hspm.org/countries/spain25062012/countrypage.aspx, acceso 2-10-15.

Padala, K. P., Padala, P. R., Malloy, T. R., Geske, J. A., Dubbert, P. M., Dennis, R. A., … Sullivan, D. H. (2012). Wii-fit for improving gait and balance in an assisted living facility: A pilot study. *Journal of Aging Research, 2012*, 1. https://doi.org/10.1155/2012/597573

Peña-Longobardo, L. M., Oliva-Moreno, J., & Garcia-Armesto, S. (2016). The Spanish long-term care system in transition: Ten years since the 2006 Dependency Act. *Health Policy, 120*, 1177–1182. https://doi.org/10.1016/j/healthpol.2016.08.012

PricewaterhouseCoopers. (2012). Estado del Arte de la Enfermedad de Alzheimer en España [State of the Art of Alzheimer's Disease in Spain]. Document prepared by PricewaterhouseCoopers for Lily.

Prince, M., Prina, M., & Guerchet, M. (2013). *World Alzheimer Report 2013*. London, UK: Alzheimer's Disease International.

Rapp, T., Grand, A., Cantet, C., Andrieu, S., Coley, N., Portet, F., & Vellas, B. (2011). Public financial support receipt and non-medical resource utilization in Alzheimer's disease results from the PLASA study. *Social Science & Medicine, 72*(8), 1310–1316. https://doi.org/10.1016/j.socscimed.2011.02.039

Rivera, B., Casal, B., & Currais, L. (2009). Provisión de cuidados informales y enfermedad de Alzheimer: valoración económica y estudio de la variabilidad del tiempo. *Hacienda Pública Española/Revista de Economía Pública, 189*(2), 107–130.

Rodríguez, M. (2013). Use of informal and formal care among community dwelling dependent elderly in Spain. *European Journal of Public Health, 24*, 668–673. https://doi.org/10.1093/eurpub/ckt088

Rodríguez, P. (2005). El apoyo informal a las personas mayores en España y la protección social a la dependencia. Del familismo a los derechos de la ciudadanía. *Revista Española de Geriatría y Gerontología, 40*(3), 5–15. https://doi.org/10.1016/S0211-139X(05)75068-X.

Rogero-García, J. (2009). Distribución en España del cuidado formal e informal a las personas de 65 y más años en situación de dependencia. *Revista Española de Salud Pública, 83*(3), 393–405.

Ruiz, M., González, C., & Mainar, A. (2015). An analysis of caregiver profile and its impact on employment situation: Primary caregivers of patients of Alzheimer's and other dementias in the South Western of Spain. *Atlantic Review of Economics, 2*, 1–18.

Shibata, T., & Wada, K. (2011). Robot therapy: A new approach for mental healthcare of the elderly - a mini-review. *Gerontology, 57*, 378–386.

Spector, A., Thorgrimsen, L., Woods, R. T., Royan, L., Davies, S., Butterworth, M., & Orrell, M. (2003). Efficacy of an evidence-based cognitive stimulation therapy programme for people with dementia. *British Journal of Psychiatry, 183*, 248–254.

State of California Department of Finance. (2017) . *Projections*. Retrieved from http://www.dof.ca.gov/Forecasting/Demographics/Projections/

Thaut, M. H. (2005). *Rhythm, music and the brain: Scientific foundations and clinical applications*. New York, NY: Taylor & Francis Group.

Torres, P., Ballesteros, E., & Sánchez, D. (2008). Programas e intervenciones de apoyo alos; cuidadores informales en España. *Gerokomos, 19*(1), 9–15.

Turró-Garriga, O., López-Pousa, S., Vilalta-Franch, J., Turon-Estrada, A., Pericot-Nierga, I., Lozano-Gallego, M., et al. (2010). Valor económico anual de la asistencia informal en la enfermedad de Alzheimer. *Revista de Neurologia, 51*(4), 201–207.

Vaquiro, S., & Stiepovich, J. (2010). Cuidado informal, un reto asumido por la mujer. *Ciencia y enfermería, 2*, 9–16.

Wengrower, H., & Chaikin, S. (2008). *La vida es Danza*. Barcelona, Spain: Gedisa Editorial.

WHO. (2014). *World Health Statistics 2014*. World Health Organization. Available from WHO Press, World Health Organization, 20 Avenue Appia, 1211 Geneva 27, Switzerland.

World Health Organization & Alzheimer's Disease International. (2012). *Dementia: A public health priority*. Geneva, Switzerland: WHO. Retrieved from http://whqlibdoc.who.int/publications/2012/9789241564458_eng.pdf

Zambrano, E., & Guerra, M. (2012). Formación del cuidador informal: relación con el tiempo de cuidado a personas mayores de 65 años. *Aquichán, 12*(3), 241–251.

Part V

Social/Public Policy and Community Perspectives

21 Community Partnerships and the Care of Latinos with Dementia: A Call for Action

Ronald J. Angel and Jacqueline L. Angel

Abstract
Latinos represent a rapidly growing segment of the population of the United States of America (USA). Because of high fertility, this population is relatively young; although like other groups, it is aging rapidly. Latino life expectancy is similar to, or higher than, that of non-Hispanic whites; however, much of the additional years of life for many Latinos are characterized by poor health and functional limitations, as well as increasing levels of dependency. In this chapter we review the institutional environment in which communities provide care to Latino infirm, aging individuals, especially those with dementia. A major focus of this chapter is the potential role of civil society organizations (CSOs) in aiding Latino communities. CSO is a term which includes non-governmental and faith-based organizations. The focus on CSOs is motivated by the fact that despite these governmental programs, the magnitude of the aging of most nations' populations, in addition to the attendant fiscal and administrative challenges, signifies that governments alone will certainly not be able to address all of the needs of older citizens, particularly those of Latino descent. As this chapter documents, demographic and social changes have reduced the family's ability to provide extensive hands-on care to its elderly afflicted by dementia. The core theoretical and practical question that this chapter's analysis raises relates to the potential role of CSOs as advocates for and complementary providers of care to the elderly.

R. J. Angel (✉) · J. L. Angel
The University of Texas at Austin, Austin, TX, USA
e-mail: rangel@austin.utexas.edu;
jangel@austin.utexas.edu

Introduction

Latinos' life expectancy at age 65 is comparable to or even superior to that of non-Hispanic whites despite Latinos' relatively unfavorable socioeconomic profiles (Arias, 2016). In conjunction with increasing longevity, low levels of education among Latinos will inevitably result in an increasing prevalence of dementia given that low levels of education are a significant risk factor (Alzheimer's Association, 2017). This increased burden of dementia will present public policy, as well as individual family caregivers, with serious new caregiving challenges. In this chapter, reviewed is what is known about the institutional context of community caregiving for older individuals with dementia and an examination of the community resources that Latino family caregivers can call upon to assist in coping with relatives who experience serious cognitive impairment.

H. Y. Adames, Y. N. Tazeau (eds.), *Caring for Latinxs with Dementia in a Globalized World*,
https://doi.org/10.1007/978-1-0716-0132-7_21

These resources range from informal, unpaid care from family members and others to formal governmental agencies that provide eldercare services. Included in this array of service providers are the non-governmental organizations (NGOs), including faith-based organizations, that are increasingly important in providing assistance to families caring for elders with dementia or to older individuals who have no family support. In addition to information on individual caregiving perspectives of dementia in the Latino population (Yeo, Gerdner, & Thompson, 2018), there are community-level studies that provide answers to such questions as to the sources of caregiver burden and their mediation in this population (Apesoa-Varano, Gomez, & Hinton, 2016; Bilbrey et al., 2018; Flores, Hinton, Barker, Franz, & Velasquez, 2009; Rote, Angel, & Markides, 2017). Drawing on a broader range of literature, apart from the clinical, this chapter frames the major questions of interest and importance for the community perspective on caregiving (Bilbrey et al., 2018). The chapter's title includes the phrase "a call to action" to emphasize the fact that, at this time, there is far less substantial empirical evidence than needed. The intention of this chapter is to identify important research and policy agendas with vitally important implications for society at large, and for families in particular, with a concern for Latinos centering this work.

Caring for individuals with advanced dementia can be especially challenging and stressful because of the behavioral and personality changes that take place (Aranda & Knight, 1999; Flores et al., 2009; Rote, Angel, & Markides, 2015; Zarit et al., 2011). This challenge is compounded by the fact that, during the latter half of the twentieth century and the beginning of the twenty-first century, various demographic and social changes including lower fertility, divorce and family disruption, geographic dispersion, and increased female labor force participation have reduced a family's ability to provide extensive support to dependent older parents. This new reality means that new combinations of formal and informal care must be explored, including the potential roles of non-governmental and faith-based organizations, in assisting families in providing care to elders with serious cognitive impairments. Knowing the various stages in the process of cognitive decline, and the institutional contexts in which families cope with an older family member's growing need for assistance, can allow professionals to more effectively target interventions, and allow them to identify older individuals at elevated risk of institutionalization, as well as assist families in accessing available community support resources.

A unique problem, however, arises from the great heterogeneity among the elderly Latino population in terms of culture, family formation, socioeconomic resources, education, and the ability to interact with formal governmental agencies. Acknowledging the service needs of various communities and groups requires comprehensive knowledge of their unique cultural, economic, and political histories. The Latino population itself is hardly homogeneous. The various nationality groups representing Latinos differ greatly in their countries of origin, their immigration histories, the political and economic situations at the time of arrival of Latinos to the United States of America (USA), and much more (Angel & Angel, 2015). In addition to standard epidemiological characteristics such as age, gender, and socioeconomic status for access to and the use of community support services, recognizing the impact of these factors is essential to the development of effective intervention strategies for individuals with dementia and other debilitating conditions (Belle et al., 2006; Gallagher-Thompson et al., 2003; Llanque & Enriquez, 2012; Napoles, Chadiha, Eversley, & Moreno-John, 2010).

As part of comprehending the institutional context of community caregiving, this chapter outlines a relatively new area in which little rigorous empirical research has been carried out. The novel approach relates to the potential role of voluntary, non-governmental, and faith-based organizations in supplementing family caregiving efforts or perhaps even substituting for the family in cases in which an older individual has no family upon whom to call. As reviewed below, these organizations hold great potential as part of a potent combination of formal and informal services. Given the community-based focus of many of these organizations, they are particularly well placed to address the needs of unique populations and communities, including various Latino groups. Of note, the advantage of this novel approach is that it is well-suited to the collectivistic nature of Latinos; however, a potential disadvantage or qualifier is that Latinos may be fearful of many organized entities because of historical patterns of exploitation by groups that may be loosely associated with government, including in their countries of origin, but especially in the USA given anti-immigrant sentiment.

The focus on NGOs in the care of elderly individuals with dementia and other debilitating conditions is motivated by various facts related to these populations and the nature of the care they require. Although medical and social service professionals are clearly vital, they are limited in their ability to assure that older individuals are not isolated and that they not only have access to needed services but that they actually use the services. In the remainder of this chapter, a review of the institutional embeddedness of family ensues, and the chapter ends with a look to the future with a proposal of a research and social policy agenda related to dementia caregiving in general and for USA-based Latino communities specifically.

The Circumstances and Scope of Caregiving

In 2016, the National Academies of Sciences, Engineering, and Medicine (NAS) issued a committee report entitled "Families Caring for an Aging America," in which the committee reviewed what is known about the consequences of caregiving for family caregivers (National Academies of Sciences, 2016). The report summarized existing research that clearly states that caring for individuals with serious physical and cognitive limitations can be demanding and can have serious negative consequences for caregivers' health and well-being. Numerous studies report that the caregivers of older adults are at risk of social isolation, depression, anxiety, and poor physical and mental health. Moreover, the more demanding the caregiving task, the greater the risk of negative outcomes for the caregiver.

Research on caregiving and caregiver burden (Dilworth-Anderson, Williams, & Gibson, 2002; Mendez-Luck, Kennedy, & Wallace, 2008; Roth, Fredman, & Haley, 2015; Taylor & Quesnel-Vallée, 2017) and particularly dementia caregiving (Cucciare, Gray, Azar, Jimenez, & Gallagher-Thompson, 2010) underscore that both the tasks involved and the burden they impose on caregivers vary greatly depending on various factors, including the severity of the physical and mental impairment of the care recipient, the availability of alternative sources of care, the caregiver's own health, and much more. As important as these findings are, a focus solely on the caregiver, or even the caregiver/care recipient dyad, ignores the likely complex institutional environment in which care is provided. This institutional environment is potentially extensive and includes a range of federal, state, and municipal services, as well as the voluntary efforts of non-governmental actors. Of course, the availability of services varies greatly depending on the city and state of residence. For example, rural residents, or the residents of seriously underserved areas such as the USA-Mexico border region, have very different levels of access to support than the residents of more affluent cities with multiple agencies and organizations.

Given the great racial and ethnic diversity of the USA, the NAS recommended the development of a deeper understanding of the role of race and ethnicity in determining the situation of both caregivers and for those for whom they provide care. For example, large racial and ethnic differences in lifetime income and wealth place older individuals and their families in very different situations when managing dementia and other infirmities of old age, with Latinos generally having lower levels of income (Angel & Angel, 2015; Silverstein & Wang, 2015). What is known is that older Latinos avoid institutionalization for as long as possible and remain in the community even when they suffer significant declines in health and functional capacity (Angel, Angel, Aranda, & Miles, 2004; Angel & Angel, 1997; S. Rote et al., 2015; Thomeer, Mudrazija, & Angel, 2015). The NAS call to action is particularly timely since there exists only a limited understanding of the community resources upon which Latinos are able to rely.

A foremost motivation for investigating the potential role of NGOs in dementia care relates to the inherent limitations that even affluent states face in addressing all of the needs of elderly individuals. The rapid aging of the populations of the USA and Latin American nations places substantial limitations on governments, at all levels, in providing care to growing numbers of dependent elders. These limitations are particularly serious in low- and middle-income nations with limited pension and social security coverage and inadequate formal eldercare systems (Beard et al., 2016; Gutiérrez, Ortega, & Lopera, 2012). In addition to limitations in state capacities, demographic and social changes increasingly limit the ability of families to provide all of the support infirm aging parents require. The increasing life spans indicate that individuals may spend many years seriously incapacitated and dependent. Evidence indicates that older Mexican-origin individuals in the USA spend over half of the years past age 65 with serious functional limitations (Angel, Angel, & Hill, 2014). Given labor force disadvantages throughout the life course, older Latinos have relatively low levels of wealth and income and are, for the most part, unable to purchase long-term support in the community (Angel & Angel, 2015).

The Relationship of Different Levels of Service Provision

The theoretical model (see Fig. 21.1) describes the relationship among different levels of service availability and emphasizes the institutional embeddedness of the family as a caregiving unit.

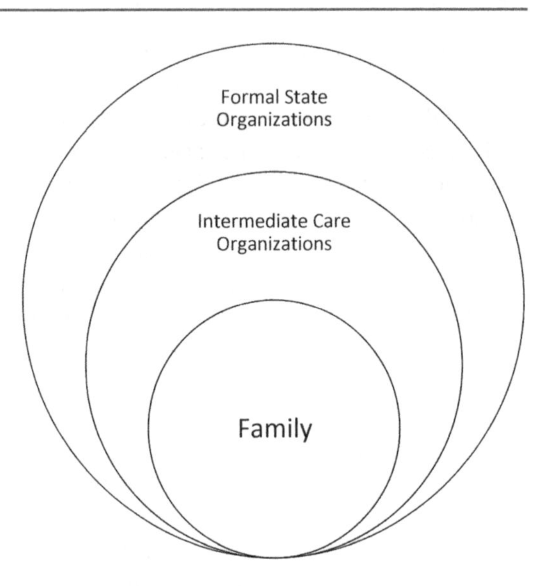

Fig. 21.1 The institutional embeddedness of family care

The model also emphasizes the intermediate role of semiformal community organizations, including non-governmental and faith-based organizations, as buffers between the family and formal sources of care. These groups are often referred to as "civil society organizations" to emphasize the fact that they differ from both the state, which provides state-sponsored services, and the market in which one can buy the services one needs. Civil society organizations include a wide range of voluntary organizations that may be part of larger social movements with various agendas. In the present context, they include organizations that support families in caring for older individuals with serious limitations. In their intermediate position, such organizations can both provide support services and act as intermediaries or advocates that place families in contact with formal state agencies.

The Family: The Basic Caregiving Institution

For most of human history, the caregiving institution of first and last resort was the family. Few other options, other than charity or the almshouse or asylum, existed (Smith & Feng,

2010). Today, however, the family's caregiver role is far more complex and demanding than in the past (National Academies of Sciences, 2016). As the NAS report notes:

> Family caregivers have always provided the lion's share of long-term services and supports, such as household tasks and self-care (getting in and out of bed, bathing, dressing, eating, or toileting) to older adults with impairment. Today, they are also tasked with managing difficult technical and medical procedures and equipment in older adults' homes, overseeing medications, and monitoring symptoms and side effects. As older adults' advocates and care coordinators, caregivers are often responsible for ensuring that care recipients obtain needed care from fragmented and complex health and social services systems. Family caregivers are often involved in older adults' decision making and may serve as surrogate decision makers when the care recipient loses the capacity to make important decisions. Many family caregivers help older adults without training, needed information, or supportive services (p. 112).

Even in the face of increasing complexity, Latino family caregivers continue to view the care of aging relatives as an honorable duty (Knight & Sayegh, 2010), yet for the reasons summarized in the NAS report, their capacity to do so, as is the case with non-Latino caregivers, is becoming increasingly strained. A demographic sea change (Rote et al., 2015) is affecting the family's old-age caring capacity. This change affects even Latino communities that have historically been highly familistic (Rote et al., 2015). For Latinos, *familismo* refers to familial ideals which inform the structures, processes, and interactions of Latino/Latina culture (Flores et al., 2009), including how adult children relate to their aging parents. Today, among all USA ethnic and racial groups, informal support systems are smaller, more dispersed, and unable to devote the time necessary to provide all of the care an aging individual might need. In the future, smaller families, divorce, the need for women to work, and migration away from an elderly individual's place of residence can only increase the number of older individuals without any source of social support in close proximity. If, as is often the case, the family lacks the financial resources to pay for in-home care services, few alternatives exist (Brown, Herrera, & Angel, 2013).

The Defamilization of Dementia Caregiving Dementia, like most other infirmities, ranges from mild early state impairment to serious advanced dementia, the latter in which an individual loses all autonomy and capacity to care for himself or herself. Advanced dementia places extreme strain on caregivers who are often simply unable to provide the constant and extensive care that is required. Although serious cognitive impairment and behavioral problems are major predictors of institutionalization (The National Coalition on Mental Health and Aging, 2017), most individuals prefer to remain in the community for as long as possible. Among Latinos, the family plays a major role in allowing them to do so (Mendez-Luck, Amorim, Anthony, & Neal, 2016). Even when an aging parent becomes seriously physically or cognitively impaired, Latino families maintain the elder parent in the community either in the parent's own home or in an adult child's home (Angel & Angel, 2015; Burcher, 2008; Gutiérrez Robledo, Ortega, & Lopera, 2012).

As important as the family has been and continues to be in providing care to elderly parents, those tasks are increasingly being "defamilized," a term which refers to a shift in responsibility from the family to the state (Angel & Angel, 2018). With industrialization and the emergence of the modern welfare state, the financial support of the elderly becomes socialized through social security systems, at least in middle- and high-income countries (Hoskins, 2010). The financing of medical care for the elderly has also increasingly shifted to the state. As a consequence, the family has been increasingly replaced by the state as the major source of financial support and medical care for elderly citizens. Yet, the state faces critical limitations in its ability to provide all of the care that elderly individuals need. Even affluent nations face fiscal limitations to the welfare state services they provide (Pierson, 2001).

The Role of Intermediate Organizations

In light of the fact that the family faces greater limitations in its capacity to provide care, other institutional arrangements perhaps including semiformal organizations take on greater significance. Although formal state and private services will always have an important role, as noted, the potential role of non-governmental and faith-based institutions, in supplementing formal governmental programs and informal family efforts in supporting and caring for seriously physically and cognitively impaired elders, raises new possibilities that must be explored. Non-governmental and faith-based organizations clearly cannot replace the state in providing financial and medical support to large numbers of infirm elderly individuals. They do, however, hold out substantial promise for supplementing the efforts of formal agencies and assisting the family in coping with the highly demanding task of caring for seriously incapacitated elders in the community. In such situations, even limited additional assistance could have a considerable positive impact on caregivers.

The potential of non-governmental and voluntary efforts arises partially from the nature of the needs of elderly individuals with special needs. For frail elders, isolation is a serious threat to well-being, and non-governmental and faith-based organizations are ideally suited to address the problem. For example, members of a local church or congregation are in an ideal position to know that an older individual is isolated, and they are in an ideal position to offer companionship. The *Meals on Wheels Association of America* (http://www.mowaa.org) is a well-known nutrition program for the elderly (Choi, Lee, & Goldstein, 2011). However, in addition to nutrition, the volunteers of the organization also provide important human and social contact.

Although most Americans are aware of *Meals on Wheels*, few are aware of *Little Brothers – Friends of the Elderly (LBFE)* (http://www.littlebrothers.org), an international network of non-profit, voluntary organizations that specifically address the problem of isolation and loneliness. Little Brothers is a member of a larger international organization known as the *Fédération Internationale des petits frères des Pauvres* (International Federation of Little Brothers of the Poor, http://www.petitsfreres.org) which addresses problems of isolation and loneliness in many countries. Another important international organization, the *Fédération Internationale des Associations de Personnes Agées* (International Federation of Associations of Older Persons (FIAPA): http://www.fiapa.org), which has its headquarters in Paris, also focuses on the prevention of isolation and on assuring better quality of life for older individuals. These organizations are particularly important in the modern world in which aging parents do not necessarily live with their children and often live far from them.

HelpAge (http://www.helpageusa.org/) is an organization that operates in many countries, including Latin America, to provide financial, medical, and emotional support to older individuals with few resources. One example of a faith-based organization that has begun addressing the needs of impaired and impoverished elderly individuals is *Hogar de Cristo* (Christ's Home: http://www.hogardecristo.cl) in Chile. The organization began in 1944 when Alberto Hurtado, a Catholic priest, committed himself to helping indigent Chileans. Although the organization's initial focus was not on the elderly, the aging of the population and high levels of elderly poverty combined with the lack of adequate social services led *Hogar de Cristo* to expand its mission to provide day care, nutritional programs, and housing to impoverished elder persons with no other means of support (Pereira, Angel, & Angel, 2007).

Community-Based Care

The importance of community care in avoiding institutionalization cannot be overstated (Gaugler, Kane, Kane, & Newcomer, 2006). As the population ages, new and innovative options will inevitably evolve. One example, discussed below, is the *Program of All-Inclusive Care for the Elderly* (PACE; (Angel, 2016)). The objective of this initiative is to improve the quality of life for older impaired individuals and their families in the community in a cost-effective manner. The program employs the services of not only professionals but also draws upon the services of community organizations that can provide companionship to isolated elders, respite for family caregivers, and services such as home and personal care, transportation, and meal preparation (Hansen & Hewitt, 2012). The program is modeled on a system of acute and long-term care services developed by *On Lok Senior Health Services* in San Francisco, California, an NGO that came into existence in 1971 (https://www.onlok.org/). Its objective was to maintain, in their own homes, the elderly Asians who lived in the San Francisco Bay Area, even as they suffered serious disabilities. The program was able to do this by bringing together teams of professionals and volunteers to provide all of the medical care, as well as daily living assistance that older individuals needed. Perhaps because of its focus on a specific group, or perhaps due to the uniqueness of the particular community, the program was a great success in reaching its objective of keeping frail elders out of nursing homes (Chatterji, Burstein, Kidder, & White, 2008; Ghosh, Orfield, & Schmitz, 2014; Gonzalez, 2017).

In the mid-1980s, that success led the U.S. Congress to pass legislation encouraging the expansion of the PACE model to other locations using federal funding through Medicare and Medicaid waiver provisions (Bloom, Sulick, & Hansen, 2011). The expansion allows states and localities to experiment with new ways of providing services in a cost-neutral manner (Ghosh et al., 2014). In 1997, PACE became a permanent option for low-income disabled and elderly individuals receiving Medicaid. PACE serves individuals 55 and older who are in such poor health that they qualify for admission to a nursing home (Chatterji et al., 2008). The program serves as a last resort, then, for maintaining the individual in the community.

PACE provides a set of comprehensive services including primary and preventive care, inpatient and outpatient medical care, and specialty services such as dentistry, podiatry, physical therapy, occupational therapy, and prescription drugs. It provides social services, including day center activities, transportation, meals, as well as physical, occupational, and recreational therapies. These services are coordinated by an interdisciplinary team of healthcare and service delivery professionals who work with family caregivers. In the event that the participant requires nursing services, PACE provides this as well (Centers for Medicare and Medicaid Services, 2011). The program also provides housing assistance for those who need it. Although PACE operates in only a few locations and largely on a trial basis, because of its comprehensiveness, it represents a promising model of community care and support.

Such a model holds particular promise for older Latinos since, on average, they have few resources and are likely to have to resort to Medicaid for long-term care. One example of a PACE program that serves a predominantly Latino population is *Bienvivir* ("Live Well" in Spanish) that operates in El Paso, Texas. This PACE program serves a largely Latino population (Angel, 2016). Approximately 90% of the participants are Spanish-speaking, and they are more frail than the average Medicare enrollee. As promising as this model may seem, however, it faces various political and financial hurdles and remains a fairly limited program because the demand for services far exceeds the available slots. In addition, a major question remains as to whether the PACE program model can be expanded to the entire state, including rural and underserved communities. While PACE may represent a useful model in terms of the package of benefits, other options must be explored.

Alternatives to PACE in Texas While PACE is only available in three cities in the state of Texas (i.e., Amarillo, El Paso, Lubbock), other community-based alternatives are offered state-wide. Since 2014, Texas' Medicaid managed care program known as STAR+PLUS has offered both acute and long-term care services for Medicaid recipients aged 21 and older (Texas Health and Human Services Commission, 2014). Low-income, older adults who are eligible for Medicaid are referred to as "dual eligibles" since they qualify for both

Medicare and Medicaid. For these individuals, Medicare covers their medical expenses, and Medicaid covers any long-term care that they receive. As of yet, it is unclear whether the change to managed long-term care will benefit Mexican-origin elders in Texas. The program could benefit rural residents by increasing access to community support for the low-income, dual-eligible population, a large portion of which is Latino. To date though, coverage remains incomplete especially in counties with large Mexican-origin populations along the Texas-Mexico border (Texas Health and Human Services Commission, 2014).

In 2014, Texas introduced another demonstration program designed to make it easier for low-income older clients to receive care and to live independently in the community (Kuhlman, 2015). Through this program, known as the "Demonstration," participants receive both Medicare and Medicaid, in addition to Part D Medicare prescription drug coverage in a single integrated Medicare-Medicaid plan (MMP). The Demonstration has been implemented in six locations, three of which have high concentrations of Latinos and include Bexar County (60.3% Latino), El Paso County (82.8% Latino), and Hidalgo County (92.2% Latino) (U.S. Census Bureau, 2018). The program benefits caregivers by providing vital services to help maintain loved ones in the community. Unfortunately, not all of the plans in Hidalgo and Bexar County offer the full range of services (STAR+PLUS Health Program, 2015). In addition, elders living in other sectors of the state do not benefit from the program.[1]

Formal Sources of Caregiver Support

For definitional purposes, caregiver and community support refers to any action or activity by a third party that simplifies or assists caregivers or that reduces the amount of time and energy they must devote to caregiving tasks, thereby hopefully reducing any negative effects to their own health and well-being. Such sources of support include non-governmental and voluntary community organizations, as well as federal and state agencies. There are hundreds of private and smaller organizations that provide specific services to individuals with dementia and their families. In this chapter, the focus is not on these organizations, rather on government-sponsored or government-funded sources of community care, as well as non-profit and non-governmental sources since these are potentially of great importance to infirm elders with few resources and their families.

As the NAS report noted, caregivers need assistance in navigating complex medical and social service environments and understanding complex technical issues. Although several sources of information are available, it is difficult to know if and how caregivers locate them or whether they are able to make optimal use of the services offered. This is particularly true for Latinos and other minority group members (Burcher, 2008; Montes-de-Oca, Ramírez, Santillanes, García, & Sáenz, 2016). The existence of programs of assistance does not signify that these reach those intended to receive help, particularly in the absence of targeted and effective outreach.

Although research on caregiver health and well-being has progressed rapidly in recent years (National Academies of Sciences, 2016), little peer-reviewed research on the use of community support services in Latino populations has been reported. It is known that older Latinos remain in the community even when they experience serious declines in health and functional status (Thomeer et al., 2015). In Latino communities, the family is clearly central in making this possible. Family members provide vital support, but increasing life spans and long periods of disability will inevitably increase the need for both formal and informal caregiving assistances.

Informal observation and previous research (e.g., Pereira & Angel, 2009) suggest that NGOs

[1] This program no longer exists.

operate largely in isolation from one another, and often with little contact with formal agencies. This fact leads to the reasonable assumption that the various agencies that might address the needs of older individuals with dementia and their caregivers engage in very little coordination of their efforts. For this sector to be of greater utility in addressing the needs of infirm elderly individuals and their families, a greater degree of communication and coordination of their efforts would be necessary. One major objective of future research should be to investigate ways in which such coordination can be encouraged and operationalized.

It is clear that although access to state-sponsored services is vital for older Latinos and their families, informal assistance remains a mainstay around the world. For example, evidence from Mexico indicates that neighbors and local community members provide a great deal of assistance to one another (Burcher, 2008). Evidence from Chile reveals that non-governmental and faith-based organizations are playing an important role in the care of dependent elderly individuals (Pereira et al., 2007). Addressing the needs of older Latinos, especially those with serious dementia and their caregivers, will require a combination of all of the formal and informal assistance available, including volunteers.

Volunteers working with non-governmental and faith-based organizations can fill a potentially vital gap in the community care safety net. They can also serve an invaluable role as advocates and guardians of the rights of all older individual, including Latinos. Given low average levels of education and a lack of familiarity with complex welfare and social security bureaucracies, such assistance is particularly important for this population. While volunteers and informal organizations cannot serve as substitutes for the state in providing financial and instrumental care to frail older individuals, they can play a very important supplemental role as advocates and providers of routine services, including companionship, and help with activities of daily living, in order to allow family members the respite that they need.

Program Awareness and Outreach

Even though home and community-based service use among ethnic minority groups has increased in recent years, minority families and particularly Mexican-origin elders and their caregivers continue to underutilize institutional and community sources of support (Herrera, Benson, Angel, Markides, & Torres-Gil, 2013; Kaye, Harrington, & LaPlante, 2010; Laditka, Laditka, & Fisher Drake, 2006; Thomeer et al., 2015; Wallace, Levy-Storms, Kington, & Andersen, 1998; Wallace & Lew-Ting, 1992). The fact that they do not use formal services signifies that Latino family members devote a greater number of hours per day to caring for their infirm relatives than other racial or ethnic groups (Ayalon & Huyck, 2002; National Alliance for Caregiving/AARP Public Policy Institute, 2015). This greater time commitment stems from the importance that Latinos place on family, as well as the lack of affordable and appropriate alternatives (Angel & Angel, 2008; Mendez-Luck et al., 2008; Silverstein & Wang, 2015).

Latinos and other minority group families tend to underutilize the entire range of institutional and community-based eldercare services. Only one-third of minority caregivers have ever used home-based services (Casado, van Vulpen & Davis, 2011). Less than 17% have used meal delivery, support, transportation, or respite care services (Casado, van Vulpen, & Davis, 2011). It is likely that this low use of community services contributes to the strain placed on family caregivers. One study of Medicaid recipients with dementia found that the low use of community-based services increases the risk of nursing home admission (Sands et al., 2008). Research also shows that caring for frail parents in the community is costly to families, particularly for Latinos who have low individual and family incomes (Rainville, Skufca, & Mehegan, 2016).

When minority group caregivers use formal services, they do so late in the aging parent's decline (Angel, Douglas, & Angel, 2003; Evercare and National Alliance for Caregiving, 2008; Kadushin, 2004; Wallace, Levy-Storms, & Ferguson, 1995). Research reveals how, even when they have adequate resources, Latino caregivers delay seeking community services until the care recipients' behavior, particularly in the case of dementia patients, has greatly worsened (Evercare and National Alliance for Caregiving, 2008). Furthermore, Latino caregivers tend to be more intensively involved in caregiving (than non-Hispanic white caregivers) in that they perform a greater number of personal care tasks for their loved ones as compared to non-Latino caregivers (Evercare and National Alliance for Caregiving, 2008).

Furthermore, individuals who would most likely benefit from caregiver support programs are among those least likely to know about them. An example relates to knowledge and use of family leave programs which allow caregivers to take paid leave for a certain period of time to care for an infant or a dependent family member (Andrew Change and Company, 2015). A study of knowledge of the Paid Family Leave program in California conducted among registered voters in 2014 revealed that only 36% of respondents knew about the program and its benefits. Awareness of the program was particularly low among Latinos, African Americans, and Asians, as well as individuals with only a high school education, low-incomes, and women (California Center for Research on Women & Families, 2015). A similar poll in New Jersey found that 60% of respondents were unaware of this potentially valuable family caregiving benefit (White, Houser, & Nisbet, 2013). Perhaps it is that caregivers can take only a limited amount of time (away from work, etc.) to care for an infirm parent who may be unlikely to regain their independence, which makes this program of limited value for those caring for older individuals with dementia. The fact that individuals are unaware of this program, though, suggests that they may also be unaware of other sources of community support.

Additional Sources of Formal Community Care

A patchwork of government-funded community-level programs can provide various forms of support to the caregivers of frail older individuals, including those with dementia, and these programs are funded by federal, state, and county governments. Various programs exist that are funded or supported by the Administration on Aging (AoA) which was established in 1965 when Congress passed the Older Americans Act (OAA) (Fritz, 1979). This law mandated the creation of state Agencies on Aging to promote delivery of social services to older Americans. The AoA is one of the units of the Administration for Community Living (ACL) in the U.S. Department of Health and Human Services Administration for Community Living (Administration for Community Living, 2017b). See Fig. 21.2 for the ACL organizational chart (2012).

The AoA is organized into five offices as shown on the left side of Fig. 21.2. These include (1) Supportive and Caregiver Services; (2) Nutritional and Health Promotion Programs; (3) Elder Rights; (4) American Indian, Alaskan Native, and Native Hawaiian Programs; and (5) Long-Term Care Ombudsman Programs. As is the case with other federal departments, AoA's structure and mission are complex, as are the needs they address. Although AoA branches potentially provide a great deal of information and access to services, it is unclear whether low-income and minority Americans are aware of these potential sources of assistance or whether they use them. Addressing the needs of Latino elders and their families will require more research and investigation into whether, and to what extent, AoA services reach their intended audience, i.e., the potential recipients of services.

Through its five branches, the AoA offers supportive services, information, and education to caregivers. Perhaps the most important of AoA's caregiver support functions are carried out by the Office of Supportive and Caregiver Services which includes programs that promote home and community-based services that provide a broad range of support services for older adults and their caregivers. In addition, this and other AoA branches offer other services including senior centers, nutrition programs such as *Meals on Wheels*, homemaker and transportation services, as well as many other services, including legal assistance, case management, disease

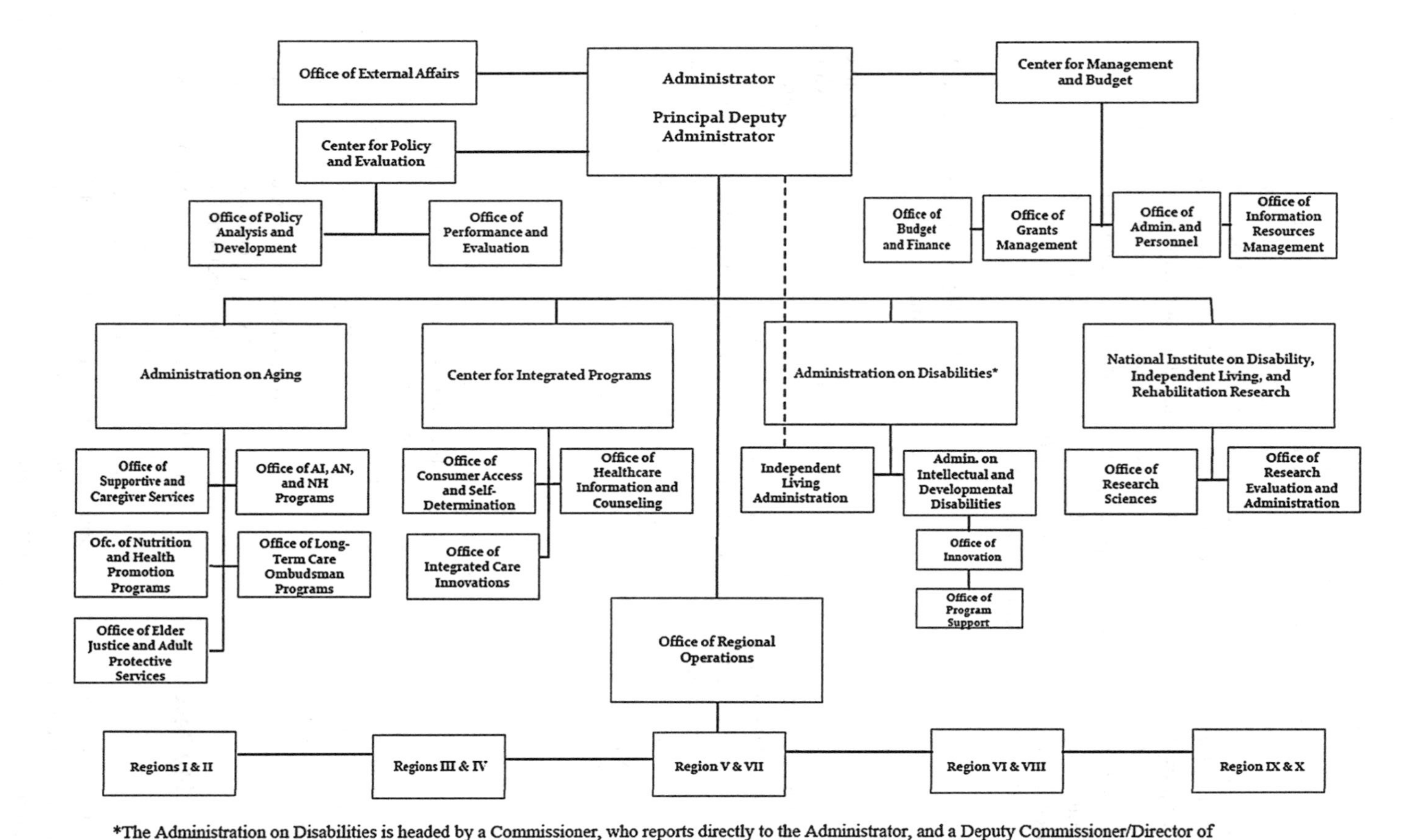

Fig. 21.2 Administration for Community Living organization chart. (Source: Administration for Community Living (2017b). Retrieved from https://www.acl.gov/about-acl/organization/organizational-chart)

prevention, health promotion, and more (Fritz, 1979; Siegler, Lama, Knight, Laureano, & Reid, 2015). As part of its mission, the AoA also provides counseling and respite services to the caregivers of frail elders through the National Family Caregiver Support Program (NFCSP) and the Native American Caregiver Support Program (NACSP), both of which are components of the AoA (Administration for Community Living, 2015). The NFCSP provides training to informal caregivers in the community, as well as respite services, adult day care, and hospice (Administration on Aging, 2012). Educational and training services are provided to adult family members who provide in-home and community care.

The Long-Term Care Ombudsman Program

An important function of AoA is carried out by the Long-Term Care Ombudsman Programs (LTCOP). This program represents an important asset to family caregivers in that it can ease concerns about potential neglect or mistreatment of elderly loved ones by providing oversight and mechanisms for correcting identified deficiencies. The program employs unpaid volunteers who respond to reports of neglect or abuse of individuals in nursing homes, board and care homes, assisted living facilities, and similar adult care organizations (Administration for Community Living 2017a). LTCOP volunteers guard against threats to health, safety, and an older individual's quality of life. Caregivers and relevant authorities are informed of any identified violations.

Aging and Disability Resource Centers

Another important source of information and referral for family caregivers are the Aging and Disability Resource Centers (ADRC; Administration for Community Living, 2017a). Since 2003, the AoA and the Centers for Medicare and Medicaid Services (CMS) have made a series of grants to states to develop these centers. The ADRC program provides information, services, and supports. The goal of the Centers is to help people of all ages and income levels to more easily access long-term services and supports through a single point of entry, to allow them to make more efficient use of care options, and to maximize the services available. The ADRC assesses a person's needs and then places them in contact with the appropriate service provider (U.S. Department of Health and Human Services, 2016).

State-Funded Home and Community-Based Programs

The Caregiver Advise, Record, Enable Act (CARE) passed by Congress in 2015 (Coleman, 2016) allows states to establish programs that involve caregivers in an elderly parent's medical discharge planning (AARP, 2015). The program informs the family caregivers when the patient is to be discharged from the hospital and provides the caregiver with education and information concerning the medical tasks the caregiver will need to perform for the patient at home. The program provides support to caregivers in 33 states (AARP, 2016). An extension of CARE in a few states, the Home Care Program for Elders, is a program designed to help caregivers of aging Connecticut residents and six other states who are at risk of nursing home placement (Connecticut Department of Social Services, 2016). The goal of the program is to allow frail seniors to remain at home in a safe environment and avoid institutionalization. The services include care management services, adult day health services, companion services, home-delivered meals, and homemaker services. In most states, the program requires a co-payment and cost sharing, based on a fee schedule if the client's income exceeds the financial eligibility criteria (Connecticut Department of Social Services, 2016).

Adult Day Services Centers

Adult day programs are intended to allow elderly and disabled persons to enjoy greater independence and to remain socially engaged while providing respite for family members (Alzheimer's and Dementia Caregiver Center, 2016). Adult day service (ADS) use is increasing (National Adult Day Service Association, 2017). The services are particularly important for individuals with Alzheimer's disease and other dementias given the high level of supervision these individuals require. This source of assistance is particularly important for family members who must work and who have other responsibilities. As of 2014, there were 5685 ADS centers in the nation (National Adult Day Service Association, 2017). In Texas, more than 60 percent of the state's 420 adult day care centers are located along the Mexican border (Zaragovia, 2014). These centers provide much-needed respite for family caregivers and allow those who are employed to go to work knowing that their impaired loved one is safe. In the absence of these centers, families would face serious challenges in providing all-day care, and many older individuals might be left alone.

Given the burgeoning older population, the market for adult day care is growing. Nevertheless, lacking is the rigorous research documenting how beneficial or effective ADS is in improving the functioning of frail elders or in reducing the burden on their caregivers (Zaragovia, 2014). Such research is essential in order to determine how best to serve the Latino population and other groups. Although certain studies have failed to demonstrate clearly positive outcomes (Gaugler, 2014; Gaugler & Zarit, 2001), others suggest that these centers may enhance the quality of life and reduce stress for caregivers (Femia, Zarit, Stephens, & Greene, 2007; Gitlin, Reever, Dennis, Mathieu, & Hauck, 2006; Siegler et al., 2015; Zarit, Kim, Femia, Almeida, & Klein, 2014).

The Potential Role of Civil Society

As noted, demographic and social changes have reduced the family's ability to provide extensive hands-on care. The core theoretical and practical question that this chapter's analysis raises then relates to the potential role of civil society organizations, including non-governmental and faith-based organizations, as advocates for and complementary providers of care to the elderly. Neoliberal economic and social policies that devolve responsibility for social services from higher levels of government to lower levels of government and to NGOs reflect the belief that local entities have certain advantages in dealing with the more routine and manageable needs of specific populations (Pereira & Angel, 2009; Pereira et al., 2007).

Although the financial support of the elderly and the often complex and expensive high-tech medicine they need can only be paid for and provided by the state, routine and relatively inexpensive services such as assistance with activities of daily living and providing companionship are often more effectively and economically provided by local groups including NGOs, congregations, and other local agencies (Angel, 2011). Such local actors are often more informed of the legal, transportation, nutritional, and other needs of the elderly in their communities.

There are many advocacy organizations in the USA and in other countries, some of which address issues of particular relevance to Latino families. For example, the *American Association of Retired Persons* (AARP) is the most visible advocate for older persons in the USA (Binstock, 2004). Other organizations, such as the *National Committee to Preserve Social Security and Medicare* (NCPSSM: http://www.ncpssm.org/), the *Alliance for Retired Americans* (ARA: http://www.retiredamericans.org/), and the *National Hispanic Council on Aging* (NHCOA: http://www.nhcoa.org/) also advocate for changes in policy to preserve and enhance programs for the elderly.

Although the literature on civil society organizations has expanded in recent years (Boli & Thomas, 1999), relatively little is known about how various NGOs frame their eldercare missions or how they focus their efforts, especially in reference to Latinos and other minorities. What is of particular importance for the Latino community is that advocacy activities be informed by an understanding of the unique needs of frail Latinos and their families. It is imperative to assure that, in addition to supporting Latino organizations, the existing and influential NGOs be made aware of the needs of the Latino population. One example is of an initiative at the University of Texas and other institutions in which the AARP supports the promotion of the next generation of emerging Latino aging scholars to investigate important issues related to population aging in the USA and Mexico: http://copp.utsa.edu/events/2016-international-conference-on-aging-in-the-americas/.

Another particularly important organization that directly addresses Latino aging issues is the *National Hispanic Council on Aging* (NHCOA: http://www.nhcoa.org/). The Council's mission includes advocacy and research, as well as the funding of community-based projects in Latino communities. The organization fosters local support networks and capacity building. One of the organization's major objectives is the support and strengthening of Hispanic, community-based organizations. Its stated mission is to "empower Hispanic community organizations and agencies, as well as Hispanic older adults and their families." It is vitally important to assess the effectiveness of the NHCOA's initiatives in order to assure that they achieve their objectives. It is also necessary for more Latinos to become involved, not only for caring for their own family members but also operationally at the level of advocacy, strategically for greater community supports, and politically to insure adequate community representation for all older individuals.

Very useful, albeit small-scale qualitative studies (Montes De Oca, Molina, & Avalos, 2008), suggest a potentially important role for local organizations in providing social support, but many questions remain. It is useful to call upon some indirect evidence. A large body of research clearly demonstrates that social support and social networks are key to the maintenance of high quality of life in old age (see Carr & Moorman, 2011). Considering the profound changes that have affected family structure and function, in the future, the elderly may find that they have no choice but to turn to such non-family sources of social support. On a positive note, the USA has historically had a very rich and active civil society (Putnam, 2000) which represents a potential major source of support for infirm elderly individuals, especially for the many who lack the resources to purchase assistance.

Future Research and Social Policy Agenda for Dementia Caregiving and the Latino Community

A key objective of this chapter is to further the discussion and to foster research ideas into the potential role of civil society in advocacy and service delivery to older individuals, especially Latinos. Described have been the eldercare functions of but a few, well-known non-governmental organizations. In today's world, many organizations for which eldercare and advocacy were not part of their original core missions have extended their focus to address the needs of the elderly and their families as life spans increase and individuals outlive their resources. A quick perusal of the Internet reveals thousands of organizations across the globe that, to some degree, address the needs of frail elderly and their families. These organizations are clearly important in preventing isolation and in improving the quality of life of seriously impaired older individuals. These groups are also important in easing the caregiving burden for family caregivers. Yet, if these organizations are to do more than solely providing palliative or respite assistance, their number one mission is to be that of advancing patient advocacy. These groups have an important role in ensuring that individuals are aware of, have access to, and apply for the available governmental programs and services to which patients are entitled.

In summary and as documented in this chapter, ethnographic and observational studies suggest that the local community and non-governmental and faith-based organizations play an important role in providing informal assistance, both in the USA and Mexico (Burcher, 2008; Montes De Oca et al., 2008). Such assistance is clearly vital for situations in which formal support is limited; still, relatively little is known about the exact sources of informal care outside of the family, nor is much known about the nature, quality, or adequacy of the care received. Determining the potential, greater role of civil society in the support of the elderly will require more research. There is limited systematic and rigorous research concerning the community support system that can be called upon for older Latinos with dementia and other disabling conditions, including at different states in the disease process. To understand the unique needs of specific populations, comparative studies will be very useful. Obtaining objective information on how families can mobilize support and reduce the barriers they face in acquiring support for the care of an aging parent is essential. The information that is necessary for the development of optimal policy concerning the community support of elderly Latinos includes information on (1) the services available in Latino communities, (2) outreach efforts by formal and informal organizations and service providers, (3) the use of those services by older Latinos and their families, and (4) the effectiveness of those services in relieving the burden that the caregivers of seriously impaired elderly experience. Given the pace of population aging in the USA, Latin America, and elsewhere in Spanish-speaking countries, harnessing the potential of non-governmental, faith-based, and other informal volunteer organization is becoming central to service availability and delivery. This third pillar of support represents a strong addition to the support efforts of federal, state, and local governments and the family and an active level of involvement and advocacy by those committed to the well-being of the elderly, especially for Latinos with dementia, yet as asserted in this chapter, civil society organizations are still no substitute for efficient and effective state policy.

References

AARP. (2015). New state law to help family caregivers. Retrieved 12/1/2016 from http://www.aarp.org/politics-society/advocacy/caregiving-advocacy/info-2014/aarp-creates-model-state-bill.html. In.

AARP. (2016). Showing support for caregivers. Retrieved 12/1/2016 from http://states.aarp.org/showing-support-kansas-caregivers/

Administration for Community Living. (2015). National aging network. Retrieved 12/1/2016 from http://www.aoa.acl.gov/AoA_Programs/OAA/Aging_Network/Index.aspx. In.

Administration for Community Living. (2017a). Long-term care ombudsman program. Retrieved 7/29/2017 from https://www.acl.gov/programs/protecting-rights-and-preventing-abuse/long-term-care-ombudsman-program. In.

Administration for Community Living. (2017b). ACL Organizational chart text version. Retrieved 7/29/2017 from https://www.acl.gov/node/760. In.

Administration on Aging. (2012). National family caregiver support program (OAA)Title IIIe. Retrieved from http://www.aoa.gov/aoa_programs/hcltc/caregiver/index.aspx http://www.caregiver.org/caregiver/jsp/content/pdfs/op_200211_10_state_texas.pdf

Alzheimer's and Dementia Caregiver Center. (2016). Adult day centers. Retrieved on 12/5/2016 from http://www.alz.org/care/alzheimers-dementia-adult-day-centers.asp#ixzz4S02HPOWA In.

Alzheimer's Association. (2017). Alzheimer's Association. 2017. Latinos and Alzheimer's. Retrieved 7/27/2017 from http://www.alz.org/espanol/about/latinos_and_alzheimers.asp. In.

Andrew Change and Company. (2015). Paid family leave market research. Retrieved 11/10/2016 from http://www.edd.ca.gov/disability/pdf/Paid:Family_Leave_Market_Research_Report_2015.pdf. In.

Angel, J. L. (2016). *A better life for low-income elders in Austin*. Retrieved from Austin, TX.

Angel, J. L., & Angel, R. J. (2008). *Caring for the elderly Mexican-origin population in Texas: Where does the burden fall? Paper presented at the Conference on Latinos and Public Policy in Texas*. Austin, Texas: The University of Texas at Austin.

Angel, J. L., Angel, R. J., Aranda, M. P., & Miles, T. P. (2004). Can the family still cope? Social support and health as determinants of nursing home use in the older Mexican-origin population. *Journal of Aging and Health, 16*, 338–354.

Angel, J. L., Douglas, N., & Angel, R. J. (2003). Gender, widowhood, and long-term care in the older

Mexican American population. *Journal of Women & Aging, 15*(2–3), 89–105. https://doi.org/10.1300/J074v15n02_06

Angel, R. J. (2011). Civil society and eldercare in post-traditional society. In R. A. Settersten & J. L. Angel (Eds.), *Handbook of sociology of aging* (pp. 549–581). New York, NY: Springer.

Angel, R. J., & Angel, J. L. (1997). *Who will care for us? Aging and long-term care in multicultural America*. New York, NY: New York University Press.

Angel, R. J., & Angel, J. L. (2015). *Latinos in an aging world: social, psychological, and economic perspectives*. New York, NY: Routledge.

Angel, R. J., & Angel, J. L. (2018). *Family, intergenerational solidarity, and post-traditional society*. New York, NY: Routledge.

Angel, R. J., Angel, J. L., & Hill, T. D. (2014). Longer lives, sicker lives? Increased longevity and extended disability among Mexican-origin elders. *The Journals of Gerontology Series B: Psychological Sciences and Social Sciences*. https://doi.org/10.1093/geronb/gbu158

Apesoa-Varano, E. C., Gomez, Y., & Hinton, L. (2016). Dementia informal caregiving in Latinos: what does the qualitative literature tell us? In W. A. Vega, K. S. Markides, J. L. Angel, & F. M. Torres-Gil (Eds.), *Challenges of Latino aging in the Americas* (pp. 141–169). New York, NY: Springer.

Aranda, M. P., & Knight, B. G. (1999). The influence of ethnicity and culture on the caregiver stress and coping process: A sociocultural review and analysis. *The Gerontologist, 37*(3), 342–354.

Arias, E. (2016). *Changes in life expectancy by race and hispanic origin in the United States, 2013–2014*. Atlanta, GA: National Center for Health Statistics.

Ayalon, L., & Huyck, M. H. (2002). Latino caregivers of relatives with Alzheimer's disease. *Clinical Gerontologist, 24*(3–4), 93–106. https://doi.org/10.1300/J018v24n03_08

Beard, J. R., Officer, A., de Carvalho, I. A., Sadana, R., Pot, A. M., Michel, J.-P., … Chatterji, S. (2016). The World report on ageing and health: a policy framework for healthy ageing. *The Lancet, 387*(10033), 2145–2154. https://doi.org/10.1016/S0140-6736(15)00516-4

Belle, S. H., Burgio, L., Burns, R., Coon, D., Czaja, S. J., Gallagher-Thompson, D., … Zhang, S. (2006). Enhancing the quality of life of dementia caregivers from different ethnic or racial groups: A randomized, controlled trial. *Annals of Internal Medicine, 145*(10), 727–738. Retrieved from http://www.ncbi.nlm.nih.gov/pmc/articles/PMC2585490/

Bilbrey, A. C., Angel, J. L., Humber, M. B., Chennapragada, L., Garcia, I., Kajiyama, B., & Gallagher, D. (2018). Working with Mexican-American families. In G. Yeo, L. Gerdner, & D. Gallagher-Thompson (Eds.), *Ethnicity and dementias* (3rd ed., p. 20). New York, NY: Taylor and Francis.

Binstock, R. H. (2004). Advocacy in an era of neoconservatism: Responses of national aging organizations. *Generations, 28*(1), 49–54.

Bloom, S., Sulick, B., & Hansen, J. C. (2011). Picking up the PACE: The affordable care act can grow and expand a proven model of care. *Generations, 35*(1), 53–55. Retrieved from http://www.ingentaconnect.com/content/asag/gen/2011/00000035/00000001/art00010

Boli, J., & Thomas, G. M. (1999). *Constructing world culture: International nongovernmental organization since 1875*. Palo Alto, CA: Stanford University Press.

Brown, H. S., Herrera, A. P., & Angel, J. L. (2013). Opportunity costs associated with caring for older Mexican-Americans. *28*, 375–389.

Burcher, J. (Ed.). (2008). *México solidario: Participación cudadana y voluntaiado*. Mexico, D.F.: Limusa.

California Center for Research on Women & Families. (2015). Just 36% of voters aware of state's paid family leave program. Retrieved 12/5/2016 from http://ccrwf.org/just-36-of-voters-aware-of-states-paid-family-leave-program/

Carr, D., & Moorman, S. (2011). Social relations and aging. In R. A. Settersten & J. L. Angel (Eds.), *Handbook of sociology of aging* (Vol. 20, pp. 145–160). New York, NY: Springer Science.

Casado, B. L., van Vulpen, K. S., & Davis, S. L. (2011). Unmet needs for home and community-based services among frail older americans and their caregivers. *Journal of Aging and Health, 23*(3), 529–553. https://doi.org/10.1177/0898264310387132

Centers for Medicare and Medicaid Services. (2011). Quick facts about programs of all-inclusive care for the elderly (PACE). Retrieved 11/14/2016 from https://www.medicare.gov/Pubs/pdf/11341.pdf. In.

Chatterji, P., Burstein, N. R., Kidder, D., & White, A. (2008). *Evaluation of the program of all-inclusive care for the elderly (PACE) Demonstration: The impact of PACE on participant outcomes*. Retrieved from Cambridge, CA.

Choi, N. G., Lee, A., & Goldstein, M. (2011). Meals on wheels: Exploring potential for and barriers to integrating depression intervention for homebound older adults. *Home Health Care Services Quarterly, 30*(4), 214–230. https://doi.org/10.1080/01621424.2011.622251

Coleman, E. A. (2016). Family caregivers as partners in care transitions: The caregiver advise record and enable act. *Journal of Hospital Medicine, 11*(12), 883–885. https://doi.org/10.1002/jhm.2637

Connecticut Department of Social Services. (2016). Connecticut home care program for elders. Retrived 12/1/2016 from http://www.ct.gov/dss/cwp/view.asp?a=2353&q=305170. In.

Cucciare, M. A., Gray, H., Azar, A., Jimenez, D., & Gallagher-Thompson, D. (2010). Exploring the relationship between physical health, depressive symptoms, and depression diagnoses in Hispanic dementia caregivers. *Journal of Aging and Mental Health, 14*(3), 274–282. https://doi.org/10.1080/13607860903483128

Dilworth-Anderson, P., Williams, I. C., & Gibson, B. E. (2002). Issues of race, ethnicity, and culture in caregiving research: A twenty – year review (1980–2000). *The Gerontologist, 42*(2), 237–272.

Evercare and National Alliance for Caregiving. (2008). Hispanic family caregiving in the U.S. Retrived 11/6/2016 from http://www.caregiving.org/data/Hispanic_Caregiver_Study_web_ENG_FINAL_11_04_08.pdf. In.

Femia, E. E., Zarit, S. H., Stephens, M. A. P., & Greene, R. (2007). Impact of adult day services on behavioral and psychological symptoms of dementia. *The Gerontologist, 47*(6), 775–788. https://doi.org/10.1093/geront/47.6.775

Flores, Y. G., Hinton, L., Barker, J. C., Franz, C. E., & Velasquez, A. (2009). Beyond familism: Ethics of care of latina caregivers of elderly parents with dementia. *Health Care for Women International, 30*(12), 1055–1072. https://doi.org/10.1080/07399330903141252

Fritz, D. (1979). The administration on aging as an advocate: Progress, problems, and prospects. *The Gerontologist, 19*(2), 141–150. https://doi.org/10.1093/geront/19.2.141

Gallagher-Thompson, D., Guy, W. W. H., Rupert, M., Argüelles, T., Zeiss, L. M., Long, C., … Ory, M. (2003). Tailoring psychological interventions for ethnically diverse dementia caregivers. *Clinical Psychologist: Science and Practice, 10*, 423–438.

Gaugler, J. E. (2014). The process of adult day service use. *Geriatric Nursing, 35*(1), 47–54. https://doi.org/10.1016/j.gerinurse.2013.10.009

Gaugler, J. E., Kane, R. L., Kane, R. A., & Newcomer, R. (2006). Predictors of institutionalization in Latinos with dementia. *Journal of Cross-Cultural Gerontology, 21*, 139–155.

Gaugler, J. E., & Zarit, S. H. (2001). The effectiveness of adult day services for disabled older people. *Journal of Aging & Social Policy, 12*(2), 23–47. https://doi.org/10.1300/J031v12n02_03

Ghosh, A., Orfield, C., & Schmitz, R. (2014). Evaluating PACE: A review of the literature. In. Washington, DC: U.S. Department of Health and Human Services, Office of the Assistant Secretary for Planning and Evaluation, Office of Disability, Aging and Long-Term Care Policy.

Gitlin, L. N., Reever, K., Dennis, M. P., Mathieu, E., & Hauck, W. W. (2006). Enhancing quality of life of families who use adult day services: Short- and long-term effects of the adult day services plus program. *The Gerontologist, 46*(5), 630–639.

Gonzalez, L. (2017). A focus on the program of all-inclusive care for the elderly (PACE). *Journal of Aging & Social Policy*, 1–16. https://doi.org/10.1080/08959420.2017.1281092

Gutiérrez, L. M., Ortega, M. L., & Lopera, V. E. A. (2012). The state of elder care in Mexico. *Current Geriatric Report, 1*(183–189).

Gutiérrez Robledo, L. M., Ortega, M. L., & Lopera, V. E. A. (2012). The state of elder care in Mexico. *Current Translational Geriatrics and Experimental Gerontology Reports, 1*(4), 183–189. Retrieved from http://link.springer.com/article/10.1007%2Fs13670-012-0028-z#

Hansen, J. C., & Hewitt, M. (2012). Pace provides a sense of belonging for elders. *Journal of the American Society for Aging*, (Spring), 37–43. Retrieved from http://www.onlok.org/Portals/0/PACE%20Provides%20a%20Sense%20of%20Belonging%20for%20Elders.pdf

Herrera, A., Benson, R., Angel, J. L., Markides, K., & Torres-Gil, F. (2013). Effectiveness and reach of caregiver services funded by the older americans act to vulnerable older hispanics and African Americans. *Home Health Care Services Quarterly*.

Hoskins, D. D. (2010). U.S. social security at 75 years: An international perspective. Retrieved 7/29/17 from https://www.ssa.gov/policy/docs/ssb/v70n3/v70n3p79.html. In (Vol. 70).

Kadushin, G. (2004). Home health care utilization: a review of the research for social work. *Health and Social Work, 29*(3), 219–244.

Kaye, H. S., Harrington, C., & LaPlante, M. P. (2010). Long-term care: who gets it, who provides it, who pays, and how much? *Health Aff (Millwood), 29*(1), 11–21. https://doi.org/10.1377/hlthaff.2009.0535

Knight, B. G., & Sayegh, P. (2010). Cultural values and caregiving: The updated sociocultural stress and coping model. *The Journals of Gerontology: Series B, 65*(1), 5–13.

Kuhlman, H. (2015). Texas medicaid and CHIP: Texas dual eligible integrated care project. Retrieved 11/15/2016 from http://legacy-hhsc.hhsc.state.tx.us/medicaid/managed-care/dual-eligible/

Laditka, S. B., Laditka, J. N., & Fisher Drake, B. (2006). Home- and community-based service use by older African American, Hispanic, and non-Hispanic white women and men. *Home Health Care Services Quarterly, 25*(3–4), 129–153. https://doi.org/10.1300/J027v25n03_08

Llanque, S. M., & Enriquez, M. (2012). Interventions for hispanic caregivers of patients with dementia: A review of the literature. *American Journal of Alzheimer's Disease and Other Dementias, 27*(1), 23–32. https://doi.org/10.1177/1533317512439794

Mendez-Luck, C. A., Amorim, C., Anthony, K. P., & Neal, M. B. (2016). Beliefs and expectations of family and nursing home care among Mexican-origin caregivers. *Journal of Women & Aging*, 1–13. https://doi.org/10.1080/08952841.2016.1222758

Mendez-Luck, C. A., Kennedy, D. P., & Wallace, S. P. (2008). Concepts of burden in giving care to older relatives: A study of female caregivers in a Mexico City neighborhood. *Journal of Cross-Cultural Gerontology, 23*(3), 2650282.

Montes-de-Oca, V., Ramírez, T., Santillanes, N., García, S. J., & Sáenz, R. (2016). Access to medical care and family arrangements among mexican elderly immigrants living in the United States. In W. Vega, K. Markides, J. L. Angel, & F. Torres-Gil (Eds.), *Challenges of latino aging in the Americas* (pp. 225–245). New York, NY: Springer Science.

Montes De Oca, V. M., Molina, A., & Avalos, R. (2008). *Migración, redes transacionales y envejecimiento: Estudio de las redes familaries transnacionales de la vejez en Guanajuato*. México, D.F.: Universidad Nacional Autónoma de México, Instituto de Investigaciones Sociales, Govierno del Estado de Guanajuato.

Napoles, A. M., Chadiha, L., Eversley, R., & Moreno-John, G. (2010). Reviews: developing culturally sensitive dementia caregiving interventions: Are we there yet? *American Journal of Alzheimer's Disease and Other Dementias, 25*(5), 389–406.

National Academies of Sciences, E., and Medicine. (2016). *Families caring for an aging America*. Washington, DC: The National Academies Press.

National Adult Day Service Association. (2017). About adult day services. Retrieved 7/29/2017 from https://www.nadsa.org/learn-more/about-adult-day-services/. In.

National Alliance for Caregiving/AARP Public Policy Institute. (2015). Caregiving in the U.S. 2015. Retrieved 11/6/2016 from http://www.aarp.org/content/dam/aarp/ppi/2015/caregiving-in-the-united-states-2015-report-revised.pdf. Retrieved from http://www.caregiving.org/wp-content/uploads/2015/05/2015_CaregivingintheUS_Care-Recipients-Over-50_WEB.pdf

Pereira, J., & Angel, R. (2009). From adversary to ally: The evolution of non-governmental organizations in the context of health reform in Santiago and Montevideo. In S. Babones (Ed.), *Social inequality and public health* (pp. 97–114). Bristol, UK: Polity Press.

Pereira, J., Angel, R. J., & Angel, J. L. (2007). A case study of the elder care functions of a Chilean non-governmental organization. *Social Science & Medicine, 64*, 2096–2106.

Pierson, P. (2001). Coping with permanent austerity: Welfare state restructuring in affluent democracies. In P. Pierson (Ed.), *The new politics of the welfare state*. Oxford and New York: Oxford University Press.

Putnam, R. (2000). *Bowling alone: The collapse and revival of American community*. New York, NY: Simon & Schuster.

Rainville, C., Skufca, L., & Mehegan, L. (2016). Family caregiving and out-of-pocket costs: 2016 report report prepared. Retrieved 12/1/2016 from http://www.aarp.org/content/dam/aarp/research/surveys_statistics/ltc/2016/family-caregiving-cost-survey-res-ltc.pdf. In.

Rote, S., Angel, J. L., & Markides, K. (2015). Health of elderly Mexican American adults and family caregiver distress. *Research on Aging, 37*(3), 306–331. https://doi.org/10.1177/0164027514531028

Rote, S. M., Angel, J. L., & Markides, K. (2017). Neighborhood context, dementia severity, and Mexican American caregiver well-being. *Journal of Aging and Health, 29*(6), 1039–1055. https://doi.org/10.1177/0898264317707141

Roth, D. L., Fredman, L., & Haley, W. E. (2015). Informal caregiving and its impact on health: A reappraisal from population-based studies. *The Gerontologist, 55*(2), 309–319. https://doi.org/10.1093/geront/gnu177

Sands, L. P., Xu, H., Weiner, M., Rosenman, M. B., Craig, B. A., & Thomas, J. (2008). Comparison of resource utilization for medicaid dementia patients using nursing homes versus home and community based waivers for long-term care. *Medical Care, 46*, 449–453.

Siegler, E. L., Lama, S. D., Knight, M. G., Laureano, E., & Reid, M. C. (2015). Community-based supports and services for older adults: A primer for clinicians. *Journal of Geriatrics, 2015*, 6. https://doi.org/10.1155/2015/678625

Silverstein, M., & Wang, R. (2015). Does familism inhibit demand for long-term care? Public policy implications of growing ethnic diversity in the United States. *Public Policy & Aging Report, 25*(3), 83–87. https://doi.org/10.1093/ppar/prv016

Smith, D. B., & Feng, Z. (2010). The accumulated challenges of long-term care. *Health Affairs, 29*(1), 29–34. https://doi.org/10.1377/hlthaff.2009.0507

STAR+PLUS Health Program. (2015). *STAR+PLUS medicare-medicaid plans for dual eligible members: Compare "Value-Added" or extra services offered by plans in Bexar, Dallas, and Hidalgo County*. Retrieved from Austin, TX: http://www.hhsc.state.tx.us/medicaid/managed-care/dual-eligible/

Taylor, M. G., & Quesnel-Vallée, A. (2017). The structural burden of caregiving: Shared challenges in the United States and Canada. *The Gerontologist, 57*(1), 19–25. https://doi.org/10.1093/geront/gnw102

Texas Health and Human Services Commission. (2014). Texas medicaid and CHIP STAR+PLUS expansion. Retrieved 12/4/2016 from http://www.hhsc.state.tx.us/medicaid/managed-care/mmc/starplus-expansion/. In. Austin, TX.

The National Coalition on Mental Health and Aging. (2017). Integrating older adult behavioral health into long-term care rebalancing- opportunities and recommendations for funders: Public health agencies and managed care organizations. Retrieved 7/29/2017 from http://www.nasuad.org/sites/nasuad/files/NCMHA%20Report%20on%20Integrating%20OABH%20into%20LTC%20-%20FINAL.PDF. In.

Thomeer, M. B., Mudrazija, S., & Angel, J. L. (2015). How do race and hispanic ethnicity affect nursing home admission? Evidence from the health and retirement study. *The Journals of Gerontology Series B: Psychological Sciences and Social Sciences, 70*(4), 628–638. https://doi.org/10.1093/geronb/gbu114

U.S. Census Bureau. (2018). *State and county quick facts*. Retrieved 6/19/19 from https://www.census.gov/quickfacts/elpasocountytexas https://www.census.gov/quickfacts/hidalgocountytexas https://www.census.gov/quickfacts/fact/table/bexarcountytexas/PST045218. Retrieved from Washington, DC:

U.S. Department of Health and Human Services. (2016). Aging & disability resource centers program/no wrong door system. In: Administration for community living, office of consumer access and self determination. Retrieved 12/5/2016 from http://www.acl.gov/Programs/CIP/OCASD/ADRC/index.aspx

Wallace, S. P., Levy-Storms, L., & Ferguson, L. R. (1995). Access to paid in-home assistance among disabled elderly people: Do Latinos differ from non-Latino whites? *American Journal of Public Health, 85*(7), 970–975.

Wallace, S. P., Levy-Storms, L., Kington, R. S., & Andersen, R. M. (1998). The persistence of race and ethnicity in the use of long-term care. *The Journals of gerontology. Series B: Psychological Sciences and Social Sciences, 53*(2), S104–S112.

Wallace, S. P., & Lew-Ting, C.-Y. (1992). Getting by at home: Community-based long-term care of Latino elders. *Western Journal of Medicine, 157*, 337–344.

White, K., Houser, L., & Nisbet, E. (2013). *Policy in action: New Jersey's family leave insurance program at age three*. Retrieved from New Brunswick, NJ.

Yeo, G., Gerdner, L. A., & Gallagher-Thompson, D. (Eds.). (2018). *Ethnicity and the dementias* (3rd ed.). New York, NY: Taylor & Francis.

Zaragovia, V. (2014). South Texas has dozens of adult day centers - and Austin has just one. Retrieved 12/4/2016 from http://kut.org/post/south-texas-has-dozens-adult-day-centers-and-austin-has-just-one

Zarit, S. H., Kim, K., Femia, E. E., Almeida, D. M., & Klein, L. C. (2014). The effects of adult day services on family caregivers' daily stress, affect, and health: Outcomes from the Daily Stress and Health (Dash) Study. *The Gerontologist, 54*(4), 570–579. https://doi.org/10.1093/geront/gnt045

Zarit, S. H., Kim, K., Femia, E. E., Almeida, D. M., Savla, J., & Molenaar, P. C. M. (2011). Effects of adult day care on daily stress of caregivers: A within-person approach. *The Journals of Gerontology Series B: Psychological Sciences and Social Sciences, 66B*(5), 538–546. https://doi.org/10.1093/geronb/gbr030

22 Latinos and Dementia: Prescriptions for Policy and Programs that Empower Older Latinos and Their Families

Valentine M. Villa and Steven P. Wallace

Abstract

As the population ages and life expectancy continues to increase, the United States (U.S.) will experience an increase in the population diagnosed with Alzheimer's Disease and Related Dementias (ADRD). The increased representation of racial ethnic minority populations among the population aged 65 and older will mean these populations will be more likely than they have in the past to be diagnosed with dementia. This is especially true for Latinos, the fastest growing population among the older adult population. The prevalence of ADRD is growing rapidly among Latinos, and this trend is expected to continue. The growing number of Latinos with dementia will increase the population requiring long-term care services and the number of Latino families becoming caregivers. In what follows, we examine policy options that meet the needs of Latino with dementia, reduce risk factors for ADRD among Latinos, as well as empower Latino family caregivers.

V. M. Villa (✉)
California State University, Los Angeles, Los Angeles, CA, USA;
UCLA Fielding School of Public Health, Los Angeles, CA, USA
e-mail: vvilla@calstatela.edu; vvilla@ucla.edu

S. P. Wallace
UCLA Fielding School of Public Health, Los Angeles, CA, USA
e-mail: swallace@ucla.edu

Introduction

As the United States (U.S.) population continues to age and experience longer life expectancy, we will continue to see an increase in the prevalence of Alzheimer's Disease and Related Dementias (ADRD) (Chin, Negash, & Hamilton, 2011; Hebert, Weuve, Scherr, & Evans, 2013). When the baby boom population (those born between 1946 and 1964) fully enters old age in 2030, one in five persons in the U.S. will be aged 65 and older (U.S. Census Bureau, 2015). At the same time, the U.S. will experience unprecedented growth in minority elderly populations. Currently, minority populations represent 20% of the older adult population; by 2050, 40% of the population aged 65 and older will be members of a minority population (Federal Interagency Forum on Aging, 2012). The fastest growing population among those aged 65 and older are Latinos.[1] At present, Latinos represent 7% of the population aged 65

[1] We use the terms Latino and Hispanic interchangeably. The bulk of the data in this chapter represents the experience of Latinos who self-identify as Mexican-American.

H. Y. Adames, Y. N. Tazeau (eds.), *Caring for Latinxs with Dementia in a Globalized World*,
https://doi.org/10.1007/978-1-0716-0132-7_22

and older, and by 2050, Latinos[2] will make up 20% of the U.S. older adult population, representing 17.5 million individuals (U.S. Census Bureau, 2015). Living longer will mean older minority populations, including Latinos, will be at greater risk of developing dementia than they have been in the past (Hebert, Scherr, Bienias, Bennett, & Evans, 2003; Mehta & Yeo, 2016).

The largest growth in prevalence of Alzheimer's Disease (AD) and other dementias is found among the Latino population (Haan et al., 2003; Harris-Kojetin, Sengupta, Park-Lee, & Valverde, 2013; Wu, Vega, Resendez, & Jin, 2016). Latinos are currently 1.5 times more likely to be diagnosed with dementia than non-Hispanic Whites (Turner et al., 2015) and have the fastest growing number of cases of dementia in the U.S. (Humes, Jones, & Ramirez, 2011). Prevalence rates for dementia among Hispanics are much higher than non-Hispanic Whites nationally, 27.9% versus 10.9% among persons aged 75–84 and 62.9% versus 30% in persons aged 85 and older (Luchsinger et al., 2015; Noble, Manly, Schupf, Tang, & Luchsinger, 2012: Thies & Bleiler, 2011). These rates are drawn from local studies and may not reflect Latino rates nationally, but they are useful since they document a common trend of excess risk faced by older Latinos. Furthermore, it is projected that among Latinos, the prevalence of AD is expected to increase from 200,000 individuals diagnosed with some form of the dementia in 2000 to 1.3 million being diagnosed by 2050 (Alzheimer's Association, 2004).

The growing number of Latinos with dementia will increase the population requiring long-term care and therefore the number of Latino families becoming caregivers (Cucciare, Gray, Azar, Jimenez, & Gallagher-Thompson, 2010). Nationally, Latinos have the highest percentage of caregivers of all groups (Turner et al., 2015). Caring for sick and disabled family members in the home is consistent with the Latino cultural value of familism which represents a strong identification and attachment to both the nuclear and extended families (Arevalo-Flechas, 2013). Some caregivers report benefits that come along with caregiving, including acknowledgment of personal strengths and mastery (Peacock et al., 2010; Zarit, 2012), but there is also evidence that caregiving can have negative economic, health, and mental health consequences for Latino caregivers including financial strain, poor health, as well as sustained feelings of burden, anxiety, and depression (Luchsinger et al., 2015; Pinquart & Sorensen, 2005; Simpson, 2010). This is particularly true for Latino dementia caregivers who are more likely than those caring for family members without dementia to experience economic strain from out-of-pocket costs for their care recipient's needs (Schulz & Eden, 2016), as well as high levels of distress and behavioral problems that Latinos with dementia are more likely to experience than other groups (Ortiz, Fitten, Cummings, Hwang, & Fonseca, 2006; Turner et al., 2015;). Moreover, there is evidence that, among Latinos, the caregiver population is becoming younger and is often faced with having to make difficult trade-offs, including placing education, career, family formation, and economic opportunities on hold (AARP, 2015, 2016; Diaz, Siskowski, & Connors, 2007) because of caring for older family members.

The nexus of AD and other dementias with the growing, graying, and increasingly diverse U.S. population will undoubtedly challenge policy makers, practitioners, and older adults and their families. Managing the impact of AD and other dementias facing the Latino population will require intervention at the policy, program, and individual level. In particular, it is essential that policies and programs be developed that meet the long-term, supportive service care needs of older Latinos with dementia, as well as empower the Latino family to preserve their own health, mental health, and economic security. In what follows, we examine (1) the role of employment policies, tax initiatives, and a possible Social Security caregiver credit for Latino family caregivers to address the costs of caregiving; (2) the expansion and integration of long-term care supportive services under Medicaid and Medicare to address access to care; and (3) the importance of increasing access to dementia care and health

[2] See footnote 1.

care to address the risk factors for dementia among Latinos.

The Cost of Caring Among Latino Dementia Caregivers: The Importance of Policies that Empower Latino Caregiver's Financial Stability

Older Latinos are less likely to utilize long-term care services than the general population and are more likely to rely on informal care in the form of family members to support their long-term care needs (Gelman, 2010; Wu et al., 2016). Given that the majority of dementia caregiving is provided by unpaid family members, dementia is often referred to as a "family disease" (Zucchella, Bartolo, Pasotti, Chiapella, & Sinforiani, 2012). Caregiving needs increase as the disease progresses, with care recipients needing assistance with basic activities including bathing, dressing, eating, walking, and toileting and ultimately 24-hour-a-day care (Alzheimer's Association, 2012). According to the Alzheimer's Association (2013), unpaid family caregivers provided an estimated 17.7 billion hours of care per year at an estimated value of $220.2 billion dollars. It has been suggested that the value of informal care in the U.S. is equal to the combined cost of medical care and long-term care for dementia.

Nationally, Latinos have the highest percentage of family caregivers for older adults of all racial/ethnic groups (Turner et al., 2015). It is estimated that 27% of Latino households provide care for a family or friend (Talamantes, Trejo, Jiménez, & Gallagher-Thompson, 2006). Of the 8 million Latino family caregivers, 1.8 million are providing care for a family member with AD (National Alliance for Caregiving and AARP, 2009). Latino caregivers for older adult family members are typically female, on average 40 years old, spend 30 hours or more per week caregiving, work either full time or part time, and are providing care for an older adult as well as taking care of a child under the age of 18 (Alzheimer's Association, 2013; Arévalo-Flechas, Acton, Escamilla, Bonner, & Lewis, 2014; National Alliance for Caregiving and AARP, 2009; Schulz & Eden, 2016). Pinquart and Sorensen (2005), in their meta-analysis of articles contrasting differences between Hispanic and non-Hispanic White caregivers, report that Hispanic caregivers are more likely to be younger, female, and less educated and have less income when compared to non-Hispanic White caregivers. This is particularly troubling given the out-of-pocket costs associated with ADRD. Because of the lack of coverage for long-term care under Medicare and the limited coverage under Medicaid (the latter of which is only for people who are impoverished), families often pay for the purchase of goods and services for the older care recipient, such as medical/pharmaceutical co-pays, meals, transportation, and housing (Schulz & Eden, 2016). The Alzheimer's Foundation of America (2013) reports that families of persons with AD can expect to pay between $41,000 and $56,000 dollars annually in costs for dementia care for their loved ones with the disease. In fact, there is ample evidence that caregivers often experience financial strain associated with their caregiving (AARP, 2015; Spillman, Wolff, Freedman, & Kasper, 2014; Wolff, Spillman, Freedman, & Kasper, 2016) with the greatest financial strain reported by dementia caregivers. Approximately 30% of the population caring for those with dementia report experiencing financial strain related to caregiving (Schulz & Eden, 2016).

Data specific to Latino caregivers finds that, when compared to other racial/ethnic groups, out-of-pocket costs related to caregiving for an older family member are greatest among Latinos because they are more likely to care for an older family member for a longer period of time and are more likely than other populations to be caring for a family member with dementia (AARP, 2016). We can expect that the cost for dementia care will outstrip the economic base of most Latino older persons and their families. For example, Wu et al. (2016) found that after adjusting for socioeconomic status and comorbid conditions associated with AD, the medical care costs for Latinos with ADRD were $7046 annually. According to AARP (2016), Latino caregivers spend about 44% of their annual income on out-of-pocket costs for caregiving, with Latino women spending the most, i.e., nearly 47% or $10,704 of their annual income. Furthermore, it is estimated that 50.8% of Latinos caregivers are employed outside the home (Schulz & Eden, 2016). A larger percentage of Latino caregivers (41%) compared to non-Latino White caregivers (29%) need to make work accommodations because of caregiving, including cutting back on their work hours, changing jobs, stopping work entirely, and taking leaves of absences (Evercare and National Alliance of Caregiving, 2008). The demand for caregiving not only negatively impacts caregivers' ability to stay in the workforce but also jeopardizes their income, job security, personal retirement savings, Social Security and retirement benefits, and long-term financial well-being (AARP, 2015; Arno, House, Viola, & Schechter, 2011; Munnell, Sanzenbacher, & Rutledge, 2015; Schulz & Eden, 2016). This has the potential to render an economically vulnerable population even more vulnerable.

Currently, Latino households have annual median incomes of $39,600, which is 52% lower than the $60,300 average household income of non-Hispanic Whites (U.S. Bureau of Labor Statistics, 2015. In 2015, 20.9% of households where the head was under the age of 65 were living in poverty compared to 10.3% of non-Hispanic White households. The group with incomes just above the poverty threshold, i.e., the near poor, includes an additional 26% of Hispanic households compared to 13.1% of non-Hispanic White households. Combining the poor and near poor shows that almost half of Hispanic households headed by a person under age 65 has a limited income (U.S. Census Bureau, 2016). Even among middle-class Latino families, it is found that 87% do not have enough assets to meet their living expenses for 3 months if their source of income is gone (Wheary, Shapiro, Draut, & Meschede, 2008).

When Latino middle-aged caregivers are no longer able to stretch their income to cover the cost of caregiving or make accommodations in their work schedules, we can expect that caregiving duties will increasingly be taken on by younger members of the Latino family. There is already evidence of an overrepresentation of Latinos among younger caregiver populations, i.e., ages 25–40 (AARP, 2015). The shift in caregiving to younger members of the Latino family has the potential to "hijack" the socioeconomic stability of the next generation of Latinos. In fact, younger millennial caregivers are more likely than older caregivers to report financial strain related to caregiving activities (AARP, 2016). They also experience stress regarding finding time for social engagements, romantic relationships, and careers (Dellmann-Jenkins, Blankemeyer, & Pinkard, 2000), have difficulty juggling the role of caregiving with being a college student (Dellmann-Jenkins et al., 2000), and experience decreases in financial stability and job growth because of the amount of time they provide for caregiving (Stephens, Franks, & Townsend, 1994). Clearly, policies that support employment and economic security are very much needed for Latino caregivers of all ages so that they can continue to provide care for their loved ones while not having to jeopardize their own financial stability.

Family Leave Policy

The best protection for working caregivers is access to paid family leave in order to continue to care for an older family member with dementia without forgoing their own economic stability. Currently, the only national family leave policy is the federal Family Medical Leave Act (FMLA) that was signed into law in 1993. The FMLA provides eligible workers with 12 weeks of *unpaid* leave for a worker's own health and time to bond with a new child or to care for a seriously ill family member, i.e., child, parent, or spouse. To be eligible, workers must have worked for the employer for at least 12 months and have worked 1250 hours. The policy only applies to worksite areas with 50 or more employees. Workers are guaranteed the ability to return to their job with the same pay, benefits, and working conditions and can keep any employer health insurance during the period of the leave (Mayer, 2013; Schulz & Eden, 2016). Because of the limits on eligibility, about two-fifths of all workers are not covered by the law (Department of Labor (DOL) Wage and Hour Division, 2012). FMLA provides a basic benefit for workers to take leave without losing their jobs, but because the leave is unpaid, low-income workers are less likely to take advantage of the policy because they are unable to forgo their wages (Feinberg, 2013). The overrepresentation of Latinos in low-paid service industry jobs makes it less likely that they would utilize FMLA benefits. Also, because the FMLA only covers care for oneself, child, parent, or spouse, those caring for a father- or mother-in-law or grandparent are not covered (Department of Labor (DOL) Wage and Hour Division, 2012). For the Latino population, this is problematic in that the bulk of Latino caregivers are women, including daughters-in-law. Moreover, because the policy does not cover grandchildren caring for a grandparent, young Latino caregivers who experience the greatest economic strain, when compared to other generations of caregivers, would not be covered. At a minimum, the FMLA should extend coverage to in-laws and grandchildren since they often provide caregiving and should benefit from the same protections as children.

Some states have enacted *paid* family leave. California began the first paid family and medical leave program in the U.S. (Wagner, 2006). To date, four states including California, New Jersey, New York, and Rhode Island have paid family leave programs for new parents and caregivers of certain critically ill family members. The state programs share some common characteristics including that they (1) are financed through an insurance model; (2) are fully funded by worker payroll deductions; (3) provide partial pay replacement for a finite period of time; (4) cover caregivers of spouses, parents, and domestic partners (California, New York, and Rhode Island include parents-in-law and grandparents); and (5) use an existing state structure to finance and administer claims (Feinberg, 2013; Schulz & Eden, 2016). Annual payroll deductions are designed to fully cover the program costs (Fiscal Policy Institute, 2014).

Paid family leave policies can improve the quality of life of Latino dementia family caregivers by providing economic security and therefore retirement security for those who need time off from work to care for a family member (Bouchey & Glynn, 2012; Lester, 2005). A recent national survey found that worry about taking leave from work was the top concern of workers who had to take time off for caregiving (Klerman, Daley, & Pozniak, 2012). By having access to paid family leave, workers are better able to continue to participate in the labor force and therefore experience greater economic security. According to Appelbaum and Milkman (2011), employees who utilized paid family leave policies to provide caregiving reported strengthened loyalty to their employer and increased the likelihood that they would return to the same employer after the leave was over. There is also evidence to suggest that employers also benefit from paid family leave policies in the form of increased productivity, reduced costs of turnover, improved employee morale, and fewer work place accidents/injuries caused by stressed employees who are worrying about losing their jobs while providing care for a seriously ill family member (Asfaw & Bushnell, 2010; Feinberg, 2013).

A drawback is the lack of public awareness of the programs, particularly among those who are most likely to need paid family leave: low-income family caregivers. Data from California finds that only 36% of registered voters knew that paid family leave programs existed, with the least knowledgeable being Latinos, African Americans, individuals with less than a high school education, low-income households, and women (Field Research Corporation and California Center for Research on Women and Families, 2015). Clearly, outreach and community education programs regarding the availability of paid family leave programs must be expanded. There is ample evidence that older Latinos and their families often are unaware of the myriad programs and services that they are, in fact, eligible for and that would greatly improve their health and economic circumstances (Wallace, Torres, Sadegh-Nobari, Pourat, & Brown, 2012). Informational campaigns that are linguistically appropriate, culturally competent, and accessible to Latinos are necessary for the population to benefit from the family leave policies available to them.

The lack of any national policy on paid family leave allows for only 12% of U.S. workers with access to paid family leave (Feinberg, 2013). Given the projected increase in dementia family caregivers, it is imperative that there be national efforts to encourage states to offer paid family leave so that caregivers can provide the care needed by their families without the risk of financial instability. Toward this end, the DOL has been taking steps to encourage the development of paid family leave policies by providing funding to states to encourage their development (Department of Labor (DOL), 2015). Since 2014, the DOL has awarded more than $2 million in grants to 12 states and localities to either evaluate their existing family leave policies or encourage their development. During a third round of Request for Proposals (RFPs) for the development of family leave programs, the DOL awarded extra points on proposals from state grantees that direct their efforts toward paid family leave for workers with elder care responsibilities (Department of Labor (DOL), 2016).

Caring for a loved one with dementia typically involves providing care for an extended period of time as the disease progresses. Older Latinos with dementia are more likely to be diagnosed when the disease is in its more advanced stages but also are more likely than other populations to live for a longer period of time with the disease (Clark et al., 2005; O'Bryant et al., 2012; O'Bryant et al., 2013;). Therefore, Latino caregivers caring for an older loved one with dementia are typically in that role for 5 years on average (Mehta et al., 2008). Thus, while the policy provides a benefit of 12 weeks, it barely makes an impact for the actual time needed for caregiving for a loved one with dementia. At best, family leave serves to allow a caregiver to take time off from work after a hospitalization when the care recipient needs a higher level of care or for a family member to take time off to provide a brief respite to the primary caregiver. Improving family leave as a workplace policy by expanding the types of families covered and by providing a cash benefit are worthwhile policy goals, but do not provide much of a comprehensive solution for Alzheimer's caregivers in Latino families.

Social Security Caregiver Credits

Social Security's Old-Age, Survivors, and Disability Insurance (OASDI) is an important source of income for older adults. The shift in private pensions away from defined benefits to defined contributions and the eroding of savings and assets have led the majority of older Americans to rely heavily on Social Security benefits in retirement. About half of older adults live in families where Social Security accounts for half or more of total income, and about one-quarter live in families where it accounts for at least 90% of their income. Among Latino older adults, close to one-third live in families where Social Security accounts for 90% or more of their income (Dushi, Iams, & Trenkamp, 2017). As noted above, Latino caregivers are more likely than other populations to care for older adults with dementia and for long periods of time and therefore are more likely to have to reduce the number of hours worked or quit working altogether. Because Social Security is based on employment history, reducing work hours or stopping work for caregiving reduces the amount of payroll taxes Latinos pay into Social Security and therefore reduces their benefits when they retire.

Developing caregiver tax credits under Social Security is one way to reduce the impact of lost wages due to caregiving on future benefits (Estes, O'Neill, & Hartmann, 2012; Morris, 2007) and has been widely adopted by European social security systems (Jankowski, 2011). A Social Security tax credit program for caregivers would prospectively credit eligible caregivers with a defined level of wages, up to a specified period of time (Schulz & Eden, 2016). For example, under such a program, full-time caregivers could receive up to 4 years of Social Security work credit equal to the individual's average wage (White-Means & Rubin, 2009). In this way, Social Security benefits in retirement would not be reduced for family caregivers who must leave the workforce because of caregiving activities. While this policy does not provide immediate economic relief to dementia caregivers, it would help reduce the retirement benefit penalty that Latino/Latinas often face later in life as a result of years of being out of the labor force and spent in caregiving.

Income Tax Credits for Caregiving

Tax credits provide an approach to assisting family members who are caregivers of dementia sufferers with their current expenses. Especially when "refundable" (i.e., eligible people can receive benefits even if they pay no taxes), tax credits can provide social welfare benefits through the tax system rather than welfare bureaucracies. Tax credits are currently used in the U.S. as a supplement to child welfare programs through the Earned Tax Income Credit (ETIC). The ETIC requires families to have earned income, with the tax credit worth 10% of the earnings up to a maximum amount and then declining as earnings continue to increase. Enacted in 1975, it has been shown to reduce poverty, improve health, and lead to better child educational outcomes (Nichols & Rothstein, 2015). Similar tax credits could be used for families caring for older adults. In a 2015 survey of caregivers of older adults, the most popular proposal for providing financial support for caregivers was paying caregivers directly for some of their hours of direct care (32%), with the proposal for a tax credit that is independent of the hours of informal caregiving being the next most popular (29%) proposal. It is telling that the popularity of the two proposals varied inversely with income – families with the lowest income strongly preferred payment for family caregiving hours, while the highest-income families strongly preferred a tax credit that could be used to offset the cost of nonfamily caregivers (National Alliance for Caregiving and AARP, 2015). Given the often-lower-income profile of Latino family caregivers compared to non-Latino White caregivers, combining these two approaches would serve Latino families the best, i.e., provide a refundable tax credit to families where the caregiver is a family member. Since current federal rules for tax advantages for caregivers (e.g., the current law of allowing the claiming of the care recipient as a dependent for tax purposes) and the ETIC are narrowly focused on those who are financially dependent on the recipient, a means to expand this to functionally dependent, but not necessarily financially dependent or co-resident, would need to be developed.

Expanding and Integrating Long-Term Care for Older Latinos with Dementia: The Role of Medicaid and Medicare

In addition to the financial costs, dementia caregiving has been associated with negative health and mental health outcomes for the caregiver as well. These impacts include depression, high blood pressure, and health problems (Gallagher, Rose, Rivera, Lovett, & Thompson, 1989; Vitaliano, Zhang, & Scanlan, 2003), as well as lower self-rated health, higher morbidity and mortality, more sleep problems, and fewer health-promoting behaviors when compared to non-caregivers (Elliott, Burgio, & DeCoster, 2010). Hispanic caregivers are more likely than non-Hispanic White caregivers to be depressed (Pinquart & Sorensen, 2005). Also, Latino caregivers who report worse physical health are also more likely to be clinically depressed as well (Cucciare et al., 2010) and to experience burden, anxiety, and stress (Mahoney, Regan, Katona, & Livingston, 2005). Moreover, Holland et al. (2011) in their study of Hispanic and non-Hispanic White dementia caregivers found that those involved with high-intensity caregiving and co-residence with the care recipient had less adaptive cortisol levels and therefore greater vulnerability to develop disease conditions. It is estimated that Latino families provide care for a longer period of time and care for individuals with higher levels of impairment and are also more likely to report higher levels of burden and distress when compared to other racial/ethnic groups (Adams, Aranda, Kemp, & Takagi, 2002; Alzheimer's Association, 2014; Arévalo-Flechas et al., 2014; Harwood et al., 2000). Clearly, Latino caregivers have high levels of supportive needs. The problem is that long-term care supportive services, for the most part, are provided through Medicaid, precisely where there is increasing political pressure to cut back spending.

Medicaid

Increasing longevity among Latinos accompanied by the increased prevalence of ADRD with age, and coupled with the need for Latino family caregivers to be employed in order to have financial stability, will increase the demand for long-term services and supports (LTSS). LTSS are a constellation of programs including home- and community-based services (HCBSs) designed to keep older adults remaining in their homes/community and out of institutions (Wallace, Satter, Padilla-Frausto, & Peter, 2009). There have been a variety of Medicaid reforms since the 1980s to encourage a shift from depending on nursing home care to providing more care in community settings; nationally, more Medicaid funds are now spent on home health and personal care than nursing homes. But in many states, Medicaid continues to spend more on nursing homes than home care (Kaiser, Henry J. Family Foundation, 2016). Medicaid has long been the largest payer of LTSS in the U.S. In addition to nursing homes, Medicaid provides HCBS that include (1) home health services and personal care services; (2) Section 1915C HCBS waivers (which allow states to provide services outside established Medicaid categories and rules); and (3) Section 1115 demonstration waivers to deliver HCBS through managed care (Wolff et al., 2016). It is estimated that in 2016, Medicaid spent $119 billion for institutional and community-based services which represents 21.4% of the total Medicaid budget of that year (Kaiser, Henry J. Family Foundation, 2017).

Medicaid only provides services to those with incomes near the poverty level and with few assets, with a few exceptions. Nevertheless, Medicaid accounts for about half of all LTSS spending nationally, in part because the high costs of LTSS cause many elderly and disabled persons to exhaust their assets and their income is insufficient to pay for their care. In 2015, Medicaid paid 43% of all LTSS expenditures, followed by Medicare at 19% and out-of-pocket spending at 13%. Private insurance covered only 9% (James, Gellad, & Hughes, 2017). The Affordable Care Act (ACA) that was signed into law in 2010 by President Obama has taken steps to improve access to LTSS in a number of ways. Additional federal funds were granted for states to expand a number of home and community services, including increased home attendant services and home health for those with chronic conditions, allowing for demonstration projects that combine Medicare acute care funding with Medicaid LTSS funding to improve care coordination, additional funding for demonstration projects that help recipients leave nursing homes (e.g., *Money Follows the Person*), and other provisions (Harrington, Ng, LaPlante, & Kaye, 2012). These programs expand Medicaid coverage, better integrate health and long-term care services for Medicaid recipients, and increase federal funding. All of these provisions benefit Latino dementia caregivers who are more likely to be represented among low-income populations and therefore meet the criteria for inclusion in these programs. There is some evidence that previous expansions of LTSS have saved Medicaid nursing home spending (Harrington, Ng, & Kitchener, 2011), but the implementation of ACA reforms is too recent to have rigorous evaluation data available. In California, the long-term care integration program (*Coordinated Care Initiative*) did not reduce total spending during the first 5 years, leading to changes in the program (Taylor, 2017).

California does provide a successful example, historically, of using Medicaid funds to support consumer-directed personal care in the form of the In-Home Supportive Services (IHSS) program, where the disabled person is able to select the caregiver of their choice and have them be paid by Medicaid. Over half (58%) of Latino elders had family who were not spouses as their paid caregiver, while the most common caregiver for non-Latino Whites was nonfamily (54%) (Newcomer & Kang, 2008). Similarly, in the four-state Cash and Counseling demonstration program, disabled people could hire family, and Latinos reported a disproportionate preference for this program because of their ability to hire family members, as well as their ability to assure that a co-ethnic was the caregiver (Mahoney, Simon-Rusinowitz, Loughlin, Desmond, & Squillace, 2004). The long-term financial support that these types of programs provide to mostly low-income family caregivers provides an important replacement wage for dementia caregivers.

The 2016 U.S. election of Donald J. Trump to the presidency, and the Republican majority in Congress, has led to a series of proposals to repeal of the ACA and make funding cuts of up to $832 billion dollars in the Medicaid program. The core of proposals to cut Medicaid, which have been discussed in U.S. Congress and conservative think tanks for many years, involves converting Medicaid from an entitlement program, where everyone who is entitled to care because they meet eligibility criteria receives benefits, to a block grant program where states receive a fixed amount of funding independent of future needs. The details vary, but to reduce projected spending – the rationale for the proposals – it is generally accepted that states will have to reduce eligibility (e.g., to the poorest of the poor), restrict services (e.g., eliminate expensive LTSS services), and limit reimbursements to providers that will end up limiting access (Carlson & Kean, 2017). Whether or not the ACA repeal passes, it is likely that pressure to reduce Medicaid funding by making it into a block grant will continue. Those most impacted by reductions that this will necessitate are those most dependent on Medicaid for services, especially Latino families with LTSS needs caused by Alzheimer's.

Medicare

Medicare is a federally funded health-care program for older persons who are eligible for Social Security benefits, some disabled individuals, and children whose parents predecease them. The program provides access primarily to acute health care, doctors' visits, hospital stays, preventive care, limited home health care, and prescription medications. Medicaid, for those eligible, pays Medicare's premiums, co-payments, and deductibles, along with LTSS that are not covered by Medicare. Medicare is financed by payroll taxes, some premiums, and general federal revenues; the program has been widely supported politically for the access it provides to almost all older adults for health care. Because the program does have both political and public supports, it has been suggested that perhaps the best way to provide universal access to long-term care supportive services is through integrating these into the Medicare program.

Developing a social insurance approach to finance long-term care might be the best option to ensure that those who need assistance with LTSS obtain the help that they need (Goldberg & Atkins, 2013). In principle, social insurance is universal and contributory and offers a benefit based on some triggering event. A sensible way to prepare for a triggering event that increases the need for LTSS is through the pooling of risk. Spreading the risk of disability across a larger population reduces the amount any one person may have to set aside in order to cover his or her expenses (Folbre & Wolf, 2012). There is ample evidence that the U.S. population supports universal coverage for long-term care and is willing to pay more in payroll taxes under Medicare to have access to that benefit (AARP, 2015). Having a universal and mandatory LTSS system also makes sense given that approximately half of older adults will need significant assistance at some point, but relatively few people think they are at risk and are resistant to the relatively high costs of coverage in the voluntary insurance market. One estimate is that coverage would require an increase in the payroll tax of 0.60–1.35%, which could be added to the Medicare tax (Favreault, Gleckman, & Johnson, 2015). By becoming part of Medicare, states and the federal government would save a significant amount of money in Medicaid which would no longer need to provide LTSS to low-income older adults, and the program would benefit from the broad-based political support for adequate access and quality of care that is created when all people benefit from the same program. The impact on Latino families caring for those with AD and other dementias would be positive because low-income families could move into the new program that has broader support, and middle-income families who cannot currently afford services would be newly eligible for these services.

Reducing Risk Factors for Dementia: Improving Access to Dementia Care and Health Care Among Latinos

The expectation that Latino older persons and their families will continue to have a high prevalence of ADRD is reasonable given the increased longevity of the population as well as the prevalence of risk factors found among the Latino population (Chin et al., 2011). According to Fitten et al. (2014), AD is a multifactorial disease that, in addition to genetic risk factors, also has modifiable risk factors including diabetes, obesity, cardiovascular disease, hypertension, and low education. There is an established body of research that supports the link between these risk factors and cognitive difficulties and the development of AD (Barnes & Yaffe, 2011; Hazzouri et al., 2012; Luchsinger et al., 2009). The Latino population has a disproportionate number of older adults with low socioeconomic status, low education levels, and certain disease conditions – especially diabetes (Sanchez-Castillo et al., 2004); an earlier onset of diabetes (O'Bryant et al., 2013); as well as suboptimal management of the diseases (Kuo et al., 2003). This combination of factors puts Latinos at a higher risk for dementia and cognitive impairment than non-Hispanic Whites (Fitten et al., 2014).

Mexican-Americans, in particular, are at greater risk for metabolic conditions and are more likely to experience multiple morbidities (Fryar, Hirsch, Eberhardt, Yoon, & Wright, 2010; Villa, Wallace, Bagdasaryan, & Aranda, 2012). Mexican-American older adults with multiple metabolic and vascular conditions exhibit greater cognitive decline than older persons with one or no conditions (Downer, Raji, & Markides, 2016). Diabetes mellitus is also associated with poorer baseline cognitive performance in memory, language processing speed, executive functioning, and visuospatial abilities (Bangen et al., 2015). Research shows that Latinos with diabetes (all ages) are more likely than other populations to not have insurance or a usual source of care (Angel, Angel, & Hill, 2008) and are much more likely to access the health-care system via the emergency room when the disease is in its more advanced stages (Wallace & Villa, 2009). As a result, improving access to care among the Latino population across the life course is one pathway for reducing disparities in diabetes and obesity and therefore risk factors for dementia. This can occur through increasing access to health insurance, improving the cultural and linguistic competence of providers, and promoting self-care programs that target the Latino community (Cersosimo & Musi, 2011).

Reducing disparities in dementia prevalence among racial/ethnic minority population is a national priority of the National Alzheimer's Plan (NAPA) and the Alzheimer's Related Dementia Conference road maps (Montine et al., 2014; U.S. Department of Health and Human Services, 2014). While these initiatives are mostly planning documents, they highlight the racial/ethnic disparities in AD risks and note the high level of informal care that is provided to those with the disease. Combined with funding augmentations to the National Institutes of Health (NIH) and National Institute on Aging (NIA), specifically for AD research that was supported by bipartisan political efforts in 2016 and 2017, the growing research and planning about AD is heartening, but constant advocacy is needed to assure that issues relevant to Latinos are included in the agenda (Alzheimer's Association, 2017).

Improving Access to Dementia Care

There is evidence that Latinos are more likely to be diagnosed with AD when it is in its more advanced stages and at younger ages (Clark et al., 2005; O'Bryant et al., 2012, 2013) yet live longer than other populations with the disease (Mehta et al., 2008). Latinos often experience delays in accessing dementia care (Chin et al., 2011). Understanding the processes though which delays occur is paramount if we are to improve the population's access to dementia care (Chin et al., 2011). Cultural values and beliefs can shape the way older ethnic minorities assign meaning to dementia and whether or not they access dementia services (Dilworth-Anderson, Williams, & Gibson, 2002). Latinos are more likely than other populations to view dementia as a normal part of aging, as well as a disease that can be diagnosed with a blood test (Grey, Jimenez, Cucciare, Tong, & Gallagher-Thompson, 2009). Other studies have found that some Latinos attribute dementia to having a difficult life or the belief that dementia is part of God's plan or caused by forces of nature including "el mal ojo," i.e., the evil eye, or by mental illness (Hinton, Franz, Yeo, & Levkoff, 2005; Talamantes et al., 2006). In addition to cultural values and beliefs, other factors also influence the lack of help-seeking behavior for dementia among Latinos including lack of knowledge about dementia and health system barriers, i.e., lack of adequate access to health care, provider discrimination, and lack of trust for the medical establishment (Chin et al., 2011; Sayegh & Knight, 2013). Overcoming barriers to dementia care among Latinos will require the establishment of culturally competent education programs that are sensitive to Latino cultural beliefs and to the varying levels of the Latino patients' and their families' acculturation, outreach messaging that clarifies the differences between normal aging and dementia, and intervention systems that are based in the communities where Latinos live in and that are conveniently located in non-threatening environments (Chin et al., 2011; Sayegh & Knight, 2013; Stansbury, Marshall, Harley, & Nelson, 2010).

Reducing risk factors for dementia among Latinos and improving brain health will require culturally and linguistically competent and accessible education programs for Latinos. There is evidence to suggest that physical activity and healthy diets may help maintain cognitive functioning and reduce the risk for AD and vascular dementia (Friedman et al., 2009). The Centers for Disease Control (CDC) through its Healthy Aging Initiative included brain health for several years and found that Hispanics and others had low levels of knowledge about the specific actions that are associated with maintaining cognitive function, e.g., there was a generalized knowledge that physical activity was good for the brain but limited specific knowledge about the duration or intensity recommended to achieve a positive effect. Their work with multiple groups determined that targeted messaging and outreach was needed to disseminate actionable information to different racial and ethnic groups (Wilcox et al., 2009). Given the diversity of Latinos in the U.S., it is also important that outreach and education programs for Cubans in Miami, Puerto Ricans in New York, Central Americans in Chicago, and Mexicans in Los Angeles or the Texas Rio Grande Valley take into account their local contexts and perspectives in trying to motivate and educate (Martinez Tyson, Arriola, & Corvin, 2016).

Improving Access to Health Care

Historically, at all ages, Latinos are less likely to have insurance coverage or a usual source of care (Gresenz, Rogowski, & Escarce, 2009; U.S. Census Bureau, 2009). It is estimated that 85% of the older Latino population have Medicare, compared with 95% of the non-Latino White population (National Council of La Raza (NCLR), 2012). For older Latinos with Medicare, it is often difficult to meet the out-of-pocket costs associated with the Medicare program including deductibles, co-insurances, and premiums (National Committee to Preserve Social Security & Medicare (NCPSSM), 2016). Latinos' likelihood of having Part D, i.e., prescription drug coverage, under Medicare is 30% lower than that of non-Hispanic Whites. Moreover, Latinos who were eligible for low-income subsidies, but were not enrolled in Medicare Part D based on their Medicaid or SSI status, were 54% less likely to have Part D coverage and 70% less likely to have other coverage under Medicare compared to non-Hispanic Whites (McGarry, Strawderman, & Li, 2014).

The pattern of poverty, poor health, and compromised access to health care found among older Latinos is also found among younger members of the Latino population as well (Villa et al., 2012). Like the generation before them, economic hardship often translates into poor health for younger Latinos. Latinos under the age of 65 have higher prevalence rates for diabetes, obesity, and uncontrolled hypertension compared to non-Hispanic White (Dominguez et al., 2015). The Latino population also has higher death rates than non-Hispanic Whites from diabetes, liver disease, essential hypertensions, and hypertensive renal disease (Dominguez et al., 2015). In spite of this health profile, Latinos are more likely to be uninsured when compared to non-Hispanic Whites, have lower use of preventive services, and experience greater delays in access to health care (Dominguez et al., 2015; National Committee to Preserve Social Security & Medicare (NCPSSM), 2016). Without access to basic health care, it will be difficult if not nearly impossible to improve the health profiles of Latinos and therefore reduce the risk of developing dementia.

Conclusion

The rapidly growing size of the older Latino population, its socioeconomic disadvantages, the particular risk factors for cognitive impairment, the high level of family involvement in caregiving, and the linguistic and cultural factors needed to effectively reach the population all combine to underscore the imperative that dementia policy and programs be adapted to the situation of Latino elders, their families, and their communities. Reducing the incidence of dementia is the best way to reduce the impact of cognitive impairment on older Latinos and their families. While the research is not yet definitive on its impact on dementia, there are multiple other benefits to measures that reduce the incidence of diabetes and control of blood sugar among diabetics. This makes diabetes prevention and control a worthwhile goal even if it results in having a limited impact on cognitive decline; and if it does prove protective, then the benefits are multiplied. Both medical and non-medical approaches have been shown to reduce diabetes and improve glycemic control. Improved health-care monitoring can also improve early detection of AD, while increased physical activity and healthy diets are also associated with both metabolic improvements and brain health.

For Latinos who suffer from dementia, new and expanded policies are crucial to support the families who are most often providers of care for them at home. Broad policies such as refundable caregiver tax credits and consumer-directed Medicaid home care programs provide long-term financial support for all caregivers of persons with dementia but will be especially helpful to Latino families who are more likely to have low incomes and provide family caregiving. These are also the types of policies that Latino caregivers are most likely to endorse as being most helpful. A universal, long-term care benefit under Medicare would provide the most equitable basis for support, especially if it includes a consumer-directed home care component. In the current political climate, advancing any of these policies will not be simple. But AD impacts people of all races and ethnicities, all incomes, and all regions of the country, and research and programs in this area have shown an unusual level of bipartisanship. Addressing the needs of Latino elders and their families will require coalitions where all families see the benefits of the new policies, but it is important to make sure that their design takes into account the needs that fall most heavily on older Latinos with dementia and their caregivers.

References

AARP. (2015). *Caregiving in the United States*. Retrieved from http://www.aarp.org/ppi/info-2015/caregiving-in-the-united-states-2015.html

AARP. (2016). *Family caregivers cost survey: What they spend & what they sacrifice*. Retrieved from http://www.aarp.org/research/topics/care/info-2016/family-caregivers-cost-survey.html

Adams, B., Aranda, M. P., Kemp, B., & Takagi, K. (2002). Ethnic and gender differences in distress among Anglo American, African American, Japanese American, and Mexican American spousal caregivers of persons with dementia. *Journal of Clinical Geropsychology, 8*(4), 279–301.

Alzheimer's Association. (2004). *Hispanics/Latinos and Alzheimer's disease*. Retrieved from https://www.alz.org/national/documents/report_hispanic.pdf

Alzheimer's Association. (2012). 2012 Alzheimer's disease facts and figures. *Alzheimer's & Dementia: The Journal of the Alzheimer's Association, 8*(2), 131–168. https://doi.org/10.1016/j.jalz.2012.02.001

Alzheimer's Association. (2013). 2013 Alzheimer's disease facts and figures. *Alzheimer's & Dementia: The Journal of the Alzheimer's Association, 9*(2), 1–71. https://doi.org/10.1016/j.jalz.2012.02.001

Alzheimer's Association. (2014). Alzheimer's disease facts and figures 2014. *Alzheimer's & Dementia: The Journal of the Alzheimer's Association, 10*(2). Retrieved from https://www.alz.org/downloads/Facts_Figures_2014.pdf

Alzheimer's Association. (2017). Congress delivers historic Alzheimer's research funding increase for second consecutive year. Alzheimer's Impact Movement. Retrieved from Alzheimer's Association website: http://www.alz.org/documents_custom/historic-funding-2017.pdf

Alzheimer's Foundation of America. (2013). *Alzheimer's disease care costs add up*. Retrieved from https://alzfdn.org/wpcontent/uploads/2017/04/Care_Costs__Alzheimer_s_Disease.pdf

Angel, R. J., Angel, J. L., & Hill, T. D. (2008). A comparison of the health of older Hispanics in the United States and Mexico: Methodological challenges. *Journal of Aging and Health, 20*(1), 3–31. https://doi.org/10.1177/0898264307309924

Appelbaum, E., & Milkman, R. (2011). *Leaves that pay: Employer and worker experiences with paid family leave in California*. The Center for Economic and Policy Research. Retrieved from http://cepr.net/publications/reports/leaves-that-pay

Arevalo-Flechas, L. C. (2013). Colombian Alzheimer's caregivers: ¿Cuidadores? *Journal of Nutrition, Health & Aging, 17*(1), OP27 206-S-5.

Arévalo-Flechas, L. C., Acton, G., Escamilla, M. I., Bonner, P. N., & Lewis, S. L. (2014). Latino Alzheimer's caregivers: What is important to them? *Journal of Managerial Psychology, 29*(6), 661–684. https://doi.org/10.1108/JMP-11-2012-0357

Arno, P. S., House, J. S., Viola, D., & Schechter, C. (2011). Social security and mortality: The role of income support policies and population health in the United States. *Journal of Public Health Policy, 32*(2), 234–250. https://doi.org/10.1057/jphp.2011.2

Asfaw, A., & Bushnell, T. (2010). Relationship of work injury severity to family member hospitalization. *American Journal of Industrial Medicine, 53*(5), 506–513.

Bangen, K. J., Gu, Y., Gross, A. L., Schneider, B. C., Skinner, J. C., Benitez, A., … Luchsinger, J. A. (2015). Relationship between type 2 diabetes mellitus and cognitive change in a multiethnic elderly cohort. *Journal of the American Geriatrics Society, 63*(6), 1075–1083. https://doi.org/10.1111/jgs.13441

Barnes, D. E., & Yaffe, K. (2011). The projected effect of risk factor reduction on Alzheimer's disease prevalence. *The Lancet Neurology, 10*(9), 819–828. https://doi.org/10.1016/S1474-4422(11)70072-2

Bouchey, H., & Glynn, S. J. (2012). The many benefits of paid family and medical leave. *Center for American Progress*. Received from: https://www.americanprogress.org/wp-content/uploads/2012/11/GlynnModelLegislationBrief-2.pdf

Carlson, E., & Kean, N. (2017, June). *Health care on the chopping block: How older Americans will suffer under senate republicans' proposal to cap Medicaid funding* (Issue Brief). Retrieved from: http://www.justiceinaging.org/wp-content/uploads/2017/06/Seniors-Will-Suffer-Under-GOP-Bill-to-Cap-Medicaid.pdf

Cersosimo, E., & Musi, N. (2011). Improving treatment in Hispanic/Latino patients. *The American Journal of Medicine, 124*(10 Suppl), S16–S21. https://doi.org/10.1016/j.amjmed.2011.07.019

Chin, A. L., Negash, S., & Hamilton, R. (2011). Diversity and disparity in dementia: The impact of ethnoracial differences in Alzheimer disease. *Alzheimer Disease and Associated Disorders, 25*(3), 187–195. https://doi.org/10.1097/WAD.0b013e318211c6c9

Clark, P. C., Kutner, N. G., Goldstein, F. C., Peterson-Hazen, S., Garner, V., Zhang, R., & Bowles, T. (2005). Impediments to timely diagnosis of Alzheimer's disease in African Americans. *Journal of the American Geriatrics Society, 53*(11), 2012–2017. doi: JGS53569.

Cucciare, M. A., Gray, H., Azar, A., Jimenez, D., & Gallagher-Thompson, D. (2010). Exploring the relationship between physical health, depressive symptoms, and depression diagnoses in Hispanic dementia caregivers. *Aging & Mental Health, 14*(3), 274–282. https://doi.org/10.1080/13607860903483128

Dellmann-Jenkins, M., Blankemeyer, M., & Pinkard, O. (2000). Young adult children and grandchildren in primary caregiver roles to older relatives and their service needs. *Family Relations, 49*(2), 177–186.

Department of Labor (DOL). (2015). *Get the facts on paid sick time*. Department of Labor, U.S.

Department of Labor (DOL). (2016). *Paid leave analysis grants*. Department of Labor, U.S.

Department of Labor (DOL) Wage and Hour Division. (2012). Family and medical leave in 2012: Executive summary. *U.S. Department of Labor, Contract #GS10F0086K TO DOLF109630906.*

Diaz, N., Siskowski, C., & Connors, L. (2007). Latino young caregivers in the United States: Who are they and what are the academic implications of this role? *Child & Youth Care Forum, 36*(4), 131–140. https://doi.org/10.1007/s10566-007-9040-4

Dilworth-Anderson, P., Williams, I. C., & Gibson, B. E. (2002). Issues of race, ethnicity, and culture in caregiving research: A 20-year review (1980–2000). *The Gerontologist, 42*(2), 237–272.

Dominguez, K., Penman-Aguilar, A., Chang, M. H., Moonesinghe, R., Castellanos, T., Rodriguez-Lainz, A., & Scieber, R. (2015). Vital signs: Leading causes of death, prevalence of diseases and risk factors and use of health services among Hispanics in the United States, 2009–2013. *MMWR. Morbidity and Mortality Weekly Report, 64*(17), 469–478.

Downer, B., Raji, M. A., & Markides, K. S. (2016). Relationship between metabolic and vascular conditions and cognitive decline among older Mexican Americans. *International Journal of Geriatric Psychiatry, 31*(3), 213–221. https://doi.org/10.1002/gps.4313

Dushi, I., Iams, H. M., & Trenkamp, B. (2017). The importance of social security benefits to the income of the aged population. *Social Security Bulletin, 77*, 1.

Elliott, A. F., Burgio, L. D., & DeCoster, J. (2010). Enhancing caregiver health: Findings from the resources for enhancing Alzheimer's caregiver health II intervention. *Journal of the American Geriatrics Society, 58*(1), 30–37. https://doi.org/10.1111/j.1532-5415.2009.02631.x

Estes, C., O'Neill, T., & Hartmann, H. (2012). *Breaking the social security glass ceiling: A proposal to modernize women's benefits*. Institute for Women's Policy Research, D502.

Evercare and National Alliance for Caregiving. (2008). *Evercare study of Hispanic family caregiving in the U.S.: Findings from a national study*. Minnetonka, MN: Evercare; Bethesda, MD: National Alliance for Caregiving.

Favreault, M. M., Gleckman, H., & Johnson, R. W. (2015). Financing long-term services and supports: Options reflect trade-offs for older Americans and federal spending. *Health Affairs, 34*(12), 2181–2191.

Federal Interagency Forum on Aging. (2012). Older Americans 2012: Key indicators of wellbeing. *U.S. Federal Interagency Forum on Aging-Related Statistics (FIFARS) with the U.S. National Center for Health Statistics (NCHS), 17*(1), 176.

Feinberg, L. (2013). Keeping up with the times: Supporting family caregivers with workplace leave policies. *American Association of Retired Persons (AARP).* Retrieved from http://www.aarp.org/home-family/caregiving/info-06-2013/supporting-family-caregivers-with-workplace-leave-policies-AARP-ppi-ltc.html

Field Research Corporation and California Center for Research on Women & Families. (2015). A survey of California registered voters about the state's paid family leave program. San Francisco: CA.

Fiscal Policy Institute. (2014). Reform of New York's temporary disability insurance program and provision of family leave insurance: Estimated costs of proposed legislation. *Fiscal Policy Institute Report.* Retrieved from http://fiscalpolicy.org/wp-content/uploads/2014/06/Reform-of-NY-TDI-and-FLI.pdf

Fitten, L. J., Ortiz, F., Fairbanks, L., Bartzokis, G., Lu, P., Klein, E., … Ringman, J. (2014). Younger age of dementia diagnosis in a Hispanic population in southern California. *International Journal of Geriatric Psychiatry, 29*(6), 586–593. https://doi.org/10.1002/gps.4040

Folbre, N., & Wolf, D. A. (2012). Long-term care coverage for all: Getting there from here. In D. Wolf & N. Folbre (Eds.), *Universal coverage for long-term care in the United States: Can we get there from here?* New York, NY: Russell Sage Foundation.

Friedman, D. B., Laditka, J. N., Hunter, R., Ivey, S. L., Wu, B., Laditka, S. B., … Mathews, A. E. (2009). Getting the message out about cognitive health: A cross-cultural comparison of older adults' media awareness and communication needs on how to maintain a healthy brain. *The Gerontologist, 49*, S50–S60. https://doi.org/10.1093/geront/gnp080

Fryar, C. D., Hirsch, R., Eberhardt, M. S., Yoon, S. S., & Wright, J. D. (2010). Hypertension, high serum total cholesterol, and diabetes: Racial and ethnic prevalence differences in U.S. adults, 1999–2006. *National Center for Health Statistics (NCHS),* (Data Brief # 36), 1–8.

Gallagher, D., Rose, J., Rivera, P., Lovett, S., & Thompson, L. W. (1989). Prevalence of depression in family caregivers. *The Gerontologist, 29*(4), 449–456.

Gelman, C. R. (2010). *"La lucha": The experiences of Latino family caregivers of patients with Alzheimer's disease.* Retrieved from http://www.socsci.uci.edu/~castellj/clfm/webdocs/Week8/Required/"La Lucha"-TheExperiencesofLatinoFamilyCaregiversofPatientswithAlzheimer'sDisease.pdf

Goldberg, L., & Atkins, G. L. (2013). Social insurance: A critical base for financing long-term services and supports. In *Shaping affordable pathways for aging with dignity*. Long Beach, CA: The Scan Foundation.

Gresenz, C. R., Rogowski, J., & Escarce, J. J. (2009). Community demographics and access to health care among U.S. Hispanics. *Health Services*

Research, 44(5 Pt 1), 1542–1562. https://doi.org/10.1111/j.1475-6773.2009.00997.x

Grey, H. L., Jimenez, D. E., Cucciare, M. A., Tong, H., & Gallagher-Thompson, D. (2009). Ethnic differences in beliefs regarding Alzheimer disease among dementia family caregivers. *The American Journal of Geriatric Psychiatry, 17*(11), 925–933. https://doi.org/10.1097/JGP.0b013e3181ad4f3c

Haan, M. N., Mungas, D. M., Gonzalez, H. M., Ortiz, T. A., Acharya, A., & Jagust, W. J. (2003). Prevalence of dementia in older Latinos: The influence of type 2 diabetes mellitus, stroke and genetic factors. *Journal of the American Geriatrics Society, 51*(2), 169–177. doi:jgs51054.

Harrington, C., Ng, T., & Kitchener, M. (2011). Do Medicaid home and community based service waivers save money? *Home Health Care Services Quarterly, 30*(4), 198–213. https://doi.org/10.1080/01621424.2011.622249

Harrington, C., Ng, T., LaPlante, M., & Kaye, H. S. (2012). Medicaid home-and community-based services: Impact of the Affordable Care Act. *Journal of Aging & Social Policy, 21*(2), 169–187.

Harris-Kojetin, L., Sengupta, M., Park-Lee, E., & Valverde, R. (2013). Long-term care services in the United States: 2013 overview. Centers for Disease Control and Prevention (CDC)/National Center for Health Statistics. *Vital Health Statistics Series 3*, (37), 1–107.

Harwood, D. G., Barker, W. W., Ownby, R. L., Bravo, M., Aguero, H., & Duara, R. (2000). Predictors of positive and negative appraisal among Cuban American caregivers of Alzheimer's disease patients. *International Journal of Geriatric Psychiatry, 15*(6), 481–487. https://doi.org/10.1002/1099-1166(200006)15:6<481::AID-GPS984>3.0.CO;2-J

Hazzouri, A., Zeki, A., Haan, M. N., Whitmer, R. A., Yaffe, K., & Neuhaus, J. (2012). Central obesity, leptin and cognitive decline: The Sacramento area Latino study on aging. *Dementia and Geriatric Cognitive Disorders, 33*(6), 400–409. https://doi.org/10.1159/000339957

Hebert, L. E., Scherr, P. A., Bienias, J. L., Bennett, D. A., & Evans, D. A. (2003). Alzheimer disease in the US population: Prevalence estimates using the 2000 census. *JAMA Neurology, 60*(8), 1119–1122. https://doi.org/10.1001/archneur.60.8.1119

Hebert, L. E., Weuve, J., Scherr, P. A., & Evans, D. A. (2013). Alzheimer disease in the United States (2010–2050) estimated using the 2010 census. *Neurology, 80*(19), 1778–1783. https://doi.org/10.1212/WNL.0b013e31828726f5

Hinton, L., Franz, C. E., Yeo, G., & Levkoff, S. E. (2005). Conceptions of dementia in a multiethnic sample of family caregivers. *Journal of the American Geriatrics Society, 53*(8), 1405–1410. doi:JGS53409.

Holland, J. M., Thompson, L. W., Cucciare, M. A., Tsuda, A., Okamura, H., Spiegel, D., … Gallagher-Thompson, D. (2011). Cortisol outcomes among Caucasian and Latina/Hispanic women caring for a family member with dementia: A preliminary examination of psychosocial predictors and effects of a psychoeducational intervention. *Stress and Health: Journal of the International Society for the Investigation of Stress, 27*(4), 334–346. https://doi.org/10.1002/smi.1375

Humes, K. R., Jones, N. A., & Ramirez, R. R. (2011). *Overview of race and Hispanic origin: 2010*. Washington, D.C.: U.S. Department of Commerce. United States Census Bureau. (C2010BR-02).

James, E., Gellad, W., & Hughes, M. (2017, March 16). *In this next phase of health reform, we cannot overlook long-term care* [Web log post]. Retrieved from http://healthaffairs.org/blog/2017/03/16/in-this-next-phase-of-health-reform-we-cannot-overlook-long-term-care/

Jankowski, J. (2011). Caregiver credits in France, Germany, and Sweden: Lessons for the United States. *Social Security Bulletin, 71*(4), 61–76.

Kaiser, Henry J. Family Foundation. (2016). Medicaid's role in meeting seniors' long-term services and supports needs. *The Kaiser Commission on Medicaid and the Uninsured.* Retrieved from http://www.kff.org/medicaid/fact-sheet/medicaids-role-in-meeting-seniors-long-term-services-and-supports-needs/

Kaiser, Henry J. Family Foundation. (2017). Distribution of Medicaid spending by service. *The Henry J. Kaiser Family Foundation.* Retrieved from http://www.kff.org/medicaid/state-indicator/distribution-of-medicaid-spending-by-service/?currentTimeframe=0&sortModel=

Klerman, J., Daley, K., & Pozniak, A. (2012). *Family and medical leave in 2012: Technical Report.* (U.S. Department of Labor Publication No. GS10F0086K). Retrieved from https://www.dol.gov/asp/evaluation/fmla/FMLA-2012-Technical-Report.pdf

Kuo, Y. F., Raji, M. A., Markides, K. S., Ray, L. A., Espino, D. V., & Goodwin, J. S. (2003). Inconsistent use of diabetes medications, diabetes complications, and mortality in older Mexican Americans over a 7-year period: Data from the Hispanic established population for the epidemiologic study of the elderly. *Diabetes Care, 26*(11), 3054–3060.

Lester, G. (2005). A defense of paid family leave. *Harvard Journal of Law & Gender, 28*, 1–83.

Luchsinger, J. A., Brickman, A. M., Reitz, C., Cho, S. J., Schupf, N., Manly, J. J., … Brown, T. R. (2009). Subclinical cerebrovascular disease in mild cognitive impairment. *Neurology, 73*(6), 450–456. Retrieved from http://www.neurology.org/content/73/6/450.short

Luchsinger, J. A., Tipiani, D., Torres-Patiño, G., Silver, S., Eimicke, J. P., Ramirez, M., … Mittelman, M. (2015). Characteristics and mental health of Hispanic dementia caregivers in New York City. *American Journal of Alzheimer's disease and Other Dementias, 30*(6), 584–590. https://doi.org/10.1177/1533317514568340

Mahoney, K. J., Simon-Rusinowitz, L., Loughlin, D. M., Desmond, S. M., & Squillace, M. R. (2004). Determining personal care consumers' preferences for a consumer-directed cash and counseling option: Survey results from Arkansas, Florida, New Jersey, and New York elders and adults with physical disabilities. *Health Services Research, 39*(3), 643–664. https://doi.org/10.1111/j.1475-6773.2004.00249.x

Mahoney, R., Regan, C., Katona, C., & Livingston, G. (2005). Anxiety and depression in family caregivers of people with Alzheimer disease: The LASER-AD study. *The American Journal of Geriatric Psychiatry, 13*(9), 795–801. doi:13/9/795 [pii].

Martinez Tyson, D., Arriola, N. B., & Corvin, J. (2016). Perceptions of depression and access to mental health care among Latino immigrants: Looking beyond one size fits all. *Qualitative Health Research, 26*(9), 1289–1302. https://doi.org/10.1177/1049732315588499

Mayer, G. (2013). *The family and medical leave act (FMLA): Policy issues*. Washington, D.C.: Congressional Research Service.

McGarry, B. E., Strawderman, R. L., & Li, Y. (2014). Lower Hispanic participation in Medicare part D may reflect program barriers. *Health Affairs, 33*(5), 856–862. Retrieved from http://content.healthaffairs.org/content/33/5/856.full. https://doi.org/10.1377/hlthaff.2013.0671

Mehta, K. M., Yaffe, K., Perez-Stable, E. J., Stewart, A., Barnes, D., Kurland, B. F., & Miller, B. L. (2008). Race/ethnic differences in AD survival in US Alzheimer's disease centers. *Neurology, 70*(14), 1163–1170. doi:01.wnl.0000285287.99923.3c.

Mehta, K. M., & Yeo, G. W. (2016). Systematic review of dementia prevalence and incidence in United States race/ethnic populations. *Alzheimer's & Dementia: The Journal of the Alzheimer's Association, 13*(1), 72–83. doi: S1552-5260(16)32677-2.

Montine, T. J., Koroshetz, W. J., Babcock, D., Dickson, D. W., Galpern, W. R., Glymour, M. M., ... ADRD 2013 Conference Organizing Committee. (2014). Recommendations of the Alzheimer's disease-related dementias conference. *Neurology, 83*(9), 851–860. https://doi.org/10.1212/WNL.0000000000000733

Morris, J. L. (2007). Explaining the elderly feminization of poverty: An analysis of retirement benefits, health care benefits, and elder care-giving. *Notre Dame Journal of Law, Ethics, & Public Policy, 21*(2), 571.

Munnell, A. H., Sanzenbacher, G. T. & Rutledge, M. S. (2015). *What causes workers to retire before they plan?* Center for Retirement Research at Boston College. Retrieved from http://crr.bc.edu/working-papers/what-causes-workers-to-retire-before-they-plan/

National Alliance for Caregiving and AARP. (2009). *Caregiving in the U.S. 2009*. Retrieved from www.caregiving.org/data/Caregiving_in_the_US_2009_full_report.pdf

National Alliance for Caregiving and AARP. (2015). *Caregiving in the U.S.* Retrieved from www.aarp.org/.../aarp/ppi/2015/caregiving-in-the-united-states-2015-report-revised.pdf

National Committee to Preserve Social Security & Medicare (NCPSSM). (2016). *Medicare and Medicaid are important to Hispanic Americans*. Retrieved from http://www.ncpssm.org/Medicare/HispanicAmericansMedicare

National Council of La Raza (NCLR). (2012). *The role of Medicare in Hispanics' health coverage*. Retrieved from http://publications.nclr.org/handle/123456789/1272

Newcomer, R., & Kang, T. (2008). Analysis of the California in-home supportive services (IHSS) plus waiver demonstration program. *U.S. Department of Health and Human Services, #HHS-100-03-0025*.

Nichols, A., & Rothstein, J. (2015). The earned income tax credit (EITC). Working paper 21211. National Bureau of Economic Research.

Noble, J. M., Manly, J. J., Schupf, N., Tang, M. X., & Luchsinger, J. A. (2012). Type 2 diabetes and ethnic disparities in cognitive impairment. *Ethnicity & Disease, 22*(1), 38–44.

O'Bryant, S. E., Johnson, L., Balldin, V., Edwards, M., Barber, R., Williams, B., ... Hall, J. (2012). Characterization of Mexican Americans with mild cognitive impairment and Alzheimer's disease. *Journal of Alzheimer's Disease: JAD, 33*(2), 373–379. https://doi.org/10.3233/JAD-2012-121420

O'Bryant, S. E., Xiao, G., Edwards, M., Devous, M., Gupta, V. B., Martins, R., ... Barber, R. (2013). Biomarkers of Alzheimer's disease among Mexican Americans. *Journal of Alzheimer's Disease, 34*(4), 841–849. Retrieved from http://mimas.calstatela.edu/login?url=http://search.ebscohost.com/login.aspx?direct=true&db=psyh&AN=2013-10218-004&site=ehost-live

Ortiz, F., Fitten, L. J., Cummings, J. L., Hwang, S., & Fonseca, M. (2006). Neuropsychiatric and behavioral symptoms in a community sample of Hispanics with Alzheimer's disease. *American Journal of Alzheimer's Disease and Other Dementias, 21*(4), 263–273. https://doi.org/10.1177/153331750628935O

Peacock, S., Forbes, D., Markle-Reid, M., Hawranik, P., Morgan, D., Jansen, L., ... Henderson, S. R. (2010). The positive aspects of the caregiving journey with dementia: Using a strengths-based perspective to reveal opportunities. *Journal of Applied Gerontology*. Retrieved from http://journals.sagepub.com/doi/abs/10.1177

Pinquart, M., & Sorensen, S. (2005). Ethnic differences in stressors, resources, and psychological outcomes of family caregiving: A meta-analysis. *The Gerontologist, 45*(1), 90–106.

Sanchez-Castillo, C. P., Velasquez-Monroy, O., Lara-Esqueda, A., Berber, A., Sepulveda, J., Tapia-Conyer, R., & James, W. P. T. (2004). Diabetes and hypertension increases in a society with abdominal obesity: Results of the Mexican national health survey 2000. *Public Health and Nutrition, 8*(1), 53–60. Retrieved from https://pdfs.semanticscholar.org/f96c/3d780fcd16471e585251303c729ae5fa96c2.pdf

Sayegh, P., & Knight, B. G. (2013). Assessment and diagnosis of dementia in Hispanic and non- Hispanic white outpatients. *The Gerontologist, 53*(5), 760–769. https://doi.org/10.1093/geront/gns190

Schulz, R., & Eden, J. (2016). *Families caring for an aging America*. The John A. Hartford Foundation. Retrieved from http://www.johnahartford.org/images/uploads/reports/Family_Caregiving_Report_National_Academy_of_Medicine_IOM.pdf

Simpson, C. (2010). Case studies of Hispanic caregivers of persons with dementia: Reconciliation of self. *Journal of Transcultural Nursing, 21*(2), 167–174. https://doi.org/10.1177/1043659609357630

Spillman, B. C., Wolff, J., Freedman, V. A. & Kasper, J. D. (2014). Informal caregiving for older Americans: An analysis of the 2011 national study of caregiving. *Urban Institute*. Retrieved from http://www.urban.org/research/publication/informal-caregiving-older-americans-analysis-2011-national-study-caregiving

Stansbury, K. L., Marshall, G. L., Harley, D. A., & Nelson, N. (2010). Rural African American clergy: An exploration of their attitudes and knowledge of Alzheimer's disease. *Journal of Gerontological Social Work, 53*(4), 352–365. https://doi.org/10.1080/01634371003741508

Stephens, M. A., Franks, M. M., & Townsend, A. L. (1994). Stress and rewards in women's multiple roles: The case of women in the middle. *Psychology and Aging, 9*(1), 45–52.

Talamantes, M. A., Trejo, L., Jiménez, D., & Gallagher-Thompson, D. (2006). Working with Mexican American families. In G. Yeo & D. Gallagher-Thompson (Eds.), *Ethnicity and the Dementias* (pp. 327–340). New York, NY: Routledge.

Taylor, M. (2017). *The 2017–18 budget: The coordinated care initiative: A critical juncture*. Legislative Analyst's Office, California Legislative Analyst's office. Retrieved from www.lao.ca.gov/reports/2017/3585/coordinated-care-022717.pdf

Thies, W., & Bleiler, L. (2011). 2011 Alzheimer's disease facts and figures. *Journal of Alzheimer's and Dementia, 7*(2), 208–244.

Turner, R. M., Hinton, L., Gallagher-Thompson, D., Tzuang, M., Tran, C., & Valle, R. (2015). Using an emic lens to understand how Latino families cope with dementia behavioral problems. *American Journal of Alzheimer's Disease and Other Dementias, 30*(5), 454–462. https://doi.org/10.1177/1533317514566115

U.S. Bureau of Labor Statistics 2015. (2015). The employment situation-December 2015. *U.S. Department of Labor, USDL-16-0001.*

U.S. Census Bureau. (2009). Income, poverty, and health insurance coverage in the United States: 2009. U.S. Census Bureau, Current Population Reports, 60–238.

U.S. Census Bureau. (2015). Millennials outnumber baby boomers and are far more diverse. *Census Bureau Reports.*

U.S. Census Bureau. (2016). Current populations survey annual social and economic supplement (CPS-ASEC). *Census Bureau Reports.*

U.S. Department of Health and Human Services. (2014). National plan to address Alzheimer's disease: 2014 update. *Census Bureau Reports.*

Villa, V. M., Wallace, S. P., Bagdasaryan, S., & Aranda, M. P. (2012). Hispanic baby boomers: Health inequities likely to persist in old age. *The Gerontologist, 52*(2), 166–176. https://doi.org/10.1093/geront/gns002

Vitaliano, P. P., Zhang, J., & Scanlan, J. M. (2003). Is caregiving hazardous to one's physical health? A meta-analysis. *Psychological Bulletin, 129*(6), 946–972. https://doi.org/10.1037/0033-2909.129.6.946

Wagner, D. L. (2006). *Families, work, and an aging population: Developing a formula that works for the workers.* Retrieved from https://www.ncbi.nlm.nih.gov/labs/articles/17135098/

Wallace, S., & Villa, V. (2009). Healthy, wealthy, wise? Challenges to income security for elders of color. In L. Rogne, C. L. Estes, B. R. Grossman, B. A. Hollister, & E. Solway (Eds.), *Social insurance, social justice and social change* (p. 165). New York, NY: Springer.

Wallace, S. P., Satter, D., Padilla-Frausto, D. I., & Peter, S. E. (2009). *Elder economic security standard index: Supplemental home and community-based long-term care service package costs.* Los Angeles, CA: UCLA Center for Health Policy Research.

Wallace, S. P., Torres, J., Sadegh-Nobari, T., Pourat, N., & Brown, E. R. (2012). *Undocumented immigrants and health care reform* (pp. 1–48). Los Angeles, CA: UCLA Center for Health Policy Research.

Wheary, J., Shapiro, T., Draut, T., & Meschede, T. (2008). Economic (in)security: The experience of the African-American and Latino middle classes. *Institute on Assets & Social Policy and Dēmos* (Report No. 2). Retrieved from: http://www.demos.org/sites/default/files/publications/EconInsecurity_Latinos_AA_Demos.pdf

White-Means, S., & Rubin, R. (2009). Retirement security for family elder caregivers with labor force employment. The National Academy of Social Insurance (NASI). Retrieved from: https://www.nasi.org/research/2009/retirement-security-family-elder-caregivers-labor-force

Wilcox, S., Sharkey, J. R., Mathews, A. E., Laditka, J. N., Laditka, S. B., Logsdon, R. G., … Liu, R. (2009). Perceptions and beliefs about the role of physical activity and nutrition on brain health in older adults. *The Gerontologist, 49*, S61–S71. https://doi.org/10.1093/geront/gnp078

Wolff, J. L., Spillman, B. C., Freedman, V. A., & Kasper, J. D. (2016). A national profile of family and unpaid caregivers who assist older adults with health care activities. *JAMA Internal Medicine, 176*(3), 372–379. https://doi.org/10.1001/jamainternmed.2015.7664

Wu, S., Vega, W. A., Resendez, J., & Jin, H. (2016). Latinos and Alzheimer's disease: New numbers behind the Crisis. *USC Edward R. Roybal Institute on Aging and the Latinos against Alzheimer's Network.*

Zarit, S. H. (2012). Positive aspects of caregiving: More than looking on the bright side. *Aging & Mental Health, 16*(6), 673–674. https://doi.org/10.1080/13607863.2012.692768

Zucchella, C., Bartolo, M., Pasotti, C., Chiapella, L., & Sinforiani, E. (2012). Caregiver burden and coping in early-stage Alzheimer disease. *Alzheimer Disease and Associated Disorders, 26*, 55–60.

Epilogue: Juntos Podemos

Yvette N. Tazeau and Hector Y. Adames

Action without vision is only passing time, vision without action is merely daydreaming, but vision with action can change the world. – Nelson Mandela (n.d., para. 32)

Where have we been and, perhaps most importantly, where are we going regarding the behavioral and psychosocial treatment of dementia for Latinxs in the United States and in countries around the world that are Spanish-speaking? For those working closely with individuals who suffer from dementia, particularly in working with Latinxs and their caregivers, the current actions employed may appear too conventional for this population, raising the question what is the overall vision of healing interventions for Latinxs regarding this debilitating health condition? *Caring for Latinxs with Dementia in a Globalized World: Behavioral and Psychosocial Treatments* presents a comprehensive portrait of interventions grounded in a blend of ethnocultural traditions and Western healthcare.

The authors of this multidisciplinary volume explore different visions about dementia care held in the United States and in various countries around the globe. The authors also question the "one-size-fits-all" approach to dementia care and unpack culturally responsive and large-scale structural programming needed to help meet the unique needs of Latinxs with dementia. Ultimately, the chapter authors critically challenge the status quo by persuading readers to develop a contextualized, scientific, and sociocultural understanding of Latinxs that explicitly underscores the heterogeneity in nationality, immigration, race, sexual orientation, gender, and political realities that exists within Latinx populations. This suggests that such "tailoring" of treatments is a path to a future that provides better care for Latinxs with dementia and their caregivers. Indeed, many of the authors in the book encourage such a perspective. From incorporating culturally responsive interventions to creating policy options that reduce risk factors for dementia, each chapter provides suggestions that can serve as "bridges" to the development of multidisciplinary approaches that, when delivered within an interdisciplinary system, can more specifically target the needs of Latinx patients with dementia as well as their families and caregivers.

The work ahead for researchers, practitioners, and policymakers is humbling. The road seems long toward developing an understanding of dementia among Latinxs that integrates the roles

Y. N. Tazeau
Independent Practice, San Jose, CA, USA
e-mail: ytaging@earthlink.net

H. Y. Adames
The Chicago School of Professional Psychology, Chicago, IL, USA
e-mail: hadames@thechicagoschool.edu

H. Y. Adames, Y. N. Tazeau (eds.), *Caring for Latinxs with Dementia in a Globalized World*,
https://doi.org/10.1007/978-1-0716-0132-7

of context, culture, and overlapping forms of historical and contemporary oppression. In fact, a less than optimistic perspective suggests that even if a "cure" were to arrive for dementia, the world's healthcare systems would not be prepared to meet the demands and financial costs associated with delivering successful treatments due to a lack of resources including that of personnel (Liu, Hlávka, Hillestad, & Mattke, 2017). However, optimists embrace a belief that developing a bold vision supported by action can propel a breakthrough treatment at any time in the near future. Elements of intrepid visions are present in this book wherein the authors help the reader plan for just such a world. *Caring for Latinxs with Dementia in a Globalized World: Behavioral and Psychosocial Treatments* is a promising resource for clinicians, researchers, students, and policymakers who want to conceive of a world with more culturally responsive choices of dementia treatment for Latinxs and, fundamentally, more quality humane care. Renowned Chicana author and cultural theorist, Gloria E. Anzaldúa (n.d., para. 1) once said, "*Caminante, no hay puentes, se hacen puentes al andar,*" which loosely translated suggests that, assuredly, as "travelers on a journey there are no bridges, only those created while walking." Our conviction is that together we can materialize bold visions through the creation of innovative "bridging" actions.

References

Anzaldúa, G. E., (n.d.). *Quotable quote: Gloria Anzaldúa.* Retrieved from https://www.goodreads.com/quotes/180475-caminante-no-hay-puentes-se-hace-puentes-al-andar-voyager

Liu, J. L., Hlávka, J. P., Hillestad, R., & Mattke, S. (2017). Assessing the preparedness of the U.S. health care system infrastructure for an Alzheimer's treatment. *Rand Corporation.* Retrieved from https://www.rand.org/pubs/research_reports/RR2272.html

Mandela, N. (n.d.). *50 Inspirational Nelson Mandela Quotes That Will Change Your Life.* Retrieved from https://www.awakenthegreatnesswithin.com/50-inspirational-nelson-mandela-quotes-that-will-change-your-life/

About the Contributors

Daisy Acosta, MD is a geriatric psychiatrist who maintains a large practice in the Dominican Republic. Since 2003, she has been the Principal Investigator of the 10/66 Dementia Research Group at the Dominican Center. She is the co-founder of the Dominican Alzheimer's Association, and past President of Alzheimer's Disease International where she is also an Honorary Vice President for Life. She is renown in the Dominican Republic for her dedication to the diagnosis, treatment and care of people with dementia, and their caregivers.

Hector Y. Adames, PsyD is a clinical neuropsychologist, and an Associate Professor at The Chicago School of Professional Psychology (TCSPP). He co-founded and co-directs the Immigration, Critical Race, And Cultural Equity Lab (IC-RACE Lab). His research focuses on race, racism, wellness, and Latinx psychology.

Paola Alejandra Andrade Calderón, PhD is a clinical neuropsychologist, and an Associate Professor at Universidad del Valle of Guatemala. Her research and academic contributions focus on neuroscience, and dementia among the elderly. As an independent clinician, she is also dedicated to assessment, diagnosis, and neuropsychological rehabilitation.

Erin E. Andrews, PsyD, ABPP is a board-certified rehabilitation psychologist, co-director of Psychology Training at the Austin VA Outpatient Clinic in the Central Texas Veterans Health Care System, and Clinical Associate Professor in the Department of Psychiatry at the Dell Medical School at the University of Texas at Austin. Her areas of clinical and research expertise are disability as diversity, and disability in families.

Jacqueline L. Angel, PhD is Professor of Sociology and Public Affairs, and a faculty affiliate at the Population Research Center at The University of Texas at Austin LBJ School. Her research and publications examine health and retirement issues in the United States with a focus on older minorities and immigrants, and the impact of social policy on the Latino population and late-life Mexican American family households.

Ronald J. Angel, PhD is Professor of Sociology at the University of Texas, Austin. He has published numerous books and journal articles that investigate the socioeconomic and health vulnerabilities of minority individuals and communities.

Lee Ashendorf, PhD, ABPP-CN is Assistant Professor of Psychiatry at the University of Massachusetts Medical School, and a staff neuropsychologist at the VA Central Western

H. Y. Adames, Y. N. Tazeau (eds.), *Caring for Latinxs with Dementia in a Globalized World*,
https://doi.org/10.1007/978-1-0716-0132-7

Massachusetts. He has served as Associate Editor of the *Archives of Clinical Neuropsychology* and *Developmental Neuro-psychology*.

Laura Barba-Ramírez, MSc is a researcher at The National Council to Prevent Discrimination in Mexico. She is a specialist in mental health. Her research focuses on the association between psychological factors and health in population-based studies.

Gregory Benson-Flórez, PhD is Assistant Professor and Associate Department Chair in the Counseling Psychology Department at The Chicago School of Professional Psychology. His research includes Latinx Psychology, social justice issues, and international psychology in Latin America.

Cristalís Capielo Rosario, PhD is Assistant Professor in the Counseling and Counseling Psychology Department at Arizona State University. Her scholarship focuses on Latinx psychology, acculturation measurement, cultural and political determinants of health disparities among Puerto Rican populations, and multiculturally informed ethical standards.

Nayeli Y. Chavez-Dueñas, PhD is a clinical psychologist and Full Professor at The Chicago School of Professional Psychology (TCSPP) where she has been the lead for the Latinx Mental Health concentration for approximately a decade. She is the co-founder and co-director of the Immigration, Critical Race, and Cultural Equity Lab (IC-RACE Lab). Her scholarship addresses issues related to race, racism, Latinx psychology, and immigration.

Jhokania De Los Santos, MA is a counseling psychology doctoral candidate at the University of Georgia. Her research interests focus on the delivery of mental health care to racial-ethnic minority groups, with an emphasis on integrated care and linguistically competent services.

Leticia E. Fernández, PhD is a senior demographer at the U.S. Census Bureau, Center for Economic Studies. Her research interests include the consistency of self-reported racial and Hispanic origin identification, and racial/ethnic disparities in health and mortality outcomes.

Milton A. Fuentes, PsyD is a clinical psychologist, and a Professor at Montclair State University (MSU). He founded and directs the Clinical and Community Studies Laboratory at MSU. His scholarship focuses on diversity, inclusion, and equity.

Joshua T. Fuller, MA is a doctoral student in clinical psychology at Boston University. He works at Massachusetts General Hospital in the Multicultural Alzheimer's Prevention Program on a study of Autosomal Dominant Alzheimer's Disease. His work focuses on elucidating the earliest cognitive symptoms of preclinical Alzheimer's Disease.

Samuel C. Gable, PhD is a clinical neuropsychologist, and independent researcher. His primary area of practice and research has been in multicultural and bilingual (Spanish-English) neuropsychology. His research utilizes machine learning to retrieve and model patient data from online medical records in order to improve diagnostic accuracy for diverse and underrepresented patients.

Dolores Gallagher-Thompson, PhD, ABPP is Emerita Professor of Research in the Department of Psychiatry and Behavioral Sciences at Stanford University's School of Medicine, and visiting Professor at the Betty Irene Moore School of Nursing, Family Caregiving Institute, at the University of California at Davis. She is a board-certified geropsychologist who is actively engaged in research on diversity and caregiving, with a particular focus on community-based participatory intervention research. She and her husband, Larry W. Thompson, Ph.D., co-founded the Optimal Aging Center, an international network of caregiving researchers.

Margaret Gatz, PhD is Professor of Psychology at the University of Southern California, where she is affiliated with the Center for Economic and Social Research. Her research focuses on the mental health of older adults, including depressive symptoms, risk and protective factors for Alzheimer's Disease, and evaluation of the effects of interventions.

Mackenzie T. Goertz, MA is a doctoral trainee in Counseling Psychology at Marquette University. She is a member of the Culture and Well-Being Lab at Marquette, in addition to working with the Immigration, Critical Race, And Cultural Equity Lab (IC-RACE Lab) in Chicago, IL. She is on the editorial board of Latinx Psychology Today (LPT), and is interested in investigating factors that engage White people in dismantling racism, including the transformative potential of mentorship.

Julie E. Horwitz, PhD, ABPP-CN is a clinical neuropsychologist at the University of Colorado Memorial Hospital (UCHealth) in Colorado Springs. She specializes in the diagnosis and treatment of neurodegenerative conditions, as well as other neurologically and psychiatrically based conditions which impact cognitive functioning in adults. She has a particular interest in cross-cultural issues related to neuropsychological assessment and intervention.

Ivonne Z. Jiménez-Velázquez, MD is an internist and geriatrician, and Professor and Chair of Medicine at the University of Puerto Rico School of Medicine, Medical Sciences Campus. She is the Founder and Director of the Internal Medicine/Geriatrics Fellowship Program, and the Geriatrics Research Center. Her research focuses on Alzheimer's Disease. She is also the Principal Investigator in Puerto Rico for the 10/66 International Research Group, the Early and Late Onset Genetic Family Study, and several other clinical trials including the DIAN 1 & 2.

Norman J. Johnson, PhD is a mathematical statistician at the U.S. Census Bureau. He is the Principal Investigator for two studies: the Mortality Disparities in American Communities Study (MDAC) and the National Longitudinal Mortality Study (NLMS). Both projects focus on the various effects of demographic, socioeconomic, occupational, and housing differentials on U.S. mortality.

Tedd Judd, PhD, ABPP-CN is a board-certified clinical neuropsychologist and academic co-Director of the Master's degree program in Clinical Neuropsychology at the Universidad del Valle de Guatemala. He is past President of the Hispanic Neuropsychological Society, and has taught neuropsychology throughout Latin America.

Juan Llibre-Rodríguez, MD, PhD is Professor of Internal Medicine and Geriatric Medicine at the Medical University of Havana, Finlay Albarran School of Medicine. He is the Principal Investigator of the 10/66 International Dementia Research Group in Cuba, as well as the Cuban Aging and Alzheimer's Study—a large prospective cohort study on chronic diseases and dementia in the elderly.

Francisco Lopera, MD is a clinical neurologist and Professor at the Medical School of Antioquia University in Colombia. He is the Director of Grupo de Neurociencias de Antioquia (GNA), and the Principal Investigator of the Alzheimer Prevention Initiative API Colombia.

Mariana López-Ortega, MPP, PhD is a researcher at the National Institute of Geriatrics (INGER) in Mexico. She is a public health and policy specialist with a focus on aging policies. Her research focuses on the social determinants of aging, care dependency, dementia care, and Long-Term Care. She also collaborates on various strategies and actions toward the creation of a National Care System in Mexico.

Claire R. Manley, MA graduated from the Latinx Mental Health concentration program at The Chicago School of Professional Psychology. She is a member of the Immigration, Critical Race, And Cultural Equity Lab (IC-RACE Lab).

Silvia Mejía-Arango, PhD is an Associate Professor at El Colegio de la Frontera Norte in Tijuana, Mexico. She is a mental health specialist with an emphasis on cognitive aging. Her research focuses on cognitive impairment, dementia and its association with health, and social determinants in population-based studies.

Julian Montoro-Rodriguez, PhD is the Director of the Gerontology Program at the University of North Carolina at Charlotte. He is a sociologist with teaching and research backgrounds in the areas of aging, health, human development, and family studies. His research examines the interrelations between formal and informal support systems, and optimal adaptation and adjustment among older adults.

Alexander Navarro, RN is a pharmacist at Grupo Neurociencias Antioquia (GNA) in Colombia. He is a member of the research team for the Colombian Alzheimer's Prevention Initiative clinical trial.

Maureen K. O'Connor, PsyD, ABPP-CN is a clinical neuropsychologist, and an Assistant Professor at Boston University School of Medicine in the Department of Neurology. She is the Director of Neuropsychology at the Bedford Veterans Hospital, and maintains a private practice in the community. Her funded research focuses on the development of interventions for individuals with memory loss.

Paula Ospina-Lopera, BA completed her undergraduate degree in psychology and currently serves as a clinical researcher at the Grupo de Neurociencias of the University of Antioquia (GNA) in Colombia where she works with Alzheimer's Disease patients and caregivers.

Viviana Padilla-Martinez, PhD is a clinical psychologist at Bay Pines VA Healthcare System. She is the 2019 President of Clinical Psychology of Ethnic Minorities (Section VI) of the American Psychological Association's Division 12 for Clinical Psychology. Her primary clinical and scholarly interests include serious mental illness, diversity, and trauma.

Shawneen R. Pazienza, PhD is a rehabilitation neuropsychologist at the Austin VA Outpatient Clinic. Her clinical and research interests include outcomes of psychoeducation, health behavior intervention, and cognitive rehabilitation among veterans with concurrent medical and psychiatric diagnoses.

Jessica G. Perez-Chavez, BA is an immigrant activist completing her doctoral degree at the University of Wisconsin-Madison. She is the graduate student Editor for *Latinx Psychology Today* (LPT), and a member of the Immigration, Critical Race, and Cultural Equity Lab (IC-RACE Lab). Her research focuses on immigration, activism, and health.

Francisco J. Piedrahita, BA is an Auxiliary Nursing Care Technician at the Group of Neuroscience of Antioquia (GNA), Colombia. He provides assistance to GNA's dementia research group participants in the form of access to and participation in medical procedures and services.

Catherine V. Piersol, PhD, OTR/L is an occupational therapist, Professor and Chair in the Department of Occupational Therapy, College of Rehabilitation Science at Thomas Jefferson University. She directs the Jefferson Elder Care Center which provides community-based clinical services; professional and family education; and agency consultation to optimize performance and quality of life in people living with dementia, and to build skills in care providers.

Yakeel T. Quiroz, PhD is a clinical neuropsychologist, and Assistant Professor in the Departments of Neurology and Psychiatry at Massachusetts General Hospital, and Harvard Medical School. She directs the Mass General Familial Dementia Neuroimaging Lab, and the Multicultural Alzheimer's Prevention Program (MAPP). She is also a Clinical Scientist at the Group of Neuroscience of Antioquia (GNA), Colombia.

Carlos A. Rodriguez, PhD is a bilingual clinical neuropsychologist, and clinical training supervisor in the Department of Clinical Neurosciences at Spectrum Health in Grand Rapids, Michigan where he completed his Neuropsychology Fellowship. He also holds an Adjunct Clinical Professor appointment at Michigan State University's College of Human Medicine. He completed his doctorate in Clinical Psychology with a focus on Geropsychology and Geriatric Neuropsychology at the University of Southern California (USC), and his internship in the Adult Neuropsychology track at the University of California, Los Angeles (UCLA) Semel Institute.

Janette Rodriguez, PsyD is a clinical psychologist. She currently serves as psychologist Program Manager at the Miami VA Healthcare System where she coordinates the recovery and rehabilitation training track for interns and postdoctoral residents.

Dinelia Rosa, PhD is a clinical psychologist, and Director of the Columbia University Teachers College (TC) Dean Hope Center or Educational & Psychological Services. She is also an Adjunct Professor (Full) in the Clinical Psychology program at TC. Her work focuses on diversity and psychotherapy. She is actively involved in advocacy at the state and national level.

Caroline Rosenthal Gelman, LCSW, PhD is an Associate Professor, and Director of the MSW Program at the Silberman School of Social Work at Hunter College. She has practiced as a clinical social worker for over 30 years, specializing in mental health issues with diverse populations, particularly Latinos. For the past 20 years she has focused her research and practice on the experiences and needs of Latino older adults living with dementia, and their caregivers.

Silvia P. Salas, MA is a counseling psychology doctoral candidate at the University of Wisconsin-Milwaukee. She is a member of the Immigration, Critical Race, And Cultural Equity Lab (IC-RACE Lab). Her scholarship focuses on Latinx psychology, and transgender Latinx immigrants.

Amanda Saldarriaga, BA is a Gerontologist at the Group of Neuroscience of Antioquia (GNA), Colombia. She is coordinator of support groups for patients with dementia, and patients with Parkinson's disease, including their respective families. Her work also involves family assessments and genealogical analysis.

Azara Santiago-Rivera, PhD is Professor Emeritus at Merrimack College. She was the founding editor of the *Journal of Latinx Psychology.* Her publications and research interests include multicultural issues in the counseling profession, bilingual therapy, Latinxs and depression, and the impact of environmental contamination on the biopsychosocial well-being of Native Americans.

Philip Sayegh, PhD, MPH is a clinical psychologist with specializations in clinical geropsychology and neuropsychology. He is also Assistant Adjunct Professor at the UCLA Psychology Department, and Associate Director of the UCLA Psychology Clinic. His research has examined cultural influences on caregiver health outcomes, as well as dementia assessment and care-seeking across cultural groups.

Juan P. Serrano Selva, PhD is an academic and research psychologist. He is Associate Professor at the School of Medicine in Albacete in Spain. His research focuses on autobiographical memory, life review, and lesbian, gay, bisexual, transgender, and intersex (LGBTI) older adults.

Amber Schaefer, MEd is a third-year doctoral student in the Counseling Psychology program at Arizona State University. Her research interests include the impacts of immigration status on mental health, with a special interest in understanding the experiences of unaccompanied immigrant minors and mixed-status siblings. She is an aspiring neuropsychologist currently on rotation at Phoenix Children's Hospital.

Christian E. Schenk-Aldahondo, MD is a behavioral neurologist, and Associate Faculty at the University of Puerto Rico School of Medicine.

He is also an investigator for the DIAN clinical trials network, a member of the American Neuropsychiatric Association, and an author on the topic of neuropsychiatric conditions, such as dementia and Tourette's Syndrome.

Shanna N. Smith, MA graduated from The Chicago School of Professional Psychology (TCSPP) with a concentration in Latinx Mental Health. Her research interests include exploring trauma-informed and strength-based interventions for Black Women and girls.

Karla Steinberg, LMSW is a social worker at the Mount Sinai Visiting Doctors Program. A graduate of Silberman School of Social work, she is committed to advocating for social and age justice, and working with older adults in underserved communities in New York City.

Yvette N. Tazeau, PhD is a licensed psychologist in independent practice in Silicon Valley. She works across the developmental life-span, with specialties including geropsychology, clinical neuropsychology, child psychology, and industrial/organizational psychology. Her interests and scholarship regarding gerodiversity topics include clinical and biopsychosocial/behavioral interventions with Latinx populations, as well as organizational development issues of geriatric mental health workforce planning.

Valeria L. Torres, MA is a doctoral student at the Florida Atlantic University. She works at the Florida Alzheimer's Disease Research Center (ADRC) and the Multicultural Alzheimer Prevention Program (MAPP) at Massachusetts General Hospital.

Mari Umpierre, LCSW, PhD is a Clinical Social Worker and Researcher at The Icahn School of Medicine at Mount Sinai in New York City.

Valentine M. Villa, PhD is Professor of social work, and Director of the Applied Gerontology Institute at California State University, Los Angeles. She is also Adjunct Professor at the University of California, Los Angeles (UCLA) Fielding School of Public Health, and Senior Scientist at the UCLA Center for Health Policy Research. Her scholarship focuses on health disparities among aging diverse populations and its implications for health and social service programs and policy.

Steven P. Wallace, PhD is Professor of Community Health Sciences at the UCLA Fielding School of Public Health. He also directs the Coordinating Center for the NIA-funded Resource Centers for Minority Aging Research. He is one of the multiple-Principal Investigators of the Coordination and Evaluation Center for NIH's largest investment in testing ways to most effectively diversify the biomedical workforce, the Diversity Program Initiative. His scholarship focuses on the social determinants of health for older adults in communities of color, the impacts of public policies on older adults, and immigrant health.

Kathleen D. Warman, EdM, MA is a clinical psychology doctoral student at The Chicago School of Professional Psychology. She earned her Master of Education and Master of Arts in counseling psychology with a concentration in bilingual Latino/a psychology at Teachers College Columbia University.

Diana Lynn Woods, PhD, GNP-BC is Associate Professor at Azusa Pacific University, a gerontological nurse practitioner, and founding faculty of the Advancement for Gerontological Nursing Science at the University of California, Los Angeles (UCLA). Her scholarship focuses on behavioral symptoms of dementia, biobehavioral correlates, and nonpharmacological interventions.

Index

H. Y. Adames, Y. N. Tazeau (eds.), *Caring for Latinxs with Dementia in a Globalized World*,
https://doi.org/10.1007/978-1-0716-0132-7

U

V

W

X

Y

9781071601303